ADVANCED TECHNIQUES IN CHROMOSOME RESEARCH

ADVANCED TECHNIQUES IN CHROMOSOME RESEARCH

edited by
Kenneth W. Adolph
University of Minnesota Medical School
Minneapolis, Minnesota

Marcel Dekker, Inc. **New York • Basel • Hong Kong**

Library of Congress Cataloging--in--Publication Data

Advanced techniques in chromosome research/edited by Kenneth W. Adolph.
p. cm.
Includes bibliographical references.
Includes index.
ISBN 0-8247-8430-8 (alk. paper)
1. Chromosomes-- --Research-- --Methodology. 2. Cytogenetics-- --Technique. I. Adolph, Kenneth W.
[DNLM: 1. Chromosomes. 2. Cytogenetics. 3. Genetics Techniques. QH 600 A2445]
QH600.A37 1991
574.87'322-- --dc20
DNLM/DLC
for Library of Congress 91--6834
CIP

This book is printed on acid-free paper.

MARCEL DEKKER, INC.
270 Madison Avenue, New York, New York 10016

Current printing (last digit):
10 9 8 7 6 5 4 3 2 1

PRINTED IN THE UNITED STATES OF AMERICA

Preface

Progress in chromosome research has greatly accelerated in recent years, thanks to the development of powerful new experimental techniques. The new approaches (molecular techniques, cytogenetic and linkage analysis, and cellular techniques) provide a variety of tools for dissection of chromosome structure and function. The molecular techniques represent the extension of molecular biology, which originated with prokaryotic and viral systems, to human genetics. The analysis of the genes and DNA sequences of human chromosomes is a central research problem that is being tackled. Improvements in procedures of cytogenetic analysis and the analysis of genetic linkage are yielding a high-resolution view of chromosome organization. Adding molecular methods to cytogenetic investigations creates an especially potent combination. Chromosome studies at the level of cell and developmental biology have not been neglected as advanced methodology has been developed. The three-dimensional arrangement of chromosomes in cells and the production of transgenic animals are two examples that illustrate the range of novel and innovative cellular techniques.

The information that is gained from these technological advances, and from additional methodology that may arise in the future, is certain to have a major impact. The greatly expanded knowledge of chromosome structure and function that accumulates will be not only intellectually interesting, but undoubtedly of immense medical importance. The clinical diagnosis of genetic disorders will be markedly improved. Increased understanding of the molecular changes in carcinogenesis may translate into knowledge that is of therapeutic value. Detailing the three-dimensional structural changes of chromosomes during cell division will provide data complementary to that gained from other studies of normal and malignant cell division. Techniques such as the production of transgenic animals are likely to have broad practical applications. Indeed, the potential for each of the advanced techniques described in this book remains to be realized to full advantage. These experimental approaches have originated only recently. But their quick acceptance and widespread use means that acronyms such as PCR, RFLP, PFGE, and YAC are now part of the standard vocabulary of biological and medical research. The rapid pace of research in these areas will be certain to give rise to improved techniques and a necessary acquaintance with a new list of acronyms.

The chapters of the book are organized into three parts dealing with molecular techniques, cytogenetic and linkage analysis, and cellular techniques. Such an arrangement indicates the range of advanced experimental approaches available to the researcher. The purpose of the volume is to provide a broad overview of the emerging genetic technology. Some chapters serve as practical guides for the application of the methods and others emphasize the research results that are obtained. If an appreciation of the power and utility of these advanced techniques in chromosome research is conveyed to the reader, the volume will be considered a success.

Kenneth W. Adolph

Contents

Contributors

Kenneth W. Adolph Department of Biochemistry, University of Minnesota Medical School, Minneapolis, Minnesota

Peter D'Eustachio Department of Biochemistry and Kaplan Cancer Center, New York University Medical Center, New York, New York

Frederick F. B. Elder Cytogenetics Laboratory, Department of Pediatrics, University of Texas Medical School at Houston, Houston, Texas

Jörg T. Epplen Department of Research for Psychiatry, Max-Planck-Institute for Psychiatry, Planegg-Martinsried, Germany

Robert P. Erickson* Departments of Pediatrics and Human Genetics, University of Michigan School of Medicine, Ann Arbor, Michigan

Jon W. Gordon Brookdale Center for Molecular Biology, Mt. Sinai School of Medicine, New York, New York

Thomas Haaf Department of Genetics, Stanford University School of Medicine, Stanford, California

---

**Current affiliation*: University of Arizona Health Science Center, Tucson, Arizona.

T. C. Hsu Department of Cell Biology, University of Texas M.D. Anderson Cancer Center, Houston, Texas

Eckart U. Meese Department of Radiation Oncology, University of Michigan, Ann Arbor, Michigan

Shinsei Minoshima Department of Molecular Biology, Keio University School of Medicine, Tokyo, Japan

Ikuo Miura* Section of Molecular Cytogenetics, Department of Medical Oncology, Fox Chase Cancer Center, Philadelphia, Pennsylvania

Joseph H. Nadeau Jackson Laboratory, Bar Harbor, Maine

Indrajit Nanda Department of Human Genetics, University of Würzburg, Würzburg, Germany

John A. Phillips III Division of Genetics, Department of Pediatrics, Vanderbilt University School of Medicine, Nashville, Tennessee

James H. Resau Department of Pathology, University of Maryland School of Medicine, Baltimore, Maryland

Stephen Scherer Department of Medical Genetics, University of Toronto, and Department of Genetics, Hospital for Sick Children, Toronto, Ontario, Canada

Michael Schmid Department of Human Genetics, University of Würzburg, Würzburg, Germany

Nobuyoshi Shimizu Department of Molecular Biology, Keio University School of Medicine, Tokyo, Japan

Jill M. Siegfried Department of Pharmacology, University of Pittsburgh School of Medicine, Pittsburgh, Pennsylvania

Greenfield Sluder Cell Biology Group, Worcester Foundation for Experimental Biology, Shrewsbury, Massachusetts

Marshall L. Summar Division of Genetics, Department of Pediatrics, Vanderbilt University School of Medicine, Nashville, Tennessee

Grant R. Sutherland Department of Cytogenetics and Molecular Genetics, Adelaide Children's Hospital, North Adelaide, South Australia, Australia

---

**Current affiliation*: Third Department of Internal Medicine, Akita University School of Medicine, Akita, Japan

Akira Suyama Department of Bio-Engineering, Nagaoka University of Technology, Nagaoka, Japan

Joseph R. Testa Section of Molecular Cytogenetics, Department of Medical Oncology, Fox Chase Cancer Center, Philadelphia, Pennsylvania

Jeffrey M. Trent Department of Radiation Oncology and Human Genetics, University of Michigan, Ann Arbor, Michigan

Lap-Chee Tsui Departments of Medical Genetics and Medical Biophysics, University of Toronto, and Research Institute, Hospital for Sick Children, Toronto, Ontario, Canada

Hiromichi Tsurui* University of Tokyo, Tokyo, Japan

Vera Verga Department of Human and Pediatric Genetics, University of Michigan School of Medicine, Ann Arbor, Michigan

Akiyoshi Wada† University of Tokyo, Tokyo, Japan

Daniel E. Weeks‡ Division of Medical Genetics, Department of Psychiatry, College of Physicians and Surgeons of Columbia University, and New York State Psychiatric Institute, New York, New York

Jorge J. Yunis Department of Neoplastic Diseases, Hahnemann University School of Medicine, Philadelphia, Pennsylvania

Current affiliation:
*Department of Pathology, Juntendo University School of Medicine, Tokyo, Japan
†Sagami Chemical Research Center, Sagamihara, Kanagawa, Japan
‡Department of Human Genetics, University of Pittsburgh, Pennsylvania

ADVANCED TECHNIQUES IN CHROMOSOME RESEARCH

I

MOLECULAR TECHNIQUES: INTRODUCTION

Versatile methods have become available in the last few years for the analysis of the human genome at the molecular level. These approaches have rapidly become established as important tools in basic research aimed at sequencing the human genome and in clinical situations for genetic diagnosis. The polymerase chain reaction (PCR) and pulsed-field gel electrophoresis (PFGE) are among the most useful and recent additions to the methodology of molecular biology. Such methodology has its foundation in procedures to sequence DNA and clone genes. Underlying these procedures has been the introduction of restriction enzyme analysis, Southern blotting, and a host of related approaches (see Chap. 1). Even more convenient and powerful techniques should soon emerge. Understanding the molecular defects for a variety of genetic diseases will be an increasingly realistic goal.

The isolation and characterization of genes involved in human diseases will require the capability to clone and propagate large DNA segments. Yeast artificial chromosome (YAC) technology is a recent advance that has permitted the cloning of large fragments of DNA in a yeast cell vector. A chapter in this section is devoted to YAC technology because of its advantages for mapping large chromosome regions, genome sequencing, and identification of disease genes.

A valuable example illustrating the application of molecular techniques is provided by a chapter concerned with chromosomal rearrangements in hematologic malignancies. Until recently, chromosomal changes associated with cancer were primarily monitored by the chromosome–staining patterns of classical cytogenetics. Molecular cytogenetics has improved the resolution of the chromosomal defects associated with malignancy to the level of individual nucleotide bases of chromosome DNA.

The isolation and amplification of specific genes by cloning makes necessary, as part of the technology, the identification of the genes with specific probes. This section of the book continues with a chapter concerned with a novel means of extracting DNA fragments through use of synthesized oligonucleotides as immobilized probes.

Nonradioactive oligonucleotide probes to chromosomes are also central to the chapter that follows. In situ hybridization of synthetic probes to chromosomes is examined. The coincidence of particular synthetic sequences with cytogenetic bands, the location of nucleolus organizer regions, and the presence of clustered repetitive sequences on Y chromosomes are demonstrated.

Flow cytometry instrumentation has proven to be extremely valuable for chromosome studies, since it allows individual, intact chromosomes to be analyzed and isolated. A chapter is therefore included which discusses the separation of human chromosomes by flow cytometry and the use of human chromosomes in gene mapping and fine structure analysis.

The section concludes with a review of pulsed-field gel electrophoresis (PFGE) and describes its application to aspects of sex determination. Large DNA molecules can be resolved by PFGE and, indeed, entire chromosomes of micro-organisms, such as yeast, can be separated. For higher eukaryotes, genetic alterations, such as translocations and deletions, are detectable. In particular, PFGE has been applied to the problem of detecting the chromosomal breakpoints giving rise to XX males and XY females in humans.

1

Molecular Techniques in Genetic and Chromosomal Diagnosis

Marshall L. Summar and John A. Phillips III

Vanderbilt University School of Medicine, Nashville, Tennesee

I. INTRODUCTION

The rate with which new diagnostic techniques and information are added to the armament of the geneticist appears to be growing exponentially. It is estimated that the human genome contains 6×10^9 base pairs (bp) of DNA and about 100,000 genes per cell; therefore, detection of mutations causing a given genetic disorder would seem to be a formidable task [1]. With the availability of extensive genetic linkage maps and the advent of DNA amplification by polymerase chain reaction (PCR), specific chemical cleavage of unpaired bases, and new probes and labeling techniques, it is difficult to keep abreast of which techniques are best suited for a particular diagnostic question. In this chapter, we will describe a variety of these techniques and illustrate how they can be used to address clinical situations. Methods of both direct and indirect detection and their clinical applications to genetic disorders will be discussed. With these techniques and others being developed, genetic diseases can now be more consistently classified by the actual genetic defect, rather than by the phenotypic expression. By classifying diseases according to their genetic defects, new insights can be gained into the function of various genetic structures and their effects on phenotypic expression, as well as into their interaction with other genetic and environmental factors.

II. SOUTHERN BLOTS AND RESTRICTION ENZYME ANALYSIS

A. Introduction

Classic genetic markers consist of observable or measurable traits in humans. These comprise physical findings such as the nail–patella syndrome, protein variants such as hemoglobin, and immunologic distinctions such as the ABO blood group or major histocompatibility complex (MHC) antigens [1]. In the last decade, a new generation of markers has been developed that consist of DNA sequence polymorphisms, referred to as restriction fragment length polymorphisms (RFLPs), which can be detected by probes for specific genes or for anonymous or uncharacterized fragments of DNA [2,3]. The DNA probes used to detect these RFLPs are derived from numerous sources, such as cloned DNA libraries or cDNA sequences, obtained from mRNA isolated from a variety of tissues. In some cases, these clones contain inserts that encode known protein products, although, usually, their sequences are unknown. The use of these probes and the RFLPs they identify as markers has revolutionized the way in which we look at inherited diseases, as well as at certain acquired diseases, such as neoplasias that are associated with chromosomal abnormalities. The use of DNA polymorphisms allows direct or indirect detection of a rapidly growing list of familial and acquired disorders that perturb the human genome.

B. Direct Methods

1. Southern Blotting

One of the most commonly used techniques in molecular genetics is Southern blotting, combined with restriction endonuclease digestion of DNA [2]. The use of sequence-specific restriction endonucleases, derived from eukaryotic organisms, has greatly expanded our ability to track copies of a gene through several generations [4]. These enzymes cleave specific DNA sequences, called recognition sites, which are most often palindromic. In the human genome, there are approximately 6×10^9 bp of DNA, throughout which the recognition sequences for restriction endonucleases are widely and, in most cases, evenly distributed [5]. In the technique of Southern blotting (Fig. 1), aliquots of genomic DNA, isolated from peripheral leukocytes or tissue, are incubated with restriction endonucleases to obtain fragments of a reproducible and manageable size that can then be separated by subjecting them to gel electrophoresis. These fragments, after denaturing to render them single-stranded, are then transferred to a membrane filter (usually composed of nylon, nitrocellulose, or a combination of the two), by either capillary, electro-, or vacuum blotting [2]. The DNA fragments are fixed to the membrane by drying or exposure to ultraviolet light, or a combi-

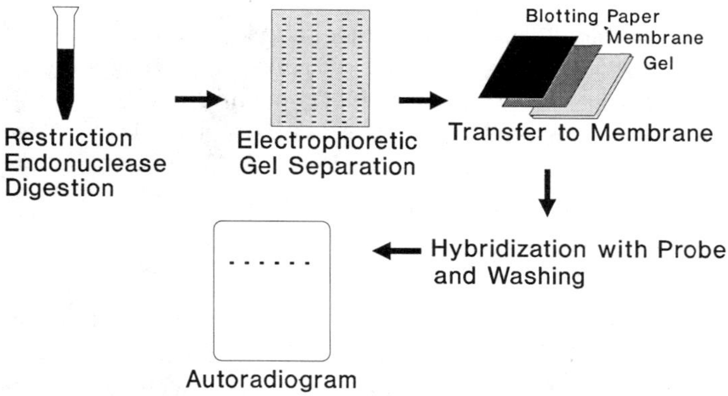

Figure 1 Schematic representation of Southern blotting.

nation. These filters are hybridized to labeled DNA probes that are specific for certain DNA sequence(s). The label incorporated into these probes can be radiolabeled or coupled to an antigen-type marker, which is then utilized in an enzymatic reaction. After the hybridization, the excess probe is removed by washing the filter, which is next either placed on film to detect the radioactive label or subjected to an enzymatically driven colorimetric reaction. The fragments that contain the complementary sequences for the probe appear as bands the size of which is determined by the relative location of restriction endonuclease sites limiting their ends.

2. Detection of Chromosome Fragments

In Turner syndrome the diagnosis is typically made by cytogenetic studies revealing a 45,X karyotype. Not all Turner patients, however, have just the 45,X karyotype; they are often mosaic for 45,X and 46,XX, or 45,X and 46,XY, or possess partial deletions of either the X or the Y chromosome, which also may be mosaic in distribution. Low-grade mosaicism is often difficult, or even impossible, to detect by current cytogenetic techniques because of the number of chromosome spreads that have to be counted to detect the mosaicism and the occurrence of tissue specificity of the mosaicism. Finally, detection of the origins of chromosomal fragments is often limited by the informativeness of stains and banding techniques. The importance of establishing the presence of a Y chromosome or Y fragment lies in the increased risk of gonadoblastoma developing in these patients [7]. By using probes specific for regions of the Y chromosome, and the techniques of Southern blot analysis, one can detect and map Y chromosome-specific fragments by hybridization with DNA that is derived from peripheral leukocytes or any other available tissue (Fig. 2) [8,9]. If one knows the portion of the Y chromosome from which a probe is derived, one can determine which part of the Y chromosome corresponds to the fragment visualized [8–10]. Southern blots are sensitive enough to detect mosaicism in the 1–2% range [8,9]. This same method can also be used to determine fetal sex at a very early gestational stage, without having to rely on cell growth for cytogenetic studies.

3. Somatic Cell Hybrids

Somatic cell hybrids have been used extensively to obtain specific probes and to regionally assign the loci for various probes [1,11–12]. By use of somatic cells that are human–rodent hybrids, cell lines have been selected that contain identified fragments of various human chromosomes, and libraries produced from these are enriched for corresponding sequences, especially if cell-sorting is used [12]. By isolating DNA from these cells and hybridizing it with the desired probe, one can determine the chromosomal origin of the probe. Somatic cell hybrids can also be used to study chromosomal deletions by incorporating chromosomes containing partial deletions, and the resulting deletion-specific hybrids may then be studied for the presence or absence of specific genes or gene regions [12].

4. Detection of Specific Gene Deletions

In one familial form of growth hormone deficiency (GH), called isolated growth hormone deficiency type IA (IGHD1A), the genetic diagnosis can be made by cleaving the DNA with restriction endonucleases and hybridizing it with the cDNA probe for GH [13,14]. Persons who have GH gene deletions do not contain genomic fragments to anneal with the GH cDNA sequences, and the resulting autoradiograms lack the fragments corresponding to those normally containing the GH gene (Fig. 3) [13]. Because other components of the GH gene family remain intact, their corresponding fragments are seen. Here, Southern blot analysis permits diagnosis of both the affected homozygote and the carrier and enables detection of a deletion defect that is not visible at the light microscope level.

5. Detection of Mutations: Sickle Cell Anemia

In sickle cell anemia, the molecular defect causes substitution of valine for glutamine in the β-globin gene, which, under certain conditions, leads to irreversible polymerization of the hemoglobin molecule and sickling of erythrocytes [13]. The DNA substitution responsible, the β^S or sickle variant, is a substitution of a thymidine for an adenine in the sixth codon of β-globin, changing GAG to GTG [15]. This substitution destroys a restriction site (CCTGAGG) for the endonuclease *Mst*II, which normally occurs at codons 5–7 of β-globin [15–17]. Loss of this restriction site leads to the production

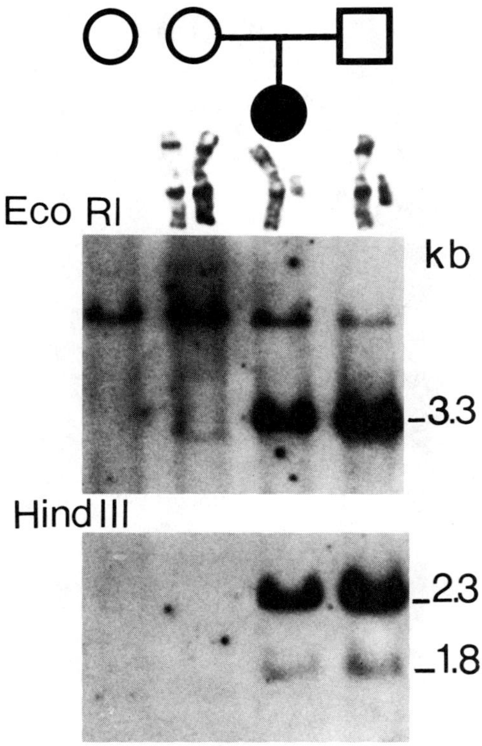

Figure 2 Chromosomal and DNA analysis of Turners' syndrome with 45,X/46,X + frag karyotype. Sex chromosomes from the parents and patient are shown beneath the pedigree. The autoradiogram beneath shows a female control (lane 1), mother (lane 2), child (lane 3), and father (lane 4) after digestion with *Eco*RI or *Hind*III and hybridization to a Y chromosome probe. Note the detection of Y-specific sequences in the patient (lane 3), indicating that the chromosome fragment is a Y derivative.

of a 1.35-kb fragment as opposed to the normal 1.15-kb fragment that results from normal β-globin genes [15–17]. This change can be detected by Southern blot analysis, identifying both the carrier and homozygous state. (Fig. 4) [15–17].

6. Detection of Mutations Using Allele-Specific Oligonucleotides: α_1-Antitrypsin

There are some mutations that do not affect a known restriction site. To detect these defects, mutation-specific oligonucleotide probes can be used which, in combination with highly specific filter-washing conditions, can

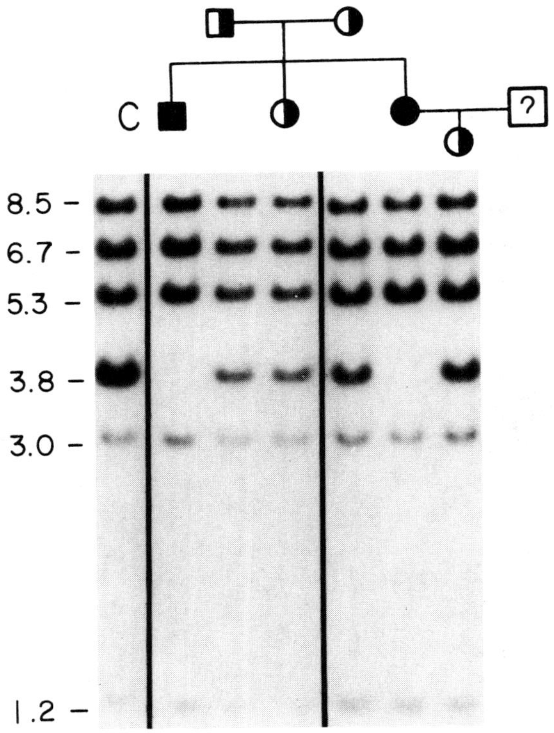

Figure 3 Pedigree and autoradiogram of DNA from a family with two affected individuals with IGHD1A. After digestion with *Bam*HI and hybridization with the GH probe, the affected individuals (solid square = son and solid circle = daughter) are missing the 3.8-kb fragments that normally contain GH1, whereas heterozygotes (half-solid symbols) show decreased amounts.

detect single-base substitutions [13,18–19]. α_1-Antitrypsin deficiency is one such disease in which the most common mutation (Z) results in a G to A substitution in codon 342 [18,19]. The oligonucleotide used to detect this mutation is 19 bases in length and codes for the normal (M) sequence [18,19]. At low-temperature hybridization and washing, the probe will bind both M and Z sequences; however, as the temperature is increased the probe eventually will appear bound to only the normal M sequence (Fig. 5) [18,19].

7. Detection of Chromosomal Rearrangements Associated with Neoplasia: Chronic Granulocytic Leukemia

A growing number of acquired forms of neoplasia are associated with characteristic chromosomal rearrangements. One of the best known, chronic

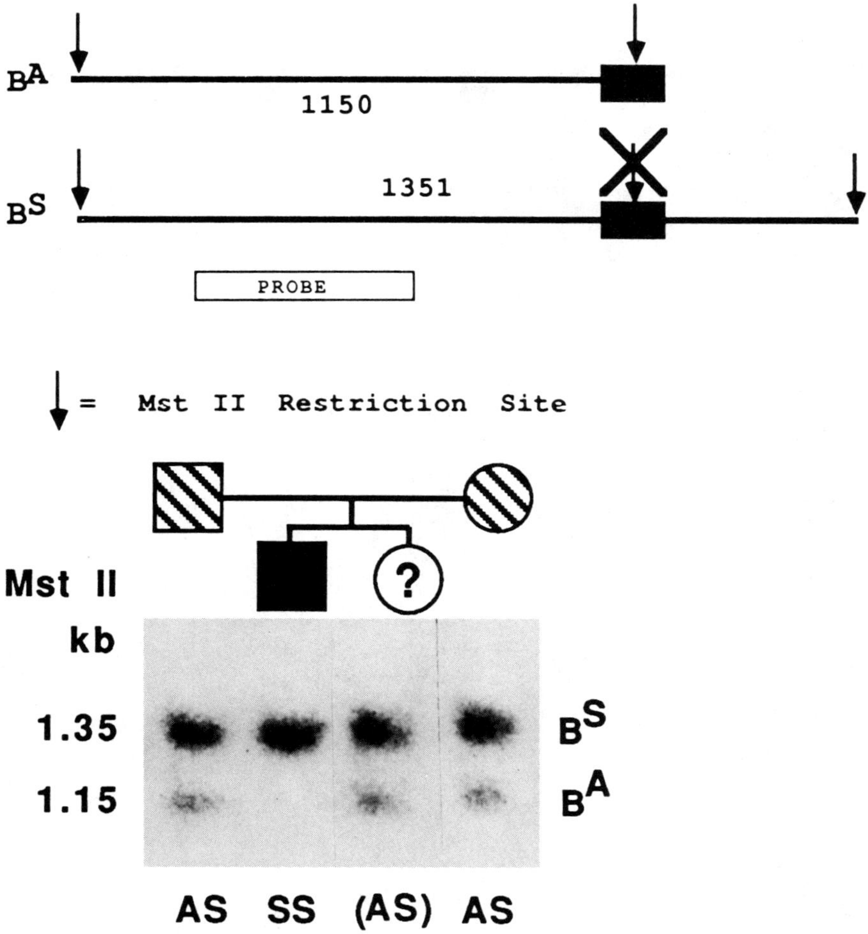

Figure 4 Diagram of *Mst*II site of β^A- and β^S-globin gene (above). Pedigree and autoradiogram of a family with the sickle cell gene (below). The affected child (solid square) is missing the 1.15-kb fragments associated with the β^A-globin gene and, instead, has only the 1.35-kb fragment derived from the β^S gene.

granulocytic leukemia (CGL), was found to be associated with the 9q34:22q11 rearrangement referred to as the Philadelphia or Ph[1] chromosome [20,21]. The Ph[1] rearrangement is now known to involve breakpoints in the *abl* protooncogene on 9q34 and the break cluster region (*bcr*) on 22q11 with use of probes derived from *bcr*, it is now possible to detect by Southern blots the altered restriction fragments that result from the *abl:bcr* fusion [22,23]. Such studies provide a sensitive and specific method to detect

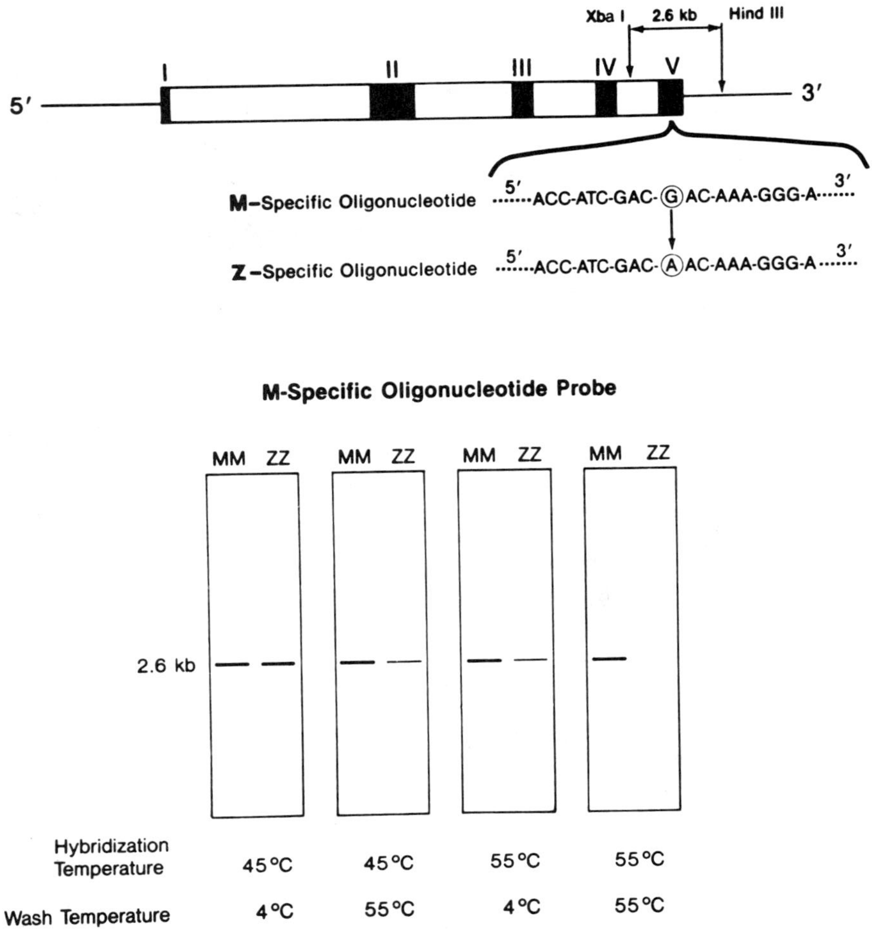

Figure 5 Detection of a single-base substitution using an oligonucleotide probe. The sequences of a portion of exon V of the normal (M) and mutant (Z) alleles of α_1-antitrypsin are shown above, and autoradiogram patterns of DNA from MM and ZZ individuals after digestion with *Xba*I and *Hind*III and hybridization to the M-specific oligonucleotide are shown below. Note the specificity of the M-specific oligonucleotide for the M allele under stringent hybridization and washing conditions.

mosaicism for Ph^1 chromosomes to monitor subjects' response to treatment and detect early relapses.

C. Indirect Methods of Detection Using Southern Blots

1. Introduction

When there is no direct method to detect a genetic defect because it does not alter a restriction site and an allele-specific oligonucleotide (ASO) probe is difficult to use, one must then rely upon indirect methods of following the transmission patterns of genes through pedigrees. These techniques involve the use of differences in the fragments generated by the restriction endonucleases, called restriction fragment length polymorphisms (RFLPs), to determine linkage relationships (4,13,24,25). This section will deal with these concepts and their use in clinical diagnosis.

2. Restriction Fragment Length Polymorphisms

It is estimated that variations in intergenic genomic DNA sequences occur about every 250–500 bp [4,5]. This variation sometimes leads to differences in the number and location of sites recognized by various restriction endonucleases. These RFLPs yield fragments of different sizes that, when separated by gel electrophoresis and hybridized to labeled probes, generate patterns which are codominant traits that can be used to trace transmission of alleles. Figure 6 illustrates how different fragment sizes might be visualized in the homozygous and heterozygous states after hybridization with a probe specific for the RFLP region. The size of the specific fragments, referred to as *alleles*, depends upon the presence or absence of a polymorphic endonuclease cleavage site. These polymorphic sites are usually located within the introns or the flanking regions of genes, rather than in the coding sequences, and they usually are unrelated to the functional qualities of the gene. By using RFLPs, one can trace the transmission of a particular gene or DNA segment through a family tree by the pattern of polymorphic alleles linked to it. The feasibility of such studies depends upon the parents being heterozygous (i.e., informative) for the RFLP, which, in turn, depends upon the frequency and distribution of the various RFLP alleles in the population under study. The informativeness of an allele, referred to as the *polymorphism information content* (PIC), within a population can be expressed mathematically by the equation, $PIC = 1 - A^2 - B^2 - N^2 - 2A^2B^2N^2$ (A = frequency of the first allele, B = frequency of the second allele, N = other alleles) [1].

The closer the PIC is to 1, the more informative the probe becomes. There are some probe–enzyme systems with over 20 distinct alleles available

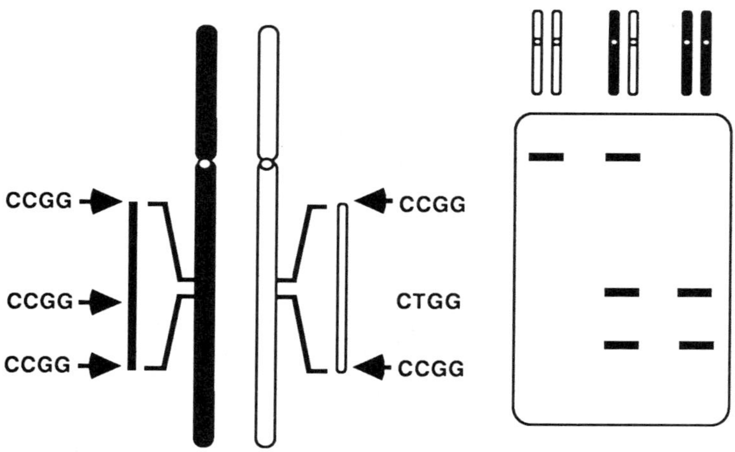

Figure 6 Schematic representation of a restriction fragment length polymorphism. Note that the substitution of a T for a C on the lighter chromosome eliminates the middle *CCGG* cutting site, leading to a larger fragment upon digestion, as shown in the autoradiogram on the right.

and PICs approaching 1; however, as the number of fragments increases, it becomes very difficult to distinguish one band from another.

3. Linkage

For many single-gene disorders there is either a polymorphic locus with few alleles, hence, a low PIC, or no RFLP at all. Here, one must rely upon closely linked markers (Fig. 7). The relationships between close markers are established through the study of statistical linkage, which evaluates the probability that two *syntenic* loci (loci on the same chromosome) will co-segregate. Mathematically, *linkage* is an expression of the percentage of recombination between two loci that occurs during meiosis I at the chiasmata of sister chromatids [26,27]. This recombination frequency is measured in centimorgans (1 cM = 1% recombination) and studies indicate that the human genome comprises approximately 3000 cM, thereby yielding about 1×10^6 bp/cM for small recombination distances (3×10^9 bp/3×10^3 cM) in the human genome.

The linkage relationships between syntenic loci can be determined by studying transmission patterns in families over several generations and noting the frequency of recombinations that occur between the loci (Fig. 8). The closer the genes are physically, the less likely they are to recombine and assort to separate gametes. The first linkage relationships were detected at the chromosomal level when it was noted that genes on the same chromo-

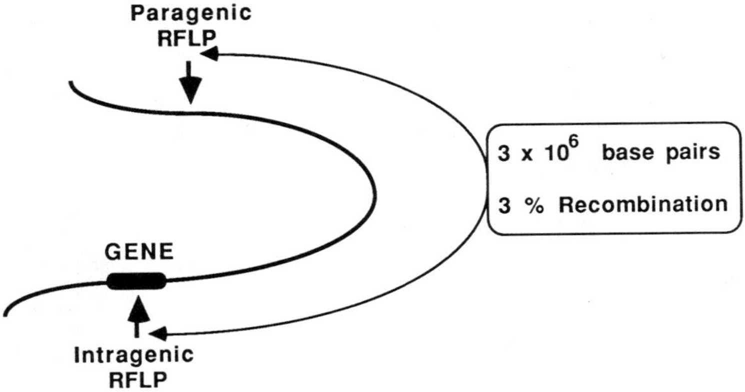

Figure 7 Diagrammatic representation of intragenic and paragenic RFLPs showing the correlation between physical distance in kilobases and recombination fraction.

some tend to travel together [26,27]. This was first described by T. H. Morgan and his colleagues at the beginning of this century in the fruit fly (*Drosophila melanogaster*) [26]. Morgan et al. observed that, occasionally, genes on the same chromosome did not cosegregate and that the closer the physical proximity of the genes, the less likely they were to recombine and assort independently [26,27]. The frequency of recombination, theta (Θ), is an expression of the percentage of all meiotic events in which recombination occurs. Therefore, a Θ of 0.5 (50% recombination) would indicate independent assortment, whereas a Θ of 0.0 would indicate that the two loci are very close together, linked, such that recombination occurs only very rarely.

4. LOD Scores

The probability of a given Θ is stated by means of the LOD or Z score. The LOD score, log of the odds, is a comparison of the probability of a particular Θ against the probability that the loci assort independently (i.e., Θ = 0.5) [26,27]. The probability of a given recombination frequency is given by the equation

$$L^{\Theta} = \Theta^{R}(1 - \Theta)^{S} L^{\Theta} = \text{likelihood of a given recombination frequency} \quad (1)$$

Θ = the given recombination frequency

R = the number of observed recombinants in a pedigree

S = the number of observed nonrecombinants

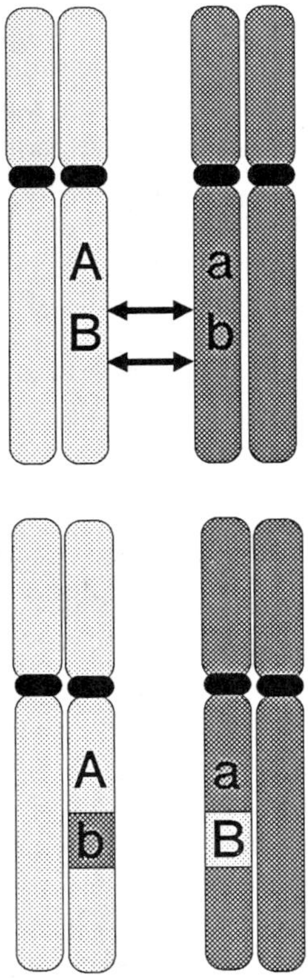

Figure 8 Recombination event during meiosis I in which sister chromatids exchange equivalent segments. Note that the recombinant chromatids have rearranged the combinations of the A(a) and B(b) alleles.

The LOD(Z) score is then determined by the equation [26,27]

$$\mathrm{LOD}(Z) = L^{\Theta}/L^{\Theta=0.5}.$$

A LOD score of 3 (odds ratio of 1000 : 1) is generally accepted as proof of linkage. The peak LOD score calculated for a range of Θs is used to select the most likely recombination frequency (method of maximum likelihood) [25,27]. A LOD score of 3, however, does not correspond to a P value of 0.001, but rather to a P value of about 0.04 because of a prior probability of linkage of 1 : 46, because there are 46 chromosomes [27]. The error range for the most likely Θ is determined by the two Θs with LOD scores corresponding to $Z^{max} - 1$, which provides a reliable confidence interval. The calculation of LOD scores rests on several assumptions: first, that the chiasmata at which recombination takes place in meiosis I are evenly distributed; second, that there is no interference of recombination between genes; and finally, that the position on the chromosome does not affect recombination. In fact, the frequency of recombination decreases as one approaches the centromere, and there are known areas of certain chromosomes where recombination is more likely to occur. Also, in males and females there is some variable degree of interference with recombination, as well as sex-specific differences in recombination rates [26,27].

To calculate the probability for a wide range of Θs for several families by hand or from tables would be prohibitively time-consuming. Fortunately, computer programs have been developed that perform these tasks quickly. One such program package, *Linkage*, calculates two-point and multipoint linkage, tests loci order, and differentiates male and female recombination rates [28,29]. With the current capacity of microcomputers, one can easily run these programs and calculate linkage with standard desktop units.

5. Linkage Maps

By combining and comparing Θs for several overlapping loci, one can construct a linkage map of a given region or linkage group of a chromosome. These linkage maps are useful for determining the relationships of markers in clinical studies, and for placing new markers in their proper relationship to established markers. Figure 9 is an example of a linkage map we constructed for the region of chromosome 17, encompassing the loci for growth hormone (GH1), protein kinase C α_1 (PKCA), collagen type 1 A1 (COL1A1), and the centromere marker D17Z1 [14]. Note that the linkage distances shown are not entirely additive, owing to the decrease in accuracy over longer distances in the recombination estimates and to variability in the random placement of chiasmata. When constructing linkage maps and performing linkage studies, several things must be considered. First, if the loci are close to the centromere, recombination events may be suppressed, resulting in Θ values, obtained for pericentromeric loci, representing greater physical distances than those derived for telomeric loci [27]. Second, recombination frequencies can vary greatly between male and female meiosis, with the female frequencies usually being greater [12,26,27]. This sex difference varies depending upon the chromosome and chromosome region being studied [27]. For routine linkage studies, however, the sex-specific recombination

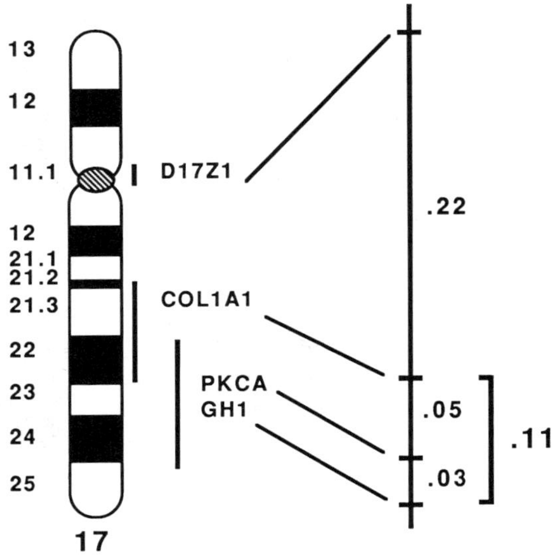

Figure 9 Linkage map of human chromosome 17 showing the relationships between chromosome morphology, regional assignment of loci, and linkage relationships.

frequencies are treated as equal, and the positional differences in Θ do not affect the statistical reliability of family studies.

6. Clinical Applications

Adult Polycystic Kidney Disease. Adult-onset polycystic kidney disease (APKD) is an autosomal dominant disorder affecting 1 : 500 to 1 : 1000 persons [30]. Affected individuals are usually asymptomatic until adulthood, at which time they develop enlarging cysts of the kidneys, as well as of the liver, spleen, and pancreas [30]. The resulting renal damage often progresses to end-stage renal failure, requiring dialysis or renal transplantation [30]. The accuracy of clinical methods for detecting the disease in the presymptomatic stage varies inversely with age, with renal ultrasound having a false-negative rate of 34% in the second decade of life [31]. More accurate presymptomatic detection of APKD could enable intervention before the development of advanced disease and provide useful information on disease status for relatives who are prospective renal transplant donors and are, themselves, at risk for having APKD.

Although the biochemical basis of APKD is unknown, Reeders et al. discovered that a clone named 3′ HVR, located near the α-globin gene on chromosome 16, demonstrated linkage at 5 cM with the APKD locus and has a high PIC value [32]. This combination makes linkage studies using 3′ HVR a useful tool in identifying affected individuals in APKD kindreds. Figure 10 shows a pedigree demonstrating how, by following transmission of the 3′ HVR probe, one can track transmission of the APKD gene. In this pedigree, it is inferred that the mother and son share the allele that cosegregates with the APKD gene. Note that the two daughters received their mothers alternate allele, therefore, leaving a 95% probability they did not inherit their mother's APKD allele (see later discussion). In addition to its diagnostic uses, the 3′ HVR probe also provides a means for identifying other useful probes as well as for testing candidate genes for recombination with the APKD locus.

Hemophilia A. Hemophilia A results from deficiency of clotting factor VIII. The factor VIII gene has been isolated, and probes that detect RFLPs have been identified for introns 18 and 22 [33]. Because the PIC value of these intragenic RFLPs are <0.5, quite often the female parents are homo-

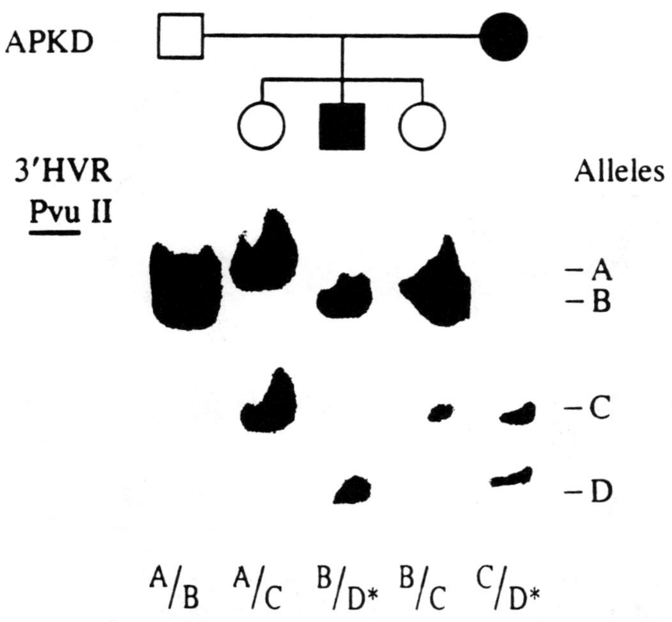

Figure 10 Tracing transmission of a gene through a closely linked probe. Shown are the autoradiogram patterns obtained using the 3′ HVR probe in a family with APKD. Note the two daughters did not inherit the same allele shared by their affected mother and brother (solid symbol) [32].

zygous for both loci, such that the transmission pattern of the mutant gene cannot be inferred from RFLP patterns (Fig. 11). In cases where the intragenic RFLPs are uninformative, one must rely upon closely linked markers or paragenic RFLPs for family studies. In these uninformative cases, there are two closely linked loci that detect RFLPs: DX13 (a two-allele system about 4 cM from factor VIII) and ST14 (a multiallele high-PIC-value probe about 3 cM from factor VIII), which can both be used to track transmission of the hemophilia gene [34,35]. By establishing the coupling phase between the mutant factor VIII gene and either the DX13 or ST14 alleles, one can infer the hemophilia status of various family members (Fig. 12). The reliability of the diagnosis derived by linkage studies can be calculated if the Θ value between the marker probe and the gene is known. A Θ of 0.04, indicating 4% recombination, leads to a decrease in reliability of 4% of the coupling phase per meiosis. If, however, several affected individuals are present, the accuracy of the coupling phase increases with each subsequent meiosis (i.e., for $\Theta = 0.04$: $n = 1$, $p = 0.96$; $n = 2$, $p = 0.9976$; $n = 3$, $p = 0.999904$), and the error is then reduced to that of the individual being studied (i.e., 4%). Figure 12 demonstrates this concept for a simple hemo-

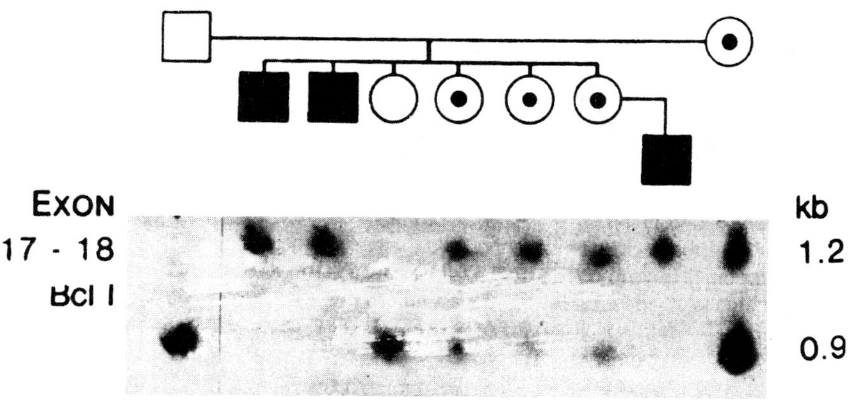

Figure 11 Autoradiogram patterns of DNAs from males with hemophilia A and their relatives following digestion with *Bcl*I and hybridization with the exon 17–18 probe derived by Gitschier [46]. Note that the −(1.2-kb) allele is in coupling with the mutant factor VIII: C gene in this family. The carrier (circles with dots) or noncarrier (open circles) status of females can be inferred from the segregation of this intragenic DNA polymorphism.

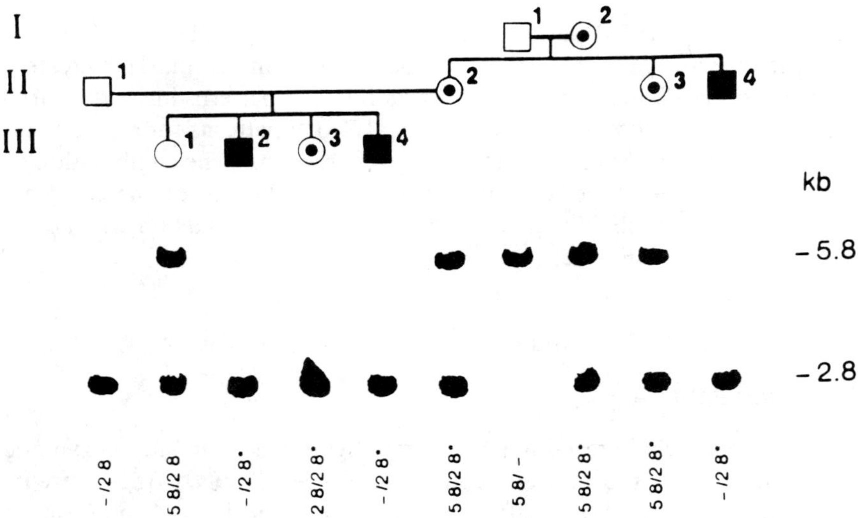

Figure 12 Pedigree of factor VIII family with DX13 probe. The statistical reliability of the coupling phase is indicated for both the affected (solid square) and study subject (light square).

philia pedigree probed with DX13 with one affected child and one unknown subject. The + allele indicates the presence of the restriction site and the − allele indicates its absence. The coupling phase appears to be with the + allele for DX13 in the affected child. The subject being studied has the − allele. The accuracy of the statement that the study subject does not have the defective factor VIII gene is as follows: For the meiosis of the affected child, the probability decreases to 96% and, with the subsequent meiosis of the study subject, the probability decreases to 92%. If there were another affected child with the + allele in this pedigree, the probability of determining the study subject's status would increase to 94.75%.

Cystic Fibrosis. Cystic fibrosis (CF) is the most common, lethal autosomal recessive disease in the United States, affecting 1 : 2000 white live births [17]. Affected individuals suffer from severe pulmonary disease, as well as gastrointestinal problems. The exact genetic defect in 70% of cases has been determined only recently, and the biochemical implications are still not fully understood [36]. However, extensive research has yielded numerous probes for diagnostic studies on chromosome 7 near the CF locus [37]. Besides their usefulness to families with affected children in linkage studies, these probes can also be used to assess carrier status-probability risks for members of the general population. The probes developed for CF are so closely linked to the

CF site that the RFLP patterns have undergone very little recombination with the putative CF ancestral gene. These tightly linked markers create a disproportionately high frequency of certain RFLP patterns in the CF carrier population. By compiling data on several RFLP patterns for an individual, one can generate haplotypes that are an expression of the combination of RFLPs. These haplotypes exist in linkage disequilibrium for the CF gene, thereby allowing probabilities to be assessed for individuals on whether or not they carry the CF gene.

III. POLYMERASE CHAIN REACTION AMPLIFICATION

A. Introduction

The polymerase chain reaction (PCR) amplification is an in vitro enzymatic process that permits large-scale replication of specific DNA fragments. Its ability to rapidly amplify specific fragments of known size and composition makes it an extremely powerful research and diagnostic tool.

B. Polymerase Chain Reaction Techniques

1. Polymerase Chain Reaction Amplification

By use of sequence-specific 15 to 25-base oligonucleotide primers, the DNA polymerase isolated from *Thermus aquaticus* (Taq polymerase) can amplify DNA fragments through successive cycles of melting, annealing, and elongation (Fig. 13) [38,39]. The thermal stability of Taq polymerase allows the enzyme to retain its activity when subjected to the 94°C temperatures required to melt dsDNA. This enzyme stability allows one to use automatic-cycling systems in which alternate heating and cooling cycles can be performed without adding additional enzyme. Selection of unique primer sequences is the basis of the specificity of PCR amplification. Primers are selected to contain unique sequences of DNA which will only pair with their complementary strands, thereby generating a single initiation site for the Taq polymerase. Considering that there are 6×10^9 bases of DNA per diploid cell and that the chances of finding a repeat of a given oligonucleotide are $(10^4)^n$, where n = the length of the primer (a 22-base length primer would have a 5.6×10^{-14} chance of an exact repeat), the chances of finding a completely unique sequence appears to be quite good, especially considering that a pair of primers is required for synthesis of a fragment. However, because cells use certain sequence patterns repeatedly, care must be taken to avoid sequences that are common among other genes such as poly-A and poly-T tracts and Alu repeat sequences, which are represented about 300,000 times [1]. The annealing temperature of the reaction is also important in the specificity of PCR amplification. The lower the annealing tem-

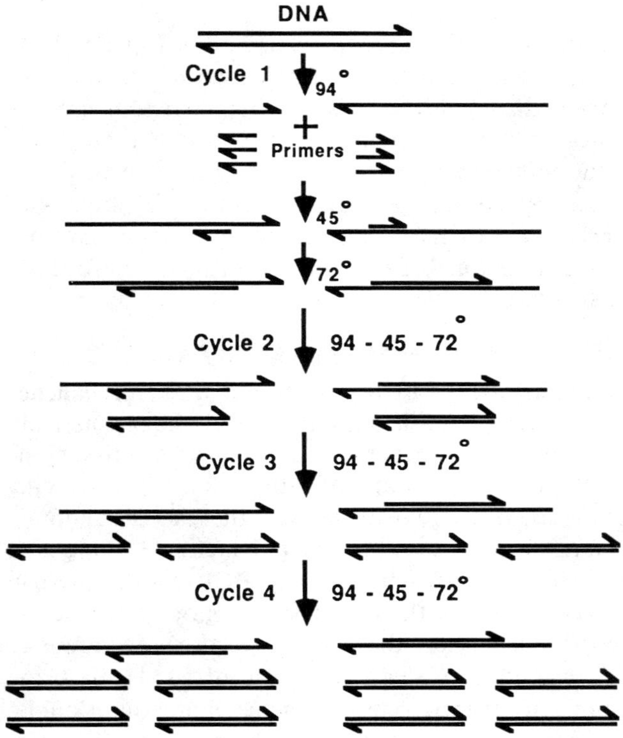

Figure 13 Schematic representation of polymerase chain reaction amplification showing complimentary genomic strands to be amplified (long arrows) and corresponding oligonucleotide primers (short arrows). Note the use of temperature in denaturation (94°C), annealing (45°C), and elongation (72°C) with the rapid accumulation of multiple PCR copies with cycles of amplification.

perature, the more nonspecific the annealing and the more likely that amplification of nonspecific fragments will result. By controlling the primer selection, the annealing temperature, and the buffer conditions, single-fragment specificity can be obtained from genomic DNA. The reaction takes only 2–3 hr to perform and allows rapid testing of specific DNA fragments. The technique has been successfully applied to DNA specimens obtained from blood, skin, spinal fluid, chorionic villi, amniotic fluid, hair, and urine, permitting a wide variety of sample collection techniques.

2. Restriction Fragment Length Polymorphism Analysis

PCR has many uses as a diagnostic tool in molecular genetics and has drastically reduced the time necessary to perform these tests. By amplifying specific fragments and cutting them with restriction endonucleases, many new RFLPs can be identified by PCR analysis. The ability to specifically amplify and detect much smaller segments of DNA than is possible with Southern blots allows one to distinguish RFLPs by using restriction endonucleases that have the more common (often 4-bp) recognition sites. These RFLPs, in some cases, detect the underlying mutation of the genetic defect involved and, in others, may be used as linkage markers for studies of transmission. Many genes, such as factor IX, which in the past have not been or were poorly informative for RFLPs, can now be analyzed for linkage with PCR techniques [40].

3. DNA Sequence Analysis Using the Polymerase Chain Reaction

One of the most discriminating uses of PCR is in the area of direct sequence analysis. Taq polymerase enables specific amplification of fragments from genomic DNA, which can then be sequenced, without having to first clone them into sequencing vectors. After PCR amplification, dideoxy sequencing is used, which involves the use of dideoxy nucleotides that, when randomly incorporated, stop 3′ elongation by removing the polymerase 3′ bond site, thereby generating fragments of different size [41]. PCR can be used in dideoxy sequencing with primers specific for a desired region. The reaction is started in its usual conditions, except that a disproportionate amount of one primer over the other is added (usually in a ratio of 50 : 1). By using disproportionate amounts of primer, one can make numerous double-stranded copies of the desired sequence, exhausting the supply of the lesser primer. The primer in excess can then be used to drive the ensuing sequence reaction. After the initial phase of amplification is complete, primer molecules specific for the single strands just produced are added along with labeled dideoxy nucleotides to randomly terminate polymerization. The label can either be radioactive, in which case four reactions are necessary, or fluorescent, in which case, with four different fluors, only one reaction is necessary. Once the reaction is complete, the fragments are subjected to separation in an electrophoretic field, and the sequences can then be exposed to film in the case of ^{32}P-labeled nucleotides, or to a fluorescent scanner during electrophoresis. The use of the fluorescent markers allows the process to be performed with an automated system that uses a photodetector device during electrophoresis, which can subsequently reduce the time needed to generate sequence data.

C. Direct Methods of Detection Using Polymerase Chain Reaction Amplification

1. Chromosome Fragment Detection

As discussed earlier in this chapter, the presence or absence of Y chromosome fragments is clinically significant in the Turner syndrome. By using Y chromosome-specific primers, Y chromosome sequences from patients with Y chromosome fragments, Y chromosome mosaics, and Y fragment mosaics can be easily detected (Fig. 14) [10]. This technique is potentially sensitive enough to detect these Y chromosome sequences at dilutions of 1 : 100,000 with female DNA [10]; It can be completed in 1–2 days and requires only a few microliters of blood [10]. This technique has been used in combination with cell-sorting to demonstrate the presence of fetal cells in maternal circulation, occasionally amplifying DNA from fewer than 10 cells [D. Bianchi, personal communication]. The major technical difficulty with this procedure involves avoiding any minute contamination by cells from males because of the process' sensitivity.

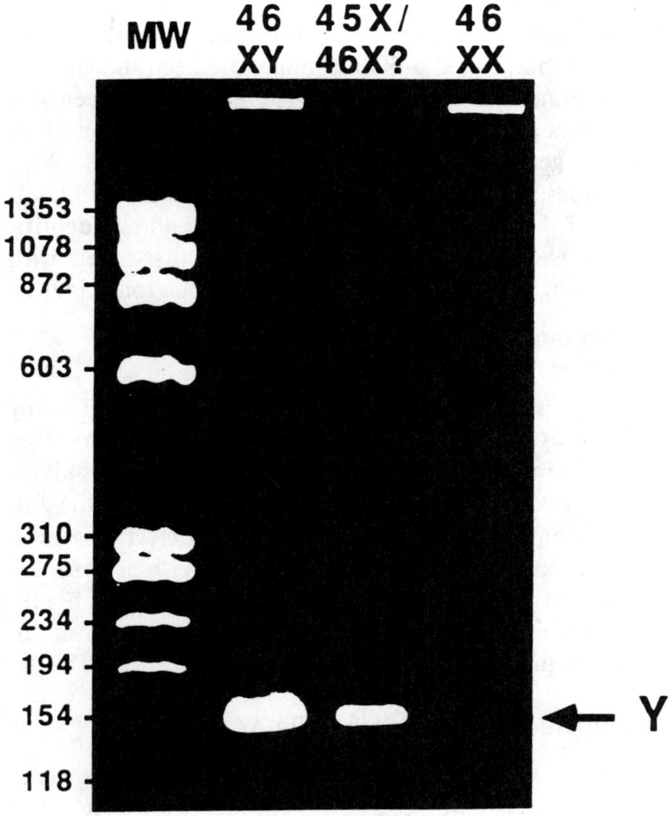

Figure 14 PCR amplification of 45,X/46,Xfrag in a Turner's syndrome patient (lane 2) with both father (lane 1) and mother (lane 3) controls.

2. Detection of Mutations by Polymerase Chain Reaction Amplification and Restriction Analysis: Sickle Cell Anemia

With use of PCR techniques, the single base change of sickle cell anemia can also be detected. By amplifying the portion of DNA encoding the defective *Mst*II site, one can test for the presence of the β^{sickle} hemoglobin gene and detect both the affected and carrier states. The amplified PCR product is subjected to enzyme cleavage by *Mst*II, and the fragments are then separated by gel electrophoresis, looking for the presence of the cleavage site (Fig. 15). The amplified fragment is 726-bp long and contains four *Mst*II recognition sites [42]. If the β^{sickle} substitution is present, a 341-bp fragment is generated that otherwise would be absent. This method can provide results within 24 hours of obtaining the sample, even from amniotic fluid, making it an effective prenatal and clinical screening tool.

3. Detection of Mutations by Polymerase Chain Reaction Amplification and DNA Sequencing: Thalassemia

Thalassemia is another defect involving the hemoglobin molecule. Those thalassemias affecting β-globin, called β-thalassemia, have been shown to be due to over 40 different mutations. Because the β-globin gene has been well characterized, PCR primers are known that can enable amplification of all segments of the β-globin gene region that spans about 3.3 kb [42]. With PCR sequencing techniques, patients can now have their mutations identified directly by analysis of the DNA sequence. This method allows identification of the exact defects and has clarified the defects associated with specific RFLP haplotypes in many different racial and ethnic groups [42].

4. Detection of Chromosomal Rearrangements Associated with Neoplasias: Certain Lymphomas

There are specific interchromosomal translocations associated with many neoplasias [43]. The most frequent in human lymphoma involves t(14;18)(q32;q21) in approximately 85% of follicular small cleaved cell lymphomas [44]. The translocation is associated with a 150-bp region in the 2.8-kb major breakpoint region (45). Southern blotting can detect this rearrangement only if the cells constitute 1–5% of the sample. With use of PCR primers from each side of the breakpoint, the cells containing the breakpoint can now be detected in minuscule quantities, thereby providing an effective monitoring technique.

D. Indirect Methods: Detection of Hemophilia A

In some mutational defects, the number of mutations, the size of the gene, or other factors make sequencing or direct detection impractical. Here, PCR can be used in linkage studies by amplifying and digesting regions generat-

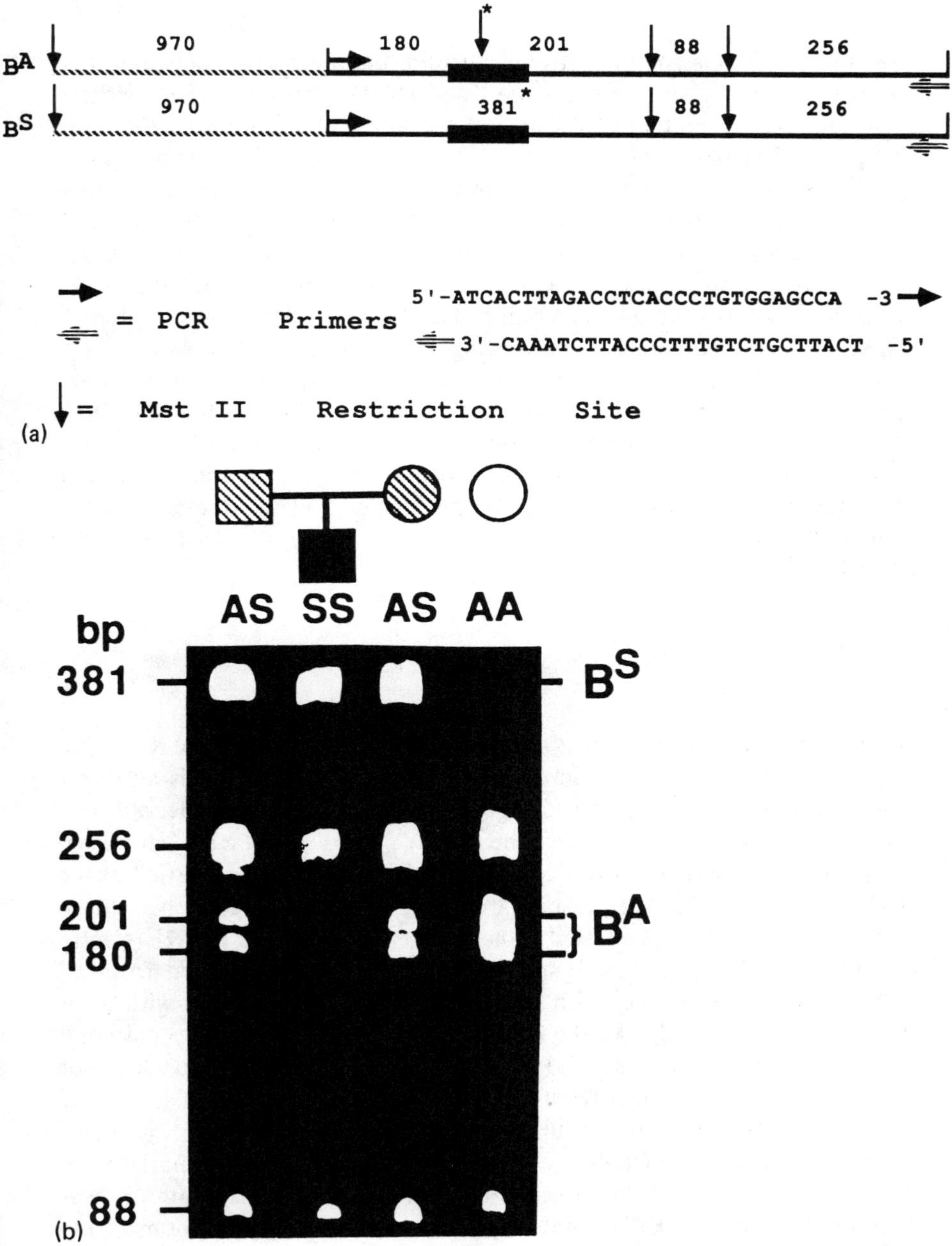

Figure 15 (a) Diagram of *Mst*II sites in PCR amplification products obtained using β^A and β^S globin templates. (b) Polyacrylamide gel analysis of digested PCR products (see Fig. 4 for symbols).

ing polymorphisms common to the general population. This data can then be used in linkage studies the same way that data generated by Southern blots are used to track mutant genes. An example of this involves the polymorphism detected in intron 18 of the factor VIII gene. By designing primers to encompass the polymorphic region of intron 18 and subjecting them to digestion with *Bcl*II, the same RFLP previously detected by intron 18 probes and Southern blotting can be accomplished on PCR-amplified products without using probes [46]. This test can be performed on DNA isolated from blood spots on scalp hair bulbs. This permits one to conduct family studies without necessitating blood being drawn from participants at a medical facility. Kits can be mailed with instructions and equipment to collect the samples, and the samples can then be returned by normal post. This approach may greatly increase participation in family studies by removing many of the geographic and personal obstacles to participation. The ease and rapidity with which PCR can be performed will make many studies previously confined to a few research facilities available on a more widespread clinical scale.

IV. NEW TECHNIQUES

A. Linkage Maps

To identify and study new genes and loci as they become available requires a reference system. The construction of the human linkage map is a project with this goal in mind. By providing linked markers at known intervals over the human genome, new genes can be rapidly placed within a known framework, and informative markers can be identified to aid in diagnostic linkage studies. To accomplish this, a standard set of family pedigrees is needed on which all markers can be tested. One such system, named CEPH (Centre d'Etude du Polymorphisme Humain), consists of DNA isolated from all members of 42 + families with three generations each, usually with eight plus children per family. All the individuals in the CEPH families have had lymphoblastoid cell lines created to produce a virtually unlimited supply of DNA. With this system, a researcher studying a gene need not test many different markers looking for linkage with the gene, but, instead, may test his marker against the CEPH system and compare it with the markers for which data have already been generated. There are currently data on over 1000 markers in the CEPH database, and the probability of demonstrating linkage with any of these is high and improves as more markers are added.

B. Pulsed-Field Gel Electrophoresis

Although techniques have been available for some time for the efficient separation of small ($<20 =$ kb) DNA fragments, large megabase-sized fragments have proved very difficult to work with, until the advent of pulsed field gel electrophoresis (PFGE) [47]. The technique of PFGE works by alternating the field directions within the gel matrix, thereby decreasing the effects of tertiary structures on the migration of the DNA by forcing it to unfold to change directions [47]. This technique permits an efficient enough separation of large DNA fragments to facilitate screening for gene locations much more quickly. With the use of restriction endonucleases that are infrequent cutters, such as *Not*I, large fragments of DNA can be generated for study by pulsed-field gels. These segments can then be further studied by inserting them into yeast artificial chromosomes (YACs) from which libraries of large fragments can be used for mapping studies [48,49].

C. Future Polymerase Chain Reaction Applications

1. Polymerase Chain Reaction: An Alternative to Cloning

The PCR amplification reaction is being used as an alternative to cloning a variety of DNA fragments [50,51]. If sequence data are known about a desired fragment, then direct amplification can be used to isolate the fragment from subjects for study. If the sequence data for a fragment are unknown, then sequencing methods such as inverted PCR can be used to provide the sequence data. The application called inverted PCR uses complimentary oligonucleotide primers [52]. After restriction enzyme digestion and self-circularization of PCR-amplified fragments with DNA ligase, the PCR primers are then used to sequence the unknown segment in the same fashion as regular PCR sequencing. By using these methods, sequence data can be generated very rapidly without risk of incorporating contaminating plasmid or bacterial sequences.

2. Detection of Mutations by Chemical Cleavage and Dissociation Gradient Gels

There are two new techniques that can be used for detection of single-base mismatches in genomic sequences. The first of these, called chemical cleavage, uses strands of DNA synthesized by PCR from a control and from an individual thought to have an alteration in a given gene. These strands are then annealed forming dsDNA heteroduplexes composed of strands from control and mutant DNA. These hybrids are then subjected to treatment with either hydroxylamine or osmium tetroxide, which on exposure to piperidine, generates cleavage of the DNA where a mismatch occurs [53,54]. Hydroxylamine detects cytosine mismatches and osmium tetroxide detects cytosine and thymidine mismatches [54]. By analyzing the number and sizes of DNA fragments generated by this process, the location and nature of single-base mismatches may be deduced.

The second process utilizes denaturing gradient gels in which hybrid sequences are run in either a urea or formamide gradient gel that favors strand dissociation as a function of the distance traveled into the gel [55]. The degree of mismatch correlates with dissociation of the complementary strands, and the resulting single-stranded DNA runs much more slowly in the gel. This method can be used as an alternative to direct sequencing for sequence differences between a control and a large group of test DNAs [55].

V. SUMMARY

Within the next few years, many of the foregoing new techniques will be available on a widespread basis. With these techniques and those already established, diagnosis of genetic and acquired chromosomal disorders will become more and more common in the clinical setting. Prenatal and neonatal diagnosis, as well as carrier detection, will provide useful data for family planning that is based upon more accurate information. Early detection of genetic disorders and possibly early detection of certain malignancies should also lead to earlier and more effective therapeutic interventions and, ultimately, to decreased morbidity and mortality. Finally, judging from the progress seen in the last 10 years, tremendous insights should be gained into how genes and chromosomal sequences are normally maintained and function. No doubt, this knowledge will, in turn, show that certain phenotypes, not now perceived to be familial, are, in many cases, due to genetic or to acquired chromosomal changes.

ACKNOWLEDGMENTS

This work is supported in part by National Institutes of Health Grants T32DK070601 (M. L. Summar), AM01434 and AM35592 (J. A. Phillips). The authors wish to thank Judy Copeland for expert preparation of this manuscript, and Valorie Summar for assistance in proofreading the text.

REFERENCES

1. McKusick, V. A. (1988). *The Morbid Anatomy of the Human Genome*. Williams & Wilkins, Baltimore.
2. Southern, E. M. (1975). Detection of specific sequences among DNA fragments separated by gel electrophoresis. *J. Mol. Biol. 98*:503–517.
3. Feinberg, A. P., and B. Vogelstein (1983). A technique for radiolabeling DNA restriction endonuclease fragments to high specific activity. *Anal. Biochem. 132*:6–13.

4. Gusella, J. F. (1986). Recombinant DNA techniques in the diagnosis of inherited disorders. *J. Clin. Invest.* *77*:1723–1726.
5. Cooper, D. N., B. A. Smith, H. J. Cooke, S. Niemann, and J. Schmidtke (1985). An estimate of unique DNA sequence heterozygosity in the human genome. *Hum. Genet.* *69*:201–205.
6. Gantt, P. A., J. R. Byrd, R. B. Greenblatt, and P. G. McDonough (1979). A clinical and cytogenetic study of fifteen patients with 45,X/46,XY gonadal dysgenesis. *Fertil. Steril.* *34*:216–221.
7. Guellaen, G., M. Casanova, C. Bishop, D. Geldwerth, G. Andre, M. Fellows, and J. Weissenbach (1984). Human XX males with Y single-copy sequence from the human Y chromosome. *Nature 307*:172–173.
8. Butler, M. G., V. G. Dev, J. A. Phillips III, S. P. Tho, J. J. Trill, J. A. Tischfield, and P. G. McDonough (1986). A child with 45,X/46,Xdel(Y)(q12) identified with a Y specific probe. *Fertil. Steril.* *46*:718–720.
9. Burk, R. D., M. A. Patrick, and K. D. Smith. (1985). Characterization and evaluation of a single-copy sequence from the human Y chromosome. *Mol. Cell Biol.* *5*:576–581.
10. Maness, K., V. Hannig, M. Summar, V. G. Dev, J. A. Phillips, D. W. Bianchi, and U. Muller (1989). Detection of Y chromosome specific fragments in Turner syndrome subjects using polymerase chain reaction amplification. *Am J. Hum. Genet.* *45*:A82.
11. Kuhn, L. C., A. McClelland, and F. H. Ruddle (1984). Gene transfer, expression, and molecular cloning of the human transferrin receptor gene. *Cell 37*: 95–103.
12. Donis-Keller, H., and D. Botstein (1988). Recombinant DNA methods: Applications to human genetics. In *Molecular Genetics in Medicine*, B. Childs, N. A. Holtzman, H. H. Kazazian, Jr., and D. L. Valle (eds.). Elsevier Science, New York, pp. 17–42.
13. Phillips, J. A. III (1988). Clinical applications of gene mapping and diagnosis. In *Molecular Genetics in Medicine*. B. Childs, N. A. Holtzman, H. H. Kazazian, Jr., and D. L. Valle (eds.). Elsevier Science, New York, p. 68.
14. Summar, M. L., J. A. Phillips, M. R. S. Krishnamani, J. Keefer, R. C. Schwartz, P. Tsipouras, H. F. Willard, and A. Ullrich. (1989). Protein kinase C: A new linkage marker for growth hormone and for COL1A1. *Genomics 5*: 163–165.
15. Wilson, J. T., P. F. Milner, M. E. Summer, F. S. Nallaseth, H. E. Fadel, R. H. Reindollar, P. G. McDonough, and L. B. Wilson (1982). Use of restriction endonucleases for mapping the allele for β^s-globin. *Proc. Natl. Acad. Sci. USA* *79*:3628–3631.
16. Chang, J. C., and Y . W. Kan (1982). A sensitive new prenatal test for sickle-cell anemia. *N. Engl. J. Med.* *307*:30–32.
17. Orkin, S. H., P. F. R. Little, H. H. Kazazian Jr., and C. D. Boehm (1982). Improved detection of the sickle mutation by DNA analysis: Application to prenatal diagnosis. *N. Engl. J. Med.* *307*:32–36.
18. Kidd, V. J., R. B. Wallace, K. Itakura, and S. L. Woo (1983). α_1-Antitrypsin deficiency detection by direct analysis of the mutation in the gene. *Nature 304*: 230–234.

19. Kidd, V. J., and S. L. C. Woo (1984). Recombinant DNA probes used to detect genetic disorders of the liver. *Hepatology 4*:731–736.
20. Rowley, J. D. (1973). A new consistent chromosomal abnormality in chronic myelogenous leukemia identified by quinacrine fluorescence and Giemsa straining. *Nature, 243*:290–293.
21. deKlein, A. A., A. Geurts van Kessel, G. Grosveld, C. R. Bartram, A. Hagemeijer, D. Bootsma, N. K. Spurr, N. Heisterkamp, J. Groffen, and J. R. Stephenson (1982). A cellular oncogene is translocated to the Philadelphia chromosome in chronic myelocytic leukemia. *Nature 300*:765–767.
22. Groffen, J., J. R. Stephenson, N. Heisterkamp, A. de Klein, C. R. Bartram, and G. Grosveld (1984). Philadelphia chromosomal breakpoints are clustered within a limited region, *bcr* on chromosome 22. *Cell 36*:93–99.
23. Collins, S. J. (1986). Breakpoints on chromosomes 9 and 22 in Philadelphia chromosome-positive chronic myelogenous leukemia (CML). *J. Clin. Invest. 78*:1392–1396.
24. Bell, G. I., M. J. Selby, and W. J. Rutter (1982). The highly polymorphic region near the human insulin gene is composed of simple tandemly repeated sequences. *Nature 295*:31–35.
25. Jeffreys, A. J., V. Wilson, and S. L. Thein (1985). Hypervariable minisatellite regions in human DNA. *Nature 314*:67–73.
26. Ott, J. (1974). Estimation of the recombination fraction in human pedigrees: Efficient computation of the likelihood for human linkage studies. *Am. J. Hum. Genet. 26*:588–597.
27. Conneally, P. M., and M. L. Rivas (1980). Linkage analysis in man. *Adv. Hum. Genet. 10*:209–266.
28. Lathrop, G. M., and J. M. Lalouel (1984). Easy calculations of LOD scores and genetic risks on small computers. *Am. J. Hum. Genet. 36*:460–465.
29. Lathrop, G. M., J. M. Lalouel, C. Julier, and J. Ott (1985). Multilocus linkage analysis in humans: Detection of linkage and estimation of recombination. *Am. J. Hum. Genet. 37*:482–498.
30. Dalgaard, O. Z. (1957). Bilateral polycystic disease of the kidneys. A follow-up of 284 patients and their families. *Acta Med. Scand. Suppl. 328*:1.
31. Bear, J. C., P. McManamon, J. Morgan, R. H. Payne, H. Lewis, M. H. Gault, and D. N. Churchill (1984). Age at clinical onset and at ultrasonographic detection of adult polycystic kidney disease: Data for genetic counseling. *Am. J. Med. Genet. 18*:45–53.
32. Reeders, S. T., M. H. Breuning, K. E. Davies, R. D. Nicholls, A. P. Jarman, D. R. Higgs, P. L. Pearson, and D. J. Weatherall (1985). A highly polymorphic DNA marker linked to adult polycystic kidney disease on chromosome 16. *Nature 317*:542–544.
33. Antonarakis, S. E., P. G. Waber, S. D. Kittur, A. S. Patel, and H. H. Kazazian, Jr. (1985). Hemophilia A: Detection of molecular defects and of carriers by DNA analysis. *N. Engl. J. Med. 313*:842–848.
34. Harper, K., R. M. Winter, M. E. Pembrey, D. Hartley, K. E. Davies, and E. G. Tuddenham (1984). A clinically useful DNA probe closely linked to haemophilia A. *Lancet 2*:6–8.

35. Oberle, I., G. Camerino, R. Heilig, L. Grunebaum, J. P. Cazenave, C. Crapanzano, P. M. Mannucci, and J. L. Mandel (1985). Genetic screening for hemophilia A (classic hemophilia) with a polymorphic DNA probe. *N. Engl. J. Med. 312*:682–686.
36. Riordan, J. R., J. M. Rommens, B.-S. Kerem, N. Alon, R. Rozmahel, Z. Grzelczak, J. Zielenski, S. Lok, N. Plavsic, J.-L. Chou, M. L. Drumm, M. C. Iannuzzi, F. S. Collins, and L.-C. Tsui. (1989). Identification of the cystic fibrosis gene: Cloning and characterization of complementary DNA. *Science 245*:1066–1073.
37. Beaudet, A. L., G. L. Feldeman, S. D. Fernbach, G. J. Buffone, and W. E. O'Brien (1989). Linkage disequilibrium, cystic fibrosis, and genetic counseling. *Am J. Hum. Genet. 44*:319–326.
38. Mullis, K. B., and F. A. Faloona (1987). Specific synthesis of DNA in vitro via a polymerase-catalyzed chain reaction. *Methods Enzymol. 155*:335–350.
39. Mullis, K., F. Faloona, S. Scharf, R. Saiki, G. Horn, and H. Erlich (1986). Specific enzymatic amplification of DNA in vitro: The polymerase chain reaction. *Cold Spring Harbor Symp. Quant. Biol. 51*:263–273.
40. Attree, O., D. Vidaud, M. Vidaud, S. Amselem, J. M. Lavergne, and M. Goossems (1989). Mutations in the catalytic domain of human coagulation factor IX: Rapid characterization by direct genomic sequencing of DNA fragments displaying an altered melting behavior. *Genomics 4*:266–272.
41. Gyllensten, U. B., and H. A. Erlich (1988). Generation of single-stranded DNA by the polymerase chain reaction and its application to direct sequencing of the *HLA-DQA* locus. *Proc. Natl. Acad. Sci. USA 85*:7652–7656.
42. Saiki, R. K., C. A. Chang, C. H. Levenson, T. C. Warren, C. D. Boehm, H. H. Kazazian, Jr., and H. A. Erlich (1988). Diagnosis of sickle cell anemia and β-thalassemia with enzymatically amplified DNA and nonradioactive allele-specific oligonucleotide probes. *N. Engl. J. Med. 319*:537–541.
43. Rowley, J. D. (1982). A new consistent chromosomal abnormality in chronic myelogenous leukemia identified by quinacrine fluorescence and Giemsa straining. *Nature, 243*:290–293.
44. Lee, M.-S., K. S. Chang, F. Cabanillas, E. J. Freireich, and J. M. Trujillo (1987). Detection of minimal residual cells carrying the t(14;18) by DNA sequence amplification. *Science, 237*:175–178.
45. Crescenzi, M., M. Seto, G. P. Herzig, P. D. Weiss, R. C. Griffith, and S. J. Korsmeyer (1988). Thermostable DNA polymerase chain amplification of t(14;18) chromosome breakpoints and detection of minimal residual disease. *Proc. Natl. Acad. Sci. USA 85*:4869–4873.
46. Kogan, S. C., M. Doherty, and J. Gitschier (1987). An improved method for prenatal diagnosis of genetic diseases by analysis of amplified DNA sequences: Application to hemophilia A. *N. Engl. J. Med. 317*:985–990.
47. Schwartz, D. C., and C. R. Cantor (1984). Separation of yeast chromosome-sized DNAs by pulsed field gradient gel electrophoresis. *Cell 37*:67–75.
48. Murray, A. W., and J. W. Szostak (1983). Construction of artificial chromosomes in yeast. *Nature 305*:189–193.

49. Burke, D. T., G. F. Carle, and M. V. Olson (1987). Cloning of large segments of exogenous DNA into yeast by means of artificial chromosome vectors. *Science* *236*:806–812.
50. Montandon, A. J., P. M. Green, F. Giannelli, and D. R. Bentley (1989). Direct detection of point mutations by mismatch analysis: Application to hemophilia B. *Nucleic Acids Res.* *17*:3347–3356.
51. Chamberlain, J. S., R. A. Gibbs, J. E. Ranier, P. N. Nguyen, and C. T. Caskey (1988). Detection screening of the muscular dystrophy locus via multiplex DNA amplification. *Nucleic Acids Res.* *16*:11141–11156.
52. Ochman, H., A. S. Gerber, and D. L. Hartl (1988). Genetic applications of an inverse polymerase chain reaction. *Genetics* *120*:621–623.
53. Grompe, M., and C. T. Caskey (1989). Mutation detection by chemical cleavage of amplified DNA in ornithine transcarbamoylase deficiency. *Clin. Res.* *37*:828A.
54. Cotton, R. G., N. R. Radriquez, and R. D. Campbell (1988). Reactivity of cytosine and thymine in single base pair mismatch with hydroxylamine and osmium tetroxide and its application to the study of mutations. *Proc. Natl. Acad. Sci. USA* *85*:4397–4401.
55. Traystman, M. D., M. Higuchi, S. E. Antonarakis, and H. H. Kazazian, Jr. (1989). Use of denaturing gradient gel electrophoresis to detect point mutations in the factor VIII gene. (submitted).

2

Cloning and Analysis of Large DNA Molecules

Stephen Scherer and Lap-Chee Tsui

University of Toronto and The Hospital for Sick Children, Toronto, Ontario, Canada

I. INTRODUCTION

As the organization of the human genome is unraveled, it becomes clear that many genes are much larger than initially realized. Some of those involved in diseases, such as hemophilia [1], retinoblastoma [2], Duchenne/Becker muscular dystrophy [3], and cystic fibrosis [4] are several hundreds or thousands of kilobase pairs (kb) in size. Isolation and physical characterization of these genes or other large DNA fragments in genome analysis are severely limited by the conventional phage- and cosmid-cloning systems. Although it is possible to extend a contiguous, cloned DNA stretch to a few hundred kilobase pairs by chromosome walking and jumping, such processes are notoriously tedious [4]. Hence, the ability to clone and propagate large DNA fragments will permit isolation and examination of many large human genes in their entirety, allow physical mapping of large chromosomal regions, provide the framework for genome sequencing, and accelerate identification of disease genes on the basis of their chromosomal location as deduced by linkage analysis.

Recent advances have been made in cloning large DNA fragments into vectors in *Saccharomyces cerevisiae* [5]. This yeast artificial chromosome (YAC) technology, together with pulsed-field gel electrophoresis (PFGE) [6], have become essential tools in physical mapping and studies of complex mammalian genomes. Large DNA molecules can be cloned as YACs with no apparent upper limit for their sizes; DNA inserts averaging at least ten times larger than those possible in a cosmid are routinely obtained. Besides its large cloning capacity, the YAC system may also allow isolation of sequences that are poorly represented in *Escherichia coli*-based libraries [7–9]. As a consequence, it will be possible to analyze chromosomal regions containing genes, gene clusters, and large functional genetic units that are otherwise impossible to study with the bacterial-cloning systems. An additional advantage of the YAC system is the ease with which large, cloned DNA fragments may be manipulated: specific nucleotide changes, insertions, and deletions can be introduced as desired with high frequency by means of homology-directed recombination [8,10]. Furthermore, YAC clones may be used in DNA-mediated gene transfer experiments aimed at studying the effect of site-directed mutagenesis and cloning of mammalian genes by biochemical selection.

The development of PFGE has made it possible to define physically the genetic entities for many microorganisms, such as *Neurospora* [11] and yeast [12], the chromosomes of which fail to condense during their growth cycle. Electrophoretic karyotyping by PFGE has revealed not only the number and size of the chromosomes in these species, but has also permitted easy assignment of cloned genes to specific chromosomes. In higher eukaryotes, PFGE has been used successfully with restriction enzymes that cut genomic DNA infrequently, for single chromosomes are much too large to be separated as bands by electrophoresis. As a result, physical (restriction) maps have been generated for many long stretches of DNA from different regions of the genome [3,13–15]. The ability to resolve large DNA fragments also allows easy detection of gross genetic alterations such as chromosomal translocations [16,17], deletions [18,19], and inversions [20].

In this chapter we will review the technical aspects of YAC cloning and illustrate how PFGE is used in the construction and characterization of YAC libraries. We will also discuss our preliminary results on some of the technical improvements and potential applications of these developing technologies concerning functional and structural studies in chromosome research.

A. The Yeast Artificial Chromosome-Cloning System

The YAC vector contains the basic functional elements necessary for replication and segregation of natural yeast chromosomes: centromeres [21–23], telomeres [21,24], and autonomously replicating sequences (ARSs), which exhibit properties expected of chromosomal origins of replication [25,26]. Yeast is the only eukaryotic organism for which these three functional units have been well-defined, to a few hundred base pairs (bp) each, and isolated as separate entities. When these elements are attached to biochemically selectable genes, the resulting DNA molecule (artificial chromosome) functions essentially as a normal chromosome [27,28]. The YAC libraries are created in vitro by ligating DNA fragments hundreds of kilobases in size between the two arms of the YAC vector (conveniently propagated as part of a bacterial plasmid). Each of these arms contains a biochemically selectable marker (e.g., *URA3* and *TRP1*) and a telomere sequence derived from *Tetrahymena*. In addition, one arm carries an autonomously replicating sequence (*ARS1*) and the other, a centromere (*CEN4*). Upon transformation into yeast, these artificial molecules can be maintained as telocentric chromosomes, at one copy per cell.

Short linear molecules (<15 kb) are, however, generally not as mitotically stable as similar, larger synthetic chromosomes (>72 kb) [28–31]. The latter observation indicates that the length of each YAC exerts a strong influence on the accuracy of chromosomal segregation; a similar conclusion can be derived from studies of deletion derivatives of natural yeast chromosomes [32]. Most YACs, however, still appear to be less stable than natural yeast chromosomes of comparable size. This difference may be attributed to the fact that multiple *ARS* elements exist at 30- to 40-kb intervals in natural yeast chromosomes [33] and that most YACs carry only three *ARS* elements—a yeast *ARS1* on the vector and the two *Tetrahymena* ARS elements associated with the telomeres [34], although additional sequences from the cloned DNA may provide heterologous *ARS* function in *S. cerevisiae* [33]. Nevertheless, the large cloning capacity and the relative stability of insert DNA sequences, together with the advanced understanding of the molecular biology of yeast [10], provide an ideal cloning system for large DNA molecules.

B. Principles of Pulsed-Field Gel Electrophoresis

In an electric field, all DNA molecules in solution move with essentially the same average velocity. Although the charge moving the molecule increases linearly with the size of the molecule, so does the friction retarding the migration, resulting in an electrophoretic mobility independent of molecular weight. In conventional gel electrophoresis, size fractionation of DNA molecules is possible owing to increased retardation conferred by the gel matrix, but not for molecules larger than 50 kb, as their effective size and shape become indistinguishable from each other. DNA molecules are, however, capable of undergoing conformational changes under the influence of an electric field to match the gel pore size, allowing penetration and migration through the matrix. As a result, extended DNA molecules snake through the gel following a leading end, with the remainder of the molecule trailing behind. The path chosen by the leading end is assumed to be a random walk, biased by the electric force applied. These properties thus form the basis for PFGE, first introduced by Schwartz and Cantor [6] who achieved size separation of large DNA molecules by periodically alternating the electric field, forcing the DNA molecules to change their direction of migration through an agarose gel. With this principle, DNA molecules from 50 kb to 10 megabase pairs (mb) in size [11] are routinely being fractionated with specially designed gel apparatuses [35–37].

During PFGE, an initial electric field is applied in one direction, and large DNA molecules respond by attempting to enter the gel at many positions along the molecules' length, resulting in the formation of extended U-shapes. One of the arms of the U-shaped molecule slips behind, allowing the opposite arm to achieve leading status. When the DNA molecule reaches its contour length (relaxed state), net motion is achieved. After a set time (pulse time), a second electric field in an alternate direction is applied, forcing the DNA molecules to distort their shape, attempting once again to achieve net migration in the direction defined by the new electric field. Because large DNA molecules spend a greater proportion of the pulse time in reorienting themselves than smaller molecules do, mobility becomes a function of size of the molecule. However, once the relaxation time exceeds the pulse time, DNA molecules no longer have time to reorient, and they all migrate through the gel at the same velocity.

Various designs of the PFGE system have been developed in attempts to improve the resolution and effective range of separation. Most modifications concern the field geometry, for example, orthogonal field alternation gel electrophoresis [38], field inversion gel electrophoresis [39], contour-clamped homogeneous electric field gel electrophoresis [35], tranverse alternating gel electrophoresis [40], crossed-field gel electrophoresis [41], rotating gel electrophoresis [42], programmable autonomously controlled electrodes [36], pulsed homogeneous orthogonal gel electrophoresis [43], and zero integrated field electrophoresis [44].

II. YEAST ARTIFICIAL CHROMOSOMAL LIBRARY CONSTRUCTION

The procedures described here will focus on the methodology used in our laboratory for the construction of an *Eco*RI partial-digest library using the vector pYAC4. These protocols are applicable for preparation of libraries generated using rare-cutting restriction enzymes and other, appropriate YAC vectors.

A. Vectors and Strains

The pYAC series of cloning vectors and the YAC-cloning system have previously been described in detail [5]. The vector used in the present discussion, shown in Figure 1, consists of two arms that supply all the elements necessary for maintenance of artificial chromosomes in *S. cerevisiae*. The left arm of the vector provides (1) a *Tetrahymena* telomere [31] that acts as a substrate for the addition of yeast telomeres and ensures chromosome stability and completion of replication of the ends of the linear artificial chromosome, the telomere also provides weak *ARS* activity [34]; (2) a centromere (*CEN4*) allowing proper meiotic and mitotic segregation [45]; (3) an autonomously replicating sequence (*ARS1*) for DNA replication [46]; and (4) the biosynthetic gene *TRP1* [46], which serves as a selectable marker in an auxotrophic *trp1* host strain. The right arm of the vector supplies a telomere defining the opposite end of the artificial chromosome and the biosynthetic marker gene *URA3* [47], which complements the host *ura3* mutation.

The cloning sites of this vector are located within a 14-bp intron of the *SUP4* (*ochre*) gene [48], which serves as a dominant selectable marker gene and divides the two arms of the vector. The *SUP4* gene encodes a tyrosine tRNA that is capable of suppressing nonsense *ochre* (UAA) stop codons [49]. Successful ligation of a foreign DNA insert at any of the cloning sites

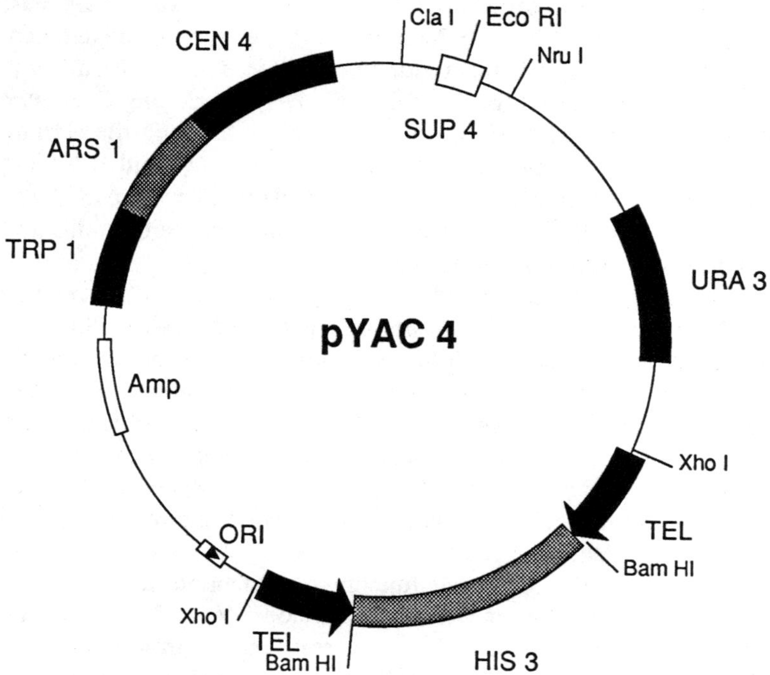

Figure 1 Example of a YAC vector.

causes interruption of the *SUP4* gene, destroying the suppressor activity that is required to release a block in the purine biosynthetic pathway conferred by the *ADE2-1* mutation in the host [50]. Blockage of this pathway leads to the accumulation of intermediates with a red hue, thus allowing YAC transformants with inserts to be identified as red colonies. Religation of the arms regenerates a functional *SUP4* gene, however, resulting in suppression of the *ADE2-1* mutation, which confers a white colony phenotype. The vector is propagated in *E. coli* as a circular plasmid with the ampicillin resistance gene from pBR322 as the selectable marker. The left arm of the vector includes the large fragment of a *Pvu*II/*Bam*HI digest of pBR322, whereas the right arm contains the small fragment of such a digest, providing end-specific probes for mapping YAC clones with inserts. Finally, another DNA segment is inserted between the two telomeres to permit separation of the two arms; this DNA segment is derived from the *HIS3* gene [31,51], of which the function per se is not important for the yeast replicon.

Different cloning sites are introduced into pYAC vectors for different purposes. The original series pYAC4, pYAC55, and pYAC3, contain *Eco*RI, *Not*I, and blunt-end restriction sites, respectively [5]. Variations of these vectors include: pYAC-RC with rare-cutting sites [52]; pYAC4-NEO [53] and pYAC neo [54] capable of conferring G418 resistance in mammalian cells; pTYACI [55], pYAC4-NEO-NOT [56] and pTV2 [57] with only one telomere for cloning telomeres by complementation in yeast [58]; pMB-1 with a *Schizosaccharomyces pombe* selectable marker and *ARS* for cloning *S. pombe* centromeres [9]; pYAC4-LLI [59] with T3 promotors flanking the cloning site for end-rescue by the polymerase chain reaction [60]; and pYACP-1 carrying *Drosophila* P-element ends and a selectable marker [8]. pJS97 and pJS98 are YAC vectors that carry each YAC arm on separate plasmids and allow plasmid rescue from both ends [61].

The host strain most commonly used for the pYAC vectors is *S. cerevisiae* AB1380 (*MATa* ψ^{+}*ura3 trp1 ade2-1 can1-100 lys2-1 his5*). AB1380 is an appropriate host not only because it harbors multiple mutations conferring the complex auxotrophic phenotype that can be complemented by the selectable markers present on the pYAC vector, but it can be transformed with vector DNA at a high frequency. To ensure that reversion of the auxotrophic phenotype (particularly *ura* and *trp*) has not occurred, however, the genotype of this host strain should initially be verified by plating on appropriate selective media [62]. When the colony color assay is to be used, it is important to ensure that the host strain has functional mitochondria (ρ^{+}) and the cytoplasmic determinant that enhances suppression is (ψ^{+}). Mitochondrial function can be confirmed by the ability of yeast isolates to grow on medium containing nonfermentable substrates such as glycerol (YPDG: see Step 1.3 for recipe) [62]. Once a transformation-efficient host strain is identified, it should be maintained as a glycerol working stock, kept at $-80°C$; it should be restreaked on YPD medium [62] no longer than 1 week before transformation experiments; prolonged storage of the host strain on YPD medium at 4°C is not recommended.

B. Size-Fractionation of High-Molecular-Weight Genomic DNA

To ensure an ample supply (>100 μg) of extremely high-molecular-weight (HMW) DNA for YAC cloning, special DNA isolation techniques are required. Carefully prepared HMW DNA can be processed through a sucrose step-gradient originally described for the isolation of intact chromosomal DNA from yeast spheroplasts [38,49]. Many sources of genomic DNA, including peripheral leukocytes [5,54,63–65], plant protoplasts [66], yeast spheroplasts [5], *C. elegans* embryos [7], *Drosophila* nuclei [8], *Myxococcus* [67], human–hamster somatic cell hybrids [61, 64–65], and animal tissues, have been prepared successfully with this procedure. Alternatively, whole cells (spheroplasts or protoplasts for cells with cell walls) can be embedded in agarose blocks [6], lysed in situ, followed by extensive proteolytic digestion, and the DNA molecules size-fractionated by PFGE [61,68]. Both methods have been used successfully in producing large quantities of HMW DNA for library construction, but the sucrose gradient method is practical only if an unlimited amount of DNA is available [5], whereas the agarose block method is simple and may be used for any DNA source. The latter method is particularly attractive when working with a limited amount of DNA (such as flow-sorted chromosomes) [61], or when precise size-fractionation is necessary (e.g., cloning a specific DNA fragment from a pulsed-field gel), or when extremely HMW DNA is required. When attempting to construct libraries with large inserts, smaller DNA molecules must be removed, regardless of how carefully the partial digestion is controlled, because they would contribute to a much higher proportion of intact ends for ligation and transform yeast at a much higher frequency than larger molecules.

Step 1.1 Preparation of High-Molecular-Weight DNA

To prepare HMW DNA in agarose blocks (as described in the PFGE section), a concentration of 2.0×10^7 cells/mL is used for lymphoblastoid cell lines or human–hamster hybrids. The general rule is to have 2–5 μg of DNA per block, regardless of the cell source; the low DNA concentration is necessary to avoid overloading problems that lead to trapping of molecules in a different region of the gel. For preparative fractionation, blocks are treated with *Eco*RI to generate ~ 100 μg of partially digested DNA. DNA molecules ranging in size from 23 kb up to 1 mb may be generated (Fig. 2A).

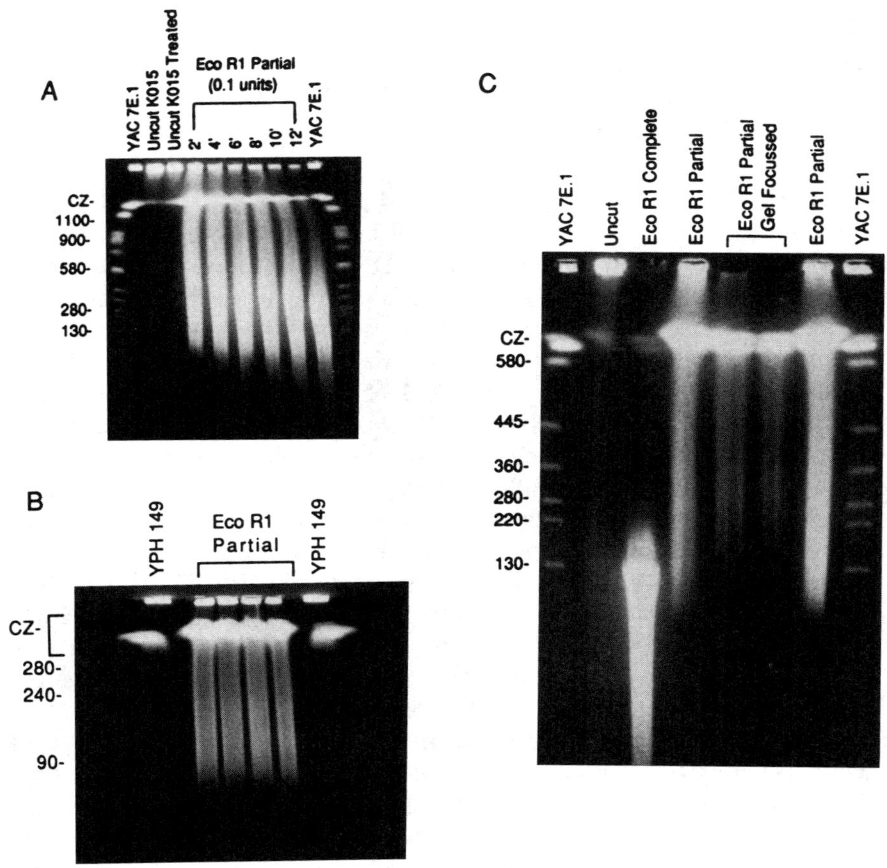

Figure 2 Preparation of high-molecular-weight DNA by PFGE fractionation: (A) Partial restriction enzyme digestions of HMW DNA contained in agarose blocks. The example shows DNA from a human–hamster somatic cell hybrid 4AF/102/K015 (which contains human chromosome 7 as its only human material). The agarose blocks were prepared as described in Step 2.1 and the partial digestions were performed with restriction enzyme *Eco*RI (see Step 1.1). The other lanes are YAC 7E.1 (a YAC clone serving as a good size marker); uncut (untreated KO15 DNA, an agarose block loaded directly on the gel); and uncut, treated (mock-treated with restriction enzyme as control for DNA degradation). (B) Concentration (focusing) of HMW DNA (partially digested with *Eco*RI) by PFGE. Four DNA-containing agarose blocks were digested with *Eco*RI for the amount of time indicated and fractionated in a 0.3% LMP agarose gel (with a 1% agarose mold as support). The three smallest yeast chromosomes of YPH 149 are shown as size markers. (C) PFGE analysis of gel-concentrated HMW DNA. The LMP gel slice containing the compression zone (CZ) from Figure 2B was processed to produce a liquid DNA sample

Before enzyme digestion, blocks are washed twice in a 20-fold excess of sterile double-distilled water (ddH_2O) for 30 min each at 4°C. At this point, blocks can be processed individually or in a group. Individually, each block is placed in an Eppendorf tube and equilibrated with 1 × *Eco*RI restriction enzyme buffer (100 mM NaCl, 10 mM $MgCl_2$, 50 mM Tris-HCl pH 8.0, 4 mM spermidine) and *Eco*RI restriction enzyme (1 μ/mL), in a final volume of 100 μL (buffer and block) for 30 min at 4°C. Enzyme digestion is initiated by incubation at 37°C. To produce desirable partial digestion products, the incubation is extended for different time intervals, up to a total of 10 min. In our experience, a 2-min incubation period gives the best results (see Fig. 2A). Digestion is terminated by the addition of 1mL of 10 mM EDTA. To avoid band migration distortion caused by high local DNA concentrations, it is important to choose a digestion condition that will yield a homogeneous and uniform distribution of different size-class DNA molecules upon PFGE fractionation. Overdigestion will cause nonuniform separation of DNA molecules and will result in trapping of small molecules in the compression zone (see the 6- to 12-min products in Fig. 2A).

PFGE may be performed in any apparatus that fractionates DNA with uniform mobility across the width of the gel, but it may be necessary to run more than one gel to prepare enough DNA with gels of small platforms. In our experience, the rotating gel system described by Southern et al. [41] produces the most desirable results for this purpose. To facilitate recovery of DNA after PFGE, low-melting-point (LMP) agarose (BRL, Ultra Pure) is used as the electrophoretic medium; an LMP agarose concentration of 0.3% is recommended, although concentration up to 1.0% is also acceptable. For gels of low percentage LMP agarose, a 1% agarose gel is cast to act as a supportive mold. The LMP agarose is also poured around the blocks to prevent them from floating away from the wells after insertion. To ensure complete solidification of the LMP agarose, the gel may be stored at 4°C for 15 min after casting at room temperature. The gel is also allowed to equilibrate with the buffer (0.5 × TBE) prior to electrophoresis, which is performed at 12°C with a predetermined voltage and pulse time to generate a compression zone capturing DNA molecules of the size range of interest (see

---

(see Step 1.1). The two lanes on either side of the gel-concentrated (focused) DNA represent DNA partially digested with *Eco*RI but not gel concentrated.

All three PFGE experiments were performed on a crossed-field gel apparatus with 0.5 × TBE at 12°C with a 110° reorientation angle. The running conditions were as follows: (A) 6 V/cm for 20 hr and a linearly ramped pulse time of 45–90 sec; (B) 4 V/cm for 10 hr at a constant switching time of 14 sec; and (C) 6 V/cm for 20 hr with a linearly ramped pulse time of 20–40 sec.

later discussion for conditions). Oligomeric λ phage DNA and yeast chromosomes are both ideal size-markers for determining the optimal conditions.

By controlling the pulse and running time, lower-molecular-weight species can be easily removed from the compression zone (see Fig. 2B). For example, we established that a voltage gradient of 4 V/cm and a pulse time of 14 sec throughout a 10-hr run is ideal for avoiding molecules <200 kb. These conditions allow the compression zone to form as a tight band; the volume of gel containing the wanted DNA is reduced to a minimum— ~4 mL when starting with 20 blocks. The compression zone is then excised without ethidium bromide staining and transferred to a polypropylene tube (e.g., Falcon 2070) containing 50 mL of 1 × TAE/100 mM NaCl/10 mM EDTA (buffer for agarase digestion). At this point, the DNA in the LMP agarose can be processed immediately or stored at 4°C for up to several months. Also, a portion of the LMP gel slice may be examined by PFGE to verify the absence of undesired, smaller molecules, and by test ligations and transformation into yeast (see Step 1.7).

The gel slice is allowed to equilibrate in the agarase buffer for >1 hr at 4°C and then transferred to a 15 mL polystyrene tube (Falcon 2057) with 300 μL of fresh buffer. The gel is slowly dissolved by increasing the temperature from 42° to 68°C over a 15-min period and holding at 68°C for an additional 5 min. Once liquefied, the molten agarose is cooled down to 37°C and agarase is added to 75 units/mL to allow complete hydrolysis of agarose to sugar oligomers [69]. [DNase-free agarase may be obtained from Calbiochem; the lyophilized enzyme can be reconstituted in 50% glycerol at 10 units/μL and stored for at least 6 months at −20°C.] To avoid rapid mixing that would shear HMW DNA, agarase is added by diluting the required amount of enzyme in 200 μL of buffer, prewarmed to 37°C, and slowly dispersing the solution throughout the molten gel. Extreme care is also taken to avoid shearing the HMW DNA that is in solution in all subsequent steps.

After agarase digestion, which usually takes 4–6 hr at 37°C, the liquefied DNA sample is transferred to an Eppendorf tube ready for phenol extraction. An equal volume of buffer-equilibrated phenol (37°C) is added slowly by pipette, down the side of the tube. The tube is laid on its side for 20 min to increase the interface between the two solutions (no rapid mixing is necessary). The tube is then spun in a microfuge for 5 sec and the aqueous layer transferred to a new tube with a broadened Eppendorf pipette tip. A further extraction with chloroform/isoamyl alcohol (24 : 1) is performed in the same way, and the final aqueous phase is transferred to an ultrathimble (molecular weight cut-off of 75,000) in a collodion bag concentrator (Schleicher & Schuell UH 100/1). The DNA solution is concentrated to a final volume of <500 μL by dialyzing against successive changes of TE placed in the outer chamber with a vacuum of ~500 mm/Hg. A portion of the HMW DNA preparation should be examined by PFGE to obtain a concentration estimate and confirm the absence of molecules <200 kb in size. Intact, HMW DNA should remain in the compression zone (see Fig. 2C).

C. Preparation of Vector DNA and Library Construction

Step 1.2 Vector DNA Preparation and Ligation

The vector DNA is prepared by the conventional cesium chloride gradient procedure [70] and is ready for further manipulation. To separate the two arms of pYAC4, ~100 μg of the purified, supercoiled plasmid is digested to completion with *Bam*HI and *Eco*RI. The double-digestion may be done sequentially, with *Bam*HI added first, followed by increasing the salt concentration and addition of *Eco*RI; the digestion is carried out for 1 hr each at 37°C [Use the appropriate enzymes and conditions for vectors other than pYAC4]. A sample of the product may be analyzed by agarose gel electrophoresis to determine if cleavage is complete. Upon complete digestion with these two enzymes, pYAC4 should yield three bands—6.0, 3.7, and 1.7 kb in size—corresponding to the left arm, the right arm, and the *HIS3* throwaway fragment. This mixture of vector arms and the *HIS3* fragment is then deproteinized by phenol and chloroform extraction, precipitated in ethanol, and dried.

To reduce transformation background, the vector arms are dissolved in a buffer containing 50 mM Tris HCl, pH 8.0, 1 mM spermidine, and 0.1 mM EDTA, and treated with calf intestinal phosphatase (CIP; Boehringer/Manheim). The reaction is carried out with 0.5 units of CIP at 37°C for 30 min (~0.01 CIP per 1 pmol of 5′ ends of DNA [70]). The reaction is terminated by heating for 10 min at 68°C, the sample extracted with phenol/chloroform, and DNA precipitated in ethanol and redissolved in 50 μL of TE. A sample of the vector arm preparation should be examined by ligation tests to ensure that it is unable to self-anneal, yet fully capable of joining foreign *Eco*RI fragments.

Because the pYAC vector may occasionally delete portions of the telomere sequence upon subculturing, individual plasmid preparations should be examined by digesting with *Hind*III to ensure that all five fragments (3.5, 3.0, 1.9, and a doublet at 1.4 kb) are present. If the 1.4-kb doublet is resolved into two bands, alteration must have occurred within the telomere segment(s).

After all the control tests are checked, the vector arms are ready for ligation with the HMW DNA. It is important to note that a vast excess of arms is used in this ligation reaction to avoid the formation of chimeric

YACs caused by ligation events involving two or more insert fragments before the addition of the YAC arms. Generally, an equal amount (by weight) of HMW DNA partially digested with *Eco*RI and dephosphorylated pYAC4 are used. For example, 50 μL of the dephosphorylated vector (100 μg) and 50 μL of 10 × ligation buffer (25 mM Tris HCl [pH7.5], 10 mM $MgCl_2$, 1 mM DTT, 0.4 mM ATP) may be gently dispersed into 400 μL (100 μg) of HMW (insert) DNA. The mixture is then allowed to equilibrate on ice for 30 min before it is brought up to 600 μL by addition of 50 units of ligase (IBI) in 100 μL of 1 × ligation buffer. After a further equilibration step on ice for 30 min the ligation mixture is left at 15°C overnight. Next day, the ligated DNA sample is gently extracted with phenol, followed by chloroform/isoamyl alcohol (24 : 1) as described previously, transferred to a UH 100/75 ultrathimble, and dialyzed/concentrated against TE for 2 hr. The final product is then transferred to an Eppendorf tube and stored at 4°C, ready for yeast transformation. A portion of the ligation product should be examined by PFGE to ensure that the DNA is intact.

D. Transformation of Yeast Spheroplasts

The ligation product is next used to transform *S. cerevisiae* AB1380 by polyethylene glycol-mediated spheroplast transformation [71,72]. The first step of this procedure is the removal of the yeast cell walls by treatment with lyticase. The spheroplasts, maintained intact in isotonic buffer, are then mixed with the ligation reaction in the presence of calcium ions, followed by polyethylene glycol. The resulting spheroplast aggregates are allowed to recover in a rich nutrient medium before plating (embedding) in a regeneration agar medium, which also selects for yeast clones that have acquired the *URA3* marker on the right arm of pYAC4. The transformation efficiency is usually higher when the right arm, rather than the left arm, is selected at this initial step. In addition, the background caused by self-ligation (circularization) of the right arm is less stable and less likely to occur, as the resulting product lacks the centromere and *ARS* sequences. All primary transformants expressing *URA3*, regardless of size of colony or color, are transferred by toothpicks to double-selection plates (SD–URA–TRP) and incubated at 30°C for 2 days. It is noteworthy that, although the color selection, based on the interruption of the *SUP4-ochre* gene, allows discrimination between recombinants and nonrecombinants, this process, in our experience, is slow and inconsistent, especially for transformants deeply embedded in the top agarose. Many of the slow-growing, white colonies eventually develop the expected $URA^{+}TRP^{+}$ red phenotype when grown on SD–URA–TRP medium in the presence of oxygen. It is also noteworthy that the red pigment development may be enhanced by placing the culture plates at 4°C.

Other transformation protocols (such as the lithium acetate [73] and the sodium alginate [54,74,75] procedures) that do not require the embedding step have been described. It is possible to transfer and replica-plate the yeast colonies directly with these methods, thereby greatly simplifying the screening step. The transformation frequency is, however, generally lower in our hands than the embedding method.

The following steps describe a modified spheroplast transformation protocol that provides a high-transformation efficiency, at 1–2 × 10^6 colonies/μg of intact vector [76]. The efficiency is much lower for ligation products; a good ligation reaction with size-selected HMW DNA generally yields more than 1000 YAC transformants per microgram vector arms, however.

***Step 1.3** Spheroplast Transformation*

1. Transfer a single large colony of AB1380 to 100 mL of YPD broth (see following) with vigorous shaking (200 rpm on a gyratory shaker) overnight at 30°C. Inoculate 100 mL of fresh YPD the next morning with 5 mL of the overnight culture and shake for 3 hr until it reaches a density of 3–5 × 10^7 cells/mL (OD_{660} of 2–4).
2. Harvest the cells by centrifugation at 400–600 *g* (2000–3000 rpm in a clinical centrifuge) for 10 min at room temperature. [All steps are carried out at room temperature unless otherwise indicated.]
3. Wash the cells by resuspension of the pellet in 20 mL of sterile ddH_2O followed by centrifugation. To reduce the centrifugation time without loss of cells, the cell suspension is divided into two aliquots in 15-mL polystyrene tubes (Falcon 2057).
4. Wash the cells again with 20 mL of 1 M sorbitol (Fisher) and resuspend them in 20 mL of SCEM (see following) containing fresh 2-mercaptoethanol (β-ME; 41 μL of 14.4 M stock solution per 20 mL) and 1000 units of lyticase (partially purified enzyme from Sigma). [The enzyme can be stored for 4 months as a 100 units/μL stock in sterile ddH_2O at 4°C.]
5. Incubate the cell suspensions in a 30°C water bath with slow shaking for 15–20 min. Monitor the rate of spheroplast formation by measuring the decrease in turbidity (at 800 nm) of a ten-fold-diluted suspension in 1 M sorbitol. When the spheroplast formation has proceeded to 80–90% completion (10–20 min), collect the cells by gentle centrifugation at 600 rpm in a clinical centrifuge for 5 min. Gently remove the supernatant from each tube without disturbing the pellet.
6. Wash each of the spheroplast pellets first with 10 mL of 1 M sorbitol and then with 10 mL of STC (see following). Gently resuspend each

pellet in 1 mL of STC and pool the suspensions before transformations.

7. Transformation of spheroplasts: Transfer 150 μL of the foregoing yeast spheroplast suspension to each of the transformation (15 mL polystyrene) tubes containing the following controls or ligation products:
   a. 1–10 ng of YCp50 (or any other control sup⁻ yeast plasmid with a centromere) plus 2 μg calf thymus DNA (red transformants)
   b. 1–10 ng of pYAC4 plus 2 μg calf thymus DNA (white transformants)
   c. 1–10 μL of ligation product from Step 1.2, with volume adjusted to 10 μL with STC
   d. 10–20 μL of LMP ligation product from Step 1.7.

   Incubate at room temperature for 10 min.
8. Add 1 mL of PEG–STC solution (see below) to each transformation sample and gently, but thoroughly, mix the contents. Incubate at room temperature for another 10 min.
9. Pellet the transformed spheroplasts by centrifugation at 600 rpm for 5 min. Gently remove the PEG–STC supernatant with a Pasteur pipette.
10. Carefully resuspend the pellet in 150 μL of SOS medium (see following) and incubate without shaking at 30°C for 30 min.
11. Gently resuspend the spheroplasts in 8 mL of prewarmed (48°C) TOP agar (see following); invert three times to mix before overlaying on selective SORB medium (dilute spheroplasts in SOS medium if necessary). Incubate agar plates at 30°C. Transformants by vectors only are visible after 2–3 days; YAC transformants grow much slower and they take 3–5 days to show.

    [*Note*: 50% of all molecules with ligated ends will contain the same vector arms making them biologically unstable (dicentric or acentric). In addition, a high proportion of the insert DNA will contain one or more ends that lack an appropriate site for ligation; there is no need to remove such molecules; they may act as carrier DNA protecting genuine YAC clones from shearing].

Media and reagents for spheroplast transformation

YPD medium: 1% yeast extract, 2% peptone, and 2% glucose. Autoclave. Add 1.5% agar for plates.

YPDG medium: 1% yeast extract, 2% peptone, 3% v/v glycerol, 0.1% glucose, 2% agar. Autoclave.

SCEM: 1 M sorbitol, 0.1 M sodium citrate (pH 5.8), 10 mM EDTA, and 30 mM β-ME. Filter-sterilize and add β-ME just before use.

STC: 1 M sorbitol, 10 mM Tris HCl (pH 7.5), and 10 mM $CaCl_2$. Filter-sterilize and store at room temperature.

PEG-STC: 10 mM Tris HCl (pH 7.5), 10 mM $CaCl_2$, and 20% polyethylene glycol 8000 (Fischer). Filter-sterilize and store at room temperature.

SOS medium: 1 M sorbitol, 6.5 mM $CaCl_2$, 0.25% yeast extract, and 0.5% peptone (supplemented with 10 μg/mL uracil for YAC transformations). Filter-sterilize and store at room temperature.

TOP agar: 1 M sorbitol, 0.67% yeast nitrogen base without amino acids (Difco), 2% glucose, and 2% agar. Autoclave. Nutrient supplements (minus uracil), filter-sterilized, are supplemented for YAC transformation as described in the yeast-cloning manual [62]. Adenine sulfate (filter-sterilized) is supplemented at 10 mg/L to allow the development of the red color associated with the *ade2-1* mutation in AB1380.

SORB medium: 1 M sorbitol, 0.67% yeast nitrogen base without amino acids, 2% glucose, and 2% agar. Autoclave. Supplement with 10 mg/L adenine sulfate and amino acids (both filter-sterilized) [minus uracil] [62]. Use 20 mL/85-mm petri plate.

SD-URA-TRP medium: 0.67% yeast nitrogen base without amino acids, 2% glucose, and 3% agar (supplemented with amino acids [minus uracil and tryptophan] and 10 mg/L adenine sulfate).

E. Small-Scale Yeast Artificial Chromosomal DNA Preparation

The following procedure for YAC chromosomal preparation allows rapid analysis of transformation products, relative to the sizes of inserts, as well as the stability of individual clones (Fig. 3). It is a scaled-down version based on the original cell-embedding procedures described for yeast chromosome preparations [6,38]. If the DNA fractionation step is properly carried out, all clones should be >200 kb in size, and it is reasonable to expect inserts of average size of 300–400 kb. The YAC chromosomal minipreps may be analyzed by PFGE together with gel-blotting and hybridization with appropriate probes. The two fragments produced by *Pvu*II/*Bam*HI double-enzyme digestion of pBR322 may be used to detect the two arms of the YAC vector, with the larger fragment specific for the left arm and the smaller specific for the right arm of the YAC vector (Fig. 3B). The yeast chromosomal preparations can also be used for analysis with restriction enzyme digestion (see Step 2.2). It is important to note that yeast has no known cytosine methylation system [77]; mammalian DNAs that resist some restriction enzyme digestion because of hypermethylation at the respective recognition sites will be susceptible after propagating in yeast.

Step 1.4 Yeast Chromosomal Preparation

1. Inoculate 5 mL of YPD broth with a single colony of a YAC transformant picked from a SD–URA–TRP plate (selective medium). Shake the culture at 30°C overnight.

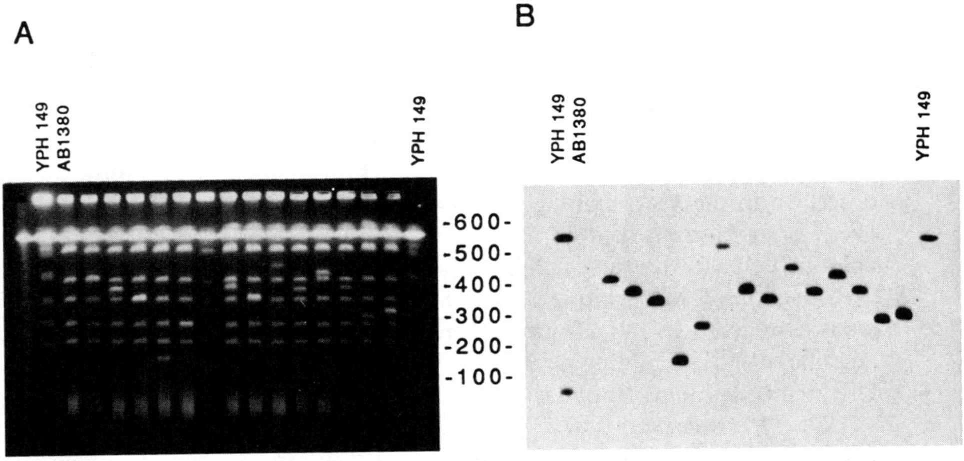

Figure 3 PFGE analysis of YAC clones: (A) An example showing 14 randomly selected YAC clones from an *Eco*RI library prepared with HMW DNA described in Figure 2. YAC chromosomes were prepared according to the small-scale method (see Step 1.4). The average size of inserts from 120 clones examined was ~375 kb. (B) Hybridization analysis of YAC DNA. The gel was blotted and hybridized with a ^{32}P-labeled probe derived from part of the vector sequence (pBR322) as described in Step 2.3. The host strain AB1380 does not contain any YAC. Hybridization to YPH 149 is due to the presence of pBR322 sequences introduced during the fragmentation process (chromosome VII is split into two fragments, 90 kb and 1010 kb) [100].

The gels shown here were run on a Bio-Rad CHEF apparatus with 0.5 × TBE at 200 V for 24 hr with a linearly ramped pulse time of 10–50 sec.

2. Collect the cells by centrifugation (2000 rpm, 5 min); resuspend the pellet in 500 μL of ddH_2O and transfer the cell suspension to a 1.5 mL Eppendorf tube.
3. Spin briefly in a microfuge (~15 sec); remove supernatant and resuspend the pellet in 180 μL of 1 M sorbitol with 50 mM EDTA.
4. Add 2 μL each of β-ME (14.4 M stock) and lyticase (100 units/μL stock); incubate at room temperature for no longer than 5 min.
5. Add 250 μL of 1% prewarmed (50°C), molten LMP agarose in 50 mM EDTA; mix gently by pipette.
6. Transfer the spheroplast suspension into a series of 1.5 × 5 × 10-mm molds to form agarose blocks. After the agarose solidifies, transfer the blocks to a 15 mL polystyrene tube (Falcon 2057) containing 3 mL of SCEM with 30 mM β-ME. Incubate at 37°C for 4–12 hr.

7. Replace the SCEM solution with 3 mL of lysis buffer (50 mM EDTA, 10 mM Tris HCl [pH 8.0], 1% Na laurylsarcosine, 1 mg/mL proteinase K [Boehringer/Manheim]); incubate at 37°C for 4–12 hr. Lysis is evident by clearing of blocks.
8. Wash the blocks extensively with 50 mM EDTA (pH 8.0). The blocks can be stored in the same solution at room temperature for several months. This protocol should yield enough sample (blocks) for 4–6 gels.

F. Yeast Colony Hybridization

Although YACs are generally present in only one copy per cell [65], much fewer than that for most bacterial plasmids, the screening techniques are adequate to allow identification of YACs with specific hybridizing probes [7–9,54–58,63,64,66,78,79]. The following protocol provides a quick and convenient method for isolating clones corresponding to a particular sequence (or sequences) from a YAC library. In addition, when constructing a library from human–rodent somatic cell hybrid line, colony hybridization with human-specific probes (e.g., labeled total human DNA) can distinguish YACs containing human sequences from those originating from the rodent background (Fig. 4), allowing clones to be selected for a specific chromosome or region of a chromosome [61,64,65].

Step 1.5 Yeast Colony Hybridization

1. Pick primary-transformants (from the selective medium lacking uracil) and inoculate them in uniform arrays on SD–URA–TRP plates, with colonies separated from each other ~5-mm apart. The plates are incubated at 30°C until there is visible growth. The colonies are then either replica-plated or reinoculated individually on sterile nitrocellulose (or nylon: Biotrans 0.2μ membrane) membranes and grown on YPD plates for 16–24 hr, at 30°C. It is important to keep the colony size smaller than 3 mm to reduce background in hybridization.
2. Transfer the membranes from YPD plates, one at a time with the colony-side up, to petri dishes lined with filter papers (Whatman 3 MM) soaked with 1 M sorbitol, 1 M sodium citrate (pH 7.0), 20 mM EDTA, 70 mM β-ME, and 50 units/mL Zymolase 100T (ICN) or 100 units/mL Lyticase. Seal the dishes in bags and store them undisturbed at 37°C for 4–16 hr.
3. Treat the membranes containing the colonies (now become spheroplasts) by successive overlaying on filter papers saturated with the following solutions for the amount of time indicated:
   a. 10% SDS, 10 min
   b. 0.5 M NaOH, 8 min

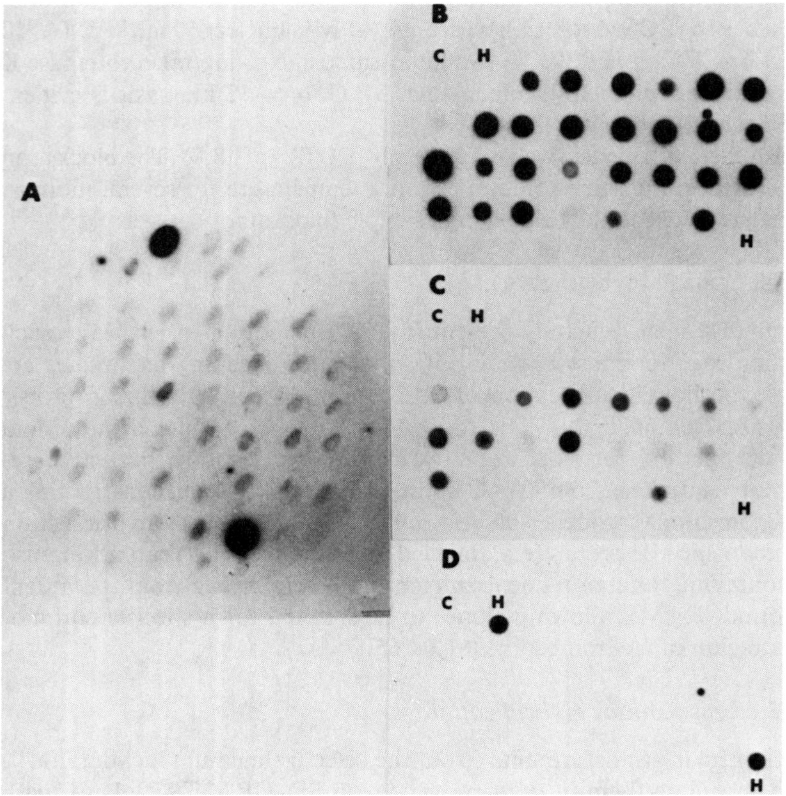

Figure 4 Hybridization analysis of YAC clones containing human genomic DNA: (A) Colony hybridization analysis (see Step 1.5). The example shows that 2 of the 54 YAC transformants in a library constructed with DNA from a human–hamster somatic cell hybrid cell line were identified to contain human genomic DNA inserts with radioactively labeled total human DNA. (B–D) Dot-blot hybridization analysis of 29 YAC clones previously identified by colony hybridization to contain human inserts. The DNA was prepared as described in Step 1.6; probe hybridization and washing were performed as described in Step 2.3. Panel B was hybridized with a human-specific *Alu* repetitive element (pBS derivative of pBLUR 8)[105]; panel C with a human *Kpn*I (LINE1) family sequence (pBSKpnI: a 4.3-kb fragment beginning at the 3′ end of the element)[106]; and panel D with Chinese hamster-specific type 1 and type 2 *Alu* repeats [107]. The absence of hybridization to both human- and hamster-specific repeats to individual YAC clones indicates that no chimeras form during the ligation process. The control samples are DNA from yeast host AB1380 (labeled "C") and two YAC clones containing hamster inserts ("H").

c. 2 × SSC, 5 min
d. 2 × SSC, 5 min
e. immerse in 2 × SSC for 1 min

4. Dry the membranes at room temperature for 30 min and bake them at 80°C for 2 hr.
5. Hybridizations are performed with conventional procedures. After hybridization, the membranes are first washed with steps of increasing stringency and exposed to x-ray film until a desirable signal/background ratio is obtained. The autoradiographs with high background noise may later be used as master templates for aligning colonies.

G. Storage and Screening of Yeast Artificial Chromosomal Libraries

In a typical experiment >95% of the clones that pass the secondary screening process (the double-selection) contain true inserts in YACs. The colonies will remain viable on plates kept at 4°C for weeks, but the stocks should be stored at −80°C. To prepare stocks, individual colonies or pools of clones are grown in 800 μL of YPD at 30°C overnight; 250 μL of a 80% glycerol solution (in YPD) is added to each culture; mix thoroughly and transfer 200 μL aliquots to individual wells of 96-well microtiter plates (with round-bottomed wells). The plates are then sealed and stored at −80°C. It is recommended that at least one set of plates be stored separately as an archival copy of the library.

Although the effort required to store individual clones is nontrivial, it is worthwhile if the library is to be used repeatedly. Furthermore, multiple screening rounds of numerous filters by colony hybridization can be avoided by subdividing the library and screening DNA from pools of colonies with the rapid and sensitive PCR [60]. For example, all 96 wells of a microtiter plate may be pooled and following identification of a desired clone, subdivided by rows or columns [80]. The DNA from individual pools can be stored for subsequent screenings.

Alternatively, YAC transformants may be stored on nitrocellulose membranes sandwiched between Whatman 3MM papers saturated with 30% glycerol (in YPD) and kept at −80°C [54]. Colonies are recovered by replica-plating from the nitrocellulose and regenerating on YPD plates at 30°C for 2 days. While the master membranes remain frozen at −80°C, the duplicate membranes are used for colony hybridization and recovery of clones.

H. Characterization of Yeast Artificial Chromosomal Clones

Rapid means to characterize YAC clones are useful not only in isolation of genes and genomic regions of interest, but also in generation of end probes for chromosomal walking with YACs to extend physical maps and link overlapping clones [8,79]. Besides the standard mapping techniques, such as subcloning DNA fragments into plasmids, phage, and cosmids, several brief characterization procedures are available.

The original pYAC series of vectors are designed to facilitate rapid recovery of the left arm by the plasmid end-rescue procedure [5], which relies on the presence of a restriction site compatible with that of *Xho*I or *Nde*I site within a reasonable distance from the left end of the insert. For example, the sticky end generated by *Sal*I is compatible with that for *Xho*I; after treating the YAC clone with these two enzymes, followed by recircularization, the resulting plasmid may be propagated in *E. coli*; the sequence created by ligation of the two ends can be cut with *Taq*I.

Another technique, based on the use of inverse PCR [81,82], allows isolation of either end of a YAC insert. The procedure involves treating the YAC clone with a frequent-cutting restriction enzyme, circularzing the resulting DNA fragments, and amplifying the end sequences by using primers derived from vector sequences located near the cloning site [8,79,83]. Although human DNA sequences in the internal regions of YAC clones may be isolated with PCR-primers directed to DNA segments containing *Alu* repeats (Alu-PCR) [84], a combination of the *Alu*-primer and a flanking vector primer will allow identification of end fragments.

The human DNA inserts may also be analyzed by a fingerprinting technique based on the presence of *Alu* and *Kpn* repetitive families. Depending on the amount and organization of these repetitive DNA elements, individual YAC clones may be identified by specific repetitive DNA patterns revealed by probe hybridization (Fig. 5). The fingerprints may thus provide a rapid means for the differentiation and identification of overlapping YAC clones; they also serve as a good method for monitoring the stability of individual YAC clones during propagation and for quantitation of the amount and distribution of various classes of repeats within individual clones.

In addition, the effort required for the isolation of YAC DNA may be substantially reduced with the glass-bead genomic-DNA isolation protocol [85], with which it is possible to perform 100 genomic DNA preparations in 1 day, producing DNA ready for restriction enzyme digestion, hybridization analysis, cloning, and PCR analysis.

Step 1.6 Rapid DNA Preparation from Yeast

1. Transfer a single colony of a YAC clone to a tube containing 10 mL of YPD; incubate the culture with shaking at 30°C overnight.
2. Harvest the cells by centrifugation at 2000 rpm for 5 min; disperse the pellet in 500 μL of ddH_2O and transfer the suspension to a 1.5 mL Eppendorf tube; centrifuge again for 10 sec, and remove the supernatant.

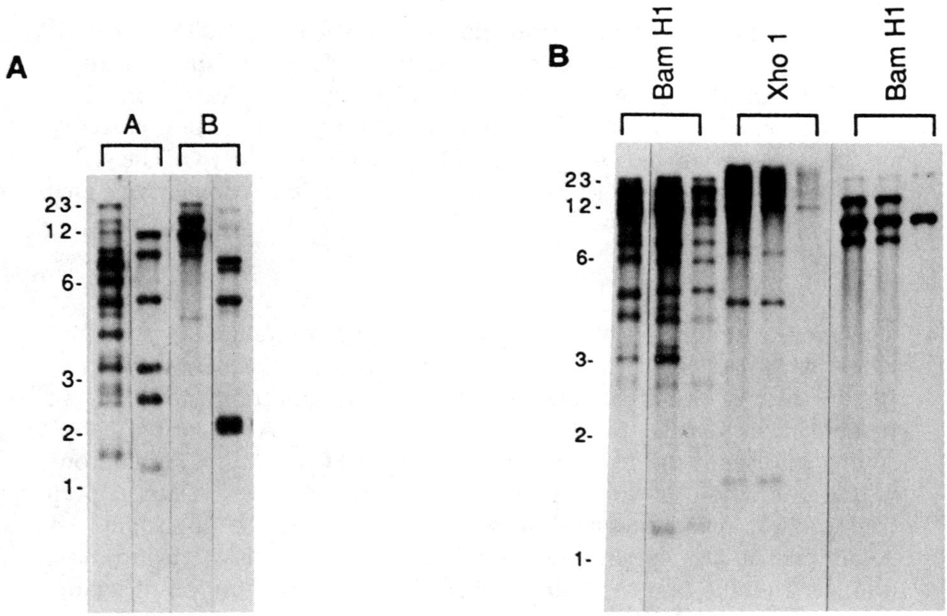

Figure 5 Fingerprinting analysis of human DNA-containing YAC clones with *Alu*- and *Kpn*I-repetitive DNA specific probes: (A) DNA from two YAC clones containing human chromosome 7 genomic DNA. The samples were digested with *Hind*III, fractionated by conventional agarose gel electrophoresis, blotted to nylon membranes, and hybridized with either (a) human specific *Alu*-repetitive sequence (pBSBlur 8) or (b) human *Kpn*1-repetitive sequence (pBSKpnI). (B) An example showing three YAC clones sharing overlapping human DNA fragments. The first two sets of lanes (from the left) were DNA samples digested with *Bam*HI and *Xho*I, as indicated, and hybridized with the *Alu*-specific probe; the last set was digested with *Bam*HI and hybridized with the *Kpn*-specific probe. The DNA samples were prepared as described in Step 1.6 and hybridization was conducted as in Step 2.3.

3. To the pellet, add 200 μL of GDIS (2% Triton X-100, 1% SDS, 100 mM NaCl, 10 mM Tris-HCl [pH 8.0] and 1 mM EDTA), 200 μL of phenol/chloroform/isoamyl alcohol (25 : 24 : 1), and 0.35 g of acid-washed glass beads (Sigma: 0.5 mm dia.). Vortex vigorously for 2.5 min. Samples may be processed in pairs.
4. Add 200 μL of sterile ddH_2O, mix, and spin 4 min in a microfuge. Transfer the aqueous layer to a new 1.5 mL Eppendorf tube and fill the tube with 100% ethanol. Invert the tube several times and allow DNA to precipitate for 2 min at room temperature. The DNA is suitable for dot-blot hybridization analysis at this stage.

5. Collect the pellet by centrifugation and dissolve it in 400 μL of sterile ddH_2O. Add 30 μg of RNase A (Sigma, DNase-free) and incubate at 37°C for 10 min. Add 8 μL of 7.5 M ammonium acetate plus 1 ml ethanol, which will precipitate the DNA at room temperature. Collect the DNA again and resuspend it in 100 μL of sterile ddH_2O. The yield is approximately 10–20 μg of DNA and, for each restriction enzyme analysis, 5 μL is sufficient.

I. A Common Problem

The most common problem encountered is the overabundance of YAC clones with sizes smaller than expected. An obvious explanation for this is the presence of contaminating small DNA molecules trapped in the compression zone during the PFGE fractionation step. As discussed earlier, trapping is usually due to too much DNA per well (overloading) or improper partial restriction enzyme digestion leading to high local DNA concentrations. The pulse time and duration of PFGE are also important. The conditions discussed here are optimized for a 0.3% LMP agarose gel; a longer running time is required when a higher concentration of agarose is used.

A rapid and practical method to test for good starting HMW DNA is by direct cloning from LMP agarose after PFGE. In this method, vector DNA is diffused into agarose slices excised from the LMP agarose gel and the resulting ligation product is used to transform yeast spheroplasts. If YAC clones of the expected size are obtained, the remainder of the DNA sample in the gel slice may be used for subsequent, large-scale ligation reactions in solution. As expected, the cloning efficiency is much lower when vector ligation is performed directly in LMP agarose, but this protocol may be considered when attempting to clone specific DNA fragments resolved by PFGE.

Step 1.7 Cloning in LMP Agarose

1. Dialyze the LMP gel slice twice against 20 mL of sterile ddH_2O at 4°C for 30 min each. Equilibrate the gel slice with 20 mL of 25 mM NaCl/10 mM Tris HCl (pH 7.5) for 1 hr at 4°C.
2. Transfer the LMP gel slice to an Eppendorf tube and add an equal amount (by weight) of YAC vector arms in the above buffer (in same volume as the gel slice).
3. Slowly heat the tube from 42° to 68°C (over a 15 min period) until the LMP agarose is liquefied. Equilibrate to 37°C and add 0.1 volume of ice-cold 10 × ligation buffer and T4 DNA ligase (IBI). Gently mix the contents by tapping the tube and incubate the mixture at 15°C overnight.

4. Heat the ligation reaction for 10 min at 65°C, cool to 37°C for 5 min, and use 10–20 μL to transform yeast spheroplasts prepared as described in Step 1.3.

III. PULSED-FIELD GEL ELECTROPHORESIS

A. Sample Preparation and Analysis

The procedures for preparing HMW genomic DNA in agarose blocks, restriction enzyme digestion of DNA in agarose, and blot-hybridization analysis are described below and followed by discussions on the general conditions used in PFGE and the various DNA size standards.

Step 2.1 *Preparation of HMW Genomic DNA in LMP Agarose Blocks*

1. Harvest cells using standard tissue culture techniques and pellet by centrifugation at 1000 rpm for 10 min at room temperature. Wash once in SE buffer (75 mM NaCl; 25 mM EDTA, pH 8.0). Count cells and resuspend them in fresh SE buffer to final concentration of 2×10^7 cells/mL.
2. Warm the cell suspension to 37°C for 3 min and quickly mix with an equal volume of 1% warm (50°C), molten LMP agarose (BRL, Ultra Pure). Immediately pour the mixture on a small glass plate (microscope slide or projector slide cover glass) to a thickness of ~1.5 min. The cell suspension is allowed to solidify at room temperature, cooled at 4°C for 10 min, and then cut into blocks (5 × 5 × 1.5 mm) with a microscope cover slip.
3. Transfer the blocks to a 250 mL flask containing 10 volumes of fresh lysis buffer (50 mM EDTA; 1% lauryl sarcosyl, pH 8.0), and 0.5 mg/mL proteinase K (Boehringer/Manheim). Incubate the flask at 50°C overnight with gentle shaking.
4. Remove the lysis buffer by decanting and rinse the deproteinized blocks with several changes of TE (10 mM Tris HCl, 10 mM EDTA) at 37°C. The blocks are stored in 50 mM EDTA at 4°C ready for restriction enzyme digestion.

Step 2.2 *Restriction Enzyme Digestion in Agarose Blocks*

1. Wash the agarose blocks twice in 20 volumes of sterile ddH_2O at 4°C for 30 min each and equilibrate with appropriate 1 × restriction enzyme buffer (as recommended by the supplier) supplemented with 4 mM spermidine HCl (Sigma).
2. Transfer individual blocks (each containing 5×10^5 cells) to Eppendorf tubes and add fresh restriction enzyme buffer to a total volume of 100 μL (block + restriction enzyme buffer).

3. Add 1–5 units of restriction enzyme for each microgram of DNA (agarose blocks containing YAC clones require slightly more enzyme; i.e., 10–20 unit μg DNA). Mix the contents and let the tube sit on ice for 30 min to allow the enzyme to diffuse into the agarose. Initiate the digestion by transferring the tube to 37°C (or other specified temperature for the enzyme). Terminate the digestions after 3–4 hr by the addition of 10 mM EDTA (pH 8.0) and the digested samples may be stored indefinitely at 4°C.

Step 2.3 Gel-Blot Hybridization Analysis

1. Visualize the DNA in the gel by staining with ethidium bromide (0.5 μg/mL) in 0.5 × TBE for 45 min (use just enough buffer to cover the gel).
2. Following destaining in ddH_2O or 0.5 × TBE and photography, treat (depurinate) the DNA by soaking the gel in 250 mL of 0.25 N HCl for 10 min. This amount of time corresponds to a 100–150 mL 0.7–1.2% agarose gel of a thickness of 5 mm; vary the time for gels of different sizes.
3. Denature the DNA by soaking the gel in 250 mL of 0.4 N NaOH/0.6 M NaCl for 30 min and neutralize with the same volume 0.5 M Tris HCl (pH 7.5)/1.5 M NaCl for 30 min.
4. Transfer the DNA overnight to either Zetaprobe (Bio-Rad) or Nytran (Schleicher & Schuell) in 10 × SSC.
5. After transfer, treat the membrane by laying it DNA-side up on a filter paper (Whatman 3MM) saturated with 50 mM sodium phosphate buffer (pH 6.6) for 2 min. Air-dry the filter and then bake it for 2 hr under vacuum at 80°C.
6. Prehybridization treatment and hybridization may be performed in CGH solution (0.5 M sodium phosphate buffer [pH 7.2] with 7% SDS, 1 mM EDTA, and 1% BSA) at 65°C [86]. Following hybridization, wash blots three times, consecutively, in 2 × SSC, 1 × SSC, and 0.2 × SSC (all with 0.1% SDS) for 20 min each at 65°C.
7. For reuse, the membranes may be deprobed by treating with 0.4 N NaOH for 30 min at 42°C, followed by neutralization in 0.2 M Tris HCl (pH 7.5)/0.5% SDS/0.1 × SSC.

B. Conditions for Pulsed-Field Gel Electrophoresis

Many studies have been conducted to investigate the parameters influencing the separation of DNA molecules by PFGE [87–90]. Such extrinsic factors include the pulse time, applied field strength, ionic environment, type and concentration of agarose, temperature, field inhomogeneity, and reorientation angle. Other factors that are intrinsic to the DNA samples include the amount of DNA per lane, the sequence-dependent topology [12], and the conformation of the DNA molecule [12,89,91].

An ideal PFGE running condition should be optimized for maximal resolution and speed of separation in the size range of interest. The condition is relatively simple for DNA molecules smaller than 1.5 mb in size; different size-class molecules can be separated with various pulse intervals; some examples are shown in Figures 6A–D. For molecules approaching 2 mb, however, the conditions become more difficult to control. Separation of DNA molecules of the latter size-class requires a substantial reduction in the voltage and an increase in the pulse time (see Figs. 6E–H). Although long pulse time is apparently critical in resolving large DNA molecules, a low voltage prevents them from forming diffused or smeared bands [6,92]. Unfortunately the optimal balance between voltage and pulse times is difficult to predict, and it remains purely empirical. In our experience, a tight compression zone is a good indication that a reasonable pulse length/voltage balance has been achieved.

Attempts to compensate for this delicate balance by using other parameters have met some success [36,87,88]. For example, reducing the agarose concentration from 1 to 0.7% and decreasing the reorientation angle from 110° to 104° allow separation of molecules greater than 5 mb in less than 96 hr. The crossed-field [41], rotating gel [42], ED [93], and PACE systems [36] are particularly suited for this purpose because the reorientation angle can be controlled. Most of the other commercially available PFGE systems (Bio-Rad CHEF-DR II, LKB Pulsaphor) have a fixed 120° reorientation angle and run with a homogeneous electric field. Beckman's Geneline apparatus uses variable reorientation angles, but it varies only from the top to the bottom of the gel. It is, however, possible to resolve DNA molecules larger than 2 mb with the latter systems, but it requires minimum run times of 4–7 days.

The field-inversion gel electrophoresis (FIGE) system [39], provides rapid separation of DNA segments smaller than 500 kb, but an elaborate pulse sequence is required for the analysis of larger molecules [37]. The conditions for FIGE have recently been described [94,95], and they are not discussed here.

C. DNA Size Standards

Three DNA markers are often used as size standards for PFGE: (1) concatamers of bacteriophage λ DNA and (2) chromosomes from *S. cerevisiae* serve as relatively accurate markers for the 50- to 1600-kb size range, and (3) chromosomes from *S. pombe* provide reference points of 5.7, 4.7, and 3.5 mb (see Figs. 6G,H) [96]. Other organisms (such as *Hansenula wingei* and

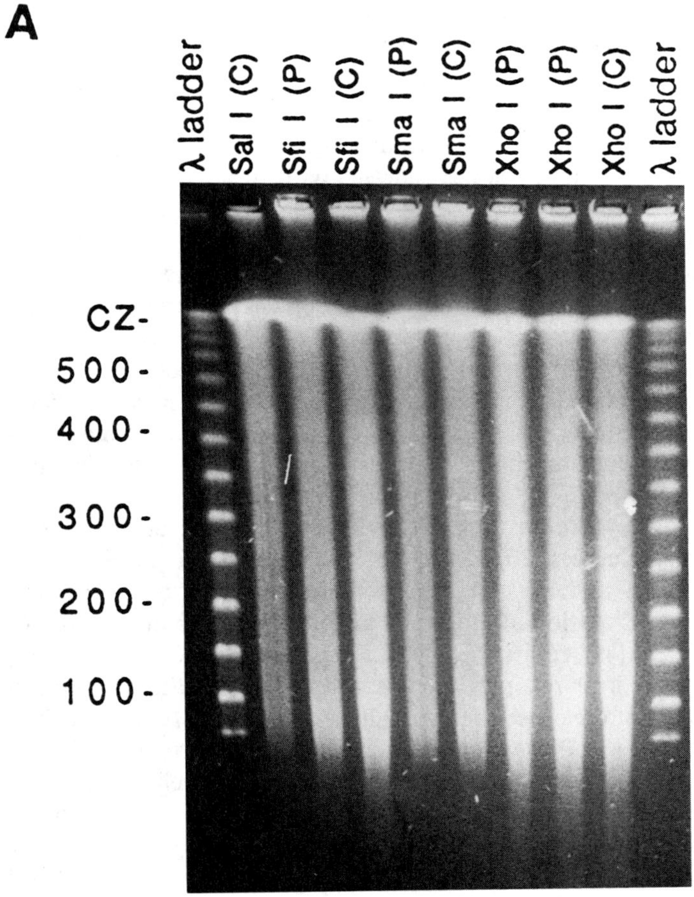

Figure 6 Optimized conditions for separation of DNA molecules in the 50- to 6000-kb size range using crossed-field gel electrophoresis: All gels were run in 0.5 × TBE at 12°C under the following conditions (the number in parentheses represents the

Candida ischiwadae) with chromosome sizes between the largest *S. cerevisiae* and smallest *S. pombe* chromosomes are available, but they are based on interpolation of the foregoing markers [90,97, and our unpublished data]. In addition, a complete *Not*I digestion product of the chromosomes from *S. pombe* provides additional molecules in the 3- to 6-mb size range [96,98]. There are, however, large variations in both the size and number of chromosomes in different laboratory strains of *S. cerevisiae* and *S. pombe* [97]. The smaller *S. cerevisiae* chromosomes may be calibrated against λ

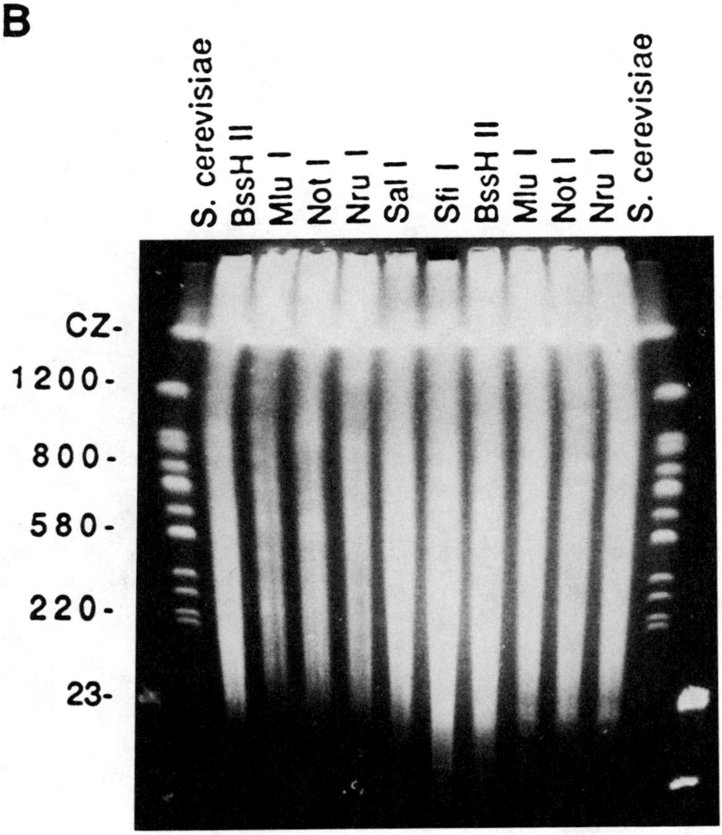

distance in centimeters from the origin to the smallest marker indicated on the gel): (A) 6 V/cm for 22 hr, 10- to 40-sec linearly ramped pulse time with a reorientation angle of 120°, 1% agarose, and a gel length of 12 cm; (B) 6 V/cm, 24 hr, 45- to 90-sec

oligomers, but it is advisable to use strains for which the chromosome sizes have been established.

Bacteriophage λ oligomers may be prepared by annealing the sticky ends within the agarose blocks [99], and yeast chromosomes may be prepared with a scaled-up version of Step 1.4 [6,38]. Chromosomes from strains such as *S. cerevisiae* YNN295 and YPH149, and *S. pombe* 972h are commercially available. YPH149 is particularly attractive because it contains 16 well-resolved yeast chromosomes, as well as a 90-kb derivative of chromosome

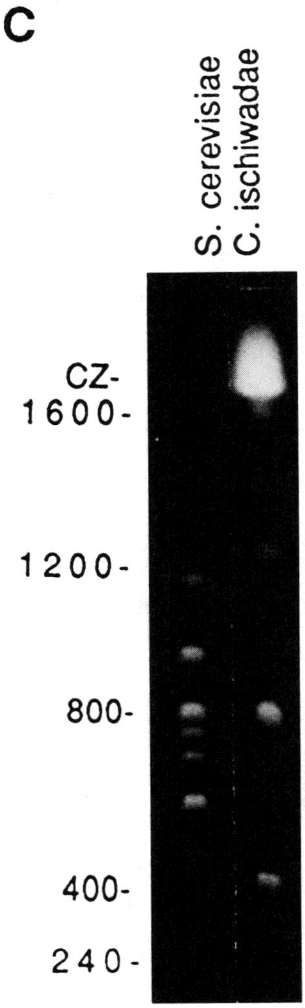

Figure 6 Continued

pulses, 120° reorientation angle, and 1% agarose (10 cm); (C) 6 V/cm, 24 hr, 60- to 120-sec pulses, 120° angle, and 1% agarose (11 cm); (D) 6 V/cm, 24 hr, 60- to 120-sec

VII, generated from fragmentation [100]. The relative mobility of the two largest yeast chromosomes (IV and XII) varies with electrophoretic conditions; therefore, inclusion of other standards in the 1- to 3-mb range is recommended.

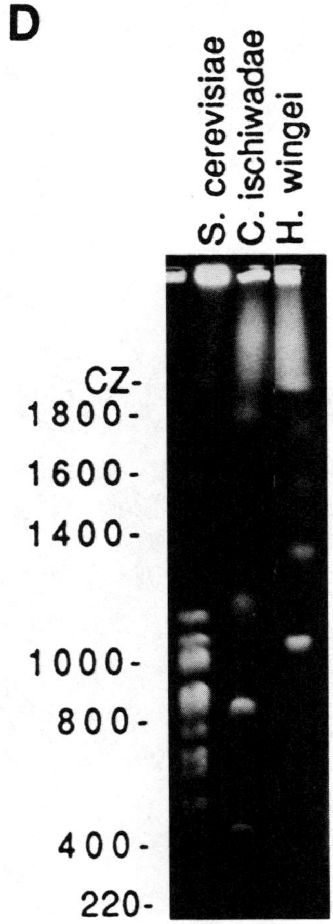

pulses, 110° angle, and 1% agarose (15 cm); (E) 4 V/cm, 72 hr, 100- to 200-sec pulses, 120° angle, and 1% agarose (15 cm); (F) 2.5 V/cm, 72 hr, 200- to 400-sec

IV. DISCUSSION

With PFGE and YAC, it is now possible to perform long-range physical mapping studies and to isolate large DNA fragments. PFGE maps of the entire genome of simple organisms like *E. coli* [101], *S. pombe* [96], and *S. cerevisiae* have been completed; studies in *Drosophila* [8,83], *C. elegans* [7], and *Myxococcus* [67] have illustrated the utility of YAC cloning as a general tool for physical mapping in less complex genomes. In combination, these two technological improvements in molecular biology should provide the

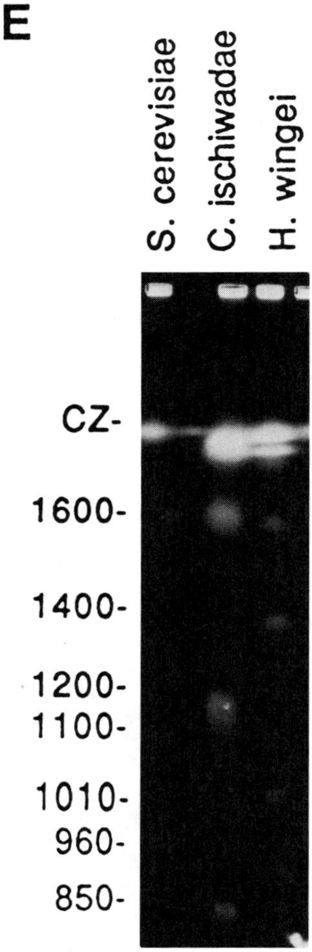

Figure 6 Continued

pulses, 104° angle, and 0.8% agarose (13 cm); (G) 2 V/cm, 30- to 45-min pulses for the first 120 hr followed by 5-min constant switching time for the next 24 hr, 110°

basic strategies for any detailed characterization of complex genomes. Given the size of the human genome (3×10^9 bp), it would require only 17,000 clones in a YAC library, with an average insert size of 400 kb, to achieve a library of 3 × genome equivalent (i.e., >90% chance of finding any given sequence). Because of the manageable numbers, it is also feasible to prepare chromosome-specific libraries with flow-sorted chromosomes or from hu-

F

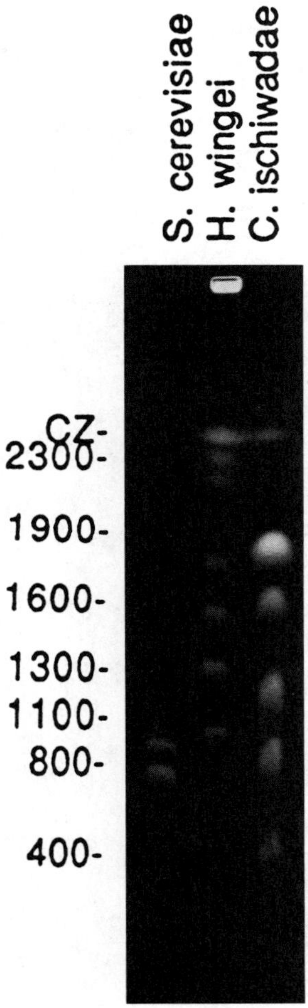

angle, and 1% agarose (7 cm); and (H) 2 V/cm, 48-, 72-, 96-hr runs (as indicated), 30- to 45-min pulses, 104° angle, and 0.7% agarose (9 cm).

man–hamster somatic cell hybrids; as illustrated above, <1000 clones are needed for a 3 × library of an average human chromosome.

Although the mammalian genome is full of repetitive sequences, preliminary data suggest that most of the YAC clones do not rearrange after long-term culture [63,65, and our unpublished data]. Therefore, YAC cloning

G

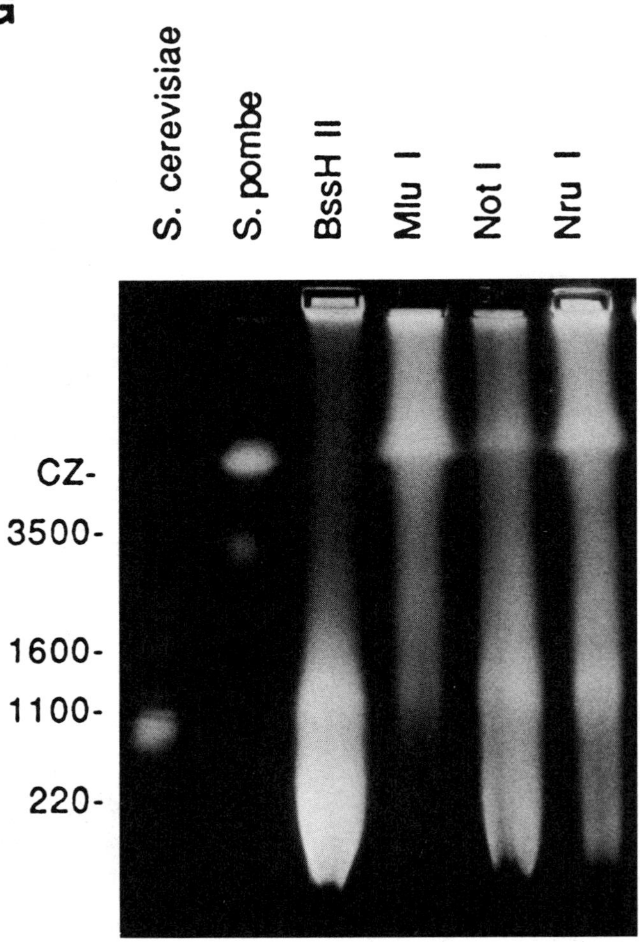

Figure 6 Continued

Note the position of the compression zone (CZ) for the *Not*I digestion; CZ for the *Sal*I digestion is apparent on the original gel. [Yeast strain *S. cerevisiae* SC252 was a

may provide an alternative means to study chromosomal regions that are unstable in *E. coli*-based cloning vectors [4]. Not all mammalian sequences are stable in YAC, however [unpublished data]. For these latter sequences, physical mapping by PFGE or other methods, such as direct genomic sequencing [86], must be used. Besides its application in chromosomal walk-

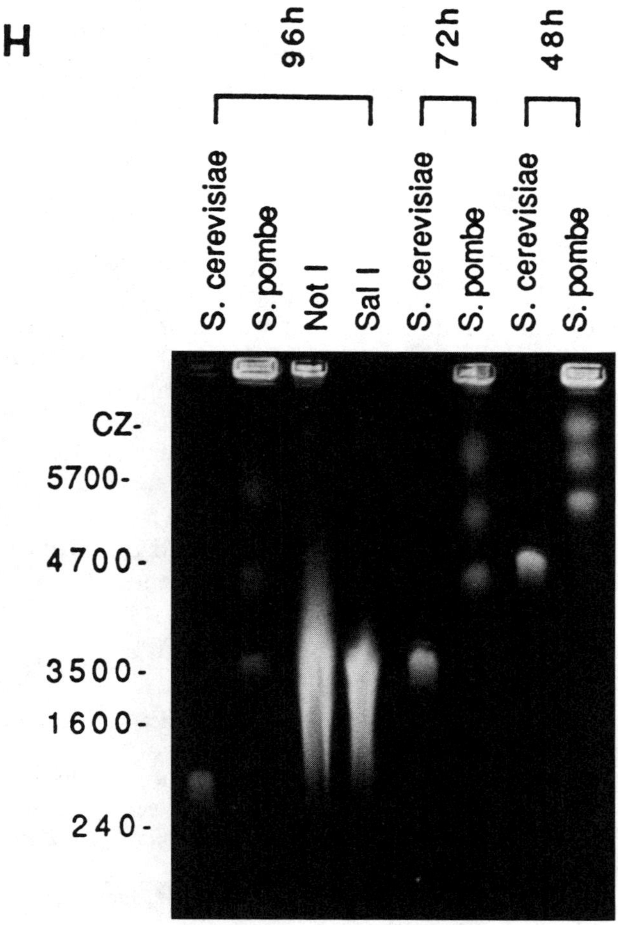

gift of J. Friesen; *S. cerevisiae* YPH 149 from P. Hieter; *S. pombe* 972 from C. Smith; *C. ischiwadae* from V. Seligy; and *H. wingei* 201 from T. Dowhanick. The λ ladder (gift of J. Rommens) was prepared from *E. coli* W3110 ($\lambda cI_{857}S_7$)].

ing, YAC cloning may also serve as a shuttle system in cloning and studying large mammalian genes by DNA-mediated gene transfer. Genes of interest may be manipulated in yeast and reintroduced into mammalian cells by techniques such as spheroplast fusion [102] or microinjection [103,104]. The ability to manipulate a large DNA segment and examine the consequence may permit identification of sequences (or sequence configurations) that are important in gene expression and gene regulation during development. Furthermore, genes of biological significance, especially those involved in DNA repair, may be identified by means of growth or biochemical selection in mammalian cell cultures.

ACKNOWLEDGMENTS

The authors thank Professor Edwin Southern and Martin Johnston (Department of Biochemistry, Oxford University) for sharing the original drawing plan of the "crossed-field" gel apparatus, Ted Warren and Arthur Tateishi (Research Development Corporation, The Hospital for Sick Children, Toronto) for technical assistance on construction of the gel apparatus, Johanna Rommens (Department of Genetics, The Hospital for Sick Children, Toronto) for helpful discussions, and David Burke (Department of Molecular Biology, Princeton University) for providing the YAC vectors and yeast strains as well as helpful advice on the YAC cloning system. This research is supported by the Canadian Cystic Fibrosis Foundation and the Medical Research Council (MRC) of Canada. L.-C. T. is an MRC Scientist.

REFERENCES

1. Antonarakis, S. E., and H. H. Kazazian (1988). The molecular basis of hemophilia A in man. *Trends Genet. 4*:233–237.
2. Bookstein, R., H. P. Lee, H. To, L. J. Young, T. W. Sery, R. C. Hayes, T. Friedmann, and W. H. Lee (1988). Human retinoblastoma gene: Genomic organization and analysis of heterozygous intragenic deletion mutants. *Proc. Natl. Acad. Sci. USA 85*:2210–2214.
3. Monaco, A. P., and L. M. Kunkel (1987). A giant locus for the Duchenne and Becker muscular dystrophy gene. *Trends Genet. 3*:33–37.
4. Rommens, J. M., M. C. Iannuzzi, B. Kerem, M. L. Drumm, G. Melmer, M. Dean, R. Rozmahel, J. L. Cole, D. Kennedy, N. Hidaka, M. Zsiga, M. Buchwald, J. R. Riordan, L. C. Tsui, and F. S. Collins (1989). Identification of the cystic fibrosis gene: Chromosome walking and jumping. *Science 245*: 1059–1065.
5. Burke, D. T., G. F. Carle, and M. V. Olson (1987). Cloning of large segments of exogenous DNA into yeast by means of yeast artificial chromosome vectors. *Science 236*:806–812.
6. Schwartz, D. C., and C. R. Cantor (1984). Separation of yeast chromosome-sized DNAs by pulsed-field gradient electrophoresis. *Cell 37*:67–75.
7. Coulson, A., R. Waterston, J. Kiff, J. Sulston, and Y. Kohara (1988). Genome linking with yeast artificial chromosomes. *Nature 335*:184–186.
8. Garza, D., J. W. Ajioka, D. T. Burke, and D. L. Hartl (1989). Mapping the *Drosophila* genome with yeast artificial chromosomes. *Science 246*:641–646.

9. Hahnenberger, K. H., M. P. Baum, M. Polizzi, and L. Clarke (1989). Construction of functional artificial minichromosomes in the fission yeast *Schizosaccharomyces pombe*. *Proc. Natl. Acad. Sci. USA 86*:577–581.
10. Struhl, K. (1983). The new yeast genetics. *Nature 305*:391–397.
11. Orbach, M. J., D. Vollrath, R. W. Davis, and C. Yanofsky (1988). An electrophoretic karyotype of *Neurospora crassa*. *Mol. Cell. Biol. 8*:1469–1473.
12. Carle, G. F., and M. V. Olson (1985). An electrophoretic karyotype for yeast. *Proc. Natl. Acad. Sci. USA 82*:3756–3760.
13. Lawrance, S. K., C. L. Smith, R. Srivastava, C. R. Cantor, and S. M. Weissman (1987). Megabase-scale mapping of the HLA gene complex by pulsed field gel electrophoresis. *Science 235*:1387–1390.
14. Rappold, G., and H. Lehrach (1988). A long range restriction map of the pseudoautosomal region by partial digest PFGE analysis from the telomere. *Nucleic Acids Res. 16*:5361–5377.
15. Rommens, J. M., S. Zengerling-Lentes, B. Kerem, G. Melmer, M. Buchwald, and L. C. Tsui (1989). Physical localization of two DNA markers linked to the cystic fibrosis locus by pulsed-field gel electrophoresis. *Am. J. Hum. Genet. 45*:932–941.
16. Higgins, M. J., M. F. Hansen, W. K. Cavenee, and M. Lalande (1989). Molecular detection of chromosomal translocations that disrupt the putative retinoblastoma susceptibility locus. *Mol. Cell. Biol. 9*:1–5.
17. Smith, C. L., P. W. Warburton, A. Gaal, and C. R. Cantor (1986). Analysis of genome organization and rearrangements by pulsed field gradient gel electrophoresis. *Genet. Eng. 8*:45–70.
18. Compton, D. A., M. W. Weil, C. Jones, V. M. Riccardi, L. C. Strong, and G. F. Saunders (1988). Long range physical map of the Wilm's tumour—Aniridia region on human chromosome 11. *Cell 55*:827–836.
19. van Omnen, G. J. B., J. M. H. Verkerk, M. H. Hofker, A. P. Monaco, L. M. Kunkel, P. Ray, R. Worton, B. Wieringa, E. Bakker, and P. L. Pearson (1986). A physical map of 4 billion bp around the Duchenne muscular dystrophy gene on the human X-chromosome. *Cell 47*:499–504.
20. Hermann, B. G., D. P. Barlow, and H. Lehrach (1987). A large inverted duplication allows homologous recombination between chromosomes heterozygous for the proximal t complex inversion. *Cell 48*:813–825.
21. Blackburn, E., and J. Szostak (1984). The molecular structure of centromeres and telomeres. *Annu. Rev. Biochem. 53*:163–194.
22. Carbon, J. (1984). Yeast centromeres: Structure and function. *Cell 37*:351–353.
23. Clarke, L., and J. Carbon (1980). Isolation of a yeast centromere and construction of functional small circular chromosomes. *Nature 287*:504–509.
24. Szostak, J., and E. Blackburn (1982). Cloning yeast telomeres on linear plasmid vectors. *Cell 29*:245–255.
25. Hsiao, C. L., and J. Carbon (1979). High frequency transformation of yeast by plasmids containing the cloned yeast *Arg4* gene. *Proc. Natl. Acad. Sci. USA 76*:3829–3833.

26. Stinchcomb, D. T., K. Struhl, and R. W. Davis (1979). Isolation and characterization of a yeast chromosomal replicator. *Nature 282*:39–43.
27. Dawson, D. S., A. W. Murray, and J. W. Szostak (1983). An alternative pathway for meiotic chromosome segregation in yeast. *Science 234*:713–717.
28. Murray, A. W., and J. W. Szostak (1983). Construction of artificial chromosomes in yeast. *Nature 305*:189–193.
29. Dani, G. M., and V. A. Zakian (1983). Mitotic and meiotic stability of linear plasmids in yeast. *Proc. Natl. Acad. Sci. USA 80*:3406–3410.
30. Hieter, P., C. Mann, M. Snyder, and R. W. Davis (1985). Mitotic stability of yeast chromosomes: A colour assay that measures nondisjunction and chromosomal loss. *Cell 40*:381–392.
31. Murray, A. W., N. P. Schultes, and J. W. Szostak (1986). Chromosome length controls mitotic chromosome segregation in yeast. *Cell 45*:529–536.
32. Surosky, R. T., C. S. Newlon, and B. K. Tye (1986). The mitotic stability of deletion derivatives of chromosome III in yeast. *Proc. Natl. Acad. Sci. USA 83*:414–418.
33. Newlon, C. S. (1988). Yeast chromosome replication and segregation. *Microbiol. Rev. 52*:568–601.
34. Kiss, G. B., A. A. Amin, and R. E. Pearlman (1981). Two separate regions of the extrachromosomal ribosomal deoxyribonucleic acid of *Tetrahymena thermophila* enable autonomous replication of plasmids in *S. cerevisiae*. *Mol. Cell. Biol. 1*:535–543.
35. Chu, G., D. Vollrath, and R. W. Davis (1986). Separation of large DNA molecules by contour-clamped homogeneous electric fields. *Science 241*: 1582–1585.
36. Clark, S. M., E. Lai, B. W. Birren, and L. Hood (1986). A novel instrument for separating large DNA molecules with pulsed homogeneous electric fields. *Science 241*:1203–1205.
37. Turmel, C., and M. Lalande (1988). Resolution of *Schizosaccharomyces pombe* by field inversion gel electrophoresis. *Nucleic Acids Res. 16*:4727.
38. Carle, G. F., and M. V. Olson (1984). Separation of chromosomal DNA molecules from yeast by orthogonal-field-alternation gel electrophoresis. *Nucleic Acids Res. 12*:5647–5664.
39. Carle, G. F., M. Frank, and M. V. Olson (1986). Electrophoretic separation of large DNA molecules by periodic inversion of the electric field. *Science 232*: 65–68.
40. Gardiner, K., W. Laas, and D. Patterson (1986). Fractionation of large mammalian DNA restriction fragments using vertical pulsed-field gradient gel electrophoresis. *Somatic Cell Mol. Genet. 12*:185–195.
41. Southern, E. M., R. Anand, W. A. Brown, and D. S. Fletcher (1987). A model for the separation of large DNA molecules by crossed field gel electrophoresis. *Nucleic Acids Res. 15*:5925–5943.
42. Serwer, P. (1987). Gel electrophoresis with discontinuous rotation of the gel: An alternative to gel electrophoresis with changing direction of the electric field. *Electrophoresis 8*:301–304.

43. Bancroft, I., and C. P. Wolk (1988). Pulsed homogeneous orthogonal field gel electrophoresis (PHOGE). *Nucleic Acids Res. 16*:7405–7418.
44. Turmel, C., G. Slater, and J. Noolandi (1989). High resolution zero integrated field electrophoresis (ZIFE) of DNA. (Submitted)
45. Mann, C., and R. W. Davis (1986). Structure and sequence of the centromeric DNA of chromosome 4 in *Saccharomyces cerevisiae. Mol. Cell. Biol. 6*:241–245.
46. Tschumper, G., and J. Carbon (1980). Sequence of a yeast DNA fragment containing a chromosomal replicator and the *TRP1* gene. *Gene 10*:157–166.
47. Rose, M., P. Grisafi, and D. Botstein (1984). Structure and function of the yeast *URA3* gene: Expression in *Escherichia coli. Gene 29*:113–124.
48. Goodman, H. M., M. V. Olson, and B. D. Hall (1977). Nucleotide sequence of a mutant eukaryotic gene: The yeast tyrosine-inserting ochre suppressor *SUP4-o. Proc. Natl. Acad. Sci. USA 74*:5453–5457.
49. Olson, M. V., K. Loughney, and B. D. Hall (1979). Identification of the yeast DNA sequences that correspond to specific tyrosine-inserting nonsense suppressor loci. *J. Mol. Biol. 132*:387–410.
50. Koshland, D., and P. Hieter (1987). A visual assay for chromosome ploidy. *Methods Enzymol. 155*:351–372.
51. Struhl, K. (1985). Nucleotide sequence and transcriptional mapping of the yeast *pet56-his3-ded1* gene region. *Nucleic Acids Res. 13*:8587–8601.
52. Marchuk, D., and F. S. Collins (1988). pYAC-RC, a yeast artificial chromosome vector for cloning DNA cut with infrequently cutting restriction endonucleases. *Nucleic Acids Res. 16*:7743.
53. Cooke, H., and S. Cross (1989). pYAC 4-Neo, a yeast artificial chromosome vector which codes for G418 resistance in mammalian cells. *Nucleic Acids Res. 16*:11817.
54. Traver, C. N., S. Klapholz, R. W. Hyman, and R. W. Davis (1989). Rapid screening of a human genomic library in yeast artificial chromosomes for single-copy sequences. *Proc. Natl. Acad. Sci. USA 86*:5898–5902.
55. Riethman, H. C., R. K. Moyzis, J. Meyne, D. T. Burke, and M. V. Olson (1989). Cloning telomeric DNA fragments into *Saccharomyces cerevisiae* using a yeast-artificial-chromosome vector. *Proc. Natl. Acad. Sci. USA 86*: 6240–6244.
56. Cross, S. H., R. C. Allshire, S. J. McKay, N. I. McGill, and H. J. Cooke (1989). Cloning of human telomeres by complementation in yeast. *Nature 338*:771–774.
57. Brown, R. A. (1989). Molecular cloning of human telomeres in yeast. *Nature 338*:774–776.
58. Cheng, J. F., C. L. Smith, and C. L. Cantor (1989). Isolation and characterization of a human telomere. *Nucleic Acids Res. 17*:6109–6127.
59. de Jong, P. J., C. Chen, and J. Garnes (1989). Applications of PCR for the construction of vectors and isolation of probes. In *Polymerase Chain Reaction: Current Communications in Molecular Biology*. H. A. Erlich, R. Gibbs,

and H. G. Kazazian (eds.). Cold Spring Harbor Laboratory Press, Cold Spring Harbor, N.Y., pp. 225–231.
60. Mullis, K., F. Faloona, S. Scharf, R. Saiki, G. Horn, and H. Erlich (1986). Specific enzymatic amplification of DNA in vitro: The polymerase chain reaction. *Cold Spring Harbor Symp. Quant. Biol. 51*:263–273.
61. McCormick, M. K., J. H. Shero, M. C. Cheung, Y. W. Kan, P. A. Hieter, and S. E. Antonarakis (1989). Construction of human chromosome 21-specific yeast artificial chromosomes. *Proc. Natl. Acad. Sci. USA 86*:9991–9995.
62. Sherman, F., G. R. Fink, and J. B. Hicks (1986). *Methods in Yeast Genetics: A Laboratory Manual*. Cold Spring Harbor Laboratory Press, Cold Spring Harbor, N.Y.
63. Brownstein, B. H., G. A. Silverman, R. A. Little, D. T. Burke, S. J. Korsmeyer, D. Schlessinger, and M. V. Olson (1989). Isolation of single-copy human genes from a library of yeast artificial chromosome clones. *Science 244*:1348–1351.
64. Little, R. D., G. Porta, G. Carle, D. Schlessinger, and M. D'Urso (1989). Yeast artificial chromosomes with 200- to 800-kilobase inserts of human DNA containing HLA, V_K, 5S, and Xq24-28 sequences. *Proc. Natl. Acad. Sci. USA 86*:1598–1602.
65. Wada, M., R. D. Little, F. Abidi, G. Porta, T. Labella, T. Cooper, G. Della Valle, M. D'Urso, and D. Schlessinger (1990). Human Xq24-Xq28: Approaches to mapping with yeast artificial chromosomes. *Am. J. Hum. Genet. 46*:95–106.
66. Guzman, P., and J. R. Ecker (1988). Development of large DNA methods for plants: Molecular cloning of large segments of *Arabidopsis* and carrot DNA into yeast. *Nucleic Acids Res. 16*:11091–11105.
67. Kuspa, A., D. Vollrath, Y. Cheng, and D. Kaiser (1989). Physical mapping of the *Myxococcus xanthus* genome by random cloning in yeast artificial chromosomes. *Proc. Natl. Acad. Sci. USA 86*:8917–8921.
68. Anand, R., A. Villasante, and C. Tyler-Smith (1989). Construction of yeast artificial chromosome libraries using fractionation by pulsed-field gel electrophoresis. *Nucleic Acids Res. 17*:3425–3433.
69. Yaphe, W. (1957). The use of agarase from *Pseudomonas atlantica* in the identification of agar in marine algae. *Can. J. Microbiol. 3*:987–993.
70. Maniatis, T., E. F. Fritsch, and J. Sambrook (1982). *Molecular Cloning: A Laboratory Manual*. Cold Spring Harbor Laboratory Press, Cold Spring Harbor, N.Y.
71. Beggs, J. D. (1978). Transformation of yeast by a replicating hybrid plasmid. *Nature 275*:104–109.
72. Hinnen, A., J. B. Hicks, and G. R. Fink (1978). Transformation of yeast. *Proc. Natl. Acad. Sci. USA 75*:1929–1933.
73. Ito, H., Y. Fukuda, K. Murata, and A. Kimura (1983). Transformation of intact yeast cells treated with alkali cations. *J. Bacteriol. 153*:163–168.
74. Lai, E., and C. Cantrell (1989). Rapid colony screening of YAC libraries by using alginate as matrix support. *Nucleic Acids Res. 17*:8008.

75. Lambie, E. J., and S. G. Roeder (1986). Repression of meiotic crossing over by a centromere (CEN 3) in *Saccharomyces cerevisiae. Genetics 114*:769–789.
76. Burgers, P. M., and K. J. Percival (1987). Transformation of yeast spheroplasts without cell fusion. *Anal. Biochem. 163*:391–397.
77. Proffitt, J. H., J. R. Davie, D. Swinton, and S. Hattman (1984). 5-Methylcytosine is not detectable in *Saccharomyces cerevisiae* DNA. *Mol. Cell. Biol. 4*: 985–988.
78. Little, R. D., and D. C. Braaten (1989). Genomic organization of human 5S rDNA and sequence of one tandem repeat. *Genomics 4*:376–383.
79. Silverman, G. A., R. D. Ye, K. M. Pollock, J. E. Sadler, and S. J. Korsmeyer (1989). Use of yeast artificial chromosome clones for mapping and walking within human chromosome segment 18q21.3. *Proc. Natl. Acad. Sci. USA 86*: 7485–7489.
80. Heard, E., B. Davies, S. Feo, and M. Fried (1989). An improved method for the screening of YAC libraries. *Nucleic Acids Res. 17*:5861.
81. Ochman, H., A. S. Gerber, and D. L. Hartl (1988). Genetic applications of an inverse polymerase chain reaction. *Genetics 120*:621–623.
82. Ochman, H., M. M. Medhora, D. Garza, and D. L. Hartl (1990). Amplification of flanking sequences by inverse PCR. In *PCR Protocols: A Guide to Methods and Applications*. M. A. Innnis, D. H. Gelfand, J. Sninsky, and T. J. White (eds.). Academic Press, San Diego, Calif., pp. 219–227.
83. Garza, D., J. W. Ajioka, J. P. Carulli, R. W. Jones, D. H. Johnson, and D. L. Hartl (1989). Physical mapping of complex genomes. *Nature 340*:577–578.
84. Nelson, D. L., S. A. Ledbetter, L. Corbo, M. F. Victoria, R. Ramirez-Solis, T. D. Webster, D. H. Ledbetter, and C. T. Caskey (1989). Alu polymerase chain reaction: A method for rapid isolation of human-specific sequences from complex DNA sources. *Proc. Natl. Acad. Sci. USA 86*:6686–6690.
85. Hoffman, C. S., and F. Winston (1987). A ten-minute DNA preparation from yeast efficiently releases autonomous plasmids for transformation of *Escherichia coli. Gene 57*:267–272.
86. Church, G. M., and W. Gilbert (1984). Genomic sequencing. *Proc. Natl. Acad. Sci. USA 81*:1991–1995.
87. Birren, B. W., E. Lai, S. M. Clark, L. Hood, and M. Simon (1988). Optimized conditions for pulsed field gel electrophoretic separation of DNA. *Nucleic Acids Res. 16*:7563–7582.
88. Lai, E., B. W. Birren, S. M. Clark, M. I. Simon, and L. Hood (1989). Pulsed field gel electrophoresis. *Biotechniques 7*:34–42.
89. Mathew, M. K., C. L. Smith, and C. R. Cantor (1988). High-resolution separation and accurate size determination in pulsed-field gel electrophoresis of DNA. 1 and 2. *Biochemistry 27*:9204–9226.
90. Sor, F. (1988). A computer program allows the separation of a wide range of chromosome sizes by pulsed field gel electrophoresis. *Nucleic Acids Res. 16*: 4853–4863.
91. Sobrol, B. W. S., and A. G. Atherly (1989). Pulse time and agarose concentration affect the electrophoretic mobility of cccDNA during electrophoresis in CHEF and in FIGE. *Nucleic Acids Res. 17*:7359–7369.

92. Smith, C. L., T. Matsumoto, O. Niwa, S. Klco, J. Fan, M. Yanagida, and C. R. Cantor (1987). An electrophoretic karyotype for *Schizosaccharomyces pombe* by pulsed field gel electrophoresis. *Nucleic Acids Res. 15*:4481–4489.
93. Schwartz, D. C., L. C. Smith, M. Baker, and M. Hsu (1989). ED: pulsed electrophoresis instrument. *Nature 342*:575–576.
94. Bostock, C. J. (1988). Parameters of field inversion gel electrophoresis for the analysis of poxvirus genomes. *Nucleic Acids Res. 16*:4239–4252.
95. Heller, C., and F. Pohl (1989). A systematic study of field inversion gel electrophoresis. *Nucleic Acids Res. 17*:5989–6003.
96. Fan, J. B., Y. Chikashige, C. L. Smith, O. Niwa, M. Yanagida, and C. R. Cantor (1988). Construction of a *NotI* restriction map of the fission yeast *Schizosaccharomyces pombe* genome. *Nucleic Acids Res. 17*:2801–2818.
97. Johnston, J. R., and R. K. Mortimer (1986). Electrophoretic karyotyping of laboratory and commercial strains of *Saccharomyces* and other yeasts. *Int. J. Syst. Bacteriol. 36*:569–572.
98. Chickashige, Y., N. Kinoshita, Y. Nakaseko, T. Matsumoto, S. Murakami, O. Niwa, and M. Yanagida (1989). Composite motifs and repeat symmetry in *S. pombe* centromeres: Direct analysis by integration of *Not*I restriction sites. *Cell 57*:739–751.
99. Waterbury, P. G., and M. J. Lane (1987). Generation of lambda concatamers for use as pulsed field electrophoresis size markers. *Nucleic Acids Res. 15*: 3930.
100. Vollrath, D., R. W. Davis, C. Connelly, and P. Hieter (1988). Physical mapping of large DNA by chromosome fragmentation. *Proc. Natl. Acad. Sci. USA 85*:6027–6031.
101. Smith, C. L., J. G. Econome, A. Schutt, S. Klco, and C. R. Cantor (1987). A physical map of the *Escherichia coli* K12 genome. *Science 236*:1448–1453.
102. Allshire, R. C., G. Cranston, J. R. Godsen, J. C. Maule, N. D. Hastie, and P. A. Fantes (1987). A fission yeast chromosome can replicate autonomously in mouse cells. *Cell 50*:391–403.
103. Couto, L. B., E. A. Spangler, and E. M. Rubin (1989). A method for the preparative isolation and concentration of intact yeast artificial chromosomes. *Nucleic Acids Res. 17*:8010.
104. O'Connor, M., M. Peifer, and W. Bender (1989). Construction of large DNA segments in *Escherichia coli. Science 244*:1307–1312.
105. Deininger, P.O., D. J. Jolly, C. M. Rubin, J. Friedmann, and C. W. Schmid (1981). Base sequence studies of 300 nucleotide renatured repeated human DNA clones. *J. Mol. Biol. 151*:17–33.
106. Fanning, T. G., and M. F. Singer (1987). Line-1: A mammalian transposable element. *Biochem. Biophys. Acta 910*:203–212.
107. Haynes, S. R., T. P. Toomey, L. Leinwald, and W. R. Jelinek (1981). The Chinese hamster Alu-equivalent sequence: A conserved, highly repetitious, interspersed deoxyribonucleic acid sequence in mammals has a structure suggestive of a transposable element. *Mol. Cell. Biol. 1*:573–583.

3

Molecular Delineation of a Chromosomal Sub-Band Commonly Rearranged in Hematologic Malignancies

Jorge J. Yunis

Hahnemann University School of Medicine, Philadelphia, Pennsylvania

I. INTRODUCTION

The combined use of refined chromosomal banding, somatic cell genetics, and molecular genetics, for the first time, is making it possible to map large regions of the human genome and to facilitate the discovery of genes and markers that are involved in human cancer [1–3]. These complementary technologies have yielded valuable information about recurrent genomic rearrangements found in a large number of patients with leukemia and non-Hodgkin's lymphoma [1,4,5]. Such information is also proving to be prognostically important in a growing number of hematologic malignancies [4–7]. Among the more common adult solid tumors, such as colorectal and small-cell lung carcinoma, the use of polymorphic markers has made it possible to uncover discrete patterns of chromosomal deletion at the molecular level which, in turn, confirm information derived from chromosomal analysis [8–12]. The missing DNA segment (e.g., del 17p13 in colorectal carcinoma and del 3p21 in small-cell lung carcinoma) can now be pinpointed by molecular analysis, and the critical missing gene characterized by reverse genetics methodology, as has been accomplished in retinoblastoma [1].

Because the human haploid genome has 3×10^9 nucleotides and 2000 prophase bands [13], a given prophase band contains an average of 1300 kilobases (kb) of DNA. Advances made through recombinant DNA technology, particularly by restriction enzyme mapping of megabase DNA using pulse-field gel electrophoresis, now makes it possible to resolve DNA fragments from hundreds to a few thousands of kilobases in length [14–16]. This opens the way for the detailed characterization of individual chromosomal prophase bands. A complete macrorestriction map of the chromosomal band Xp21, where the Duchene muscular dystrophy gene resides, has recently been constructed by using this technology [16]. We have recently focused on the chromosomal sub-band 11q23.3 using long-range molecular-mapping, cloning, and somatic cell techniques [17]. Our studies of this sub-band, which is frequently found rearranged in hematologic malignancies, illustrates the continuing progress made towards "bridging the gap" between genes and chromosomes. We believe such a combined strategy may serve as a model approach for mapping chromosomal band (sub-band) involvement in human malignancy.

Other recently developed tools for large-scale mapping of chromosomal bands and sub-bands that could be considered for use include linkage analysis of numerous chromosome-specific single-copy [18] or polymorphic DNA probes [19], generation of chromosomal jumping libraries [20], and enzyme amplification of libraries from microdissected chromosomal bands [21].

II. MOLECULAR ANALYSIS OF CHROMOSOMAL REGION 11q22-q-25: A GENERAL STRATEGY FOR MAPPING OF CHROMOSOMAL SUB-BANDS INVOLVED IN HUMAN MALIGNANCY

Microscopic analysis of breakpoints of chromosomal rearrangements, Southern blot analysis of somatic cell hybrids containing a rearranged human chromosome, and pulse-field gel electrophoresis, are tools used by investigators interested in detailed studies of a region of the human genome involved in neoplastic disease. We have used these techniques to map the order of 12 genes, taking advantage of breakpoints found in cancer and constitutional chromosomal rearrangements that involve the chromosomal band region 11q22-q25 [17]. Band q23 of chromosome 11 is a region of particular interest to cancer geneticists because it is the site of a chromosomal breakpoint region common to hematologic malignancies [17]. In addition, five cell proliferation and differentiation genes belonging to the immunoglobulin superfamily [22–24] and the protooncogene *ets*-1 [25] have been mapped to this band. It also harbors a heritable [26] as well as constitutive fragile site [27] that is sensitive to a wide variety of mutagens and carcinogens [28], the ataxia telangiectasia gene [29], and the breakpoint of a common familial reciprocal translocation, t(11;22)(q23;q11) [30]. A t(11;22) (q23q24;q11), with a breakpoint distal to that present in hematologic malignancies, is also found in Ewing's sarcoma and neuroepithelioma [31].

We had previously reported *ets*-1 oncogene amplification in a patient with acute myelomonocytic leukemia (AMMoL) that had a homogeneous-staining region (HSR) that included band 11q23 [32]. As it is known that neighboring genes can be coamplified during the amplification process [33], we initiated our investigation of band 11q23 by attempting to determine whether genes neighboring *ets*-1 were amplified in this same patient and whether they are physically linked on a relatively short DNA segment.

A. Southern Blot Analysis of a Homogeneous-Staining Region of Chromosomal Segment 11q23-q25

Molecular investigation of an HSR on the patient with AMMoL showed that 11 genes were coamplified with *ets*-1 (Fig. 1) [17]. Southern blot analysis followed by serial dilutions and densitometric analysis showed a tenfold amplification together with rearrangement of *ets*-1 following digestion with the restriction enzymes *Bam*HI, *Eco*RI, *Xba*I, and *Hind*III. In addition, a tenfold amplification (with no rearrangement) was also found for the neutral cell adhesion molecule (N-CAM); the cell-differentiation antigens, CD3δ, γ, ϵ; the T-cell surface antigen 1 (Thy-1); porphobilinogen deaminase (PBGD); the signal recognition particle receptor (SRPR); and DNA probes for the apolipoprotein AI, CIII, and AIV gene cluster (see Fig. 1) [17].

Previously, it was found that an amplified chromosomal region can be 200- to 3000-kb long [34,35] and sometimes three to five genes have been identified as being coamplified [34]. Since we found coamplification of 12 genes mapped to two or three metaphase bands (q23 and q24 or q23, q24, and q25; each metaphase band with 8000–10,000 kb), it is possible that HSR gene amplification may involve several dozens of genes.

Whether a rearranged gene, such as *ets*-1 or a neighboring gene, may have precipitated the amplification process awaits further investigation. This has been suggested as a mechanism for N-*myc* amplification [36], which is frequently found in the advanced stages of neuroblastoma and is generally associated with a poor prognosis [37]. Amplification of chromosome 2p in neuroblastoma has been found to involve coamplified and rearranged loci adjacent to an amplified N-*myc* in different cell lines [36]. Although the amplicon region could be much smaller if a recombinational event took place [36], for example, between the rearranged *ets*-1 oncogene and *Thy-1*, there is no evidence that such a recombination took place in our patient.

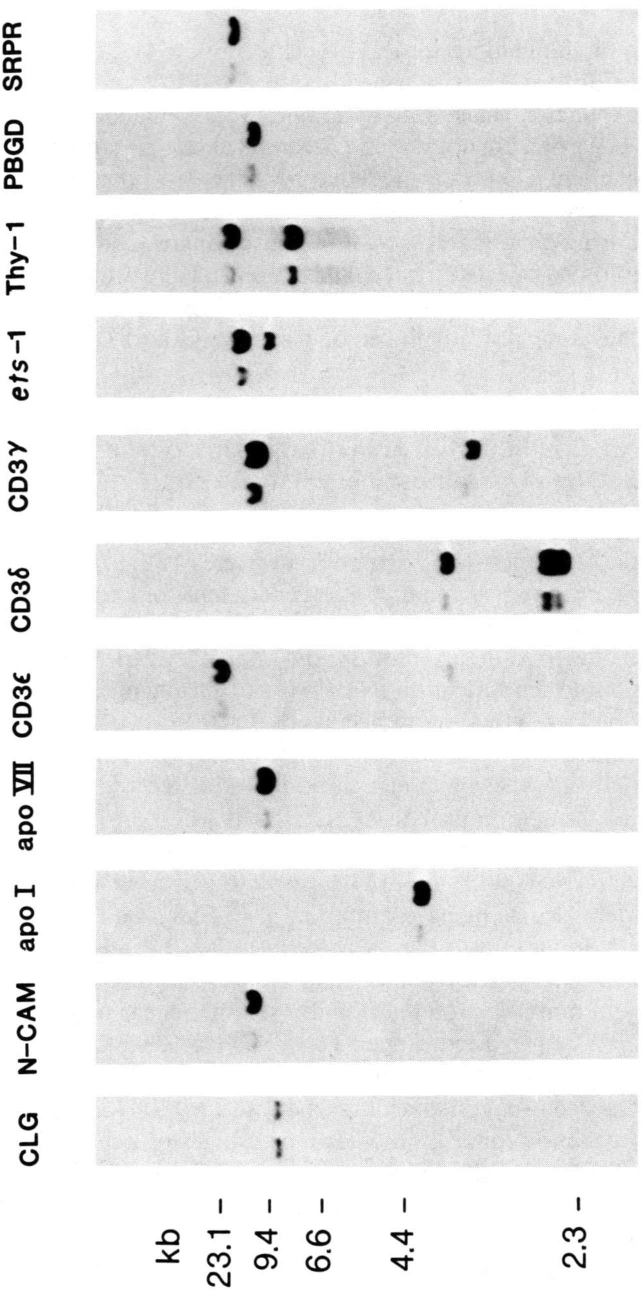

CLG
N-CAM
apo I
apo VIII
CD3ε
CD3δ
CD3γ
ets-1
Thy-1
PBGD
SRPR
kb
23.1 -
9.4 -
6.6 -
4.4 -
2.3 -

Figure 1 Southern blot illustrating the coamplification of N-CAM, apolipoprotein probes I and VII, CD3ϵ, δ, and γ genes, *ets*-1, Thy-1, porphobilinogen deaminase (PBGD), and signal recognition particular receptor (SRPR) in bone marrow DNA from a patient with acute myelomonocytic leukemia and a large homogeneous-staining region (HSR) at 11q23qter (lane 2): Bone marrow DNA from the same patient in complete clinical remission after chemotherapy is shown in lane 1 as a control. Normal and HSR DNA (e μg each) were digested with 30 units of *Bam*HI. The Southern blot was first hybridized with ^{32}P-labeled collagenase (CLG), used as one of the control gene probes, and then dehybridized and rehybridized successively with ^{32}P-labeled N-CAM, apo probes I and VII, CD3ϵ, CD3δ, CDγ, *ets*-1, Thy-1, PBGD, and SRPR. *bcl*-1, mapped to 11q13, progesterone receptor (PGR), mapped to 11q21q22, as well as actin and N-*ras*, which were used as controls. The actin and N-*ras*, like the PGR gene probe mapped to 11q21q22, showed similar signal intensity in the control and HSR DNA. A λ-DNA–*Hind*III fragment marker was used to determine DNA migration in kilobases, as indicated on the left. The signal for each probe was found in the HSR DNA to be amplified approximately ten times that of the normal DNA signal by use of serial dilutions (data not shown). (From Ref. 17, with permission)

B. Southern Blot Analysis of Somatic Cell Hybrids with 11q21-q23.3 Breakpoints

The general order of DNA probes representing 14 genes was determined by Southern blot analysis of DNA from rodent–human somatic cell hybrids [17]. Each of these hybrids contained a single human chromosome 11, derived from six different chromosomal defects, with breakpoints at bands 11q21, q23.2, and q23.3 (Table 1, Fig. 2). The defects included an inversion insertion (X;11)(q24;q23.3q21) from a patient with AMMoL; a t(11;14)(q23.3;q32.3) from a diffuse B–large-cell lymphoma cell line; a constitutional deletion (11)(q22.3qter) from a patient whose mother and brother have a heritable fragile site 11q23.3; and a constitutional translocation t(X;11)(q26;q23.2). A Chinese hamster–human somatic cell hybrid containing a single normal human chromosome 11 was used as control.

Thirteen different chromosome 11-specific DNA probes were hybridized to the DNA containing an intact chromosome 11 used as the control (see Table 1 and Fig. 2). Nevertheless, various derivative chromosomes containing different segments of the long arm of chromosome 11 showed that certain genes were present and others absent, depending on the location of the breakpoint. The somatic cell hybrid with a constitutional del(11)(q23.3qter), for example, showed that all chromosome 11-specific genes tested were present, except *ets*-1, *Thy-1*, *PBGD*, and *SRPR* (see Table 1,F) [17]. In contrast, hybrids with a derivative chromosome 4, X, and 14 from hematologic malignancies with an inv ins (X;11)(q24;q23.3q21), a t(4;11)(q21;q23.3), and t(11;14)(q23.3;q32.3), respectively, showed the opposite (see Table 1, C–E). The counterpart derivative X chromosome from the inv ins (X;11), which includes the chromosomal segment 11q21q23, tested positive for the progesterone receptor (*PGR*), collagenase (*CLG*) and *N-CAM* genes, the *apo* probes, and the *CD3* genes (Table 1,B). A Southern blot illustrating some of these findings is shown in Figure 2 [17].

This approach allowed the establishment of a provisional order of genes and breakpoints on the long arm of chromosome 11 to read as follows: *bcl*-1 (q13)--inv ins (X;11)(q24;q23.3q21) breakpoint--*PGR*, *CLG*--del(11)(q22.3qter) breakpoint--*apo*I, *apo*VII, CD3ϵ,δ,γ--t(4;11)(q21;q23.3), t(11;14)(q23.3;q32.3), inv ins (X;11)(q24;q23.3q21) and del (11)(q23.3qter) breakpoints--*ets*-1--,*Thy-1*--,*PBGD*--, *SRPR* (see Table 1 and Fig. 2). Our findings are consistent with the general observations by other investigators that apolipoprotein AI and CIII, *CD3*ϵ, δ, and γ--*ets*-1, *PBGD*, and *Thy-1* are located on chromosome 11 distal to the constitutional t(11;22)(q23;q11) and proximal to the Ewing's sarcoma t(11;22)(q23q24;q11) breakpoints [31].

Table 1 Presence or Absence of Specific Genes in Various Chromosome Defects Involving Varying Breakpoints in the Long Arm of Chromosome 11[a]

Chromosome 11 breakpoint	Probes	A (box)	A	B (box)	B	C (box)	C	D (box)	D	E (box)	E	F (box)	F	G (box)	G	H (box)	H
q21 →	bcl 1		+	X	−	11	+		−		−		+		+		+
	PGR		+		+		−		−		−		+		+	11	+
q22.3 →	CLG		+		+		−		−		−		+	11	+		+
q23.2 →	N-CAM		+		+		−		−		−		+		+		−
	apo I		+		+		−		−		−		+		−		−
	apo VII	11	+	11	+		−	4	−	14	−	11	+	X	−		−
	CD3ϵ		+		+		−		−		−		+		−		−
	CD3δ		+		+		−		−		−		+		−		−
q23.2 →	CD3γ		+		+		−		−		−		+		−		−
	ets-1		+		−		+		+		+		−		−		−
	Thy-1		+	X	−	11	+	11	+	11	+		−		−		−
	PBGD		+		−		+		+		+		−		−		−
	SRPR		+		−		+		+		+		−		−		−

[a]All DNA samples represent rodent–human somatic cell hybrids with a single normal or derivative human chromosome 11 schematically represented in boxed areas on the left of each column, with corresponding presence or absence of specific genes on the right (A–H). Southern blot analysis of DNA digested with *Bam*HI, *Eco*RI, and *Hind*III restriction enzymes. A, normal chromosome 11; B to H represent various specific rearrangements involving chromosome 11; B, der X with insertion 11q; C, del 11q21q23.3 from a leukemia with inv ins(X;11)(q24;q23.3q21); D, der 4 from a leukemia t(4;11)(q21;q23.2); E, der 14 from a lymphoma t(11;14)(q23.3;q32.3); F, constitutional del(11) (q23.3qter); G, der 11 from constitutional t(X;11)(q26;q23.2); H, constitutional del(11)(q22.3qter). For illustration of a selected Southern blot, see Figure 2.
Source: Ref. 17, with permission.

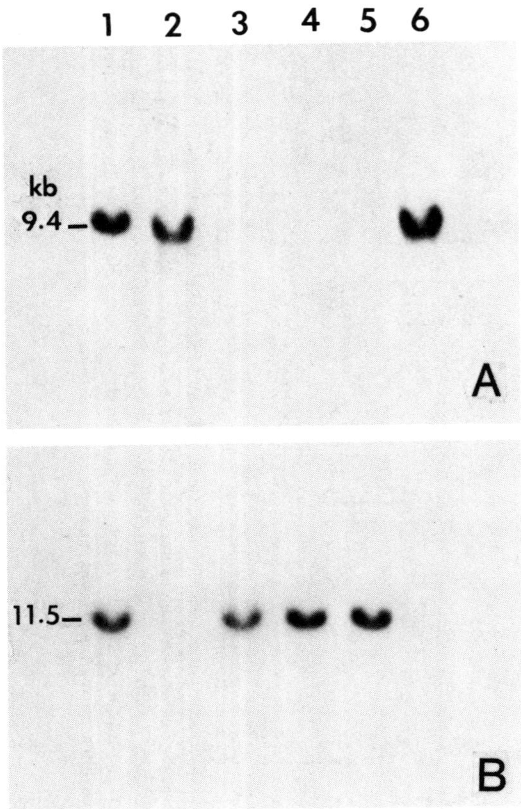

Figure 2 Southern blots of somatic cell hybrids containing a single normal or derivative human chromosome 11: 4 μg of DNA digested with *Eco*RI were used in each lane. The figure shows DNA from a hybrid containing a single normal chromosome (lane 1), a derivative 11 (lane 2), and derivative X (lane 3) from a leukemia with an inv ins(X;11)(q24;q23.3q21), a derivative 4 from a t(4;11)(q21;q23.3) (lane 4), a derivative 14 from a t(11;14)(q23.3;q32.3) (lane 5), and a del(11)(q23.3qter) (lane 6). (A) results of DNA hybridization with CD3γ; (B) results with *ets*-1. (From Ref. 17, with permission)

C. Long-Range Mapping of Genes in Chromosomal Segment 11q23-q25 by Pulse-Field Gel Electrophoresis

A mapping study with transverse alternating-field gel electrophoresis (TAFE), a variant of pulse-field gel electrophoresis, was carried out in an attempt to determine the approximate distance separating genes in the 11q22qter region. A set of seven different rare-cutting restriction enzymes (*Sal*I, *Sac*II, *Not*I, *Cla*I, *Mlu*I, *Nru*I, and *Sfi*I) and nine different DNA probes (N-*CAM*; *CD3*ε, δ, and γ; *ets*-1; *Thy-1*; *PBGD*; and *apo I* and *VII*) were used in an analysis of DNA from normal cultured human fibroblasts and lymphocytes.

As shown in Table 2, the *apo* probes and *CD3*γ genes comigrated on DNA cut with two or more enzymes (e.g., the *apo* probes and *CD3* genes comigrated with *Sal*I and *Sac*II restriction enzymes). The distance between genes found to comigrate was calculated by determining the fastest-migrating labeled band (smaller relative molecular mass) observed between two probes and accepting it as the maximum physical distance. This approach allowed us to estimate, for example, the maximum distance as 370 kb between the *apoI* probe and the *CD3*γ gene. In addition, comparison of fragment comigration patterns using different restriction enzymes and probes yielded a tentative order of genes found to comigrate by TAFE. Comigration of N-*CAM*, *ets*-1, *Thy-1*, and *PBGD* with each other or with the other genes analyzed was not found.

On the basis of our results, a proposed order of genes in the 11q21qter region is illustrated in Figure 3. All the 11q23.3 breakpoints found in hema-

Table 2 Differential Migration and Comigration of Genes in Chromosomal Band 11q23, Using Transverse Alternating-Field Electrophoresis[a]

DNA Probes	*Sal*I	*Sac*II	*Mlu*I	*Not*I	*Nru*I	*Cla*I	*Sfi*I
N-CAM	150	400 270	1100	1350 550	1300	80	410
apo I	425	370	890	410	420	320	50
apo VII	425	370	890	410	420	320	50
CD3ε	425	370	290	290	1350	280	50
CD3δ	425	370	290	290	1350	280	350
CD3γ	425	370	290	290	1350	280	350
ets-1	740 580	610	1550	1350	1550	300	260
Thy-1	520 380	170	1350	410	650	580	140
PBGD	240	30	1200	600	260	560	50

[a]For each probe, the size of the DNA fragments found with a given restriction enzyme digestion is given in kilobases. (± 20 kb for DNA segments over 200 kb). Results were based on hybridization of different DNA probes on the same autoradiograms with DNA from *Saccharomyces cerevisiae* yeast chromosomes and a λ ladder DNA as molecular markers. Genes found to comigrate with a given restriction enzyme cut are boxed in.
Source: Ref. 17.

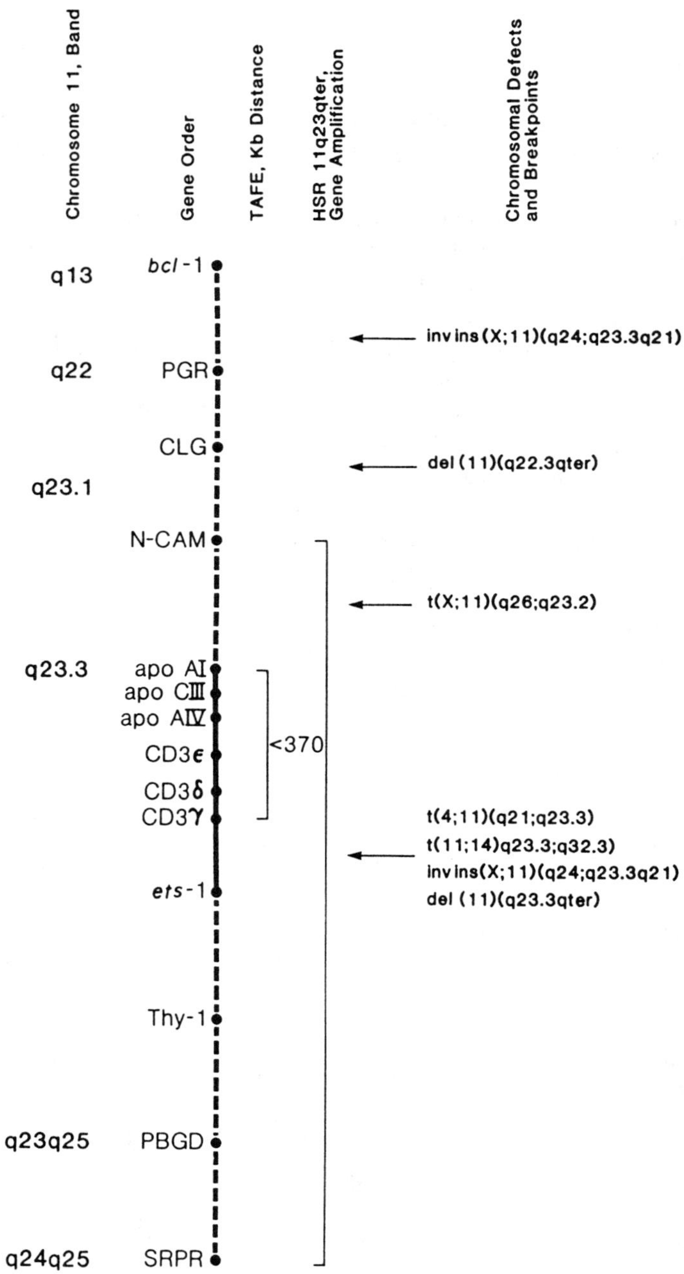
Chromosome 11, Band
Gene Order
TAFE, Kb Distance
HSR 11q23qter, Gene Amplification
Chromosomal Defects and Breakpoints
q13
bcl-1
q22
PGR
CLG
q23.1
N-CAM
q23.3
apo AI
apo CIII
apo AIV
CD3ε
CD3δ
CD3γ
ets-1
Thy-1
q23q25
PBGD
q24q25
SRPR
<370
inv ins(X;11)(q24;q23.3q21)
del(11)(q22.3qter)
t(X;11)(q26;q23.2)
t(4;11)(q21;q23.3)
t(11;14)q23.3;q32.3)
inv ins(X;11)(q24;q23.3q21)
del(11)(q23.3qter)

tologic malignancies, the breakpoint(s) of the t(11;14)(q23.3;q32.3) in diffuse large-cell lymphoma, the t(4;11)(q21;q23.3) frequently found in patients with childhood acute lymphocytic leukemia, and an inv ins (X;11)(q24;q23.3q21) in AMMoL, occur proximal to the *ets*-1 oncogene and distal to the *CD3*γ gene (see Fig. 3).

III. CLONING AND SEQUENCING OF AN 11q23.3 CANCER CHROMOSOMAL BREAKPOINT

Numerous patients with non-Hodgkin's lymphoma and leukemia have a reciprocal chromosomal translocation with precise breakpoints in their malignant cells [1,2]. Among these reciprocal translocations, two are particularly common and involve either chromosome sub-band 14q32.3 or 11q23.3, along with a second, but varying chromosome. Rearrangement of the immunoglobulin heavy-chain genes (*IgH*) at band 14q32.3 is found in approximately 65% of patients with non-Hodgkin's lymphoma and in some with acute or chronic lymphocytic leukemia [1]. Molecular investigation has shown that *IgH* rearranges with an oncogene or putative oncogene in the reciprocal chromosome, more often a chromosome 18 [1,2,4].

Rearrangements of sub-band 11q23.3 with one of several discrete regions of other chromosomes have been found in numerous hematopoietic malignancies with different morphologies. These include approximately 10% of patients with adult acute myelomonocytic or myeloproliferative syndrome, 25% with childhood acute lymphocytic leukemia, many with acute leukemia or common lineage, approximately 10% of patients with small lymphocytic cell lymphoma, and several patients with diffuse large-cell lymphoma [1,22].

Because *IgH* regions involved in recurrent breakpoints found in lymphoid malignancies are well characterized [1,2], a cell line with a t(11;14)(q23.3;q32.3) from a patient with diffuse large-cell lymphoma [38] present-

Figure 3 Proposed gene order, amplification, and distance in kilobases of genes on the long arm of human chromosome 11, using breakpoints of selected chromosomal rearrangements: The order of the *apo* genes and of the *CD3* genes, among themselves, is based on published data. It is not clear, however, whether *apoAI* or *AIV* is proximal to *CD3*ϵ. The gene order of *Thy*-1, *PBGD*, and *SRPR* is also unclear. These genes have been mapped to 11q23, q23q25, and q24q25, respectively (HHMI, 1988). The length of the amplicon region could prove smaller than indicated, if a recombinational deletion event took place between some of the genes. (From Ref. 17, with permission)

ed a unique opportunity for molecular investigation of the 11q23.3 breakpoint in the t(11;14)(q23.3;q32.3), which, in turn, may facilitate investigation of 11q23.3 rearrangements found in the various hematologic malignancies [22].

A. Rearrangement of an Immunoglobulin Heavy Chain Constant Gene in a 14q+ Chromosome from the t(11;14)(q23.3;q32.3) Lymphoma Cell Line

In our initial investigation of the t(11;14)(q23.3;q32.3) diffuse large-cell lymphoma cell line, we constructed three λ libraries from DNA of human–Chinese hamster somatic cell hybrids derived from the t(11;14). One hybrid contained the 14q+ derivative chromosome, a second contained the derivative 11q− chromosome, and a third the chromosome 11 not involved in the t(11;14). The 11q− and the 14q+ libraries were initially screened with a J_H and a Cγ3 cDNA probe. This approach enabled us to identify the breakpoint occurring in the *IgH* genes between J_H and Cγ in the t(11;14) [39]. The Cγ3 cDNA probe, known to cross-hybridize to all constant γ genes [40], was found to detect a weakly labeled and abnormally short 4.3-kb *Hind*III band by Southern blot analysis of t(11;14) cells and the somatic cell hybrid with the 14q+ chromosome (Fig. 4). This short band was not evident in Cγ germline DNA from fibroblasts or in productively rearranged Cγ DNA from antibody-producing B lymphocytes [41]. The abnormal Cγ gene was subsequently cloned from the genomic library containing, as a single human chromosome, the 14q+ human-specific chromosome involved in the t(11;14)(q23.3;q32.2).

The screening of recombinant phage plaques with the Cγ probe disclosed two DNA clones with a large *Eco*RI–*Xho*I segment positive for Cγ. These were purified and characterized by restriction map analysis (Fig. 5B, upper two clones). Chromosomal walking, using as a probe the 2.15-kb *Sac*I–*Bam*HI DNA fragment 5′ distal to the *Eco*RI–*Xho*I Cγ-positive region, disclosed two new partially overlapping subclones. These, in turn, revealed a second and weakly positive *Hind*III–*Xho*I Cγ DNA fragment (see Fig. 5B, third and fourth clones). A 12.2-kb DNA segment 5′ to the weakly positive Cγ DNA (see Fig. 5B, lowest clone) was isolated after additional walking, by using a single-copy 0.5-kb *Hind*III segment as a probe.

When the known germline restriction map of the Cγ4 and Cγ2 gene region (see Fig. 5A) [41] is compared with the map generated from the 14q+ chromosome (see Fig. 5B), it was shown that the breakpoint in chromosome 14 occurs next to a partially deleted Cγ2 gene. In fact, DNA sequence analysis of the first 259 base pairs (bp) in the 5′ portion of the *Hind*III–

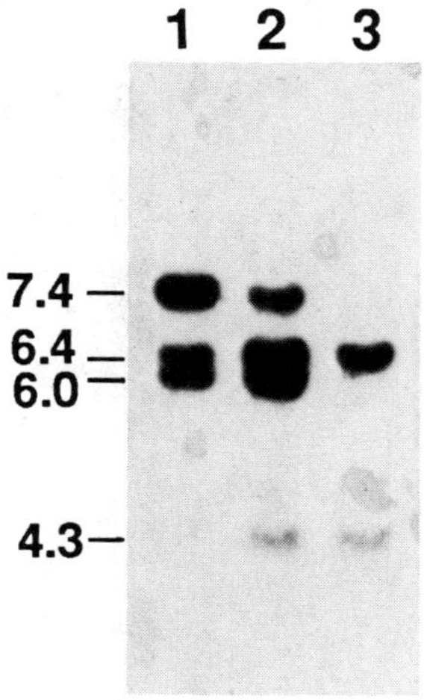

Figure 4 Southern blot hybridization with DNA digested with *Hind*III restriction enzyme and hybridized with a $C\gamma$ DNA probe: Samples of 3 μg of DNA were loaded on the gel. Lanes contain DNA from (1) normal fibroblasts, (2) the t(11;14)(q23.3;q32.3) lymphoma cell line, (3) a 14q+ X CHO hybrid derived from the t(11;14) cell line. Size of DNA fragments is given in kilobases. DNA was transferred from an 0.8% agarose gel to a Zetabind membrane (Cuno Inc., Meridan, Conn.) and DNA probes were ^{32}P labeled. (From Ref. 39)

*Xho*I $C\gamma$ region indicated that most of the $C\gamma$ gene was deleted, with only 43 bp of the last exon being retained [39]. It is possible to speculate that the unusually large deletion involving most of the $C\gamma2$ gene may have occurred as a result of heavy chain disease, which frequently involves a deletion of *IgH* constant gene sequences [40].

Immediately adjacent and 5′ to the abbreviated $C\gamma$ gene (see Fig. 5B), a 1.34-kb *Hind*III fragment was found and subdivided by a *Stu*I restriction enzyme site into a 0.63-kb *Hind*III–*Stu*I and a 0.71-kb *Stu*I–*Hind*III fragment. Southern hybridization, with a switch γ-2 ($S\gamma2$) probe, as well as sequence analysis of the larger 0.71-kb fragment, showed it to include $S\gamma2$ DNA. In contrast, the 0.63-kb *Hind*III–*Stu*I DNA fragment that was hybridized with DNA from a somatic cell hybrid contained only a normal human chromosome 11 and was negative with chromosome 14 DNA. Thus, the 0.63-kb *Hind*III–*Stu*I DNA fragment was chromosome 11-specific and the 1.34-kb *Hind*III fragment contained the junction of the t(11;14) translocation [39].

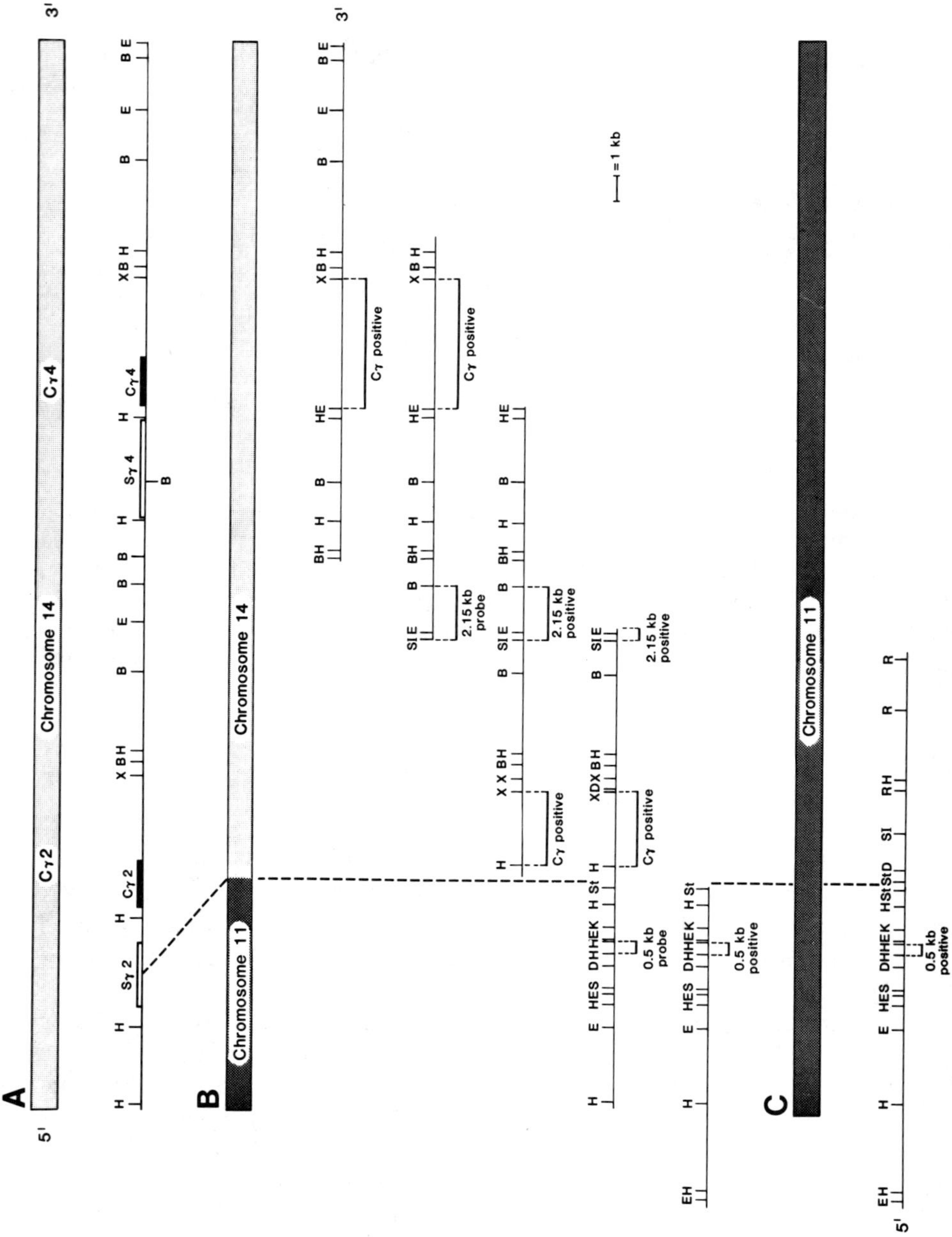
A
5'
3'
Cγ2
Cγ4
Chromosome 14
Sγ 2
Cγ 2
Sγ 4
Cγ 4
B
Chromosome 11
Chromosome 14
Cγ positive
2.15 kb probe
2.15 kb positive
0.5 kb probe
0.5 kb positive
= 1 kb
C
Chromosome 11

B. Rearrangement and Sequencing of the 11q23.3 Breakpoint in the t(11;14) Lymphoma Cell Line

To test for rearrangement of the 11q23.3 breakpoint in the t(11;14), the 12.2-kb DNA segment 5′ to the Cγ DNA (see Fig. 5B, lowest clone) was analyzed by Southern gel electrophoresis of somatic cell hybrids that contained human chromosome 11 or 14 as a single human chromosome. All restriction DNA fragments of the 12.2-kb clone obtained were specific for chromosome 11. With this clone, the single-copy 0.5-kb *Hind*III DNA fragment detected rearrangement in the t(11;14) cell line and in the somatic cell hybrid with the derivative 14q+ chromosome. Hybridization with the 0.5-kb *Hind*III fragment in Southern blots of DNA, digested with the restriction enzyme *Dra*I, showed a single germline band migrating at 3.7 kb in normal human total genomic DNA, whereas the germline band and an abnormal 6.7-kb band were found in the t(11;14) DNA. The abnormal 6.7-kb band was also found in DNA from the 14q+ chromosome (Fig. 6 and Table 3) [39]. Rearrangement was confirmed by inversed-field electrophoresis with *Sac*I and *Bam*HI restriction enzyme digestion: germline DNA migrated at 18.5 and 33 kb, respectively, and the abnormal band migrated at 25.5 and 25 kb, respectively [39].

To characterize the 11-specific DNA region 5′ and 3′ of the 11q23.3 breakpoint, we used the 0.5-kb *Hind*III probe on a λ genomic library of the normal human chromosome 11 of the t(11;14) cell line, selected in a Chinese hamster somatic cell hybrid. This approach yielded a 21.2-kb clone from chromosome 11. The restriction map of the 5′ half of this clone was identi-

---

Figure 5 (A) Schematic representation and restriction map of germline Cγ2 and Cγ4 DNA segment of chromosome 14. The switch and exon regions of Cγ2 and Cγ4 are indicated by open and solid bars, respectively. (B) Schematic representation of the 14q+ chromosome of t(11;14)(q23.3;q32.3) and of clones isolated from a λ library of a somatic cell hybrid containing the 14q+ chromosome as a single human chromosome. The two upper clones were isolated using a Cγ probe and DNA digested with the restriction enzymes *Bam*HI (B), *Eco*RI (R), *Hind*III (H), *Sac*I (S), and *Xho*I (X). For chromosomal walking, the 5′ most distal 2.15-kb *Sac*I–*Bam*HI segment found in one of the two isolated clones was used as a probe, and two new and partially overlapping subclones were isolated (third and fourth clones). The new clones showed a second Cγ gene, and the fourth clone also showed a single copy 0.5-kb *Hind*III DNA segment that was chosen to detect a fifth clone that only contained 11-specific DNA. Additional restriction enzymes used in the two lowest clones included *Dra*I (D), *Kpn*I (K), *Sal*I (S), and *Stu*I (St). (C) The 0.5-kb *Hind*III fragment was also used to clone the corresponding region on the normal chromosome 11 of t(11;14). The t(11;14) breakpoint was found between the two *Stu*I restriction sites (dotted line). (From Ref. 39)

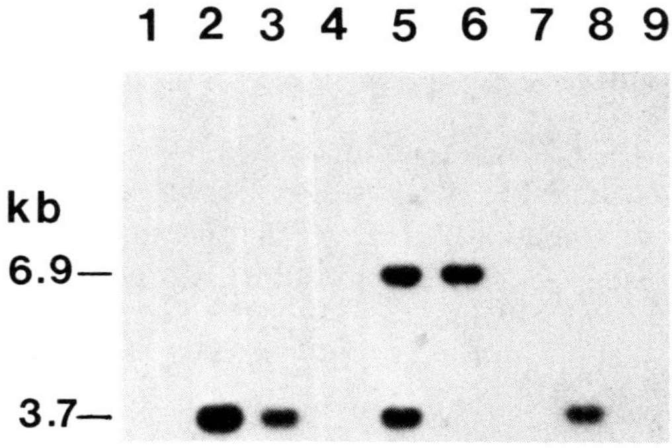

Figure 6 Southern blot hybridization using DNA digested with *Dra*I restriction enzyme and hybridized with a chromosome 11-specific 0.5-kb *Hind*III probe: Samples of 3 μg of DNA were loaded on the gel. Lanes contain DNA from (1) CHO cells derived from Chinese hamster DNA; (2) normal human DNA; (3) human X–CHO hybrid cells containing only a normal human chromosome 11 control; (4) human X–CHO hybrid cells containing only a normal human chromosome 14 control; (5) DNA from t(11;14)(q23.3;q32.3) lymphoma cell line; (6) 14q+ X–CHO hybrid derived from the t(11;14) cell line; (7) 11q− X–CHO hybrid derived from the t(11;14) cell line; (8) human X–CHO hybrid containing the normal chromosome 11 of the t(11;14) cell line; (9) human X–CHO hybrid containing the chromosome 14 of the lymphoma cell line, not involved in the t(11;14). For interpretation of results, see Table 1. (From Ref. 39)

cal with the 11-specific segment of the 14q+ chromosome map (see Fig. 5C). The restriction map of the 3′ half of the 21.2-kb clone, however, was different from the segment of the 14q+ chromosome starting with a unique *Stu*I–*Stu*I 0.29-kb DNA fragment. This suggested that the chromosome 11 translocation breakpoint occurred between two *Stu*I restriction sites. DNA sequencing of this small region and of the 5′ end of the *Stu*I–*Hind*III fragment adjacent to the Sγ2/Cγ2 in the 14q+ chromosome uncovered the breakpoint (Fig. 7) [39].

Rearrangement of 14q32.3-IgH lymphoid malignancies generally follows one of two basic mechanisms. An error can occur in the normal recombinational process of a diversity (D_H), joining (J_H), and variable (V_H) region, most often in a chromosome 14 of a pre-B cell [1,2]. In the more mature B cells, an aberrant switching mechanism is often involved [1,2]. Our analysis of the t(11;14) (q23.3;q32.3) lymphoma cell line, showed that the rearrangement error involved the switch segment of the *IgH* Cγ2 gene but was unexpectedly preceded by an *alu* repeat.

Table 3 Germline and Rearranged Locus in the t(11;14)(q23.3;q32.3) Lymphoma Cell Line and in Derived Somatic Cell Hybrids Containing the Pertinent Chromosomes 11 and 14[a]

	Human chromosomes present				11q23.3 Breakpoint locus	
	11	11q−	14	14q+	Germline	Rearranged
Pertinent Chromosomes of t(11;14)						
t(11;14) DNA	+	+	+	+	+	+
14q+ chromosome	−	−	−	+	−	+
11q− chromosome	−	+	−	−	−	−
Normal chromosome 11	+	−	−	−	+	−
Other chromosome 14	−	−	+	−	−	−
Controls						
Chinese hamster DNA	−	−	−	−	−	−
Normal human DNA	+	−	+	−	+	−
Normal human chromosome 11[b]	+	−	−	−	+	−
Normal human chromosome 14[b]	−	−	+	−	−	−

[a]*Dra*I, *Sac*I, and *Bam*HI were used as restriction enzymes in Southern blots in which the 0.5-kb *Hind*III probe was used. See Figure 6 for results obtained with *Dra*I.
[b]Somatic cell hybrids containing a single human chromosome.
Source: Ref. 39.

C. Molecular Studies of 11q23.3 Sub-Band Rearrangements in Different Types of Hematologic Malignancies

There are more than 15 recurrent chromosomal subtypes of hematologic malignancies, in addition to the t(11;14) (q23.3;q32.3) diffuse large-cell lymphoma, that have a breakpoint in the 11q23.3 region (1,22). In a preliminary study, we used the 0.5-kb *Hind*III probe derived from the t(11;14) (q23.3;q32.3) breakpoint region against DNA from patients with other 11q23.3 rearrangements. These included t(4;11)(q21;q23.3) acute lymphocytic leukemia and t(9;11)(p22;q23.3), t(11;17)(q23.3;q25), and inv ins (X;11)(q24;q23.3q21) acute myelomonocytic leukemia. DNA digestion with restriction enzymes *Dra*I, *Sac*I, and *Sfi*I showed band migration at 3.7, 18.5, and 100 kb, respectively, the same as found in DNA from normal controls [39]. This indicated that the breakpoint(s) of the 11q23.3 rearrangements is

```
Chromosome 11     CGAAGATGAGGGAATATGAGGCCGAAGGGGGCGGCAAGGACTTAGGGGATGAT
                  |||||||||||||||||||||||||||||||||||||||||||||||||||||
Chromosome 14q+   CGAAGATGAGGGAATATGAGGCCGAAGGGGGCGGCAAGGACTTAGGGGATGAT
                                        ↓
Chromosome 11     GGTGGAGGTTATAGGGTACGAGGCGGCCTATAGGGGGCGGCAGGGACATGCTGT
                  |||||||||||||||||||||||
Chromosome 14q+   GGTGGAGGTTATAGGGTACGAGGAAATTATACATTTTATATAAGAAATGTAAC

Chromosome 11     ATTGGAGGGCTCGGTTTAGAAGCAAACCGGCCGCTGCTTGATTGC

Chromosome 14q+   GGGGTTGCTGGAGGTGGTGGCTCACACCTGTAATCCAGCACTTTG
```

Figure 7 DNA base sequences surrounding the breakpoint (arrow) of the reciprocal translocation t(11;14)(q23.3;q32.3) cell line: Sequences of the normal chromosome 11 and the 14q+ are compared. Base identity is shown with vertical lines. (From Ref. 39)

not at the same site as found for the t(11;14)(q23.3;q32.3) lymphoma cell line. A breakpoint heterogeneity of this type has been found in Ph^1-positive acute lymphocytic leukemia, in which the *bcr* gene breakpoint at the chromosome band 22q11 in the t(9;22)(q34;qll) translocation and the genomic breakpoint within the *abl* gene on chromosome 9 can occur over a 200-kb region [42]. Thus, it is possible that a given putative oncogene at 11q23.3 may be involved in several of the 11q23.3 rearrangement breakpoints.

We have termed the putative 11q23.3 breakpoint gene in the t(11;14) lymphoma cell line *bcl*-4 (for B-cell lymphoma leukemia 4). If *bcl*-4 is found to be involved in other 11q23.3 rearrangements besides the t(11;14)(q23.3;q32.3) diffuse large-cell lymphoma, this would open the way to determining the location and function of a gene(s) in the breakpoint of a second, but varying, chromosome involved in the rearrangement [e.g., chromosome band 4q21 in the t(4;11)(q21;23.3) and 9p22 in the t(9;11)(p22;q23.3)]. As chromosomal rearrangements of sub-band 11q23.3 are associated with a pre-B, B, myelomonocytic, or monocytic cell phenotype in hematologic malignancies, it is possible that genes in the breakpoint of the varying chromosome may be associated with marrow stem cell differentiation [43].

IV. CONCLUSION

Several complementary technologies are now available for mapping and studying genes in specific chromosomal bands (sub-bands) involved in human disease. The combined application of refined chromosomal banding, Southern blot analysis of somatic cell hybrids, and pulsed-field gel electrophoresis is proving to be particularly useful in delineating at the molecular level, chromosomal bands and sub-bands frequently rearranged in hematologic malignancies. We have employed this approach to show the general order of genes in the q22–q25 region of human chromosome 11. In addition, 11 genes in this region were found to be amplified in a patient with acute myelomonocytic leukemia and an HSR 11q23qter, providing evidence that HSR gene amplification may occur over a larger domain than previously believed. Interestingly, the breakpoint of a constitutional deletion from a patient whose mother and brother have a heritable 11q23.3 fragile site was found to occur between the *CD3*γ gene and the *ets*-1 oncogene, in the same region as the breakpoint(s) of 11q23.3 sub-band rearrangements frequently observed in acute leukemia and, sometimes, in non-Hodgkin's lymphoma.

Molecular investigation of the sub-band 14q32.3-IgH rearrangements found in most patients with non-Hodgkin's lymphoma and some with acute or chronic lymphocytic leukemia, has shown that an *IgH* gene rearranges with an oncogene or putative oncogene in the reciprocal chromosome, frequently a chromosome 8 or 18. Similarly, rearrangement of sub-band 11q23.3 with one of several discrete bands of other chromosomes has been observed in numerous hematologic malignancies with different phenotypes. We analyzed a diffuse large-cell lymphoma cell line with a t(11;14) translocation involving both sub-band 14q32.3 and sub-band 11q23.3. Cloning and sequence analysis of the t(11;14) breakpoint showed that the $S\gamma2/C\gamma2$ gene of *IgH* rearranged with a putative oncogene at sub-band q23.3 of chromosome 11. Characterization of the 11q23.3 breakpoint gene in the t(11;14) (q23.3;q32.3) diffuse large-cell lymphoma cell line will facilitate molecular investigation of other 11q23.3 cancer chromosomal rearrangements.

ACKNOWLEDGMENTS

The excellent editorial assistance and critique of William Hoffman are gratefully acknowledged. This work was supported in part by NIH grant CA33314.

REFERENCES

1. Yunis, J. J. (1989). Genes and chromosomes in the pathogenesis and prognosis of human cancers. *Adv. Pathol. 2*:147–188.
2. Nowell, P. C., and C. M. Croce (1988). Chromosomal approaches to oncogenes and oncogenesis. *FASEB J. 2*:3054–3060.
3. Bishop, J. M. (1987). The molecular genetics of cancer. *Science 235*:305–311.
4. Yunis, J. J., M. Lobel, M. A. Arnesen, M. M. Oken, M. G. Mayer, R. E. Rydell, and R. E. Brunning (1988). Refined chromosome study helps define prognostic subgroups in most patients with primary myelodysplastic syndrome and acute myelogeneous leukemia. *Br. J. Haematol. 68*:189–194.

5. Yunis, J. J., M. G. Mayer, M. A. Arnesen, D. P. Aeppli, M. M. Oken, and G. Frizzera (1989). *bcl*-2 and other genomic alterations in the prognosis of large-cell lymphoma. *N. Engl. J. Med. 320*:1047–1054.
6. Williams, D. L., J. Harber, S. B. Murphy, et al. (1986). Chromosomal translocations play a unique role in influencing prognosis in childhood acute lymphoblastic leukemia. *Blood 68*:205–212.
7. Dreazen, O., F. Rassool, R. S. Sparkes, et al. (1987). Do oncogenes determine clinical features in chronic myeloid leukemia? *Lancet 1*:1402–1405.
8. Vogelstein, B., E. R. Fearon, S. E. Kern, S. R. Hamilton, A. C. Preisinger, Y. Nakamura, and R. White (1989). Allelotype of colorectal carcinomas. *Science 244*:207–211.
9. Hansen, M. F., and W. K. Cavenee (1988). Tumor suppressors: Recessive mutations that lead to cancer. *Cell 53*:172–173.
10. Kok, K., J. Osinga, B. Carritt, et al. (1987). Deletion of a DNA sequence at the chromosomal region 3p21 in all major types of lung cancer. *Nature 330*:578–581.
11. Zbar, B., J. Brauch, Talmadge, C. et al. (1987). Loss of alleles of loci on the short arm of chromosomes 3 in renal cell carcinoma. *Nature 327*:721–724.
12. Yunis, J. J. (1987). Multiple recurrent genomic deletions and duplications are a common feature in most human neoplasias. In *UCLA Symposium: Recent Advances in Leukemia and Lymphoma*, Gale, R. P., and D. W. Golde (eds.). Alan R. Liss, New York, pp. 3–21.
13. Yunis, J. J. (1981). Mid-prophase human chromosomes: The attainment of 2000 bands. *Hum. Genet. Cell 56*:293–298.
14. Carle, G. F., M. Frank, and M. V. Olson, (1986). Electrophoretic separations of large DNA molecules by periodic inversion of the electric field. *Science 232*: 65–68.
15. Patterson, M., M. Bell, C. Schwartz, and K. Davies (1988). Pulse-field gel mapping studies in the vicinity of the fragile site at Xq27.3. *Am. J. Med. Genet. 30*:581–591.
16. Burmeister, M., A. P. Monoco, E. F. Gillard, G. J. B. Van Ommen, N. A. Affara, et al. (1988). A 10-megabase physical map of human Xp21, including the Duchene muscular dystrophy gene. *Genomics 2*:189–202.
17. Yunis, J. J., C. Jones, M. Madden, D. Lu, and M. G. Mayer (1989). Gene order, amplification and rearrangement of chromosome band 11q23 in hematologic malignancies. *Genomics 5*:84–90.
18. Smith, D. I., R. Mangrulker, R. Geist, J. Gilbert, K. Kinsman, H. Drabkin, and W. Golembeiski (1989). Saturation of human chromosome 3 with unique sequence hybridization probes. *Genomics 4*:453–459.
19. Donis-Keller, H., et al. (1987). A genetic linkage map of the human genome. *Cell 51*:319–337.
20. Poustka, A. M., and H. Lehrach (1986). Jumping libraries and linking libraries: The next generation of molecular tools in mammalian genetics. *Trends Genet. 2*:174–179.

21. Ludecke, H. J., G. Senger, U. Claussen, and B. Horsthemke (1989). Cloning defined regions of the human genome by microdissection of banded chromosomes and enzymatic amplification. *Nature 338*:348–350.
22. Yunis, J. J., and R. D. Brunning (1986). Prognostic significance of chromosomal abnormalities in acute leukemias and myelodysplastic syndromes. *Clin. Hematol. 15*:597–620.
23. Gold, D. P., J. J. M. Van Dongen, C. C. Morton, G. A. Bruns, P. van den Elsen, et al. (1987). The gene encoding the E subunit of the t3/T-cell receptor complex maps to chromosome 11 in humans and to chromosome 9 in mice. *Proc. Natl. Acad. Sci. USA 84*:1664–1678.
24. Seki, T., N. Spurr, F. Obata, S. Goyert, P. Goodfellow, and J. Silver (1985). The human *Thy*-1 gene: Structure and chromosomal location. *Proc. Natl. Acad. Sci. USA 82*:6657–6661.
25. De Taisne, C., A. Gegonne, D. Stehelin, A. Bernheim, and R. Berger. (1984). Chromosomal localization of the human proto-oncogene c-*ets*. *Nature 310*:581–583.
26. Sutherland, G. R., P. B. Jacky, E. Baker, and A. Manuel (1983). Heritable fragile sites on human chromosomes: New folate-sensitive fragile sites: 6p23, 9p21, 9q32 and 11q23. *Am. J. Hum. Genet. 35*:432–437.
27. Yunis, J. J., and A. L. Soreng (1984). Constitutive fragile sites and cancer. *Science 226*:1199–1204.
28. Yunis, J. J., A. L. Soreng, and A. E. Bowe (1987). Fragile sites are targets of diverse mutagens and carcinogens. *Oncogene 1*:56–69.
29. Gatti, R. A., I. Berkel, E. Boder, et al. (1988). Localization of an ataxia-telangiectasia gene to chromosome 11q22-23. *Nature 336*:557–580.
30. Iselius, L., J. Lindsten, A. Aurias, M. Fraccaro, et al. (1983). The 11q;22q translocation: A collaborative study of 20 new cases and analysis of 110 families. *Hum. Genet. 64*:343–355.
31. Burdaf, M., B. Sellinger, C. Griffin, and B. S. Emanuel (1989). Comparative mapping of the constitutional and tumor-associated 11;22 translocations. *Am. J. Hum. Genet. 45*:128–139.
32. Rovigatti, U., D. K. Watson, and J. J. Yunis (1986). Amplification and rearrangement of Hu-*ets*-1 in leukemia and lymphoma with involvement of 11q23. *Science 232*:398–400.
33. Alitalo, K., and M. Schwab (1987). Oncogene amplification in tumor cells. *Adv. Cancer Res. 47*:235–281.
34. Debatisse, M., I. Saits, G. Buttin, and G. R. Stark (1988). Preferential amplification of rearranged sequences near amplified adenylate deaminase genes. *Mol. Cell. Biol. 8*:17–24.
35. Montgomery, K. T., and P. W. Melera (1988). Analysis of a novel locus frequently coamplified with N-*myc* in human neuroblastoma. *Prog. Clin. Biol. Res. 271*:71–88.
36. Shimatake, H., T. Kikuchi, C. Abe-Nakagawa, N. Kanda, and T. Tsuchida (1988). Rearrangement dynamics involved in gene amplification in human neuroblastoma cells. *Prog. Clin. Biol. Res. 271*:121–131.

37. Seeger, R. C., G. M. Brodeur, H. Sather, et al. (1986). Association of multiple copies of the N-*myc* oncogene with rapid progression of neuroblastomas. *N. Engl. J. Med. 313*:1111–1116.
38. Kubonishi, I., K. Niya, M. Yamashita, S. Yano, T. Abe, Y. Ohsuki, and I. Miyoshi (1986). Characterization of a new human lymphoma cell line (RC-K8) with t(11;14) chromosome abnormality. *Cancer 58*:1453–1460.
39. Yunis, J. J., M. G. Mayer, C. Jones, and D. Lu (1990). Molecular cloning of the 11q23.3 chromosome breakpoint in diffuse lymphoma. *Science* (In press).
40. Alexander, A., I. Anicito, and J. Buxbaum (1988). Gamma heavy chain disease in man. *J. Clin. Invest. 82*:1244–1252.
41. Flanagan, J. G., and T. H. Rabbitts (1982). Arrangement of human immunoglobin heavy chain constant region genes implies evolutionary duplication of a segment containing gamma, epsilon, and alpha genes. *Nature 300*: 709–713.
42. Westbrook, C. A., et al. (1988). *Blood 71*:697–702.
43. Rowley, J. D. (1984). Biologic implications of consistent chromosome rearrangements in leukemia and lymphoma. *Cancer Res. 44*:3159–3168.

4

High Base-Sequence-Resolution Extraction of DNA Fragments by Immobilized Probe for High-Efficiency Gene Cloning

Akiyoshi Wada* and Hiromichi Tsurui†

The University of Tokyo, Tokyo, Japan

Akira Suyama

Nagaoka University of Technology, Niigata, Japan

I. INTRODUCTION

The affinity and specificity of hybridization between complementary DNA (or RNA) strands are extraordinarily high compared with other types of biological macromolecular interaction. This is because of the long-range nature of cooperativity along a nucleotide chain in double-helix formation. Thus, even one base mismatching among several tens of bases in a double helix greatly reduces its stability [1–5]. This distinctive molecular characteristic of the hybridization reaction has enabled us to devise a highly effective method for detecting and extracting a DNA (or RNA) fragment of a preselected base sequence by using a probe strand having the sequence complementary to it. Recent advances in machine-aided synthesis of sequence-designed polynucleotides have made a sequence-tunable detection–extraction possible, and this method has gained much popularity among researchers in molecular biology and biochemical genetics [reviewed in Refs. 6–9].

The present chapter is intended to give a systematic account of a modern DNA (or RNA) base-sequence-probing method with a synthesized short (10–50-bp) oligonucleotide as an immobilized probe. The primary aim here is to describe the details of the preparation and the characteristics of our

Current affiliation:
*Sagami Chemical Research Center, Sagamihara, Kanagawa, Japan
†Juntendo University, Tokyo, Japan

newly developed probe system, in which probe DNA fragments are so immobilized on a high-performance liquid chromatograph (HPLC) hard-gel surface that its whole base sequence is in full exposure for access by sample DNA strands without undesirable steric hindrance of the solid support. We hope that the device described in this chapter will remove some of the burden from researchers now responsible for tedious manual gene-cloning work.

In addition to the functional descriptions, we would especially like to stress the theoretical basis of the method. Namely, because the mechanism of hybridization has been elucidated by the theory of the helix-coil transition of the DNA double-strand, we can physically and quantitatively characterize the base–sequence-resolving ability of the present system in terms of the molecular thermodynamic properties of DNA. This is analogous to describing the wavelength resolution of light or the atomic-resolving power of a mass spectrometer by the electromagnetic theory and the mechanism of the spectrometer. Both theoretical study and experimental results in this chapter should demonstrate that the system described should be useful as a high-precision tool for the detection and extraction of a fragment of a desired base sequence. The design principle and apparatus constants of this DNA probe system have a definite molecular physical base; thus, the system has a high potential to be a "sequence spectrometer" in DNA analysis.

II. IMMOBILIZE SAMPLE OR IMMOBILIZE PROBE?

The detection or extraction of a DNA fragment having a specific base sequence is performed, directly or indirectly, through double-helix formation with its antistrand (i.e., a complementary probe fragment). In general, there can be two types of hybridization systems: sample-strand immobilization and probe-strand immobilization. The former is now commonly used in gene cloning. Namely, colonies of bacteria transformed with recombinant plasmids or bacteriophage plaques that contain a stretch of sample DNA are replicated in situ onto a nitrocellulose or nylon membrane. The DNA is then liberated, denatured, and bound to the membrane. A specific DNA fragment is then applied to the immobilized DNA as a probe for an aimed sequence. The hybridization, which may be detected with autoradiography or fluorescence signal, indicates recombinants containing complementary sequences, which are searched for gene cloning. As only a small portion of the DNA tethers it to the membrane, even though it may be attached to the surface at many places along the length of its long strand, most of its bases are freely accessible by a whole stretch of a short probe DNA fragment. Therefore, a high base-sequence resolution can be secured [10,11].

In such a system, however, because the sample fragments are immobilized, when one tries to extract and clone a gene from a billion base pair (bp) genome such as that of higher organisms, as many as 10^6–10^7 colonies or plaques should be investigated with appropriate probe(s) [7–9]. This screening procedure requires tedious and time-consuming manual work, which requires a month or more and is a primary rate-limiting step in the current gene-cloning process.

An alternative system (i.e., probe immobilization) is, therefore, an ideal form for searching and "fishing out" a fragment of a desired base sequence directly from a pool of genomic DNA fragments. Unfortunately, in a classic device for the immobilization of a DNA fragment to a membrane such as a nitrocellulose filter, masking of a substantial part of the short probe base sequence is inevitable because of the steric hindrance of the solid bed to which the probe is immobilized; only a small part of the strand is available for hybridization, and a considerable reduction in the hybridization rate and of the base-sequence resolution results [12,13]. The reduction of resolution is also brought about by the diversity of duplex stability coming from the variety of its (G + C) content [14,15, and references therein]. Consequently, even though the lengths of all the probes are made uniform, the individual probes differ considerably in (G + C) content, making it difficult to find suitably stringent and selective hybridization conditions for all probe members. A large number of false-positives can occur when screening highly complex genes of low abundance [11]. These difficulties in the probe immobilization system are surmountable with the strategy described in the following sections.

III. HOW TO GET HIGH BASE-SEQUENCE RESOLUTION

A high base–sequence-tuning ability in a hybridization system may be achieved by setting experimental conditions such that the stability of a hybridized double helix depends upon the degree of base mismatching alone.

The molecular parameters affecting the stability, or melting temperature, of a DNA–DNA or DNA–RNA double helix with a full base-sequence complementarity are the length and the (G + C) content of the fragment. The first of the basic requirements for a good probe is, therefore, a probe of homogeneous length, which is easily obtained by virtue of today's DNA synthesis instrumentation.

The second is the solvent that suppresses the difference between hydrogen-bonding strengths of G–C and A–T base pairs [11,16]. Several solvents, such as the tetraalkyl ammonium salts, are known to equalize the stability of these base pairs. In these solvents, the ions are bound more tightly to the A–T-rich region than to the G–C-rich region of a DNA, and stabilize bound regions [17]. As a result, at appropriate concentrations of these salts, a double strand melts within a very narrow temperature range regardless of its

(G + C) content [18,19]. For instance, the melting of ColE1 DNA in a standard saline citrate (SSC) buffer starts at 64°C from the less stable A–T-rich region of the DNA, propagates to regions of higher stability, and is completed at 81°C. This melting temperature ranges as wide as 17°C and is narrowed into an approximate 1°C width in a 2.4 M tetraethylammonium (TEA) chloride buffer [20]; 3.0 M tetramethylammonium (TMA) chloride has the same effect as TEA. Thus, all DNA double helices of homogeneous length with a perfectly matched (complementary) base sequence give a melting temperature independent of the base sequence; the difference depends upon only the degree of base mismatching. The usefulness of these reagents has already been demonstrated in gene screening with a sample immobilizing system [11] and in studying the single-copy DNA sequence difference between individual sea urchins among local populations [21].

The final requirement for a good probe is the absence of interference from the probe-immobilizing bed, to guarantee free access by sample DNAs to the entire base sequence of the probe strand and to realize hybridization characteristics similar to DNA in solution [22,23]. For this purpose, we adopted 5′ chain terminal immobilization in which hydroxyl groups on the silica gel surface or cellulose matrices are activated by carbonyldiimidasole, and then covalently bound with carbodiimide to the 5′ end of a single-stranded DNA probe through a linker such as hexaethylenediamine. This kind of 5′ end-mooring system for practical use has been developed by several researchers [22–24 and references therein], and its hybridization properties as a probe system have been extensively explored [12,13,20, and references therein]. The present study is based on these pioneering efforts.

We examined probe-supporting gels according to the following criteria: (1) chemical and physical stability, (2) easy accessibility for sample DNA to the immobilized probe strand, (3) ease and feasibility of chemical modification, (4) capacity for immobilization of the probe, and (5) absence of a nonspecific interaction to sample DNA (i.e., low background noise) [25]. LiChrospher Si 1000 (Merck Co.), which is a hard gel originally designed for HPLC, satisfied the foregoing conditions [26]; that is, it is stable enough to endure vigorous stirring, centrifugation, and autoclaving. It is also stable chemically except that it dissolves in an alkaline solution (pH >7.5). The pore size of the gel is about 100 nm: a moderate size suitable for our purpose. This is large enough for the DNA sample to diffuse into a gel particle and reach the immobilized probe quickly [25], whereas a gel of this pore size has enough surface area (20 m^2/g) for ligand immobilization. The rapid access of the sample DNA to the immobilized strand is essential for the isolation of a particular nucleotide sequence, especially when there are complementary strands that compete with the sample–probe duplex formation reaction. The particle diameter of 10 μm is smaller than that of normal soft gels and is preferable for rapid access by the sample to the strand immobilized in the gel, whereas it is not too small to prohibit treatment of the gel on a sintered glass filter or drying in a rotary evaporator. The hydroxyl group on the surface of the gel is not end-capped; hence, we can make use of this functional group for chemical modification. The gel is hydrophilic and, when converted to diol-silica, has no electrostatic charges on its surface, assuring that the sample DNA cannot adhere to the gel. We can easily add an amino group at the 5′ terminus of an oligonucleotide with an amino link at the final step of the synthesis. The coupling procedure of the oligonucleotide and tresyl-activated gel is very simple. We prepared several gels with different probe densities, from 4 to 60 OD/g dry gel, which is high enough for the isolation of a particular DNA sequence. We used this gel system in the examination of the probing ability of the present system described in Section V. The best gel system examined to date in our laboratory and the probe strand-attaching protocol are described in detail in the Appendix.

As stated, the use of 2.4 M TEA or 3.0 M TMA salt enabled the establishment of a hybridization elution procedure depending only on the probe length and not on its (G + C) content. Moreover, TEA or TMA salt accelerates the double-helix–coil transition reaction [27].

IV. THEORETICAL EVALUATION OF BASE-SEQUENCE RESOLUTION AND THE DESIGN PRINCIPLE FOR A DESIRABLE PROBE SYSTEM

A DNA double strand is equivalent physically to a one-dimensional crystal lattice, the melting of which is thus formulated within the framework of statistical thermodynamics [14,15, 28–31]. The Poland–Fixman–Freire algorithm [28,29] has enabled us to calculate the details of the helix-coil transition of a sequence-determined DNA. In this algorithm, the stability difference between the G–C and A–T base pairs in a normal solvent can be rigorously taken into account. In the present solvent system, G–C and A–T base pairs have equal stability as described earlier, and a common set of stability parameters may be applied. In contrast, zero bonding energy is instead assigned to the mismatched base pair(s) to evaluate the melting temperature depression by a defect in a double helix. The melting temperature and the temperature width of the helix–coil transition thus calculated present a theoretical basis on which to draw a diagram of the resolving power of the base sequence as shown in Figure 1, which delineates the probe length effect and the position effect of mismatching. It is evident from this diagram that the shorter the probe length is, the better the resolution becomes. However, this correlation is not particularly strong.

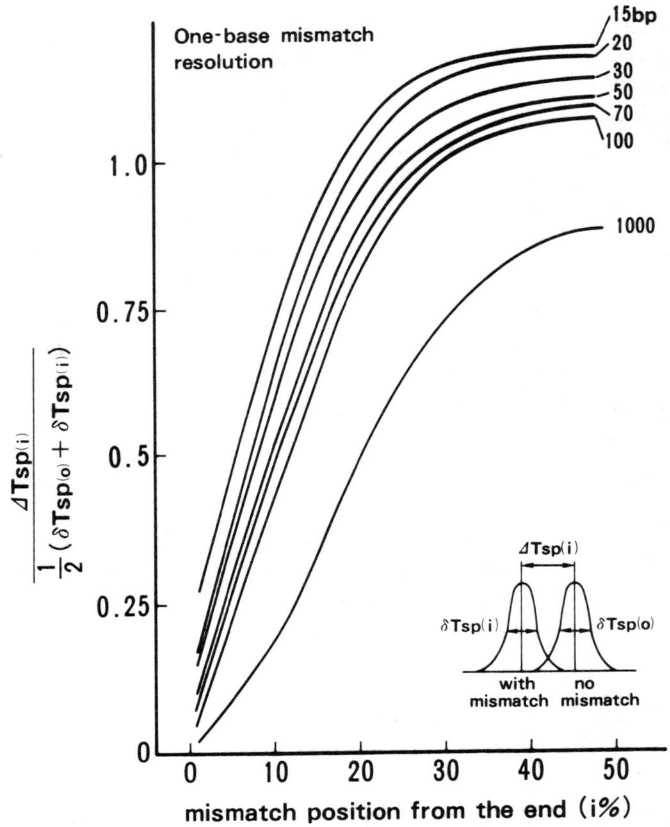

Figure 1 Theoretical diagram showing the base-sequence resolution of DNA probe system as a function of probe length and mismatch location: The latter is shown by percentage length from the end. Definitions of the separation temperature difference between the elution bands of probe–complementary DNA hybrid and probe–mismatched DNA hybrid, Δ Tsp, and their band width, δ Tsp, are illustrated at bottom right. The separation temperature is defined as that giving 50% chain separation of the hybrid. It is almost equal to the melting temperature in hybrids of short length.

The lower and upper limits of the probe length are determined as follows: The lower limit depends upon the complexity of the sequence under study. It is given by the condition that the probability of a random occurrence of the probe sequence in the DNA library being screened is much smaller than unity; accordingly, the redundancy in the selection (i.e., the probability of fishing out a spurious fragment) is negligible. The condition for a DNA with an equal base composition of the four bases with their random distribution is represented as

$$1 - [(1 - 1/4^{l_p})^{l_f - l_p + 1}]^{L/l_f} << 1 \quad (1)$$

where l_p is the probe length, l_f the average length of sample fragments, and L the total length of the library to be screened. In practice $l_f >> l_p >> 1$, so that the equation is equal to $L/4^{l_p} << 1$; thus we have

$$l_p >> 1.66 \log L. \quad (2)$$

This condition, however, is not sufficient for an optimal DNA probe because of the low detectability of a mismatched base that is located near the end of the probe as shown in Figure 1. Therefore, it is desirable for l_p to have two or three additional nucleotides of a length indicated by Equation 2.

Apart from a trivial technical difficulty in synthesizing long-probe nucleotide chains, the accuracy in base sequence discrimination, as revealed in Figure 1, determines the upper limit of the probe length. As shown, the stability reduction of hybrid duplexes from base mismatch(es) depends on their position. The reduction also depends on the probe length. When the base-mismatch number is kept constant, longer DNA probes result in less stability reduction, although the theoretical resolution does not change drastically, as shown in the figure, because longer hybrid duplexes have sharper melting transitions. The DNA probe length for practical use should be fewer than 50 nucleotides if single-base mismatch detection is required.

In addition, the following base sequences should be avoided in designing a probe: base sequences that easily form stable secondary structures, and those consisting partially of repetitive sequences. When base sequences of the particular sample fragments to be detected are already known, local base sequences to be adopted for the probe, which are involved in that sequence, and do not easily form stable secondary structures, can be easily selected by examining the hairpin stability along the chain by using the dynamic programming method developed for calculation of the optimal RNA secondary structure [32,33].

V. EXPERIMENTS AND RESULTS

There are two types of experimental procedures in a probe-immobilized system: batch processing and continuous temperature-gradient elution processing. Batch processing is carried out in 100 μL–1 mL in an Eppendorf tube, in which a series of annealing, separation, washing, and extraction steps are performed, as shown in Figure 2.

The temperature-gradient elution experiment comprises four processes: sample injection and denaturation, annealing, washing, and temperature-gradient elution. In the first process the temperature of the DNA probe column is raised to the denaturation temperature, T_d, and then the sample

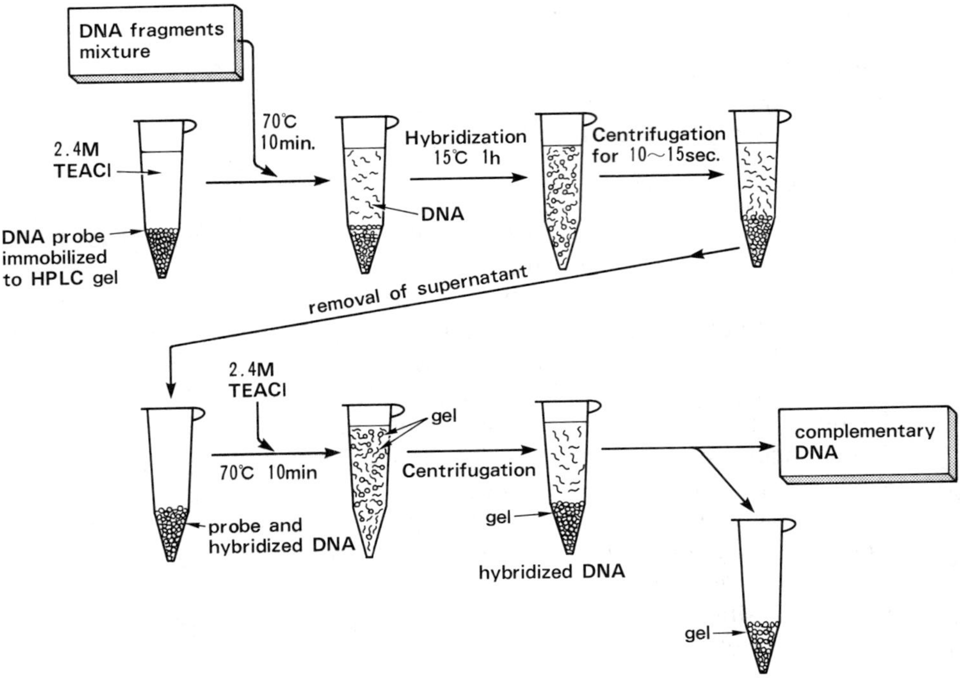

Figure 2 Experimental steps in the batch hybridization procedure.

solution is injected into the column. The injected double-stranded sample DNA is thus denatured and separated into single-stranded random coils. Secondary structures of the injected single-stranded DNA or RNA samples, if they exist, are unfolded as well, owing to the breakage of intramolecular base pairings. T_d is usually determined in reference to the length of samples, because the sample DNA is usually longer than DNA probes, and longer duplexes have a higher melting temperature. When DNA samples of more than several hundred base pairs in a double-stranded form are involved, the highest temperatures required for T_d are 65°C in 2.4 M TEACl and about 95°C in 3.0 M TMACl. These T_d values are about 5°C higher than T_m of the double-stranded long sample DNAs. In the second process, the column temperature is decreased to T_a, the annealing or hybridization temperature. Our system is designed to decrease the column temperature at the rate of −1°C/min. The column is then kept at T_a for an adequate time interval to hybridize the sample to the probe DNA. Tetraalkylammonium salt accelerates hybridization [19,26] so that 20 min is normally long enough for the annealing time interval. T_a is adjusted to be T_m − (15–25°C), where T_m is the melting temperature of the hybrid duplex formed, because at this temperature the annealing rate is most accelerated [27,34]. In the third process, the column temperature is raised to T_w, the wash temperature. The sample DNAs (or RNAs) that do not hybridize to DNA probes are washed away at this temperature. Those DNAs (or RNAs) without homologous base sequences to DNA probes form a self-annealed double helix with their partner strand in the sample DNA. In the last process, the column temperature is raised gradually, and samples are eluted: the temperature-gradient elution process. The elution buffer containing tetraalkylammonium salt also accelerates the melting reaction rate, so that a relatively fast temperature-elevating speed of 1°C/min is applied to the column.

A. Characterization of Probe–Sample DNA or RNA Interaction with a Hard-Gel Support

We studied the basic characteristics and the performance of the present DNA–probe system with batch processing from both equilibrium and dynamic aspects, using an adenovirus gene [35].

The oligonucleotide 18-mer in the antisense strand of the adenovirus type 2 *E1A* gene was used as the probe DNA attached to the HPLC gel. The *Puv*II–*Xbal* fragment (positions from 624 to 1336) of the adenovirus *E1A* gene with the ^{32}P end-labeled sense strand (Figure 3A) was used as a sample DNA to examine this probe system. The time course of the total radioactivity change of the unhybridized (supernatant) and hybridized DNAs was measured in a 2.4 M TEACl + 0.1 × SSC buffer in Eppendorf tubes as follows (see Fig. 2): 5 mg of the probe-immobilized gel was suspended in 100 μL of the buffer, and the sample DNA was added. The tube was kept at 70°C for 10 min to denature sample DNA. The mixture was then incubated at 15°C for different lengths of time to hybridize the sample DNA to the immobilized probe. The tubes from different incubation times were centrifuged, and the supernatants were collected. After washing the precipitated gel with the same buffer, 100 μL of the buffer was added to each tube and incubated at 70°C for 10 min to extract the hybridized DNA from the immobilized probe. The total radioactivity of the unhybridized (supernatant) and probe-hybridized (extract) DNAs was determined by liquid scintillation counting. Hybridization of RNA to the immobilized oligonucleotide probe was carried out using the same procedure, as just described, except the gel was autoclaved at 120°C for 40 min to inactivate RNase. The following are the typical characteristics surveyed:

1. Nonspecific adsorption of the sample DNA on the gel surface was examined (blank test) using gel particles without probe DNA immobilized. The result in Figure 3B shows that the immobilizing bed used in the present

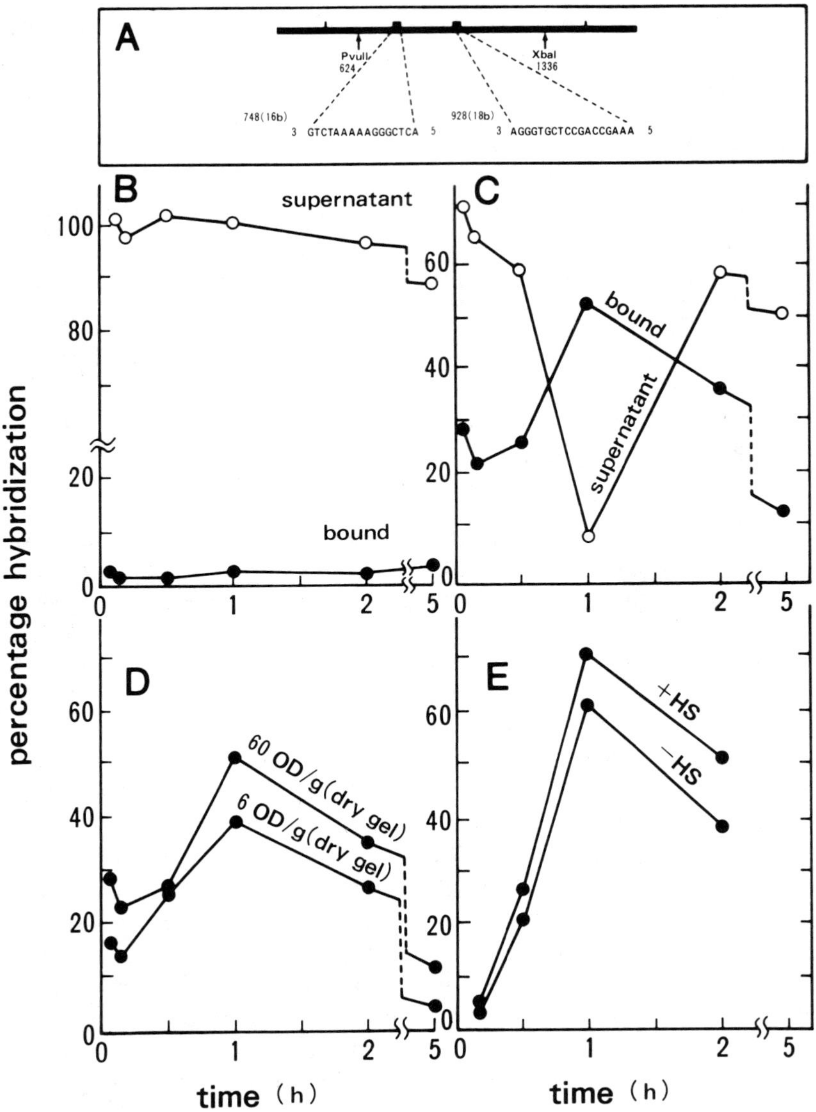

Figure 3 Hybridization characteristics of an 18-bp probe system: (A) The location of the probe fragments and sample fragment in the E1a adenovirus gene; (B) Absence of nonspecific adsorption of sample fragment to solid support (without probe DNA immobilized); (C) change of the amount of sample DNA bound to the immobilized probe DNA. (For the mechanism of the reduction of bound DNA over a long elution period see text); (D) Effect of probe DNA density on the sample/probe hybridization efficiency; (E) Effect of foreign DNA fragments (HS, herring sperm) on the sample–probe hybridization efficiency.

system was free of undesirable adhesion of sample DNA, which guarantees low background noise operation.

2. Time course of hybridization: the hybridization kinetics revealed an unexpected but reproducible characteristic (see Fig. 3C). Namely, the amount of the probe-hybridized sample DNA first increases, reaches a maximum after about 1 hr and then decreases. This distinctive time course is reproducible, and the same result was also obtained with a 16-mer probe, the sequence of which is shown in Figure 3A. This decrease in the amount of hybridized DNA is accounted for by a pulling out of a once-hybridized sample DNA by its complementary strand in solution (withdrawal effect) as follows: During the initial stage of the reaction, the formation of a probe–sample hybrid is much faster than double-helix formation between sample fragments because of the excess of probe strands over those of the sample DNA. After the probe–sample hybridization has proceeded to some extent, however, a sample DNA hybridized to an immobilized probe strand is pulled out into the solution phase by forming a fully complementary double strand with its antistrand of sample DNA. This results because the sample double helix is longer and, thus, much more stable than the short probe–sample hybrid. As a matter of fact, a single-stranded RNA, which lacks complementary strands, shows a binding curve that simply increases with time and shows no declining curve from the withdrawal effect. This is an important factor to consider when determining the optimal hybridization conditions for cloning a gene with a short or medium-sized probe.

3. The effect of the probe concentration on the probing efficiency was studied by two kinds of gels with high (60 OD/g dry gel) and low (6 OD/g dry gel) immobilized DNA densities (see Fig. 3D). A fixed amount of the labeled DNA (12,000 cpm, $\sim 10^{10}$ molecules) was added to these gels and hybridization was carried out in a TEACl + 0.1SSC buffer. In both cases, the amount of the hybridized sample DNA reached a maximum after 1 hr and then decreased. The difference in the hybridization profiles between these two gels of different probe densities was smaller than that expected, presumably because of a large excess concentration of the probe DNA over that of the sample DNA, and the withdrawal effect by complementary strands.

4. The effect of an excess amount of unrelated DNA on the hybridization efficiency was studied by mixing the labeled sample DNA with a 10^4-fold concentration of herring sperm (HS) DNA (see Fig. 3E). The hybridization profiles obtained with and without HS DNA were similar, whereas the hybridization efficiency with HS DNA was even higher than that without HS DNA. The reproducibility of this difference was confirmed. We speculate that the presence of a nonspecific DNA promotes the hybridization

between the probe and sample by the volume-exclusion effect and also by suppressing the withdrawal through partial hybrid formation between the sample and nonspecific DNA.

B. Resolving Power of One Base Pair Mismatching

The discrimination and separation of a DNA single strand, which has a one-base difference from a strand that has perfect complementarity with the probe base sequence, is possible when the melting temperature depression produced by the defect is comparable with, or larger than, the width of the melting range (see Fig. 1). As discussed in the preceding section, the melting temperature depression depends upon the location of the defect in the helix. This is clearly shown experimentally with model oligomers, 21 nucleotides long which have a different base-mismatch placement (Fig. 4), when a set of temperature-gradient elution profiles of the oligomers from a probe-immobilized column are plotted against column temperature (i.e., chain separation temperature T_{sp}). The one-base mismatch is detectable, unless it is located near the chain end. The melting temperature change, produced by single-base mismatches, demonstrated in this experiment is found to be much larger than the theoretically predicted one (7°C versus 4°C). Probably, this base mismatch gives a more positive double-helix destabilizing effect than the zero base-pairing energy assumed in the theoretical calculation. Actually, the destabilizing effect depends upon the type of the mismatched pair [1,35].

C. Results of RNA or DNA Extraction

1. Extraction of $tRNA^{Ser}$

Extraction of a specific tRNA from a crude mixture of mitochondrial RNA is demonstrated by the temperature-gradient elution with a 21-mer probe (cellulose support) having a part of the base sequence of bovine mitochondrial $tRNA^{Ser}$ (Fig. 5A).

Figure 5B shows a typical elution profile of the tRNA mixture [16]. The elution buffer is 0.1 × SSC and contains 3.0 M TMACl. Temperature is gradually raised from 25°C to 85°C at the rate of 1°C/min and the flow rate is 0.05 mL/min. The numbers in parentheses are those of the fraction. The peak eluted at about 62°C (fractions 7, 8, and 9) is expected to contain the targeted substance, serine tRNA. The peak eluted above 70°C (fractions 11 and 12) may be due to longer RNAs that adhere nonspecifically to the support. Figure 5C shows bands from gel electrophoresis of column-loaded crude RNA and fractionated samples. The gel was 15% polyacrylamide and the solvent was 7 M urea–1 × TBE. The numbers indicate the fractions of

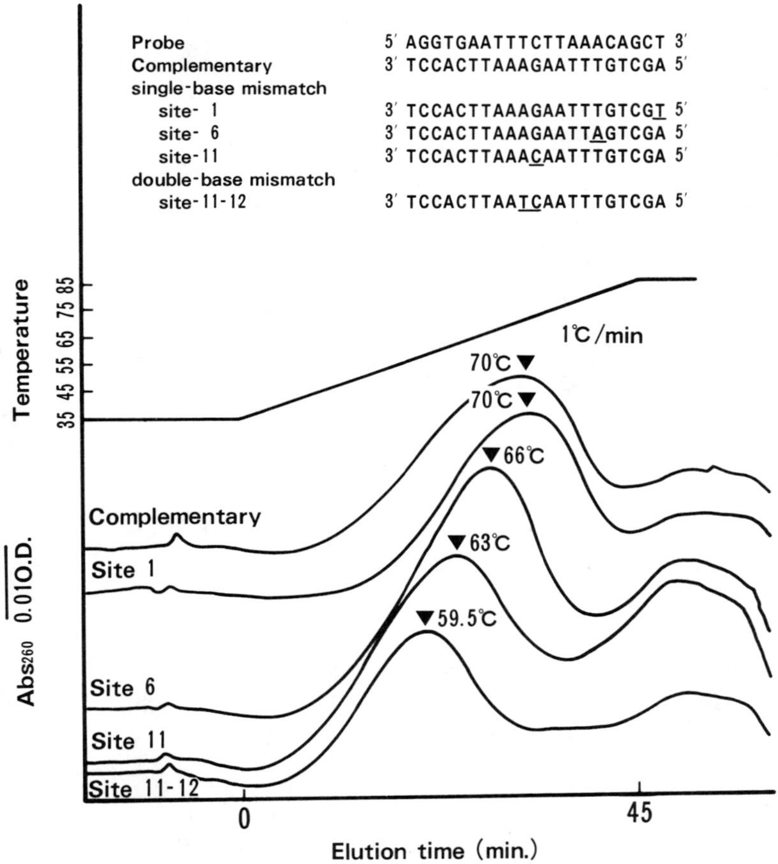

Figure 4 Temperature-gradient elution profiles of several oligonucleotides (21 mer) having one or two mismatches at different portions of the chains, which are listed at the top. The time course of programmed temperature change is plotted in the middle. Difference in separation temperature reduction is shown according to the difference of the mismatched base position.

the temperature-gradient chromatography, M refers to the mixture of mitochondrial tRNA, and S the temperature-gradient chromatographed sample. The latter is obtained by refractioning M through benzyl–DEAE cellulose and RPC-5 to enrich $tRNA^{Ser}$ concentration. S contains a small amount of $tRNA^{Ser}$ (indicated by the horizontal arrow), whereas fraction 8 is exclusively composed of $tRNA^{Ser}$.

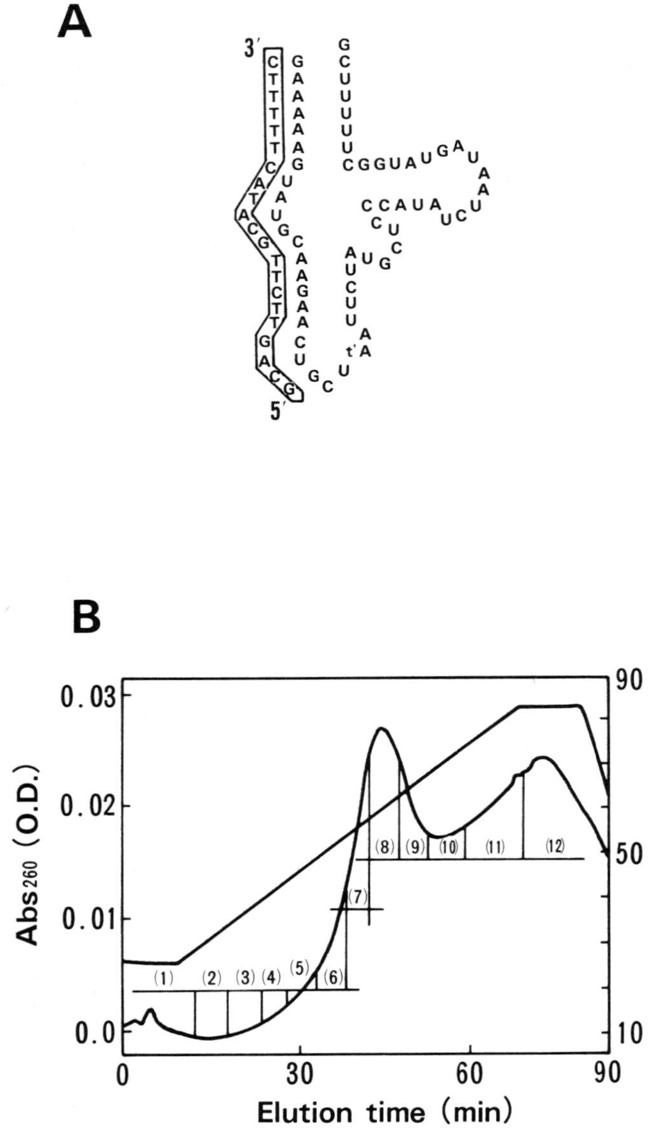

Figure 5 Temperature-gradient elution profile of mitochondrial tRNA: (A) Base sequence of $tRNA^{Ser}$ and that of probe DNA; (B) Elution profile of tRNA mixtures using the probe with the base sequence shown above. Numbers in parenthesis indi-

C.

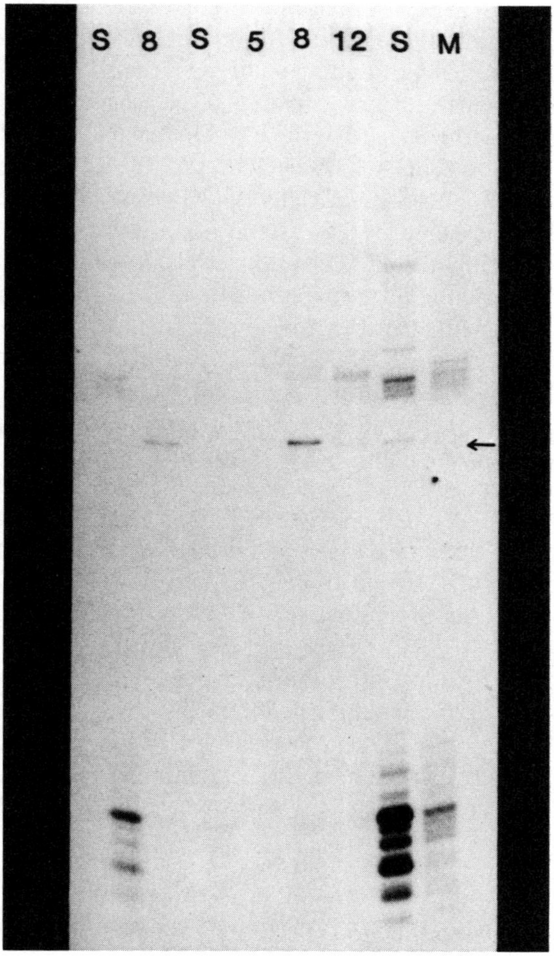

cate fractions. The time course of programmed temperature elevation is superposed; (C) Gel electrophoresis of whole mixture of mit-tRNA (M), loaded sample (S), and fractions 5, 8, and 12. The whole mixture is preliminarily fractionated through benzyl–DEAE cellulose and RPC-5 column to yield the sample that was loaded onto the temperature-gradient column. Fraction 8 appears to be exclusively composed of $tRNA^{Ser}$.

2. Extraction of the Fibronectin Gene and Its Flanking Region

To demonstrate the high performance of the present system, we tried to "fish out" the 5′ flanking region of the human fibronectin (*FN*) gene directly from a pool of total genomic DNA of human embryonal carcinoma NEC 15 cells [36]. The batch process with LiChrospher Si 1000 gel was used. Two oligonucleotides of 27 bases near the 5′ end of the FN cDNA from positions 15 to 41 (5′-TGTGCTGCACAGGGGGAGGAGAGGGAA-3′) and 304 to 330 (5′-GCCGTCCTGTGCCTGGGGACAGCGGTG-3′) were synthesized according to the sequence measured by Dean et al. [36], and their 5′ ends were moored to the gel. A fragmented DNA mixture of the carcinoma cell, each of which is about 10-kb long on average, was hybridized to the immobilized probes and then extracted according to the process illustrated in Figure 2. The hybridized single-stranded DNA thus concentrated was made double-stranded by DNA polymerase I using random primers, and was cloned in a Charon 4A vector after the ligation of *Eco*RI linkers and packaged into λ phage in vitro. The recombinant phages were plated with *Escherichia coli* LE392, and the existence of the *FN* gene sequence was screened with ^{32}P-labeled human FN cDNA clone pFH134 [37] after a transfer of plaques on nylon membrane. The clone pFH134 contains an *FN* cDNA sequence about 2-kb long, and its 5′ end is located about 650 bases downstream from the 5′ end of the *FN* cDNA. About 3000 plaques were formed, and 1000 of these were positive clones.

The result clearly demonstrated the extremely high efficiency of our probe system: One-third of the human genomic DNA fragments extracted in reference to the *FN* gene base sequence were found to contain the 5′ flanking region of the *FN* gene, which is a high 30% extraction efficiency. The details are reported in a separate article [38].

VI. DISCUSSION

The extremely high base-sequence selectivity of our system obviates the laborious, repetitive, and time-consuming labor required to screen a target genetic base sequence of low abundance from a complex genomic DNA pool. The extent of concentrating a gene is extremely high, and weeks or months of sometimes fruitless labor can be dramatically reduced to a routine operation of several hours.

The high performance and uncomplicated procedure of this system will make it possible to develop, in the near future, an innovative method for automated gene mapping by combining the present probe system with an automated DNA sequencer and DNA-probe synthesizer. A series of processes: (1) fishing out a DNA fragment with a probe DNA, followed by its amplification; (2) sequencing several tens of base pairs near the ends of the fragment; (3) synthesis of the DNA probe having the determined sequences; and (4) fishing out another fragment, which should be contiguous to the first fragment, can be automated without any difficulty and will constitute a machine-operated system that may justly be termed an "automatic gene-walker."

As we have stressed, modern biological research involves a multistep series of operations from sample preparation, physical and chemical measurements, to information processing [39]. The desire to accelerate a particular process in this series of operations usually pressures researchers to automate adjoining units which limit the overall rate of processing. A whole system (e.g., a genome analysis system) can thus be developed into one that is well balanced.

It is interesting to compare the present end-moored probe (EMP) method with the polymerase chain reaction (PCR) method [40,41], which is a clever and highly effective device to amplify a DNA fragment in the flank of a primer fragment preselected by a researcher. They are broadly classified and evaluated in Table 1. These methods are complementary in performance, and both can be used properly according to the purpose of the research.

In addition to performing the function of gene screening, the future prospects for this method becoming an analysis instrument are bright: the

Table 1 Comparison of the PCR and EMP Methods for DNA Sequencing

Evaluation	PCR	EMP
Features	Direct in vitro amplification	Extension of current device: in vivo amplification after fishing out a DNA fragment
Preselected sequence	Primer sequence	Probe sequence
Obtained fragment	Sequence sandwiched by two primer sequences (~kb)	Fragments (kb ~ 100-kb) carrying probe sequence
Sensitivity	Very high	Depend upon the recombination efficiency of screened DNA with vector DNA
Efficiency and versatility	Very good	Very good; appropriate for seeking new sequences and concentrated library construction

base–sequence-resolving power will be drastically improved if the probe density is substantially increased and a longer column is used. By doing so, the chromatographic effect, which is brought about by a multiple-step sample–probe binding and is equivalent to multiple screening, throughout the long probe-interacting passage will contribute to sharper separation of DNA bands. The improvement of the resolving power of the system thereby achieved will eventually realize a high-precision mechanism enabling us to measure the *sequence homology spectrum* of genes. The degree of homology of DNA sequences to a standard sequence is displayed according to the order of the elution of the fragments having different degrees of base mismatching with the standard probe DNA. The high-precision extraction and analysis of DNA-binding proteins will be performed as well by making the probe DNA into a double strand by appropriate primer and polymerase, or by annealing a synthesized complementary strand.

ACKNOWLEDGMENTS

We thank Prof. K. Watanabe of Tokyo Institute of Technology and Prof. K. Oda of Science University of Tokyo, and the members of their laboratories, for their generous cooperation in our study which was briefly reviewed in the present chapter. We also thank Mr. Yasutomi Mitoma of Tosoh Co. for his technical advice in preparing probe-bound gel supports. Our studies were supported by a Grant-in-Aid for Scientific Research from the Ministry of Education, Science and Culture, and by a Specially Coordinated Fund from the Science and Technology Agency of Japan.

REFERENCES

1. Gillam, S., K. Waterman, and M. Smith (1975). The base-pairing specificity of cellulose-pdT9*. *Nucleic Acids Res. 2*:625–634.
2. Wallace, R. B., J. Shaffer, R. F. Murphy, J. Bonner, T. Hirose, and K. Itakura (1979). Hybridization of synthetic oligodeoxyribonucleotides to phi X-174 DNA: The effect of single base pair mismatch. *Nucleic Acids Res. 6*:3543–3557.
3. Wallace, R. B., M. J. Johnson, T. Hirose, T. Miyake, E. H. Kawashima, and K. Itakura (1981). The use of synthetic oligonucleotides as hybridization probes. II. Hybridization of oligonucleotides of mixed sequence to rabbit β-globin DNA. *Nucleic Acids Res. 9*:879–894.
4. Lerman, L. S., S. G. Fischer, I. Hurley, K. Silverstein, and N. A. Lumelsky (1984). Sequence-determined DNA separations. *Annu. Rev. Biophys. Bioeng. 13*:399–433.
5. Myers, R. M., S. G. Fischer, L. S. Lerman, and T. Maniatis (1985). Nearly all single base substitutions in DNA fragments joined to a GC-clamp can be

determined by denaturing gradient gel electrophoresis. *Nucleic Acids Res. 13*: 3131–3145.
6. Meinkoth, J., and G. Wahl (1984). Hybridization of nucleic acid immobilized on solid supports. *Anal. Biochem. 138*:267–284.
7. Hames, B. D., and S. J. Higgins (eds.) (1985). *Nucleic Acid Hybridisation*. IRL Press, Oxford.
8. Lerman, L. S. (ed.) (1986). *DNA Probes*. Cold Spring Harbor Laboratory, Cold Spring Harbor, NY.
9. Davies, K. E. (ed.) (1986). *Human Genetic Diseases*. IRL Press, Oxford.
10. Szostak, J. W., J. I. Stiles, B. K. Tye, P. Chiu, F. Sherman, and R. Wu (1979). Hybridization with synthetic oligonucleotides. *Methods Enzymol. 68*:419–428.
11. Wood, W. I., J. Gitschier, L. A. Lasky, and R. M. Lawn (1985). Base composition-independent hybridization in tetramethylammonium chloride: A method for oligonucleotide screening of highly complex gene libraries. *Proc. Natl. Acad. Sci. USA 82*:1585–1588.
12. Buenemann, H. (1982). Immobilization of denatured DNA to macroporous supports: II. Steric and kinetic parameters of heterogeneous hybridization reaction. *Nucleic Acids Res. 10*:7181–7196.
13. Gingeras, T. R., D. Y. Kwoh, and G. R. Davis (1987). Hybridization property of immobilized nucleic acids. *Nucleic Acids Res. 15*:5373–5390.
14. Wada, A., S. Yabuki, and Y. Husimi (1980). Fine structure in the thermal denaturation of DNA: High temperature-resolution spectrophotometric studies. *CRC Crit. Rev. Biochem. 9*:87–144.
15. Wada, A., and A. Suyama (1986). Local stability of DNA and RNA secondary structure and its relation to biological functions. *Prog. Biophys. Mol. Biol. 47*: 113–157.
16. Tsurui, H., A. Suyama, and A. Wada (1988). Temperature gradient DNA-probe chromatography: As tools for detection and purification of particular DNAs and RNAs. *Nucleic Acids Res. Symp. Ser. 19*:49–52.
17. Shapiro, J. T., J. M. Stannard, and G. Felsenfeld (1969). The binding of small cations to deoxyribonucleic acid. Nucleotide specificity. *Biochemistry 8*:3233–3241.
18. Melchior, W. B., and P. H. von Hippel (1973). Alteration of the relative stability of dA–dT and dG–dC base pairs in DNA. *Proc. Natl. Acad. Sci. USA 70*: 298–302.
19. Chang, C.-T., T. C. Hain, J. R. Hutton, and J. G. Wetmur (1974). Effect of microscopic and macroscopic viscosity on the rate of renaturation of DNA. *Biopolymers 13*:1847–1858.
20. Suyama, A., H. Tsurui, M. Yoneyama, N. Baba, and A. Wada (1989). Temperature gradient DNA-probe column chromatography: A new method for detection and purification of particular DNAs and RNAs. In *Recent Progress of Life Science Technology in Japan*. Academic Press, New York.
21. Britten, R. J., A. Cetta, and E. H. Davidson (1978). The single-copy DNA sequence polymorphism of the sea urchin *Strongylocentrotus purpuratus*. *Cell 15*:1175–1186.

22. Wolf, S., L. Haines, J. Fisch, J. Kremsky, J. Dougherty, and K. Jacobs (1987). Rapid hybridization kinetics of DNA attached to submicron latex particles. *Nucleic Acids Res. 15*:2911–2926.
23. Ghosh, S. S., and G. F. Musso (1987). Covalent attachment of oligonucleotides to solid supports. *Nucleic Acids Res. 15*:5353–5390.
24. Kremsky, J., J. L. Wooters, J. P. Dougherty, R. E. Meyers, M. Collins, and E. L. Brown (1987). Immobilization of DNA via oligonucleotides containing an aldehyde or carboxylic acid group at the 5′ terminus. *Nucleic Acids Res. 15*: 2891–2909.
25. Fassina, G., and I. M. Chaiken (1988). Analytical high-performance affinity chromatography. *Adv. Chromatogr. 27*:248–297.
26. Larsson, P. O., M. Glad, L. Hansson, M.-O. Mansson, S. Ohlson, and K. Mosbach (1983). High-performance liquid affinity chromatography. *Adv. Chromatogr. 21*:41–84.
27. Orosz, J. M., Jr., and J. G. Wetmur (1977). DNA melting temperatures and renaturation rates in concentrated alkylammonium salt solutions. *Biopolymers 16*:1183–1199.
28. Poland, D. (1974). Recursion relation generation of probability profiles for specific-sequence macromolecules with long-range correlation. *Biopolymers 13*:1859–1871.
29. Fixman, M., and J. J. Freire (1977). Theory of DNA melting curves. *Biopolymers 16*:2693–2704.
30. Gotoh, O. (1983). Prediction of melting profiles and local helix stability for sequenced DNA. *Adv. Biophys. 16*:1–52.
31. Wartell, R. M., and A. S. Benight (1985). Thermal denaturation of DNA molecules: A comparison of theory with experiment. *Phys. Rep. 126*:67–107.
32. Zuker, M., and P. Stiegler (1981). Optimal computer folding of large RNA sequences using thermodynamics and auxiliary information. *Nucleic Acids Res. 9*:113–148.
33. Suyama, A. (1984). RNA secondary structure and its relation to biological functions. In *Proceedings of the 10th Taniguchi International Symposium*. A. Wada (ed.). Taniguchi Foundation, Kyoto, pp. 162–172.
34. Marmur, J., and P. Doty (1961). Thermal denaturation of deoxyribonucleic acids. *J. Mol. Biol. 3*:585–594.
35. Aboul-ela, F., D. Koh, I. Tinoco, Jr., and F. H. Martin (1985). Base–base mismatches. Thermodynamics of double helix formation for dCA_3XA_3G + dCT_3YT_3G (X,Y = A,C,G,T). *Nucleic Acids Res. 13*:4811–4824.
36. Dean, D. C., L. C. Bowlus, and S. Bourgeois (1987). Cloning and analysis of the promoter region of the human fibronectin gene. *Proc. Natl. Acad. Sci. USA 84*:1876–1880.
37. Kornblihtt, A. R., K. Umezawa, K. Vive-Pederson, and E. F. Baralle (1985). Primary structure of human fibronectin: Differential splicing may generate at least 10 polypeptides from a single gene. *EMBO J. 4*:1755–1759.
38. Tsurui, H., E. Hara, K. Oda, A. Suyama, and A. Wada (1989). A rapid and efficient cloning method with solid phase DNA probe: Application for cloning of 5′ flanking region of the human fibronectin gene. *Gene 88*:233–239.

39. Wada, A., S. Kidokoro, and S. Endo (1989). Expanding roles of computers and robotics in biological macromolecular research. *Annu. Rev. Biophys. Biophys. Chem. 18*:1–24.
40. Saiki, R. K., S. Scharf, F. Faloona, K. B. Mullis, G. T. Horn, H. A. Erlich, and N. Arnheim (1985). Enzymatic amplification of β-globin genomic sequences and restriction site analysis for diagnosis of sickle cell anemia. *Science 230*: 1350–1354.
41. Saiki, R. K., D. H. Gelfand, S. Stoffe, S. J. Scharf, R. Higuchi, G. T. Horn, K. B. Mullis, and H. A. Erlich (1988). Primer-directed enzymatic amplification of DNA with a thermostable DNA polymerase. *Science 239*:487–491.

APPENDIX

A. Activation of Diol Silica with Tresyl Chloride

1. Diol silica can be purchased from Nagel (Nucleosil 1000-7-OH) or prepared according to Larsson et al. [26].
2. Wash 1 g of diol silica with acetone and dry acetone briefly on a glass filter.
3. Put the gel into a 10-mL messflask with 1.2 mL of dry acetone, 100 μL of dry pyridine, and a small stirrer tip.
4. Add 100 μL of tresyl chloride dropwise (1 min) with vigorous stirring under a slow stream of dry nitrogen gas to ensure unhydrous condition. Keep the reaction flask at 0°C with a vinyl bag filled with crushed ice.
5. Slow the stirrer speed and maintain even speed for 20 min.
6. Briefly wash the gel on the glass filter with acetone, acetone–5 mM HCl (1 : 1), 5 mM HC1, and acetone.
7. Dry the gel under vacuum and keep it at 4°C.

B. Immobilization of Probe DNA(s) to Support Matrix

1. Synthesize the probe DNA(s) with Aminolink 2 at the 5′ terminus according to the principles described in the text.
2. After deprotection, dissolve the DNA in 180 μL of coupling buffer (0.2 M $NaHCO_3$, 0.5 M NaCl, pH 7.5).
3. Mix the DNA with 100 mg of tresyl-activated gel. Keep the conjugate at 23–25°C for 24 hr with gentle stirring.
4. After the coupling reaction, wash the gel with distilled water.
5. Suspend 5 mg of saturated sucrose solution and measure the concentration of immobilized DNA spectroscopically [26]. Several OD per gram dry gel is adequate for EMP.

C. Fractionation of the Object DNA Fragments from Genomic DNA

1. Put 100 mg of probe-immobilized gel, sample DNA, and 400–500 μL of hybridization solution into an Eppendorf tube.
2. Vortex the conjugate and keep the tube at 70°C for 10 min to denature sample DNA.
3. Keep the tube at 15°C for 1 hr to hybridize the object DNA fragments with probe DNA. Vortex the suspension every 5 min.
4. Centrifuge the tube for 10–20 sec and remove the supernatant.
5. Add 400–500 μL of hybridization solution to the gel, vortex, centrifuge for 10–20 sec, and remove the supernatant.
6. Add 200 μL of hybridization solution to the gel and keep the tube at 70°C for 10 min to allow the hybridized DNA fragments to unfold.
7. Centrifuge the tube for 10–20 sec and collect the supernatant, which contains the targeted DNA fragments.

D. Construction of a Highly Concentrated Library

1. Dialyze the fractionated DNA fragments against Tris-EDTA (TE) buff-er.
2. Add random primer (Amersham, mixture of hexaoligonucleotides) to the sample, keep the tube at 60°C for 20 min to anneal.
3. Synthesize second-strand DNA at 67 mM potassium phosphate buffer (pH 7.4), 6.7 mM $MgCl_2$, 1 mM 2-mercaptoethanol, 1 mM each dATP, dGTP, dTTP, 20 μCi [$\alpha-^{32}P$] dCTP (3000 Ci/m mol), 20 units DNA polymerase I at 12°C for 12 hr.
4. Divide the DNA into three portions and treat with 50, 125, and 375 units of S1 nuclease, respectively, in 30 mM sodium acetate buffer (pH 4.6), 100 mM NaCl, 1 mM $ZnSO_4$ at 37°C for 30 min.
5. Combine the three fractions and treat with 20 units *Eco*RI methylase in 100 mM Tris-HCl (pH 8.0), 2 mM DTT, 10 mM EDTA, 80 mM *S*-adenosylmethionine at 37°C for 15 min.
6. Treat the DNA with the Klenow fragment of DNA polymerase I at 37°C for 30 min to make blunt ends, ligate to *Eco*RI linker, and cleave with *Eco*RI.
7. Fractionate the reaction mixture by gel filtration through a Biorad A 100 column and collect the high-molecular-weight DNA.
8. Ligate the DNA to 0.1 μg of *Eco*RI-cut Charon 4A vector DNA at 15°C for 16 hr.
9. Package the recombinant DNA vector using the Amersham in vitro packaging kit No. 334Y.

5

In Situ Hybridization of Nonradioactive Oligonucleotide Probes to Chromosomes

Indrajit Nanda and Michael Schmid

University of Würzburg, Würzburg, Germany

Jörg T. Epplen

Max-Planck-Institute for Psychiatry, Planegg-Martinsried, Germany

I. INTRODUCTION

The conventional chromosomal banding techniques, flow cytometry, and somatic cell genetic approaches have greatly refined the classical method of chromosomal analysis. Although the aforementioned strategies have been instrumental in gene mapping, the limitations for regional localization of the gene(s) of interest have been circumvented only with the availability of suitable DNA probes in conjunction with the advent of in situ hybridization techniques that allow detection and localization of specific nucleic acid sequences directly within the cell or tissue. This technique, originally developed by Gall and Pardue [1] and John et al. [2], involves the annealing of polynucleotide probes to chromosomal or cellular preparations, the DNA–RNA of which has been denatured to allow hybridization with the exogenous labeled probe. The label is subsequently detected by autoradiography or by the more recently developed staining techniques. Because this technique yields both molecular and morphological information about individual cells, its usefulness has been proved in many areas of biomedical research. For example, the detection of RNA in situ identifies those cells that contain particular messages in a heterogeneous cell population. Similarly, the genomes of infectious agents can be demonstrated in cells, tissues, or body fluids. Thus, in situ hybridization has become an increasingly important molecular tool in recombinant DNA technology for characterization of the basic features of gene structure and gene expression. Here, our primary interest has to be confined to cytogenetic analyses. It is impossible, however, to provide a comprehensive outline of the development of the technique [for a thorough review see Ref. 3].

Initially, hybridization in situ was employed to localize the chromosomal sites of numerous repeated sequences on both polytene and diploid chromosomes in several animal species and in humans [4–6]. Usually, cloned DNA probes, ^{3}H-labeled by nick translation, were hybridized to high-quality preparations, which had been previously treated with ribonuclease and denatured in the presence of dextran sulfate. Subsequently, several modifications have been introduced in the in situ hybridization methodology for mapping genes on chromosomes, to attain higher sensitivity. The detection of single-copy sequences on mitotic chromosomes was achieved by enhancing the specific activities of the probe molecules with random-prime-labeling technique [7] or RNA polymerase-based transcription [8]. Following autoradiography, chromosomal preparations are G-banded directly by several cycles of staining–destaining–restaining with Wright's stain to record the precise grain location on a specific band [9]. Obviously, in situ hybridization techniques with radioactive probes suffer from the serious disadvantages of prolonged autoradiographic exposure times (several weeks) and limited spatial resolution. In addition, the use of radioisotopes excludes the methodology for many laboratories, purely for safety reasons. These problems have gradually been overcome by introducing a variety of different immunogenic, fluorescent, and enzymatic labels [10–12]. However, the application of such methods has hithertofore been restricted by sensitivity limitations for the detection of high-copy number or comparatively large single-copy sequences [13]. In addition to the widespread use of cloned probes in in situ hybridization studies, alternative approaches with synthetic DNA sequences have recently been explored to localize ubiquitously interspersed, tandemly arranged, repetitive sequences in eukaryotic genomes. Numerous different tandem, simple-repetitive sequences are organized extensively in a head-to-tail fashion and undergo recombination or other mutational events far more frequently than unique sequences. This results in a considerable diversity at these DNA loci [14,15]. Thus, the simple-repeat probe $(CAC)_5$, for example, generates individual-specific DNA fingerprints in humans [16]. The chromosomal locations for certain simple repeats have been reported when using cloned probes that simultaneously evoke unspecific signals at many unrelated sites. Because the simple-repeat probes are short, one approach to study their organization is by constructing specific synthetic probes. Synthetic oligonucleotides are particularly useful for localizing specific mRNA in histological preparations of nervous and endocrine tissues [17], as they reach target sites easily and thus are advantageous over long cloned cDNA probes. Because such short probes can be synthesized in a few hours by a fully automated apparatus and because known sequences may be adapted for many different species, it seems promising to employ this approach also at the chromosomal level. In the following sections, some of our experiences and results are presented as obtained with synthetic oligonucleotide probes in in situ hybridizations onto metaphase chromosomes. Nonradioactive nucleic acid probes are of considerable importance because of both simplicity and flexibility. Two different nonradioactive reporter groups (biotin and digoxigenin) were used here for labeling the short oligonucleotides. The sensitivities obtained in the detection of target DNA will be compared.

II. METHODOLOGY

Essentially, absolute hybridization specificity can be achieved by synthetic oligonucleotide probes under appropriate conditions. By selecting suitable stringency conditions of hybridization and washing, a perfectly matched duplex between oligonucleotide probes and complementary target DNA stretches can be obtained, whereas mismatches at one or more positions in the duplexes will prevent annealing and signal generation. On the basis of the aforementioned principles, hybridization of oligonucleotides has the potential to provide a means for detecting single-base changes within genomic DNA [18]. The most common method for labeling oligonucleotides has been the incorporation of the isotope γ-^{32}P. For in situ hybridization purposes, we have attached two types of organic molecules to the 5′ termini of the deprotected oligodeoxyribonucleotides. The following section outlines the experimental details of the in situ hybridization protocol for the nonradioactive approach.

A. Preparation of Chromosomes

Lymphocyte cultures from healthy humans are set up by adding 0.8 mL of heparinized blood to 10 mL of RPMI 1640 supplemented with 10% fetal calf serum and 0.1 mL phytohemagglutinin M (Gibco, FRG). To obtain long chromosomes, cell cultures are incubated for 72 hr at 37°C and then exposed to a low dose of demecolcine (Colcemid; 0.08 μg/mL) for a relatively short time (60 min). The cells are centrifuged and resuspended in 0.075 M hypotonic KCl solution for 10 min at 37°C, recentrifuged, and fixed in cold methanol/acetic acid (3 : 1) for at least 30 min. After three changes with fixative, the cell suspension is dropped onto cold, wet slides and air-dried.

For preparing chromosomes from mouse tissues (C3H strain), animals are injected with colchicine intraperitoneally. Usually 0.05 mL of a 1 mg/mL stock solution is adequate for a normal-sized mouse. One to two hours after colchicine injection, suitable tissues are removed from the anesthetized animal. Bone marrow cells are flushed, by a hypodermic syringe with a small amount of hypotonic solution, into a centrifuge tube. The cells are briefly agitated mechanically to obtain a single-cell suspension and are then exposed to the hypotonic solution for 10 min at 37°C. For meiotic stages, testis tubules are carefully isolated from the tissue and minced in phosphate-buffered saline (PBS) to release the cells. After pelleting by centrifugation, cells are resuspended in 0.075 M KCl for 10 min at room temperature. Fixation and preparation of both mitotic and meiotic chromosomes follow the same technique as just described for short-term leukocyte cultures. For hybridization in situ, slides are aged for about 2 weeks at 4°C.

B. Labeling of the Probes

The purified oligonucleotides can be radiolabeled by one of the three following methods, depending upon the specific activity of the oligonucleotide probe required:

1. 5′ End-labeling with γ-[^{32}P]dATP using T4 polynucleotide kinase: To make certain that virtually all probe molecules are carrying the label, the kinase-treated oligonucleotide can be purified by denaturing polyacrylamide gel electrophoresis [18].
2. 3′ End-labeling with α-labeled dNTP using terminal transferase.
3. Enzymatic synthesis of antisense oligonucleotides using radioactive precursors by extension of a short antisense primer annealed to a longer sense-strand oligonucleotide (template) with the Klenow fragment of DNA polymerase I.

In the nonradioactive approach, a suitable hapten molecule can be attached to the oligonucleotides by either method 2 or 3. Usually, 5′ end-labeling can be achieved by a phosphoramidite or a phosphodiester linkage. Here, however, only a single nonradioactive reporter group is usually attached to the probe, whereas multiple attachment can be accomplished by enzymatic means using terminal transferase or the Klenow fragment of DNA polymerase I. The major disadvantages of multiple-reporter groups include the reduced stability of the probe–target sequence duplex, thereby reducing the specificity of the hybridization.

In the present experiments, nonradioactive labeling of synthetic oligonucleotides is carried out employing the 5′ end-labeling method with both biotin (vitamin H) and a steroid hapten, digoxigenin. A 15-mer $(CAC)_5$ and two 16 mers, $(GATA)_4$ and $(GACA)_4$, are synthesized on an Applied Biosystems Synthesizer (model 380A) by the solid-phase phosphoramidite method. At the 5′ end of the oligonucleotide, an aminoethylphosphate linker

(Aminolink; Applied Biosystems, FRG) is added as described by the manufacturer. After deprotection, the highly reactive primary amino group is subsequently allowed to react with the digoxigenin succinimide ester (Boehringer-Mannheim, FRG) or *N*-hydroxysuccinimidobiotin (Sigma, FRG) overnight, at room temperature, in sodium bicarbonate buffer (pH 9.0) [19]. The nonradioactive derivatives are purified subsequently by reversed-phase high-performance liquid chromatography (HPLC) on a commercially available column (RP18, 5-μm particle size; Bischoff, FRG). The respective peak fractions are tested for biotin or digoxigenin content by spot-blotting onto nitrocellulose, followed by colorimetric detection using streptavidin–alkaline phosphatase (Blue Gene; BRL, FRG). The positive fractions are dried under vacuum. The pellet is dissolved in bidistilled water, and the concentrations can be determined by absorbance measurement at 260nm.

C. In Situ Hybridization

Because the sizes of the probes are small (15–16 bases), the temperatures for duplex stability can be calculated with the formula: $T_d = 4(G + C) + 2(A + T)$ in 1 M Na^+[18]. The exact hybridization temperatures are then standardized usually by decreasing the calculated temperature by 5°–10°C.

Chromosomal preparations are treated with DNase-free preboiled RNase for 1–2 hr at a concentration of 100 μg/mL in 2 × SSC at 37°C. After removal of coverslips in several washes of 2 × SSC, slides are dehydrated in an ethanol series for 2 min in each dilution. The chromosomes are denatured in 70% formamide–2 × SSC at 70°C for 1–2 min, washed several minutes in a cold ethanol series, and then air-dried. In our experience, meiotic chromosomes (metaphase I) require a relatively short time for denaturation. On each slide 20 μL of hybridization mixture containing 5 × SSPE, 0.5% SDS, 5 × Denhardt's solution, 50 μg/mL denatured *Escherichia coli* DNA, and 1–2 pmol of probe are placed under a coverslip and sealed with a rubber solution. Hybridization is carried out in a moist chamber at the respective hybridization temperatures, 45°C $(CAC)_5$, 43°C $(GACA)_4$, and 35°C $(GATA)_4$, for 5–8 hr. The hybridization reaction is terminated by washing the slides three times for at least 1 hr at room temperature in 6 × SSC. The stringent wash at the hybridization temperature is performed for 2 min. Ethanol dehydration at this stage should be avoided, as the slides are immediately subjected to signal detection.

D. Cytochemical Detection

For biotinylated oligonucleotides, the detection kit of ENZO renders reproducible signals with the exception that diaminobenzidine (DAB; Sigma,

FRG) is used as a developing agent, instead of the developer provided in the kit. After the stringent wash, slides hybridized with biotinylated probes are rinsed for 3 min in PBS, including 0.05% Tween 20, as well as in two changes of PBS at room temperature. Then, 200 μL of the peroxidase–streptavidin conjugate (ENZO) at a dilution of 1 : 200 in the DETEK buffer is applied to each slide, covered with a 24 × 60 mm coverslip and incubated at room temperature for 20–40 min. After a wash in PBS, containing 0.05% Tween, followed by three more washes in PBS, the slides are stained with 0.5 mg/mL 3,3′-diaminobenzidine tetrahydrochloride (DAB)–H_2O_2 (0.003%) in PBS, in the dark, for about 10 min at room temperature, which results in the precipitation of insoluble brownish complexes over the chromosomal locations. It is essential to prepare the DAB solution immediately before use. The reaction is terminated by rinsing in PBS and air-drying the slides. Good contrast in biotin hybridizations can be achieved by a brief Giemsa staining and phase-contrast microscopy.

The hybridization of digoxigenated oligonucleotide probes is visualized by a polyclonal antibody conjugate directed against the hapten. Slides removed from 6 × SSC are transferred to buffer I (100 mM Tris–HC1, pH 7.5, 900 mM NaC1) at room temperature for 3–5 min, followed by a 30 min incubation in 0.5% blocking reagent (Boehringer-Mannheim, FRG) prepared in buffer I. After washing repeatedly in buffer I, Dig–AP conjugate (polyclonal sheep antidigoxigenin–Fab fragments conjugated to alkaline phosphatase) is applied in a 1 : 1000 dilution for 1 hr at room temperature. Unbound antibody is subsequently washed off by rinsing several times in buffer I. The slides are transferred to color-developing solution (buffer II: 100 mM Tris–HCl, pH 9.5, 900 mM NaCl, and 50 mM $MgCl_2$) for equilibration. Finally, slides are incubated overnight, in the dark, in staining solution containing 0.17 mg/mL 5-bromo-4-chloro-3-indolyl phosphate (BCIP) and 0.33 mg/mL nitroblue tetrazolium salt (NBT) dissolved in buffer II. The color reaction is stopped with TE (10 mM Tris–HCl, 1 mM EDTA; pH 8.0) and, finally, in water. The location of the color precipitation over metaphases or interphase nuclei can be observed, without counterstaining, by phase-constrast microscopy.

III. RESULTS

Three different simple-repetitive oligonucleotide probes, either biotinylated or digoxigenated, have been hybridized to mouse and human metaphase chromosomes. Depending on the probes used, multiple-banding patterns or localized staining could be observed on human metaphase chromosomes. Repetitive DNA consisting of simple quadruplet repeats (sqr) GATA/GACA are sex-specifically organized in mice [20,21]. Therefore, these two probes are extremely useful in identifying the Y chromosome of mice in normal and aberrant situations. The results are discussed in the following sections.

A. Coincidence of $(CAC)_n$ Loci and R-Bands in Humans

Recently, a generally applicable method has been devised to individualize the genomes of any animal species yet investigated by oligonucleotides specific for simple-repetitive DNA [22]. For example, the probe $(CAC)_5$, or its complementary sequence $(GTG)_5$, reveals a highly variable-banding pattern, with differentiating signals, distributed over a broad range of fragment sizes (4–30 kilobase; kb) after in-gel hybridization of restriction enzyme-digested human DNA [22]. Accordingly this particular probe displays a great potential, along with the hitherto constructed oligonucleotide probes, for discriminating between individuals. Both radioactively labeled as well as the digoxigenated $(CAC)_5$ reveal an identical banding pattern [23]. Because of the high number of bands generated by $(CAC)_5$, it seems possible that $(CAC)_n$-related simple repeats are distributed all over the human genome. The latter speculation can be validated by in situ hybridization to human metaphase chromosomes with the digoxigenated $(CAC)_5$ probe. Figure 1 displays a karyotype of the haploid set of human chromosomes (right-hand side) obtained by in situ hybridization. In the 50 metaphases analyzed, the hybridization sites coincide preponderantly with the R-band-positive regions (left). Those R-bands that are frequently identified with the $(CAC)_5$ probe are depicted with dots in Figure 1; particularly, the R-bands over chromosomes 13, 14, 19, 21, and 22 can be regularly demonstrated with this probe. The probe $(CAC)_5$ and its complement contain 67% G or C. R-bands are fairly enriched with G–C-rich sequences [24]. Hence, our probe reveals most G–C-rich R-band regions of human chromosomes. In this context, it is worth mentioning that, recently, with use of the radiolabeled minisatellite core sequence, clustering of silver grains has been recorded around chiasmata regions of meiotic bivalents [25]. Chiasmata in meiosis I or crossingover events are more frequent in the R-band regions of the mitotic chromosomes [26]. Therefore, hybridization of $(CAC)_5$ in R-band regions further supports the concept that both types of repetitive sequences are subject to, or even play a more causative role in, recombination.

Only occasionally can a weak $(CAC)_5$ signal be marked on the Y chromosome. This could be further corroborated from the mouse–human hybrid cell line (3E7) that contains Y as the only human chromosome. No Y-specific band larger than 2 kb could be resolved when the $(CAC)_5$ hybridization pattern was compared between DNA from the hybrid cell line and normal male genomic DNA (see Fig. 4). Thus, accumulation of long $(CAC)_n$ stretches on the Y chromosome can be excluded.

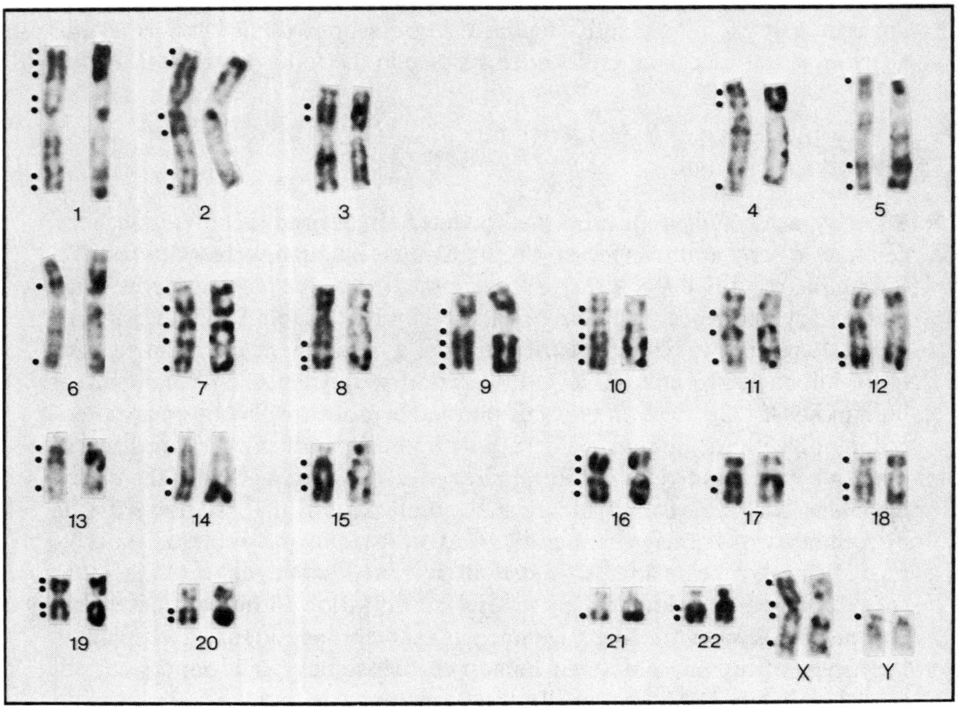

Figure 1 Human chromosomes after $(CAC)_5$ oligonucleotide hybridization in situ (right): The hybridization sites coincide with a subset of the R-bands (dots) as shown for each banded chromosome (left).

B. Characterization of Human Chromosomes Bearing Nucleolus Organizer Regions

Human pericentromeric alphoid-repetitive DNA sequences (α-repeats) are organized as chromosome-specific subfamilies that can be used as chromosome-specific markers [27]. Recent evidence points to these markers as being useful for studying chromosomal arrangements [28] and for the construction of centromere-based genetic linkage maps of human chromosomes [29]. The organization of centromeric alphoid-repetitive DNA is mostly studied with rodent–human hybrid cells. Here, precise assignments in specific chromosomes are not apparent in routine blot hybridizations; but the in situ hybridization approach, with cloned probes, has been successful for this purpose.

Among our simple-repeat probes tested, the hybridization results obtained with $(GACA)_4$, particularly, are quite specific for the short acrocentric chromosomes. Both biotin and digoxigenated $(GACA)_4$ elicit prominent hybridization signals at the centromeric regions and short arms of all five pairs of the human acrocentric chromosomes (Fig. 2a,c). The nucleolus organizer regions (NORs), which are selectively stained by silver ($AgNO_3$), are located in the short arms of the acrocentric chromosomes. The number of active and Ag-stained NORs varies from one person to another [30]. Furthermore, the silver reaction of a given NOR is generally constant from one cell to another [31]. The hybridization patterns on these acrocentric chromosomes are nearly identical with those observed with NORs. Figure 2b shows that in different metaphases the signal obtained from this particular person is quite identical, and a characteristic feature is that one of the acrocentric chromosomes always bears two hybridization signals. In another person, the centromeric regions and short arms of all five pairs of acrocentric chromosomes are positive with this probe (see Fig. 2c), supporting the concept of significant individual variation. We will investigate if these probes recognize spacers of ribosomal DNA sequences (rDNA). Certainly, alphoid-repetitive sequences could be closely associated with the NOR regions. This was further analyzed by in situ hybridization in the chimpanzee (see Fig. 2e) and gorilla for which the $(GACA)_4$ signals are confined to those chromosomal regions in which the NORs are located.

Several α-satellite clones described in the literature are specific to the NOR-bearing chromosomes. However, in no instance were all acrocentric chromosomes characterized simultaneously by in situ hybridization [32,33]. The use of chromosomal-specific α-satellite DNA for the direct detection of trisomy 21 has not yet been successful, because it has been impossible to isolate a chromosome 21-specific α-satellite DNA sequence. Therefore, the probe $(GACA)_4$ presented here may provide a useful tool for the molecular investigation of these chromosomal regions and their associated disorders. The human centromeric regions and the short arms of acrocentric chromosomes have been indistinctly labeled, along with many chromosomes [34], when using a mixed uncloned satellite DNA, the BKm probe. Both $(GATA)_n$ and $(GACA)_m$ are fairly enriched in this satellite DNA. Therefore, with the "pure" $(GACA)_4$ synthetic oligonucleotide, one can specifically characterize the acrocentric chromosomes.

C. Clustered Simple Repetitive Sequences on Vertebrate Y Chromosomes

The heterochromatic nature of the Y chromosome in numerous vertebrates makes it seem probable that the bulk of the DNA in this chromosome is highly repetitive. Therefore, repetitive probes have been sought as investigative tools for the analysis of the structure of Y chromosomes. Among many

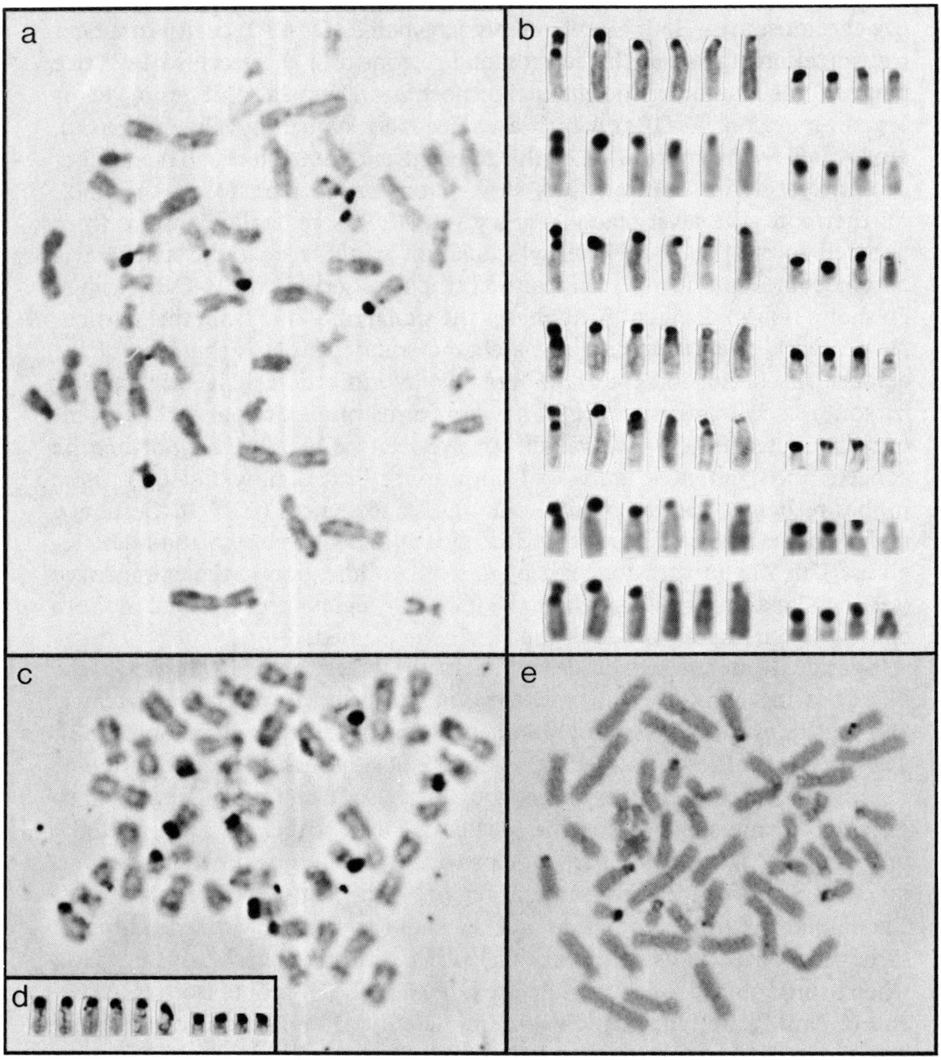

Figure 2 Photomicrographs showing in situ hybridization of the oligonucleotide probe $(GACA)_4$ to metaphase chromosomes of human (a, b, c, d) and chimpanzee (e): The hybridization is carried out with both biotinylated (a, b, e) and digoxigenated probes (c, d). Nucleolus organizer regions (NORs) of the chromosomes are selectively labeled with this probe. Variation of the hybridization signals over acrocentric chromosomes of humans can be observed by comparing individuals (a, c). However the signals on metaphases of the same person are more-or-less identical (b).

repetitive DNA sequence families, the simple GATA/GACA repeat, originally isolated from female banded krait (BKm, [35]), has received attention because it is represented in fair amount in every eukaryotic species yet investigated by DNA hybridization analysis [36]. Besides the universal interspersion, these sequences have been shown specifically accumulated on the mouse Y chromosome and on the W chromosome of snakes. In many animal species, they generate a male-specific signal in blot hybridization analysis (e.g., see Fig. 4). This has led to speculation on the relevance of these sequences for sex determination in the heterogametic sex. With use of cloned BKm sequences, Y chromosome-specific hybridization has been reported in the mouse, with substantial signals over the whole chromosomal complement [37]. Interestingly, with the same cloned probe, a completely different chromosome (17) was found to be positive [38]. Therefore, the sex chromosomal distribution of GATA/GACA simple repeats will be outlined, from synthetic probes, in the extensively studied mouse.

In situ hybridization of the biotinylated $(GACA)_4$ probe to male metaphase chromosome spreads of *Mus musculus* (C3H) always demonstrates a concentration of the signal in the paracentromeric region, including the sex-determining region of the Y chromosome (Fig. 3a,b). This specific signal is also demonstrable in interphase nuclei (see Fig. 3d,e). Hybridization with the $(GATA)_4$ probe also reveals identical signals over the Y chromosome. Even for the situation in which the Y chromosome has undergone inversion (resulting in a metacentric Y chromosome), the signal is present on the short arm, including the centromeric region (see Fig. 3f,g). Singh et al. [39], using the Bkm-related probe (^{3}H-2(8)RNA), could demonstrate signal only at the centromeric region of this rearranged Y chromosome. Even in meiotic bivalents, the Y chromosome can be identified with both of these probes (see Fig. 3h). As signals specific for the Y chromosome can be detected in 8–9 hr (total procedure), it is possible to determine the Y-bearing sperm in random samples and, likewise, determine if the Y chromosome is present in a particular embryo.

In contrast with the highly differentiated Y chromosome in mice, the heteromorphic sex chromosomes could not be identified by simple karyologic analysis among most lower vertebrates. In some species, the Y chromosome can be identified by conventional C-banding [40]. The evolution of differentiated sex chromosomes often correlates with a significant accumulation of simple-repetitive sequences. Male-specific hybridization owing to the accumulation of simple repeats on the Y chromosome is illustrated by in situ hybridization in mice and snakes. Therefore, the present approach that uses simple repeats may provide a fast complementation to identify the sex

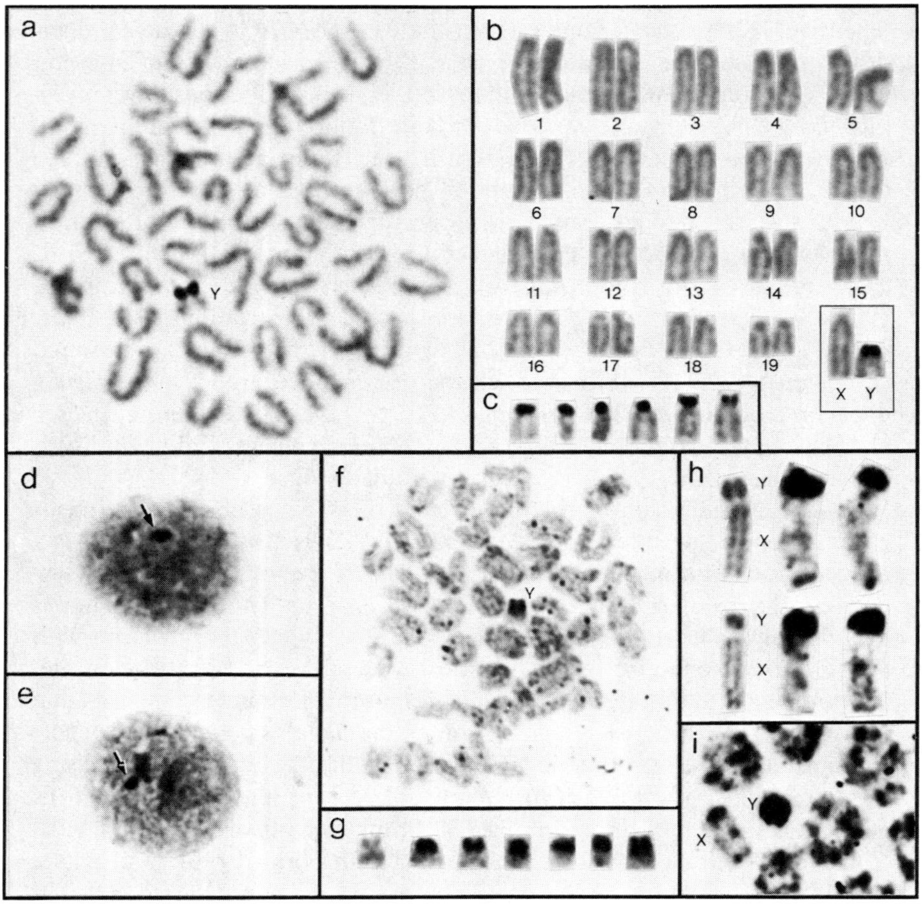

Figure 3 The Y chromosome of the mouse can be identified by in situ hybridization using either the biotinylated or the digoxigenated $(GACA)_4$: (a) In metaphase spreads, the Y chromosomes are exclusively marked with biotinylated $(GACA)_4$. (b) The constructed karyotype (b) and the isolated Y chromosomes from different metaphase plates (c) display the reproducibility of the signal specific to the Y chromosome. (d,e) In interphase nuclei, the Y chromosome can be marked using the digoxigenin-labeled probe. (f,g) In a mutated Y chromosome (pericentromeric inversion), the signal of the digoxigenated $(GACA)_4$ probe is predominantly spread over the short arm. (h,i) In the meiotic Y chromosome the association and orientation of the $(GACA)_4$-positive short arm can be recognized in X–Y bivalents in metaphase I.

chromosomes status in gonochoristic fish species. With ^{32}P-labeled synthetic $(GACA)_4$, male-specific DNA hybridization patterns could be generated for the guppy fish *Poecilia reticulata* (Fig. 4), in which the presence of the Y chromosome has been established through noncytogenetic approaches. With use of the $(GACA)_4$ probe in males, characteristic peroxidase staining at the telomeric end of the Y chromosome was noticed after hybridization onto metaphase chromosomes (Fig. 5b). This allows differentiation between the homomorphic XY sex chromosomes.

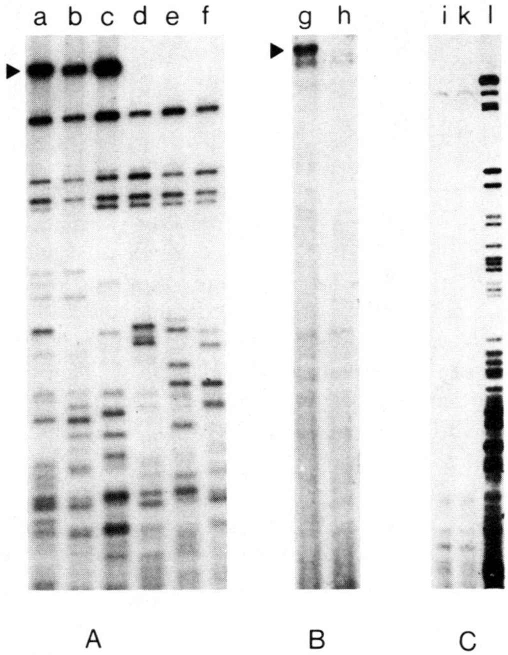

Figure 4 Different radiolabeled simple-repeat probes hybridized in the gel provide a useful means to investigate the distribution of simple sequences on the Y chromosome. (A) Male-specific signal (arrow) in the high-molecular-weight range can be obtained after probing male (a, b, c) and female (d, e, f) guppy DNA with the $(GACA)_4$ oligonucleotide. (B) With the $(GATA)_3(GACA)_2$ probe, hybridization signals can be visualized specific for the Y chromosome in the mouse (arrow). (C) The lack of long stretches of $(CAC)_n$ repeats on the human Y chromosome is illustrated by comparing the hybridization pattern between the mouse–human hybrid cell line 3E7 harboring human Y chromosomes (i), the mouse control cell line 7/2 (k), and normal human male DNA (l). Interspersion of long $(CAC)_n$ stretches is limited to the autosomes (and the X ?).

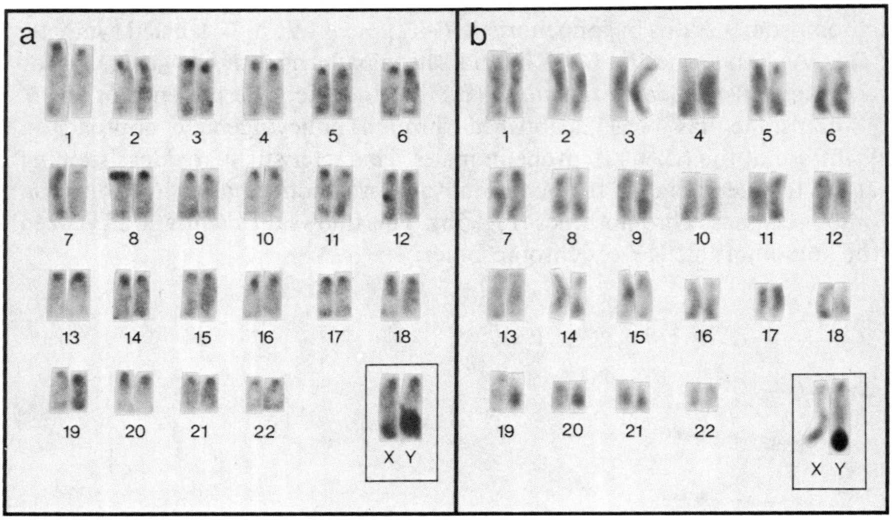

Figure 5 The Y chromosome of the fish *Poecilia reticulata* (guppy) can be detected by conventional C-banding technique. This reveals conspicuous Y-specific staining (a) in addition to the staining on the X as well as the centromeres of other chromosomes. However, because of experimental variations, Y-specific staining could not be distinguished from the rest of the C-banded autosomes, particularly on the homomorphic sex chromosomes. Hence, in situ hybridization with the simple-repeat probe $(GACA)_4$ provides a more informative, alternative approach. The guppy karyotype constructed from male metaphases hybridized with biotinylated $(GACA)_4$ denotes exclusively a signal on the Y chromosome (b).

IV. CONCLUSIONS

Simple-repetitive DNA sequences are widespread in all eukaryotes. This phenomenon can be easily demonstrated by hybridization in situ. With three synthetic oligonucleotide probes, the present study covered three unrelated aspects in different biological subjects: coincidence of R-bands and $(CAC)_n$ loci, association of $(GACA)_n$ sequences and NORs, as well as the identification of the guppy Y chromosome. Nevertheless, the biological meaning of simple-repetitive DNA still remains an enigma.

In most instances, oligonucleotides labeled with γ-^{32}P are not suitable for hybridization to chromosomes in situ because the energy of the emitted β-particles is too high. This causes difficulties in relating the far-tracking signals to the morphological structure itself. Obviously, nonradioactive probes avoid this shortcoming. The specificity of the probes used here was ascertained from control slides that were treated identically to the other slides, with the exception that the hybridization mixture lacked the labeled probe. No signals were noticed in these slides. In addition, areas not covered with the hybridization mixture did not show any dye precipitate when processed by the detection system. During the present investigation, several aspects of the hybridization with biotinylated and digoxigenated oligonucleotide probes could be compared. For example, hybridization signals as obtained with digoxigenin probes are always more specific, reproducible, and prominent, compared with those evoked by biotin. On the other hand, much of the drudgery of hybridization in situ with digoxigenated probes results from the long incubation in the color-developing solution at elevated pH (9.5), which often ends in a poor structural appearance of the chromosomes. The advantage of nonradioactive in situ hybridization with biotin-labeled probes lies in the better microscopic resolution of the label and the absolute lack of background interference. Furthermore, as a result of the counterstaining with Giemsa, the biotinylated oligonucleotide pictures emerge much sharper. Another major advantage with biotin-based signal detection is the reduced time-consumption. Because strong hybridization signals could be elicited by the digoxigenin, detection systems based on horseradish peroxidase-coupled antidigoxigenin antibodies might be useful for oligonucleotide hybridization in situ.

In situ hybridization with a 42-bp oligomer probe to metaphase chromosomes has been performed recently using a nonisotopic method to characterize telomeric sequences [41] through multiple attachment of biotin (biotin-11-dCTP) at the 3′ ends by terminal deoxynucleotide transferase. Signals were detected by the fluorescein–avidin system [42]. This high-contrast fluorescent detection system can replace the usual colorimetric procedure, and the individual chromosomes can be identified simultaneously. Moreover, the hybridization signal for oligoprobes specific to low-copy sequences can be amplified by performing several rounds of F-avidin/antiavidin incubation. However, on the basis of the time consumed for the whole procedure and also because of material costs, simple end-labeling of oligonucleotides and single-step detection offers a fast and easy alternative. In the future, the advantages for cellular studies of oligonucleotide probes with multiple nonradioactive reporter groups versus a single reporter group should be compared in detail.

ACKNOWLEDGMENTS

We gratefully acknowledge the invaluable assistance of Mr. H. Zischler and Ms. C. Deubelbeiss for discussions, as well as for providing data from their own experiments. We thank Dr. H. Winking (Lübeck) for providing us the laboratory strains of mice. We also thank our colleagues M. Guttenbach, H. Seyschab, and C. Steeg for their help. This work was supported by the Deutsche Forschungsgemeinschaft (Schm 484/4-1) and the VW-Stiftung. The oligonucleotide probes are subject to patent applications and commercial enquiries should be directed to FRESENIUS AG, Obersursel, FRG.

REFERENCES

1. Gall, J. G., and M. L. Pardue (1969). Formation and detection of RNA–DNA hybrid molecules in cytological preparations. *Proc. Natl. Acad. Sci. USA 63*: 378–383.
2. John, H., M. L. Birnstiel, and K. W. Jones (1969). RNA–DNA hybrids at the cytological level. *Nature 223*:582–587.
3. Szabo, P., and D. C. Ward (1982). What's new with hybridization in situ? *Trends Biochem. Sci. 7*:425–427.
4. Pardue, M. L., and J. G. Gall (1975). Nucleic acid hybridization to the DNA of cytological preparations. In *Methods in Cell Biology, Vol. 10*. D. Prescott (ed.). Academic Press, New York, pp. 1–16.
5. Steffensen, D. M. (1977). Human gene localization by RNA:DNA hybridization in situ. In *Molecular Structure of Human Chromosomes*. J. J. Yunis (ed.). Academic Press, New York, pp. 59–88.
6. Jones, K. W. (1970). Chromosomal and nuclear location of mouse satellite DNA in individual cells. *Nature 225*:912–915.
7. Feinberg, A. P., and B. Vogelstein (1983). A technique for radiolabeling DNA restriction endonuclease fragments to high specific activity. *Anal. Biochem. 132*:6–13.
8. Krieg, P. A., and D. A. Melton (1984). Functional messenger RNAs are produced by SP6 in vitro transcription of cloned cDNAs. *Nucleic Acids Res. 12*: 7057–7070.
9. Harper, M. E., and G. F. Saunders (1981). Localization of single copy DNA sequences on G-banded human chromosomes by in situ hybridization. *Chromosoma 83*:431–439.
10. Langer-Safer, P. R., M. Levine, and D. C. Ward (1982). Immunological method for mapping genes on *Drosophila* polytene chromosomes. *Proc. Natl. Acad. Sci. USA 79*:4381–4385.
11. Bauman, J. G., J. Wiegant, P. Borst, and P. Van Duijn (1980). A new method for fluorescence microscopical localization of specific DNA sequences by in situ hybridization of fluorochrome-labeled RNA. *Exp. Cell Res. 128*:485–490.
12. Manuelidis, L., P. R. Langer-Safer, and D. C. Ward (1982). High resolution mapping of satellite DNA using biotin labeled DNA probes. *J. Cell Biol. 95*: 619–625.
13. Ambros, P. F., M. A. Matzke, and A. J. M. Matzke (1986). Detection of a 17 kb unique sequence (T-DNA) in plant chromosomes by in situ hybridization. *Chromosoma 94*:11–18.

14. Jeffreys, A. J. (1987). Highly variable minisatellites and DNA fingerprints. *Biochem. Soc. Trans. 15*:309–317.
15. Studer, R., and J. T. Epplen (1990). On the organization of animal genomes: Ubiquitous interspersion of repetitive DNA sequences. In *Proceedings of the Congress on Genome Analysis in Domestic Animals*. H. Geldermann, F. Ellendorff, and G. Stranzinger (eds.). VCH, Weinheim (in press).
16. Schäfer, R., H. Zischler, U. Birsner, A. Becker, and J. T. Epplen (1988). Optimized oligonucleotide probes for DNA fingerprinting. *Electrophoresis 9*:369–374.
17. Lewis, M. E., T. G. Sherman, and S. J. Watson (1985). In situ hybridization histochemistry with synthetic oligonucleotides: Strategies and methods. *Peptides 6* (Suppl. 2):75–87.
18. Thein, S. L., and R. B. Wallace (1986). The use of synthetic oligonucleotides as specific hybridization probes in the diagnosis of genetic disorders. In *Human Genetic Diseases A Practical Approach*. K. E. Davies (ed.). IRL Press, Oxford, pp. 33–50.
19. Zischler, H., I. Nanda, R. Schäfer, M. Schmid, and J. T. Epplen (1989). Digoxigenated oligonucleotide probes specific for simple repeats in DNA fingerprinting and hybridization in situ. *Hum. Genet. 82*:227–233.
20. Epplen, J. T., J. R. McCarrey, S. Sutou, and S. Ohno (1982). Base sequence of a cloned snake W-chromosome DNA fragment and identification of a male specific putative mRNA in the mouse. *Proc. Natl. Acad. Sci. USA 79*:3798–3802.
21. Singh, L., I. F. Purdom, and K. W. Jones (1981). Conserved sex-chromosome-associated nucleotide sequences in eukaryotes. *Cold Spring Harbor Symp. Quant. Biol. 45*:805–813.
22. Zischler, H., I. Nanda, C. Steeg, M. Schmid, and J. T. Epplen (1990). How to demonstrate genetic individuality in any human and animal subject. In *Proceedings of the Congress on Genome Analysis in Domestic Animals*. H. Geldermann, F. Ellendorff, and G. Stranzinger (eds.). VCH, Weinheim (in press).
23. Zischler, H., R. Schäfer, I. Nanda, and J. T. Epplen (1990). Towards an optimal method for DNA fingerprinting. In *Advances in Molecular Genetics Vol. 1*. J. Henke, J. Kömpf, and A. J. Driesel (eds.). Hüthig, Heidelberg, pp. 125–138.
24. Holmquist, G., M. Gray, T. Porter, and J. Jordan (1982). Characterization of Giemsa dark- and light-band DNA. *Cell 31*:121–129.
25. Chandley, A. C., and A. R. Mitchell (1988). Hypervariable minisatellite regions are sites for crossing-over at meiosis in man. *Cytogenet. Cell Genet. 48*: 152–155.
26. Holmquist, G. P. (1989). Evolution of chromosome bands: Molecular ecology of non coding DNA. *J. Mol. Evol. 28*:469–486.
27. Willard, H. F. (1985). Chromosome-specific organization of human alpha satellite DNA. *Am. J. Hum. Genet. 37*:524–532.
28. Cremer, T., J. Landegent, A. Brückner, H. P. Scholl, M. Schardin, H. D. Hager, P. Devilee, P. Pearson, and M. van der Ploeg (1986). Detection of chromosome aberrations in the human interphase nucleus by visualization of specific target DNAs with radioactive and non-radioactive in situ hybridization techniques: Diagnosis of trisomy 18 with probe L1.84. *Hum. Genet. 74*:346–352.

29. Waye, J. S., G. M. Greig, and H. F. Willard (1987). Detection of novel centromeric polymorphisms associated with alpha satellite DNA from human chromosome 11. *Hum. Genet.* *77*:151–156.
30. Mikelsaar, A. V., M. Schmid, W. Krone, H. G. Schwarzacher, and W. Schnedl (1977). Frequency of Ag-stained nucleolus organizer regions in the acrocentric chromosomes of man. *Hum. Genet.* *37*:73–77.
31. Mikelsaar, A. V., H. G. Schwarzacher, W. Schnedl, and P. Wagenbichler (1977). Inheritance of Ag-stainability of nucleolar organizer regions. *Hum. Genet.* *38*: 183–188.
32. Jørgensen, A. L., C. J. Bostock, and A. L. Bak (1987). Homologous subfamilies of human alphoid repetitive DNA on different nucleolus organizing chromosomes. *Proc. Natl. Acad. Sci. USA* *84*:1075–1079.
33. Jørgenesen, A. L., S. Kolvraa, C. Jones, and A. L. Bak (1988). A subfamily of alphoid repetitive DNA shared by the NOR-bearing chromosomes 14 and 22. *Genomics* *3*:100–109.
34. Singh, L., and K. W. Jones (1986). Bkm sequences are polymorphic in humans and are clustered in pericentric regions of various acrocentric chromosomes including the Y. *Hum. Genet.* *73*:304–308.
35. Singh, L., I. F. Purdom, and K. W. Jones (1980). Sex chromosome-associated satellite DNA: Evolution and conservation. *Chromosoma* *79*:137–157.
36. Epplen, J. T. (1988). On simple repeated GATA/GACA sequences in animal genomes: A critical reappraisal. *J. Hered.* *79*:409–417.
37. Schäfer, R., S. Ali, and J. T. Epplen (1986). The organization of the evolutionarily conserved GATA/GACA repeats in the mouse genome. *Chromosoma* *93*:502–510.
38. Kiel-Metzger, K., and R. P. Erickson (1984). Regional localization of sex-specific Bkm-related sequences on proximal chromosome 17 of mice. *Nature* *310*:579–581.
39. Singh, L., H. Winking, K. W. Jones, and A. Gropp (1988). Restriction fragment polymorphism in the sex-determining region of the Y chromosomal DNA of European wild mice. *Mol. Gen. Genet.* *212*:440–449.
40. Schmid, M. (1983). Evolution of sex chromosomes and heterogametic systems in amphibia. In *Mechanisms of Gonadal Differentiation in Vertebrates.* U. Müller, and W. W. Franke (eds.). *Differentiation* *23*(suppl):13–22.
41. Moyzis, R. K., J. M. Buckingham, L. S. Cram, M. Dani, L. L. Deaven, M. D. Jones, J. Meyne, R. L. Ratliff, and J.-R. Wu (1988). A highly conserved repetitive DNA sequence, $(TTAGGG)_n$, present at the telomeres of human chromosomes. *Proc. Natl. Acad. Sci. USA* *85*:6622–6626.
42. Pinkel, D., T. Straume, and J. W. Gray (1986). Cytogenetic analysis using quantitative, high-sensitivity, fluorescence hybridization. *Proc. Natl. Acad. Sci. USA* *83*:2934–2938.

6

Gene Mapping and Fine-Structure Analysis of the Human Genome Using Flow-Sorted Chromosomes

Nobuyoshi Shimizu and Shinsei Minoshima

Keio University School of Medicine, Tokyo, Japan

I. INTRODUCTION

The human genome consists of 22 pairs of autosomes (numbered 1–22) and a pair of sex chromosomes (X and Y). Previously, these chromosomes have been identified individually by conventional chromosomal-banding techniques. Recently, flow cytometry has provided us with a fast and accurate method to measure the minute quantity of DNA in each human chromosome. Almost all of the 24 different types of chromosomes can now be resolved, and individual variation of the flow karyotype has been recognized [1,2]. Flow cytometry has enabled us to distinguish even the subtle difference between two chromosome homologues [3].

The sorting capabilities of most flow machines have provided a method for the fractionation and collection of individual chromosomes. Small quantities of the flow-sorted chromosomes deposited on nitrocellulose membrane filters have facilitated human gene mapping with specific DNA probes [1,4,5]. Relatively large quantities of the sorted chromosomes have been used to produce a chromosome-specific clone library of numerous DNA segments [6]. These clone libraries are being used as marker probes for genetic linkage analysis, physical mapping of the individual chromosomes, and the molecular analysis of genetic diseases.

More recently, mapping and sequencing of the entire human genome has become a major interest in medical genetics and human molecular biology

[7]. High-quality flow-sorted human chromosomes have become one of the essential components for this challenging project. In this chapter, we will describe advanced flow-cytometry techniques for the preparation of highly purified intact human chromosomes and some molecular biological techniques necessary for genome analysis.

II. HIGH-RESOLUTION FLOW ANALYSIS AND SORTING OF HUMAN CHROMOSOMES

We have established a method to sort human chromosomes at high resolution using a conventional cell sorter, equipped with a single argon ion laser [8]. The method for preparation and staining of chromosomes was improved, and the flow system was optimized, as described in detail in Section VIII of this chapter. With these improved techniques, flow-karyotypes of the human B-lymphoblastoid line GM130B (46,XY) was established after Hoechst 33258 (H33258) or propidium iodide (PrI) staining (Figs. 1A and 2A). Chromosomes stained with H33258 were separated into 17 fractions, whereas chromosomes stained with PrI were separated into 18 fractions. The chromosome types contained in each fraction were determined by spot-blot hybridization, using chromosome-specific DNA markers as probes. The separation pattern of chromosomes stained with the two different fluorescent dyes significantly differed from each other. By appropriate use of either dye, 17 among 24 types of human chromosomes can be sorted as single components. Chromosomes 3, 8, 13, 18, 20, 21, and 22 are well separated and can be obtained in a considerably pure form.

A dual-laser flow system can be used to sort chromosomes at higher resolution after staining with both chromomycin A3 and H33258 (Fig. 3). All the human chromosomes except four types (chromosomes 9 to 12) can be clearly separated. The dual-laser system, however, requires much more attention to the sorting procedure, and chromomycin A3 staining has a relatively long incubation time. Therefore, the single-laser system is convenient, and its resolution is sufficient for many experimental purposes.

III. GENE MAPPING BY SPOT-BLOT HYBRIDIZATION WITH A PANEL OF FLOW-SORTED HUMAN CHROMOSOMES

Usually, 50,000 chromosomes each are collected as small spots on nitrocellulose membrane filters and hybridization is performed with a human gene probe of interest. In many cases, single strong hybridization signals are obtained and assignment of the gene to that particular chromosome is relatively straightforward. Figure 4 shows the assignment of 80K-H protein gene to chromosome 19 [9]. However, multiple minor signals are occasionally

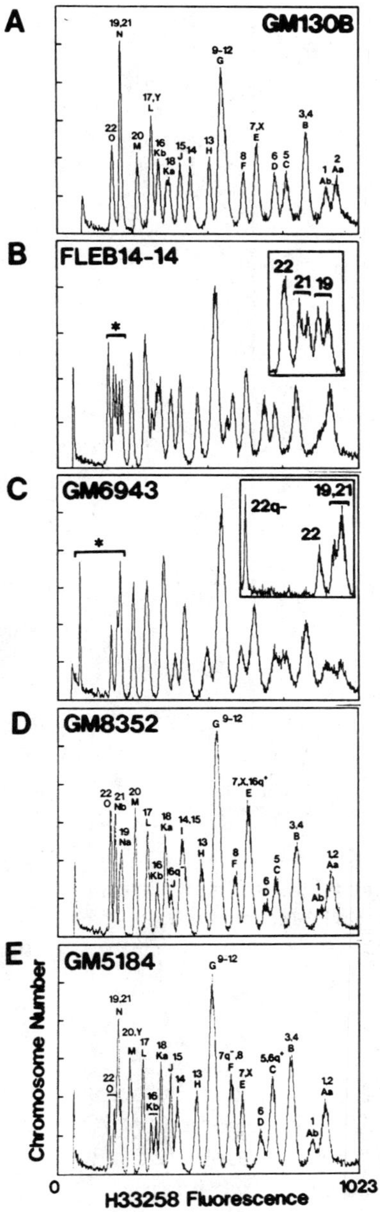

Figure 1 Flow-karyotypes of five human B-lymphoblastoid lines: Metaphase chromosomes were stained with Hoechst 33258 (H33258) and analyzed by a FACS440 sorter equipped with a single argon laser. The asterisked areas are enlarged in the inserted figures.

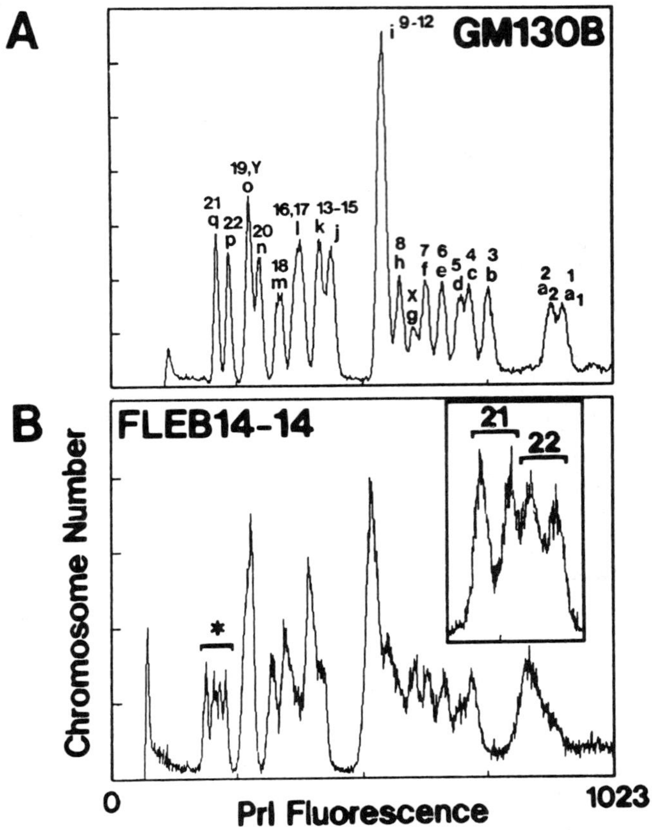

Figure 2 Flow-karyotypes of two human B-lymphoblastoid lines: Metaphase chromosomes were stained with propidium iodide (PrI) and analyzed by a FACS440 sorter equipped with a single argon laser. The asterisked area is enlarged in the inserted figure.

detected in different fractions because of the presence of homologous sequences. Figure 5 shows such an example for the assignment of ATP synthase β-subunit (*ATPSB*) gene [5]. Spot-blot hybridization with flow-sorted chromosomes stained with H33258 revealed a strong hybridization signal in fraction G (chromosomes 9–12) and weak signals in fraction Aa (chromosome 2) and fraction L (chromosome 17 and Y). Thus, at least three human chromosomes carry the homologous sequences to the *ATPSB* gene. Hybridization with sorted chromosomes with PrI also revealed a strong signal in fraction i (chromosomes 9–12) and two weak signals in fraction a2 (chromo-

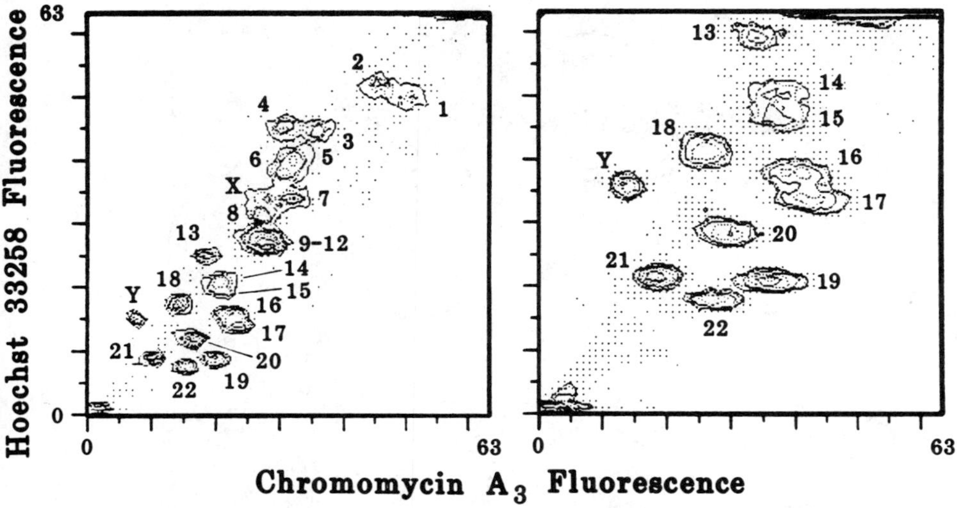

Figure 3 Flow-karyotypes of the human B-lymphoblastoid line GM130B: Metaphase chromosomes were stained with both chromomycin A_3 and Hoechst 33258, and analyzed by a FACS440 sorter with a dual-laser system. The right figure shows a selected area of the left figure.

some 2) and fraction 1, which does not contain the Y chromosome (located in fraction o), but does contain chromosomes 16 and 17. These results confirm the location of one of the homologous sequences on at least one of chromosomes 9–12 and the other on chromosome 2 and indicate that the location of another sequence is on chromosome 17. Further analysis with a somatic cell hybrid panel eliminated chromosomes 9–11 and assigned the actual *ATPSB* gene to chromosome 12 [5].

IV. REGIONAL MAPPING USING TRANSLOCATION CHROMOSOMES

Confirmation of the chromosomal assignment and regional localization of the gene can be performed by using flow-sorted translocation chromosomes. Figure 6 shows the flow-karyotype of GM7173 cells, which carry a t(5;7) translocation. The gene for interleukin 4 (*IL4*) has been assigned to chromosome 5 [10]. When an IL-4 DNA probe was applied to chromosomes from GM7173 cells, hybridization signals appeared in fraction A (chromosomes 1 and 2), fraction B (chromosomes 3 and 4), and fraction CD (chromosome 5 and 6). The translocation chromosome 5q+, which is larger than the nor-

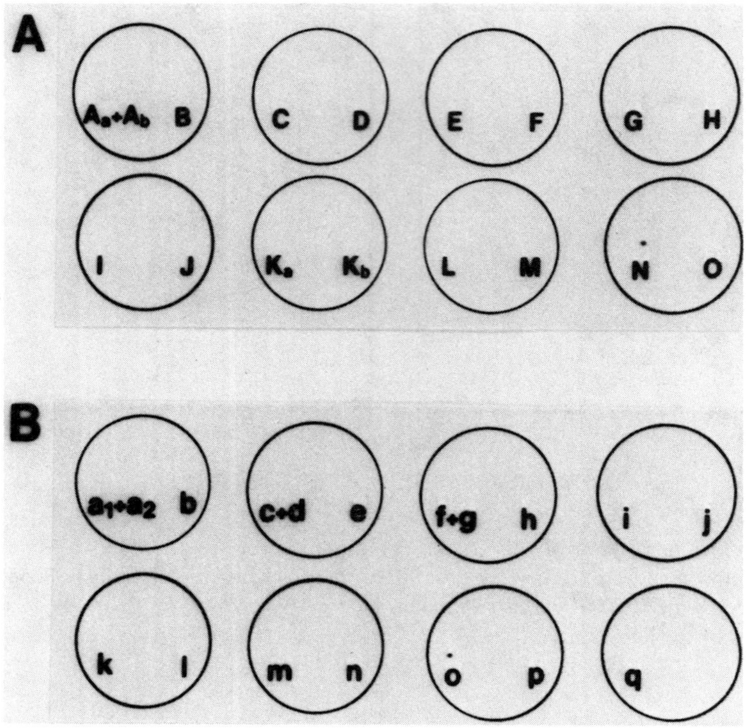

Figure 4 Assignment of the human 80K-H protein gene to chromosome 19 by spot-blot hybridization: 50,000 chromosomes each were collected on nitrocellulose membrane filters, and hybridization was performed with the cDNA probe for 80K-H protein. (A) Chromosome fractions from Figure 1A. Fractions Aa and Ab were collected as a mixture. (B) Chromosome fractions from Figure 2A. Fractions a_1 and a_2, c and d, and f and g were collected as mixtures, respectively.

mal chromosome 5, is sorted at a position between A and B, and the signals in fractions A and B must be generated from the translocation chromosome 5q+. These results confirm the assignment of *IL4* gene on chromosome 5 and provide further evidence that the *IL4* gene is located on the pter-q35 region of human chromosome 5.

The flow-karyotype of GM8352 cells, which carry a t(6;16) translocation, is shown in Figure 1D. Here, the translocation chromosome 16q+ is sorted in fraction E, whereas the translocation chromosome 6q− is sorted as a new peak between fractions I and Ka. Accordingly, the peak heights of normal chromosomes 6 (fraction D) and 16 (fraction Kb) are reduced. Fig-

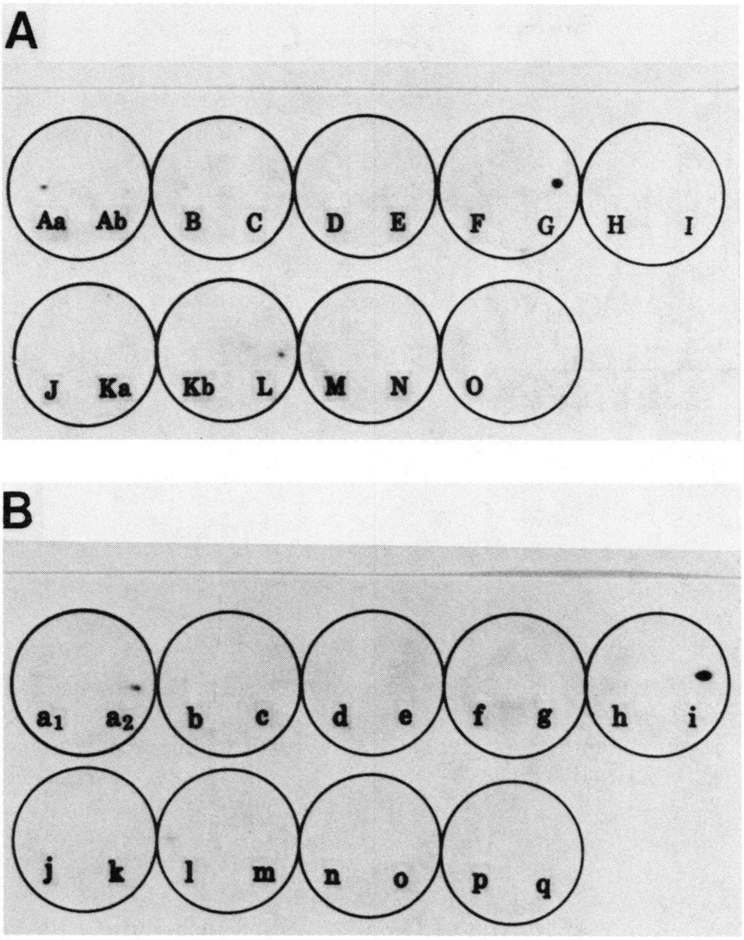

Figure 5 Spot-blot hybridization of flow-sorted chromosomes with human *ATPB* gene probe: (A) Chromosome fractions from Figure 1A. (B) Chromosome fractions from Figure 2A.

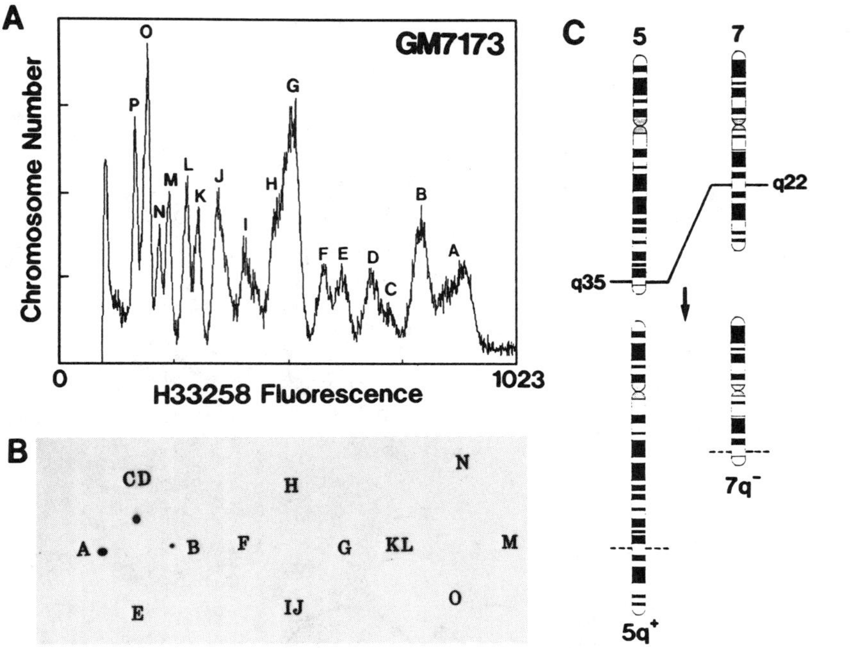

Figure 6 Regional mapping of IL-4 gene using a translocation chromosome: (A) Flow-karyotype of GM7173 after H33258 staining. (B) Spot-blot hybridization of IL-4 gene probe with chromosomal fractions from Figure 6A. (C) Breakpoints of the t(5;7) translocation in GM7173 cells.

ure 1E shows the flow-karyotype of GM5184 cells, which carry a t(6;7) translocation. The translocation chromosomes 6q+ and 7q− are sorted in the fractions of chromosomes 5 (fraction C) and 8 (fraction F), respectively. Again, the peak heights of normal chromosomes 6 (fraction D) and 7 + X (fraction E) are reduced. These translocation chromosomes are useful for regional mapping of genes. Thus, spot-blot hybridization with flow-sorted human normal and translocation chromosomes has provided evidence for chromosomal assignment and regional localization of several human genes: myeloperoxidase (*MPO*/chromosome 17) [11]; phosphoribosylpyrophosphate synthetase subunit I and II (*PRPS1*/Xq21-qter and *PRPS2*/Xpter-q21) [12]; *PRPS1*-related sequences 1 and 2 (*PRPS1L1*/7 and *PRPS1L2*/9) [12]; interleukin 4 (*IL4*/5pter-q35) [10]; ATP synthase β-subunit (*ATPSB*/12p13-qter) [5]; *ATPSB*-homologous sequences (*ATPSBL1*/2 and *ATPSBL2*/17) [5]; 80K-H protein (*G19P1*/19) [9]; subunits L1, L2, L3, and

S of calpain (*CAPN1*/11, *CAPN2*/1, *CAPN3*/15, *CAPN4/19)* [13]; and calpastatin (*CAST*/5) [14].

V. SEPARATION OF SINGLE HOMOLOGUES

Several human chromosomes, such as chromosomes 1, 9, 13, 14, 15, 21, and 22, contain heterochromatic regions, and their size often differs among individuals. This heteromorphic nature of certain chromosomes and subtle differences between two homologues can be accurately distinguished by flow-cytometry. We have improved the conditions to analyze and separate one of two chromosome homologues using B-lymphoblastoid cells bearing variations in heterochromatic regions. Heteromorphisms can be seen in the flow-karyotypes of GM130B (46,XY) (see Fig. 1A); FLEB14-14 (46,XY) (see Fig. 1B); GM6943 [46,XX, t(21;22)(p11;q11.2)] (see Fig. 1C); GM8352 [46,XX, t(6;16)(q15;q24)] (see Fig. 1D); and GM5184 [46,XY, t(6;7)(q21;q21.2)] (see Fig. 1E). One of the chromosome 1 homologues is clearly separated in GM8352 and GM5184 (see Figs. 1D, E). Chromosomes 14 and 15 are sorted at separate peaks in GM130B (see Fig. 1A), but these are in the same fraction in GM8352. Chromosomes 21 and 19 usually are not separated, but they are separated in GM8352 (see Fig. 1D). Homologues of chromosomes 19 and 21 are clearly separated in FLEB14-14 (see Fig. 1B). These flow-karyotypes were made after H33258 staining. Homologues of both chromosomes 21 and 22 can be separated in FLEB14-14 if PrI staining is used (see Fig. 2B).

Homologues can also be separated from the cells carrying unique translocation chromosomes. The cell line GM6943 carries a translocation t(21;22) (see Figs. 1C and 7). Because the translocation results in significant differences in size, normal chromosome 22 is clearly separated from the translocation chromosome 22q− after H33258 staining.

Single homologues are useful when individual DNA polymorphisms need to be considered for genome analysis. Comparison of the DNA restriction patterns between normal and translocation chromosomes provides direct information on the particular regions involved in the translocation.

VI. PULSED-FIELD GEL ANALYSIS OF FLOW-SORTED HUMAN CHROMOSOMES

There have been concerns about the intactness of flow-sorted human chromosomes. Chromosomes may be damaged by laser-irradiation and physical constraints during sorting. Previously, there was no reliable method to test for intactness. Human chromosomes sorted by our method have been shown to be intact [8]. This was accomplished by avoiding the possible problems

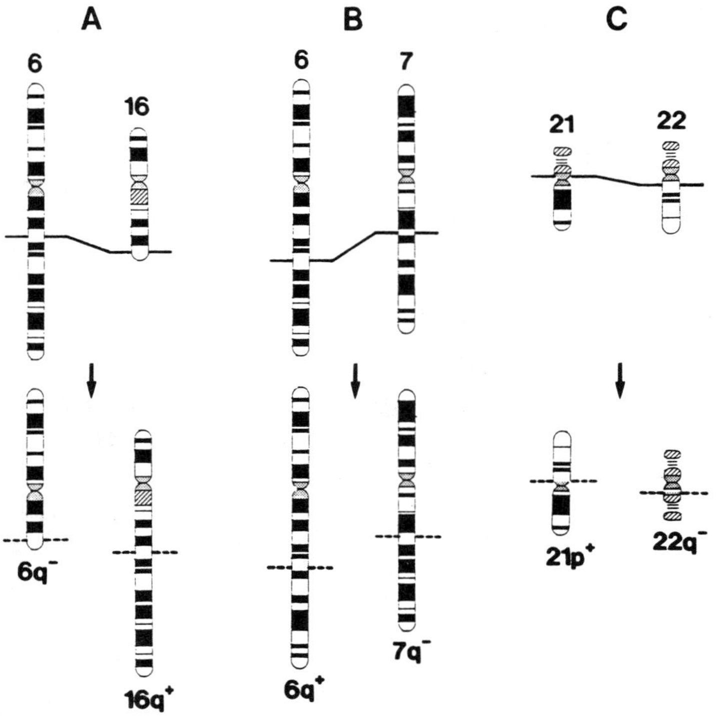

Figure 7 The translocation breakpoints in three B-lymphoblastoid lines: (A) GM8352, (B) GM5184, (C) GM6943.

that cause chromosome breakage, as described in Section VIII [8]. Analysis was done by pulsed-field gel electrophoresis, as shown in Figure 8. Seven hundred thousand chromosomes 21 and 22 were collected separately in tubes and cut with *Not*I restriction enzyme. The DNA fragments were separated by pulsed-field gel electrophoresis and analyzed by Southern blot hybridization with the human *Alu* sequence as a probe. The analysis revealed over 30 distinct giant DNA fragments, ranging from 50 kb to longer than 2.5 Mb in size. In this manner, the flow-sorted chromosomes were proved to be intact.

VII. PROSPECTS

Construction of a linkage map and a physical map of specific human chromosomes with fine resolution is required for the isolation of disease-related genes by reverse genetics [15]. A linkage map is being constructed by the use of CEPH family DNA panel and radiation hybrids with various DNA probes. Use of chromosome-specific clone libraries of DNA segments that are generated from flow-sorted human chromosomes has been inevitable. Plasmid and phage libraries containing DNA segments of 5–20 kb in size have been constructed. Cosmid libraries containing DNA segments of 50 kb in average size are being constructed. The P1 phage vector appears useful for cloning DNA segments up to 100 kb [16].

The chromosome-specific plasmid libraries are being used to construct a "linking library" by which two adjacent restriction fragments can be identified [17]. The cosmid library has been employed to construct "contig maps" of certain chromosomes [18]. These chromosome-specific libraries, together with cell hybrids containing different chromosomal segments, are utilized to construct a physical map of the human chromosome.

Pulsed-field gel electrophoresis has been used with chromosomal DNA that was directly released from intact cells in a gel matrix, but it has never been used for sorted chromosomes until recently. We have shown that flow-sorted human chromosomes are intact and that giant DNA fragments can be prepared. Cloning giant restriction fragments as large as 500 kb into yeast artificial chromosomal (YAC) vectors is now possible [19]. The YAC clones will serve as useful DNA sources for physical mapping and sequencing, as well as for functional analysis of particular chromosomal regions. Construction of region-specific clone libraries, using isolated large DNA segments and shotgun polymerase chain reaction (PCR) is now being pursued [K. Kawasaki and N. Shimizu, unpublished data].

Flow-sorted human chromosomes can be used for the construction of chromosome-specific cDNA libraries that can be screened from the tissue-specific cDNA library. Microinjection of the sorted chromosomes into eggs or intact cells of various tissue origins would provide a unique system for studying gene regulation. Thus, flow-sorted human chromosomes, particularly with high purity and intactness, are invaluable materials, and the utilization of these chromosomes will open various new avenues to study the molecular biology of the human genome.

VIII. EXPERIMENTAL PROCEDURES

A. Cell Culture with High Mitotic Index

Metaphase chromosomes are prepared by the polyamine–digitonin method according to Lebo et al. [1], with several modifications [8]. The spontaneously transformed human B-lymphoblastoid line GM130B (NIGMS Human Genetic Mutant Cell Repository, Camden) is used to establish a standard male flow-karyotype. Grow the cells in RPMI 1640 medium (Gibco) containing 15% heat-inactivated fetal calf serum (Boehringer Mannheim),

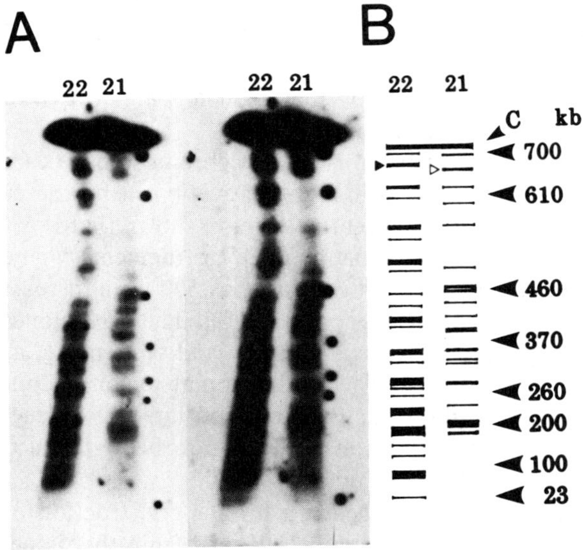

Figure 8 Analysis of the *Not*I DNA fragments of flow-sorted human chromosomes 21 and 22 by pulsed-field gel electrophoresis: 700,000 each of sorted chromosomes 21 and 22 were embedded in a low-melting-temperature agarose block. Chromosomal DNA was liberated by treating with proteinase K and Sarkosyl and digested with *Not*I restriction endonuclease. DNA fragments were separated by PFGE and analyzed by Southern blot hybridization using the *Alu* repetitive sequence as a probe. (A) Pulse time of 55 sec, (B) pulse time of 120 sec. Two panels in A and C are shorter

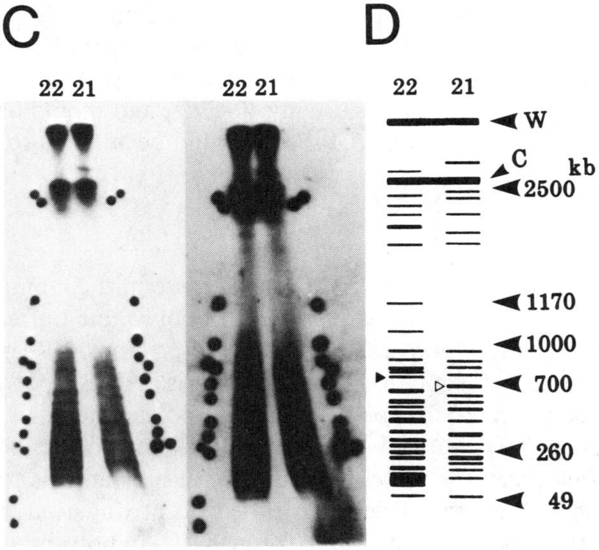

(left) and longer (right) exposures of the same autoradiogram. B and D are diagrams of A and C, respectively. C and W represent the positions of compression and the sampling well, respectively. Open and solid triangles in panels B and D indicate the corresponding DNA fragments detected under two different pulse-time conditions for chromosomes 21 and 22, respectively.

0.3 mg/mL glutamine, 100 μg/mL kanamycin (Meiji), and 1 μg/mL amphotericin B (fungizone; Squibb). Maintain cultures in T75-type flasks (Corning 25110) with a maximum volume of 30 mL until the cultures reach stationary phase ($> 1.5 \times 10^6$ cells/mL). Then, dilute each fivefold with fresh medium and replate into Falcon 3025 (15-cm diameter) dishes. Keep the maximum volume for each dish 80 mL or less to maintain the growth rate (doubling time: ~24 hr). Allow cells to grow for 24–30 hr, then add demecolcine (Colcemid) at 30 ng/mL. After 10–12 hr, harvest cells by centrifugation at 120x*g* for 5 min. Suspend cell pellets in 75 mM KCl solution (25- to 50-fold volume of cell pellet). Take a 100 μL aliquot of the cell suspension and leave at room temperature for 10 min. Fix with Carnoy's fixative and spread on a glass slide. Observe with a phase-contrast microscope (Nikon TMS) to calculate the mitotic index. The mitotic index is usually 40–50%, and should be as high as possible. The optimum culture conditions should be adapted for each cell line.

B. Chromosome Preparation

Incubate the cell suspension for 15 min at 37°C with intermittent mild agitation. Spin down cells and suspend the cell pellets in polyamine buffer [20] with the same volume as the 75 mM KCl solution used previously. Keep for 30 min on ice with intermittent agitation. Spin down cells and resuspend in polyamine–digitonin solution [27-fold pellet volume of polyamine buffer and threefold pellet volume of saturated digitonin (Merck) in polyamine buffer]*. Vortex cells four times for 30 sec on ice with 30-sec intervals to liberate chromosomes. Spin down the nuclei and debris by centrifugation at 70x*g* for 5 min. Collect the supernatant and add Triton X-100 in polyamine buffer to 1%.

C. Staining Chromosomes with Fluorescent Dyes

Metaphase chromosomes are stained with either 2 μg/mL Hoechst 33258 (H33258; Sigma) or 35 μg/mL propidium iodide (PrI; Sigma).

The stock solution of H33258 (0.5 mg/mL) is made with distilled water. The stock solution of PrI (1 mg/mL) is made with 1.2% Na_3citrate · $3H_2O$ [21] and irradiated with excitation light as follows: place 1 mL of the PrI stock solution in a 15-mL transparent plastic centrifuge tube (Sumitomo MS55150; 19-mm diameter) and irradiate for 20–30 min at room tempera-

---

*To prepare saturated digitonin solution: add 10 mg/mL digitonin to polyamine buffer, incubate at 37°C for 30 min, remove the residual solid by centrifugation at 470x*g* for 5 min, and filter the supernatant.

ture with a spotlight (2-cm diameter; 488-nm wave length) focused from an argon ion laser (Spectra Physics 164-05; 1 W).

Chromosomes stained with H33258 are stable for 2 days or longer and can be used for sorting with a good resolution. However, only 1–2 mL suspension at one time should be stained with PrI 30 min before sorting because PrI causes aggregation of chromosomes within 3 hr.

D. Flow-Karyotyping and Chromosome Sorting

Chromosomes are analyzed and sorted with a FACS 440 cell sorter (Becton-Dickinson) equipped with a single argon ion laser. Analysis and sorting are generally carried out at a flow rate of 1000–3000/sec. For H33258, ultraviolet light (150 mW) is used and, after cutting the incident light with long-pass filters LP390L and LP400, the signals are AD-converted in 1024 channels and stored in a computer (model 310, Yokogawa-Hewlett-Packard). For PrI, 488-nm light (1 W) is used, and the incident light is cut by long-pass filters LP490 and LP520. For machine adjustment with autofluorescence of chicken red blood cells, the optical filter DM530/30 is used with 488-nm light. Tris-buffered saline containing EDTA (TES; 0.15 M NaCl, 10 mM EDTA, 10 mM Tris–HCl pH7.4) is used as a sheath fluid after autoclaving. The sheath fluid pressure is 15–17 psi.

The preparation and optimization of the sorter is very critical for maximum resolution. The critical points of these procedures will be summarized:

1. Adjust the seven parameters with chicken red blood cells (Sigma) suspended in phosphate-buffered saline to obtain a low coefficient of variation (CV) for the fluorescence (at most 3.0 for 488 nm and 4.0 for UV): these parameters include the positions (X, Y, and Z) and angles (Θ and Ψ) of the nozzle (ceramic nozzle with an orifice diameter of 50 μm), the focusing of laser light on the jet, and the focusing of emission light on the photomultiplier tube.
2. Apply chromosomes and allow to flow through the system at a rate of 1000–1500/sec for at least 10 min to stabilize the flow.
3. Perform fine adjustment of the foregoing parameters.
4. Make sure that the direction of the jet from the nozzle is absolutely perpendicular and that the laser light hits the jet at exactly right angles.
5. Run analysis and sorting at a flow rate of 1000–3000/sec.
6. Clean the flow system (sample delivery tube and connecting tubings) by flushing with 1% sodium dodecylsulfate (SDS) at least three times a week.
7. Replace the sample delivery filter every day and change the sheath fluid filter (0.22 μm) at least once a week.

For spot-blot hybridization, chromosomes are sorted onto nitrocellulose filters as described previously [1,8]. Collect 50,000 chromosomes of each type as a tiny spot directly onto nitrocellulose filter disks (25-mm diameter, Schleicher & Schuell). Take pictures of these filters under illumination with a hand-held UV light to mark the location and shape of collected chromosomal spots.

For giant DNA fragment analysis, chromosomes are sorted into a sterile silicone-coated transparent centrifuge tube† (1.5-mL; IEDA). Place a pair of tubes, filled with 0.5 mL TES, 5-mm apart in an ice bath. Set it under the sorting chamber at the position where the tubes collect the drops deflected in the two directions. Keep the distance between the deflection plate and tube at 4.5 cm. Place each collecting tube along the deflected stream by tilting approximately 6°. Try to collect the drops in the center of the liquid (TES) surface. Monitor carefully and correct the angle of the deflected stream to make sure the drops alight directly on the liquid surface. Once the liquid volume becomes 1.5 mL, remove the tubes and centrifuge at 3200x*g* for 15 min. Aspirate out 1 mL of supernatant and leave the chromosomes in 0.5 mL of TES. If necessary, continue sorting using the same tube without resuspending.

E. Spot-Blot Hybridization

Fifty thousand chromosomes each are collected on nitrocellulose membrane filters as small spots and hybridization is performed with a human gene probe of interest. Treat filters once with 0.5 M NaOH/1.5 M NaCl for 5 min and then twice with 0.5 M Tris–HCl (pH 7.5)/1.5 M NaCl, and bake at 80°C for 2 hr in a vacuum oven. Carry out prehybridization and hybridization as described previously [11]. After hybridization, wash the filters twice for 30 min with 0.5x SSC (75 mM NaCl/7.5 mM sodium citrate, pH 7.0) containing 0.1% SDS, at 55°C. If background is high, use 0.2x SSC or 0.1x SSC and raise temperature up to 70°C.

F. Pulsed-Field Gel Electrophoresis

Usually, 10 million chromosomes each are collected in a sterile silicone-coated centrifuge tube. Spin down the chromosomes by centrifugation at 3200x*g* for 15 min. Aspirate out the supernatant carefully down to 50 μL. Suspend chromosomes by gentle tapping. Add 600 μL of buffer L (0.1 M EDTA, 10 mM Tris–HCl pH 7.5, 20 mM NaCl) and centrifuge as before.

---

†Immerse the tubes and pipet tips with freshly prepared 2% dimethyldichlorosilane in acetone for a moment, wash with distilled water, and autoclave.

Aspirate out supernatant down to 25 μL, resuspend chromosomes and warm at 42°C for 30 sec. Add 25 μL of 1% low-melting-temperature agarose (SEAPLAQUE; FMC Corp.) in buffer L kept at 42°C. Mix well with gentle tapping. Transfer the mixture into a slit (2 $\times$ 5 $\times$ 10 mm) in a custom-made transparent insert mold, the bottom of which is sealed with adhesive tape. (The chromosome suspension should be pipetted as gently as possible to avoid breakage by shearing forces.) Wrap the insert mold with Saran wrap and keep in ice for 15 min. Take out the agarose block and soak in 300 μL buffer L containing 1 mg/mL proteinase K and 1% Sarkosyl. Incubate at 55°C for 24 hr. Wash the block extensively in TE (10 mM Tris–HCl pH 7.5/1 mM EDTA). Apply restriction enzymes, such as *Not*I, at desired amounts and incubate the agarose block for an appropriate time at 37°C. Run the orthogonal field-alteration gel electrophoresis (OFAGE) using a Pulsaphor system (LKB) as described [22], except that the electric field is altered every 55 sec or 120 sec. Perform Southern hybridization using an *Alu* repetitive sequence (BLUR8) as a probe as described [8].

ACKNOWLEDGMENTS

The authors thank the following people who participated in the project: J. Kudoh, R. Fukuyama, K. Kawasaki, M. Hirai, K. Sakai, and M. Maekawa. They also thank Dr. T. Honjo at Kyoto University for providing us with the cell line FLEB14-14, and Dr. D. Wallace at Emory University and Dr. K. Arai at the University of Tokyo for providing us with DNA probes.

REFERENCES

1. Lebo, R. V., D. R. Tolan, B. D. Bruce, M.-C. Cheung, and Y. W. Kan (1985). Spot-blot analysis of sorted chromosomes assigns a fructose intolerance disease locus to chromosome 9. *Cytometry 6*:478–483.
2. Green, D. K., J. A. Fantes, K. E. Buckton, J. K. Elder, P. Malloy, A. Carothers, and H. J. Evans (1984). Karyotyping and identification of human chromosome polymorphisms by single fluorochrome flow cytometry. *Hum. Genet. 66*:143–146.
3. Trask, B., G. van den Engh, B. Mayall, and J. W. Gray (1989). Chromosomes heteromorphism quantified by high-resolution bivariate flow karyotyping. *Am. J. Hum. Genet. 45*:739–752.
4. Li, K., C. K. Warner, J. A. Hodge, S. Minoshima, J. Kudoh, R. Fukuyama, M. Maekawa, Y. Shimizu, N. Shimizu, and D. C. Wallace (1989). A human muscle adenine nucleotide translocator gene has four exons, is located on chromosome 4 and is differentially expressed. *J. Biol. Chem. 264*:13998–14004.
5. Neckelmann, N., C. K. Warner, A. Chung, J. Kudoh, S. Minoshima, R. Fukuyama, M. Maekawa, Y. Shimizu, N. Shimizu, J. D. Liu, and D. C. Wallace (1989). The human ATP synthase β subunit gene: Sequence analysis, chromosome assignment, and differential expression. *Genomics 5*:829–843.

6. Van Dilla, M. A., L. L. Deaven, K. L. Albright, N. A. Allen, M. R. Aubuchon, M. F. Bartholdi, N. C. Brown, E. W. Campbell, A. V. Carrano, L. M. Clark, L. S. Cram, B. D. Crawford, J. C. Fuscoe, J. W. Gray, C. E. Hildebrand, P. J. Jackson, J. H. Jett, J. L. Longmire, C. R. Lozes, M. L. Luedemann, J. C. Martin, J. S. McNinch, L. J. Meincke, M. L. Mendelsohn, J. Meyne, R. K. Moyzis, A. C. Munk, J. Perlman, D. C. Peters, A. J. Silva, and B. J. Trask (1986). Human chromosome-specific DNA libraries: Construction and availability. *Biotechnology 4*:537–552.
7. McKusick, V. A., and F. H. Ruddle (1987). Toward a complete map of the human genome. [Editorial]. *Genomics 1*:103–106.
8. Minoshima, S., K. Kawasaki, R. Fukuyama, M. Maekawa, J. Kudoh, and N. Shimizu (1990). Isolation of giant DNA fragments from flow-sorted human chromosomes. *Cytometry 11:* 539–546.
9. Sakai, K., M. Hirai, S. Minoshima, J. Kudoh, R. Fukuyama, and N. Shimizu (1989). Isolation of cDNAs encoding a substrate for protein kinase C: Nucleotide sequence and chromosomal mapping of the gene for a human 80K protein. *Genomics 5*:309–315.
10. Arai, N., D. Nomura, D. Villaret, R. DeWaal Malefijt, M. Seiki, M. Yoshida, S. Minoshima, R. Fukuyama, M. Maekawa, J. Kudoh, N. Shimizu, K. Yokota, E. Abe, T. Yokota, Y. Takebe, and K. Arai (1989). Complete nucleotide sequence of the chromosomal gene for human IL-4 and its expression. *J. Immunol. 142*:274–282.
11. Kudoh, J., S. Minoshima, K. Hashinaka, C. Nishio, M. Yamada, Y. Shimizu, and N. Shimizu (1988). Assignment of the myeloperoxidase gene *MPO* to human chromosome 17 using somatic cell hybrids and flow-sorted chromosomes. *Jpn. J. Hum. Genet. 33*:315–324.
12. Taira, M., J. Kudoh, S. Minoshima, T. Iizasa, H. Shimada, Y. Shimizu, M. Tatibana, and N. Shimizu (1989). Localization of human phosphoribosylpyrophosphate synthetase subunit I and II genes (*PRPS1* and *PRPS2*) to different regions of the X chromosome and assignment of two *PRPS1*-related genes to autosomes. *Somat. Cell Mol. Genet. 15*:29–37.
13. Ohno, S., S. Minoshima, J. Kudoh, R. Fukuyama, Y. Shimizu, S. Ohmi-Imajoh, N. Shimizu, and K. Suzuki (1990). Four genes for the calpain family locate on four distinct human chromosomes. *Cytogenet. Cell Genet.* (in press).
14. Inazawa, J., H. Nakagawa, S. Misawa, T. Abe, S. Minoshima, R. Fukuyama, M. Maki, T. Murachi, M. Hatanaka, and N. Shimizu (1990). Assignment of the human calpastatin gene (*CAST*) to chromosome 5 at region q14-q22. *Cytogenet. Cell Genet.* (in press).
15. Rommens, J. M., M. C. Iannuzzi, B.-S. Kerem, M. L. Drumm, G. Melmer, M. Dean, R. Rozmahel, J. L. Cole, D. Kennedy, N. Hidata, M. Zsiga, M. Buchwald, J. R. Riordan, L.-C. Tsui, and F. S. Collins (1989). Identification of the cystic fibrosis gene: Chromosome walking and jumping. *Science 245*:1059–1065.
16. Sternberg, N. (1990). Bacteriophage P1 cloning system for the isolation, ampli-

fication, and recovery of DNA fragments as large as 100 kilobase pairs. *Proc. Natl. Acad. Sci. USA 87*:103–107.

17. Poustka, A., and H. Lehrach (1986). Jumping libraries and linking libraries: The next generation of molecular tools in mammalian genes. *Trends Genet. 2*: 174–179.
18. Carrans, A. V., J. Lamerdin, L. K. Ashworth, B. Watkins, E. Branscomb, T. Slezak, M. Raff, P. T. de Jong, D. Keith, L. McBride, S. Meister, and M. K. Ronick (1989). A high resolution, fluorescence-based, semiautomated method for DNA fingerprinting. *Genomics 4*:129–136.
19. McCormick, M. K., J. H. Shero, M. C. Cheung, Y. W. Kan, P. A. Hieter, and S. E. Antonarakis (1989). Construction of human chromosome 21-specific yeast artificial chromosomes. *Proc. Natl. Acad. Sci. USA 87*:103–107.
20. Sillar, R., and B. D. Young (1981). A new method for the preparation of metaphase chromosomes for flow analysis. *J. Histochem. Cytochem. 29*:74–78.
21. Crissman, H. A., and J. A. Steinkamp (1973). Rapid, simultaneous measurement of DNA, protein, and cell volume in single cells from large mammalian cell populations. *J. Cell Biol. 59*:766–771.
22. Kawasaki, K., J. Kudoh, K. Omoto, and N. Shimizu (1988). Mega base map of the epidermal growth factor (EGF) receptor gene flanking regions and structure of the amplification units in EGF receptor-hyperproducing squamous carcinoma cells. *Jpn. J. Cancer Res. 79*:1174–1183.

7

Pulsed-Field Gel Electrophoresis and Its Application to Human and Mouse Sex Determination

Vera Verga and Robert P. Erickson*

University of Michigan School of Medicine, Ann Arbor, Michigan

I. INTRODUCTION TO PULSED-FIELD GEL ELECTROPHORESIS

A. Agarose Gel Electrophoresis

In the 1930s, Tiselius [1] separated human serum into several groups of its constituent proteins by passing an electric current through a solution. Today gel electrophoresis is one of the three main techniques for separation in biology, together with ultracentrifugation and chromatography. Agarose gel electrophoresis of nucleic acids is one of the most commonly used techniques in molecular biology. The negatively charged DNA molecule migrates through a solid support, such as an agarose gel, toward the positive terminal when an electric field is passed across it. The speed at which the DNA moves depends on a number of parameters such as the relative molecular mass (M_r) of the DNA, the percentage and thickness of the agarose gel, and the strength of the electric field. When ethidium bromide (EtBr) intercalates the DNA, these molecules can be visualized in the gel with ultraviolet (UV) light owing to the fluorescent properties of the DNA–EtBr complex.

The ability to directly study DNA has had a major impact in the field of molecular biology especially since the development of the Southern transfer technique by Southern [2]. This procedure allows DNA to be directly transferred from the gel matrix to a more permanent support, namely nitrocellulose or nylon membranes. By using a combination of restriction enzymes, which cleave DNA at specific sites on the DNA double helix, and agarose gel

---

**Current affiliation:* University of Arizona Health Science Center, Tucson, Arizona.

electrophoresis, large DNA can be separated into a unique set of smaller fragments for study. Conventional gel electrophoresis has been successfully used to study DNA molecules up to about 50 kilobases (kb), although resolution above 20 kb is hard to obtain, as DNA is randomly sheared during the usual extraction procedures.

The size of the haploid human genome is 3×10^9 base pairs (bp), (i.e., an average of 1.3×10^8 per chromosome) [3]. It is postulated that the DNA is arranged in a linear fashion along these chromosomes. Many human geneticists have contributed to the construction of linkage maps of each chromosome [4]. From recombinational studies, it is estimated that the human genome contains 3300 centimorgans (cM) of genetic distance. Thus, 1 cM is approximately equivalent to 1000 kb or 1 million bp. This relationship is not linear and is affected by sex, the position of the DNA relative to the centromere, the presence of repetitive sequences, and other unknown factors. Cytogenetic observations of whole chromosomes allowed the analysis of translocations and deletions that were large enough to be visible as alterations of banding patterns of the chromosomes involved. The resolution limit of banding is the size of a band that is estimated to contain about 2 million bp (2000 kb or 2 megabases (Mb)). In situ hybridization can resolve distances to one band. It can be seen that DNA technology has been limited by a gap in resolution between cytological and standard DNA fractionation techniques.

B. Pulsed-Field Gel Electrophoresis

A new type of DNA electrophoresis, pulsed-field gel electrophoresis (PFGE), was developed by Schwartz and co-workers [5,6]. They described a technique whereby they successfully separated the chromosomes of the yeast, *Saccharomyces cerevisiae*, producing 12 distinct chromosomal bands in a size range of up to 2000 kb. They varied the direction of the electric field during electrophoresis. After Southern hybridizations, the different bands were verified as distinct chromosomes by their hybridization to previously assigned genes [5].

Under standard conditions, DNA molecules that are larger than 20 kb cannot be separated according to size in agarose because they are too large to be sieved by the agarose pores. However, these larger DNA molecules can move in agarose because of their ability to alter their shape when influenced by the electric field. They migrate in an end-on fashion that has been called *reptation* [7]. Once this occurs, the molecules are no longer separated according to size, and all move with the same speed through the gel. By altering the field around the DNA molecule, the molecules are continuously forced to change the direction in which they are moving and, hence, change their configuration. The rate at which a DNA molecule can change its configuration is dependent on its molecular mass; consequently, PFGE allows size separation in a higher M_r range. Recently, Schwartz and Koval [8] designed an experiment that has allowed the observation of the movement of individual DNA molecules when an electric field is applied. The DNA was labeled with a fluorescent dye and a gel apparatus was set up, such that the DNA motion could be observed by a microscope. The DNA moves in cycles of elongation and contraction under the influence of the electric field. These results were confirmed by Smith et al. [9].

C. Principles of Pulsed-Field Gel Electrophoresis

Since the discovery that large DNA molecules can be separated by alternating the electric fields, several different separation methods have been developed and are summarized in Table 1.

The PFGE that was developed by Schwartz and Cantor, and modified by Carle and Olson, has become known as OFAGE (orthogonal field alteration gel electrophoresis) [5,6,10,11]. This acronym is used for all methods of separation in which there are two approximately orthogonal electric fields. Many different sophisticated gel boxes have been designed and are commercially available. In OFAGE the point electrode configuration is known as double inhomogeneous. This inhomogeneous electric field causes the DNA to migrate with a mobility and trajectory that depend on its loading position on the gel. After hybridization, the resulting bands in the different lanes cannot be accurately compared, as they are not in a straight line (Fig. 1).

Field-inversion gel electrophoresis (FIGE) was developed by Carle et al. [12]. They discovered that simply by periodically reversing the direction of the electric field 180°, larger DNA molecules could be separated in a size-dependent manner. The "net" migration of the DNA is in the forward direction because the greater portion of time is with a field orientated for the forward direction. This method was simpler in terms of electrical geometry, and could separate DNA effectively in the size range of 20–700 kb.

Finally, a different electric field manipulation scheme was developed by Chu et al. [13] and has become known as CHEF (contour-clamped homogenous electric field). This system has multiple electrodes arranged on a closed contour and electrically clamped to predetermined potentials. The DNA alternates in direction at a fixed angle of 120°. DNA larger than 5 Mb has been successfully separated with this system [14].

II. METHODOLOGY

To perform PFGE, it is extremely important to first isolate the DNA to be examined in as intact a form as possible. The DNA needs to be protected against shear due to motion when it is in a liquid. Schwartz and associates [5] developed a unique isolation method for DNA in agarose. The DNA is

Table 1 A Comparison of the Different Types of PFGE

Name	Acronym	Methods	Ref.
Orthogonal-field altered gel electrophoresis	OFAGE	Point electrode configuration; double inhomogeneous field; nonuniform electric field; migration dependent on position of DNA in the gel.	5,6,11,12
Contour-clamped homogeneous electric field	CHEF	Two alternating fields in a fixed reorientation angle of 120°	13–14
Field-inversion gel electrophoresis	FIGE	Electric field in forward and reverse direction in a 3 : 1 ratio; good separation up to 1 Mb	12
Other Systems			
Programmable autonomously controlled electrode gel electrophoresis	PACE	24 electrodes arranged in closed hexagonal contour; homogeneous electric fields; good separation 270 kb to 5 Mb	15
Pulsed-homogeneous orthogonal-field gel electrophoresis	PHOGE	Homogeneous, orthogonal pulsed-fields; reorientation angle at 90°; straight lanes	16
Gel Rotation System			
Rotary pulsed-field gel electrophoresis	RPFGE	Gel on a platform that rotates 60° to the left and 60° to the right, forming an angle of 120°	17–19

protected from shear by embedding it in a solid support, yet it is accessible by enzymes for further analysis. The DNA is extracted from cells after they have been formed into a block of agarose that is the correct size for a gel slot (gel inserts). This method of DNA extraction avoids any harsh organic chemicals, such as phenol and chloroform, and harsh forces caused by centifugation or precipitation, which are standard procedures for conventional DNA analysis and which all cause a great amount of DNA shearing. DNA, in these gel inserts, is very stable in the presence of 0.5 M EDTA. It appears that it is important not to use rapidly dividing cells for DNA preparation, as the replication forks might lead to extra bands [20]. It is also suggested that one not use overconfluent cells; they may have begun the production of enzymes that can degrade DNA. Finally, this method of DNA

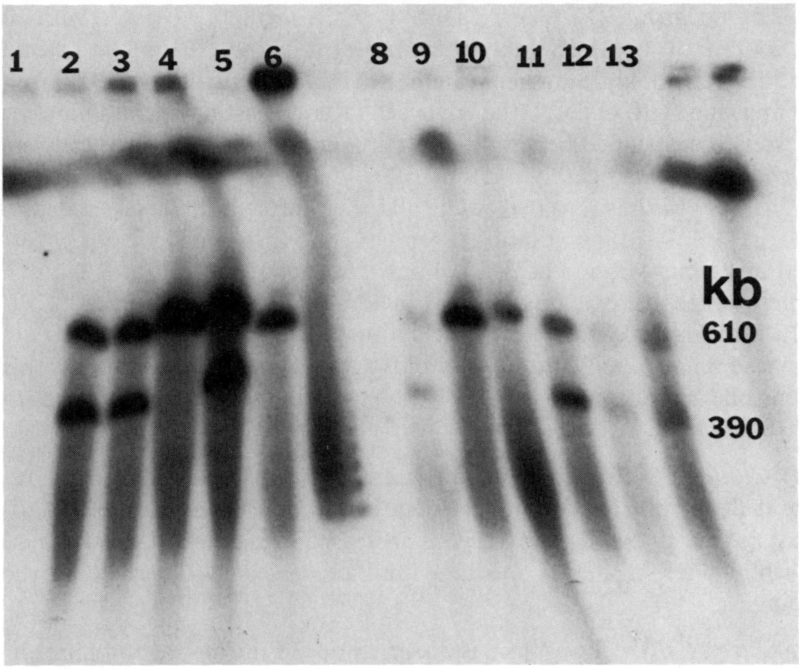

Figure 1 A Southern hybridization of DNA probe 47Z (DXYS5) to an OFAGE pulsed-field gel: The bands in the outer lanes (2, 3, 12, and 13) can be seen to be farther from the origin than those in lanes 5, 6, and 8. The 610-kb fragment is the X-specific copy, whereas the 390-kb fragment is Y-specific one. The DNA samples are from normal females (lanes 3, 5, 9, and 10), normal males (lanes 8 and 11), XY females (lanes 1, 2, 12, and 13) and an XX male (lane 4). The gel was prepared using an LKB Pulsaphor with a switch time of 60 sec, at 350 V for 18 hr. Weak bands corresponding to the λ-concatemer can be seen in lane 6.

preparation is relatively easy, and the resulting DNA is so stable that the inserts can even be mailed at room temperature.

A. Preparation of DNA in Agarose Blocks

As the DNA must be prepared from fresh cells, several different cell types can be used. Only the initial preparation of these cells differs. We have successfully prepared large-molecular-mass, unsheared, intact DNA from whole blood, cultured fibroblasts and lymphoblasts, and mouse spleen cells.

Noncoagulated Human Blood. Twenty milliliters of human blood, isolated using heparin, EDTA, or ACD as the anticoagulant, is mixed with an equal volume of 1 × PBS (phosphate-buffered saline); 20 mL is layered onto a Ficol/Histopaque gradient (Sigma). After centrifugation at 2000 rpm for 15 min, the "buffy coat," (the white cell layer consisting of mononuclear leukocytes) is suspended between the plasma and Ficol. The white cells are removed, resuspended gently in 50 mL of 1 × PBS, and repelleted at 2000 rpm. The PBS wash is repeated twice, and then the cells are resuspended in 10 mL of PBS. After an additional centrifugation, the cells are again suspended at a final volume of 3 × 10^7/mL. An equal volume of 2% LMT (low-melt temperature agarose; FMC) in 125 mM EDTA (kept at 50°C) is mixed with the cells, which were quickly prewarmed to 50°C. The cells are then placed immediately into block molds that form 100 μL blocks. At this dilution, one block contains about 1.5 × 10^6 cells, which will be equal to about 10 μg of DNA.

Freshly Cultured Cells. (1) Lymphoblasts: Cells are harvested by centrifugation at 2000 rpm for 5 min and then rinsed in PBS as before. (2) Fibroblasts: After trypsinization the cells are rinsed in Hanks buffer twice and then kept on ice in culture medium until processed as described in the foregoing.

Mouse Spleen Cells. The spleen is dissected out of the mouse and immediately placed into PBS on ice. The spleen is very gently homogenized with a glass-on-glass homogenizer by hand until it is broken up. The connective tissue, which does not break down, is removed, and the cells are mixed with PBS and then treated as described in the foregoing.

After the cells are poured into block molds, the molds are placed briefly into a freezer (−20°C) until the agarose has hardened, and then they are pushed out, using a Pasteur pipette (sealed end), into a solution of 1% NDS (sodium *N*-lauroyl sarcosine, 1 mg/mL proteinase K, 0.5 M EDTA), and incubated at 50°C for 2 days. The tubes are gently rocked, and the lysis buffer is changed at least once. After this, the blocks containing DNA can be stored indefinitely in the NDS solution at 4°C.

The blocks can be removed from the solution with Pasteur pipettes that have been bent and sealed at the tips. About ten blocks are prepared for use at a time. They are placed into 1 × TE, pH 8.0, and rinsed by rocking gently for 2 hrs. The first rinse contains 1 mM phenylmethyl sulfonyl fluoride (PMSF), which inactivates any of the proteinase that is still active. (Fresh PMSF is prepared in isopropanol when needed). The TE is replaced twice at 2-hr intervals, and the blocks are then placed into a fresh solution of TE and stored at 4°C.

The DNA can be examined for size and purity. One block is removed and cut into three parts with a sterile scalpel. One-third is treated as for a normal digestion (see later), the second part is incubated with all the components of the restriction reaction, except the enzyme, and the final section is placed in a tube containing only the TE buffer that the block was stored in. They are incubated at 37°C (or the recommended temperature for the enzyme), as for a normal digestion, and then run on a PFGE apparatus. No DNA or only very small amounts, should move out of the lane in the latter two conditions, but if a substantial amount does enter the gel, the DNA might be contaminated with microorganisms producing DNases or residual cellular enzymes and, thus, not be useful for PFGE.

The preparation of DNA in agarose has been a successful method of DNA isolation for experiments other than PFGE. Large M_r DNA in agarose can be used for conventional gel electrophoresis, and the preparation method gives a very high yield of DNA because there are no steps during which the DNA can be lost. Finally, the purity of the agarose is crucial. We have used FMC (LMT) successfully, but batch variation can occur and cause degradation; consequently, it is important to first check the agarose.

B. Restriction Endonuclease Digestions

There are a growing number of restriction enzymes that fall into a class known as *rare-cutters* (i.e., they cut the DNA infrequently and usually generate DNA fragments larger than 100 kb). (Table 2) Many of these enzymes contain CpG dinucleotides in their recognition sequences. There appears to be an underrepresentation of these sequences in mammalian genomes because the cytosine residue is methylated and readily mutates to form thymine [21]. Furthermore, it appears that there is a clustering of CpG sequences at the 5′ regions of genes, and these islands have been called *Hpa*II tiny fragments or HTF islands [22]. After PFGE technology was developed, there was a need for enzymes to study the genomes of higher organisms the chromosomes of which are too large to be resolved by the current technology. Thus, restriction enzymes that have six-base recognition sequences, containing CpG sequences that are rare in the human genome, were found to be useful for PFGE. In addition, the rare-cutters, which had eight bases in their recognition sites, were *Sfi*I and *Not*I (New England Biolabs) [23]. Partial digestions with these rare-cutters are extremely useful to map even larger regions. Also, many of these enzymes are sensitive to methylation and, hence, a site will often not be cut and thereby allow even larger regions to be studied.

A typical restriction reaction is performed in a standard polypropylene Eppendorf tube, in a reaction volume of 300 μL containing 0.1 volume of 10X reaction buffer (recommended by the manufacturer), 100 μg/mL of BSA, 6 mM spermidine, and 10–30 units of restriction enzyme. The blocks

Table 2 Restriction Enzymes Useful for PFGE

Enzyme	Recognition sequence	Fragment size[a]
*BssH*II	GCGCGC	1
*Cla*I	ATCGAT	1
*Eag*I	CGGCCG	1
*Mlu*I	ACGCGT	2
*Nae*I	GCCGGC	1
*Nar*I	GGCGCC	1
*Not*I	GCGGCCGC	2
*Nru*I	TCGCGA	2
*Pvu*I	CGATCG	1
*Sac*II	CCGCGG	1
*Sal*I	GTCGAC	1
*Sfi*I	GGCCNNNNNGGCC	3
*Sma*I	CCCGGG	1
*Xho*I	CTCGAC	1

[a]1, Some fragments below 50 kb; 2, Few fragments below 50 kb; 3, Some fragments in 10- to 15-kb range.

are removed using a sterile (alcohol flamed) bent Pasteur pipette, and cut in half (5 μg) with a sterile blade. Reactions are incubated for only 2 hrs and then placed at 4°C for a couple of minutes. If double digestions are carried out, the enzyme requiring the lowest salt concentration is used first; the block is then removed and incubated with the second enzyme. If the reactions are not going to be electrophoresed immediately they can be stored at 4°C with 1 mL of 0.5 M EDTA. Some researchers like to first remove the restriction enzyme by a proteinase digestion before electrophoresis [24]. We do not find this step necessary.

C. Electrophoresis and Running Conditions

There are now many different types of electrophoresis apparatuses (see Table 1). We have performed all of our pulsed-field experiments using an LKB Pulsaphor (Pharmacia). The original apparatus consisted of a double inhomogeneous field with electrodes placed around the sides. The anodes were placed on the top and left sides at position 90, whereas the cathodes were placed on the right and bottom sides at positions 5, 95, and 180. The DNA is loaded into the wells at the origin, which is placed in a bottom-right line in the running tank. We have now switched to the CHEF system or hexagonal array which fits into the LKB system. The switch time is crucial for separation of different size ranges of DNA. For a range up to 1000 kb we use a 75-sec switch at 150 V for 42 hr (Figure 2). The blocks are loaded into the gel, after removing them from the Eppendorf tubes with a curved sterile Pasteur pipette. Generally, for larger DNA, longer switch times are needed to separate the DNA molecules. There are many publications concerning the optimal conditions for DNA separations [25,26]. The important parameters to consider for each run are the concentration of agarose, the temperature surrounding the gel during the run, the electric field and current, and, finally, the switch times. To maintain a constant temperature, the buffer is

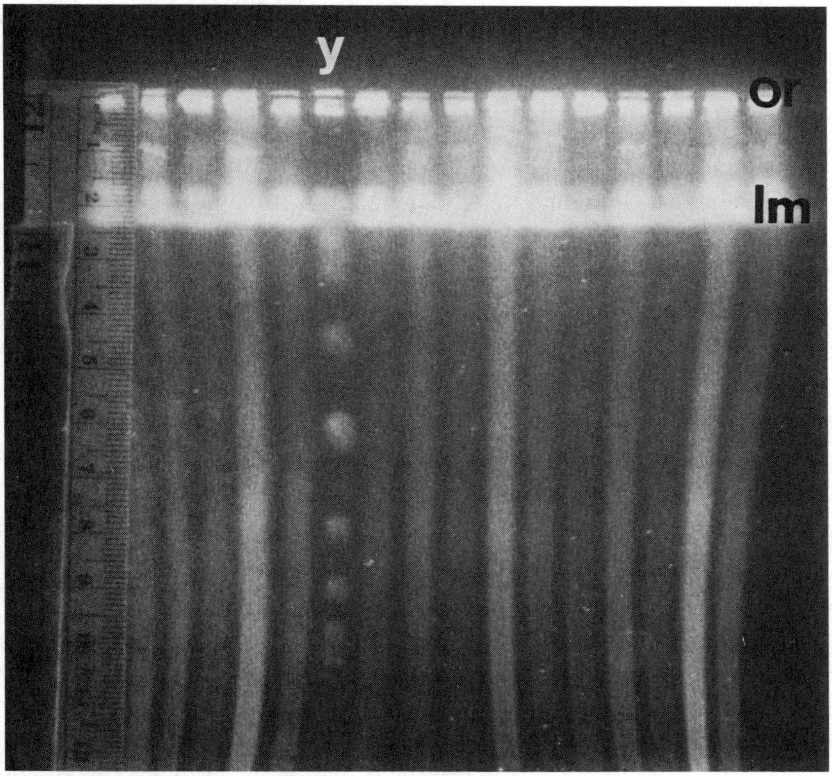

Figure 2 A CHEF-type pulsed-field gel of DNA digested with rare restriction enzymes and run at 170 V for 42 hr, with a switch time of 75 sec. The range of yeast chromosomes is 260–900 kb. The yeast marker is in lane Y. The origin (or) and limited mobility region (lm) are also indicated. The bright-staining region in the origin indicates that a large amount of DNA was still uncut or still too large to move into the gel with the conditions used.

circulated around the gel during a run. Computer programs have been developed that optimize electrophoresis conditions [27–29], and a new switching box has been marketed by LKB (Pharmacia), which permits ramping conditions during the runs (i.e., the switch times can be incremented and a more effective separation can be obtained).

D. Southern Transfers

After electrophoresis, the gel is first stained with EtBr. If EtBr is added before electrophoresis, degradation can occur that may prevent a positive signal after hybridization. The gel is photographed by conventional methods. The large molecules do not transfer efficiently unless they are first nicked into smaller fragments. We routinely nick the DNA by exposure to a 240-nm UV light source for 90 sec. Other laboratories have had success using 0.25 M HCl [30]. This method did not work in our hands, and the DNA appeared to be sheared into fragments so small that they moved through the membrane, and the filters were not useful for hybridizations.

The gel is then placed into denaturing solution (0.5 M NaOH; 1.5 M NaCl) and rinsed in the solution. An alkaline Southern transfer [2,31] is set up with denaturing solution as the transfer solution to the membrane Genescreen (NEN). The transfer is allowed to proceed overnight, and the membrane is neutralized in 1 M ammonium acetate (pH 7.5), 0.02 N NaOH for 3–5 min. The filter is air-dried, then baked at 80°C for 20 min; the DNA is then cross-linked to the filter with a 330-nm UV light source for 90 sec.

Several different types of membranes are used for PFGE; we report on Genescreen (NEN) because it has consistently given good results in our laboratory.

E. DNA Hybridizations

Prehybridizations have been performed with two types of hybridization buffers; both have given us good hybridization signals. The first solution used is Church and Gilbert's [32] hybridization solution, which consists of 7% SDS, 0.5 M sodium phosphate, pH 7.2, and 1 mM EDTA, at 65°C. The second hybridization solution is the one recommended by Collaborative Research, which contains 1 M NaCl, 30% formamide, 100 μg/mL of salmon sperm DNA, 0.1M Tris-HCl, pH7.5, 500 μg/mL heparin, 0.1% sodium phosphate, 0.05% N-Lauroylsarcosine at 50°C.

Probes are labeled with phosphorus-32 by the random priming method in low-melting-temperature agarose to a specific activity of 3×10^7 cpm (μg.mL) [33]. The unincorporated nucleotides are removed using spin columns equilibrated with STE buffer [33].

Posthybridization washes are at room temperature for 15–30 mins in 0.04 M sodium phosphate pH 7.2, 0.5% SDS, and 1 mM EDTA, and then at 65°C for 30 min with a change of wash after 15 min. Filters are exposed for 1–7 days at −70°C with two intensifying screens. To reuse filters, they are stripped with 1 × TE pH8, 1% SDS heated to 85°C, then washed at 65°C for 30 min [34].

F. Preparation of Size Markers

1. λ DNA Ladders

To determine the sizes of the fragments detected by PFGE, two types of markers are used. First, as the λ genome has 12 complementary single base pairs at each of its linear ends, a solution of any strain of λ DNA molecules will form concatemers, seen as a ladder in the gel, which increases linearly in size according to the size of the phage genome. Strain cI857 forms ladders that are useful size markers that increase in 48.5-kb jumps and are valuable for sizing DNA under 1000 kb (Fig. 3). Often λ-concatemers are not visible in the gel, but after hybridization to a small amount of labeled λ DNA, bands can be detected in the upper region of the gel.

Preparation of λ-Concatemers. Host LE392 bacteria is prepared by growing an overnight culture in NZCYM containing 0.4% maltose. Inoculate 1 million plaque forming units (PFU) of bacteriophage with 300 μL of the overnight culture of LE392. Place at 37°C for 15 min. Add to 1 L of NZCYM (no maltose) and incubate for 8–10 hr until confluent lysis occurs. Lysis is completed by the addition of 20–30 mL of chloroform. Add 29 g of NaCl (0.5 M) and shake sample for a further 10–15 min. The bacterial debris is removed by a 5000 rpm centrifugation. To precipitate the phage particles, 120 g of polyethylene glycol (PEG 6000) is added to the supernatant, and it is incubated on ice for at least 1 hr. The phage are pelleted by a 15,000 rpm spin for 20 min. The pellet is resuspended in 10 mL of SM buffer (5 mL of 2% gelatin, 5.8 g NaCl, 2 g $MgSO_4 \cdot 7H_2O$, and 50 mL of 1 M Tris–HCl pH 7.5 per liter), and extracted with an equal volume of chloroform. The aqueous phase is then centrifuged for 2 hr at 19,000 rpm, and the phage pellet is resuspended in 2 mL of SM buffer. A $CsCl_2$ step-gradient is prepared, and the phage are layered on top. They are centrifuged at 35,000 rpm for 2 hr. The phage band at a density of 1.45 g/mL is removed with a needle and dialyzed against TE buffer overnight to remove the $CsCl_2$. The phage particles, in approximately 2 mL, are mixed with an equal volume of LMT agarose, as prepared for normal block preparation, and 100-μL blocks are formed. The phage are allowed to anneal in the lysis buffer, at 37°C, for at least 2 days. One block is cut up into different sizes and run on a gel to

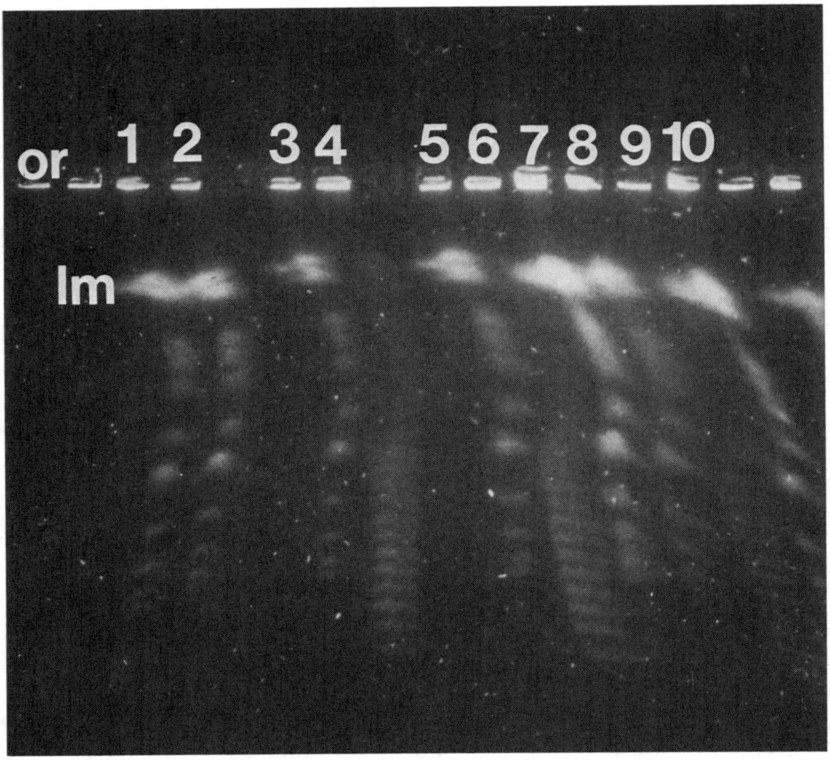

Figure 3 An OFAGE pulsed-field gel of bacteriophage λ (cI857) ladders (lanes 1, 4, 6, and 9) and yeast chromosomes (*S. cerevisiae*) in lanes 2, 3, 5, 7, 8, and 10. The chromosomes migrate at different rates, depending on their position in the gel. The outer lanes migrate farther from the origin than the center lanes. In lane 6, at least 14 bands of the λ-ladder can be detected and 12 yeast chromosomes are visible in lane 5. This gel was prepared using an LKB Pulsaphor (Pharmacia), with the anodes placed at position 95, and the cathodes at positions 5, 95, and 180. The gel was run at 350 V for 18 hr: lm, limited mobility region; or, origin.

determine the optimum amount to use (see Fig. 2). These λ-concatemers are stored at 4°C. It has also been reported that commercially available λ can be used to form concatemers for PFGE [35].

2. Yeast Chromosome Markers

Yeast chromosomes can be individually separated and provide markers of known size at the position of each chromosome from 240 kb to over 2 Mb.

The most commonly used strains are *Saccharomyces cerevisiae* and *Schizosaccharomyces pombe* (Table 3). *Saccharomyces cerevisiae* has 16 chromosomes. The size of each chromosome has been determined using λ-concatemers. Earlier estimates of the size of the largest yeast chromosome was 1.3 Mb when compared with the λ-concatemers, but recent results indicate that the largest chromosome exceeds 2 Mb [5,11]. A strain of fission yeast, *Schizosaccharomyces pombe*, was found to be a convenient source of DNA and contained three nuclear chromosomes and minichromosomes. The sizes of these chromosomes are 3, 6, and 9 Mb, and the minichromosomes are in the size range of 300–500 kb [36]. By altering extraction procedures, decreasing electric field conditions, and increasing run times the largest *S. pombe* chromosomes were effectively separated. Hence, separation of intact human chromosomes, the smallest of which is estimated to be five times larger than 9 Mb, might be possible in the future [36].

Table 3 Markers Useful for PFGE[a]

Yeast[b] (*S. cerevisiae*)	PHAGE (cI857)
2x 1290	970
2x 1080	921
980	873
940	824
830	776
2x 780	727
660	679
2x 570	630
400	582
320	533
260	485
220	436
	388
	340
	291
	242
	194
	145
	97
	48.5

[a]Sizes in kilobases (kb).
[b]Data from Refs. 6 and 10.

Preparation of Yeast Chromosome Markers. A fresh plate of yeast is prepared using YPD (YPD consists of 1 g yeast extract, 1 g dextrose, and 2 g bactopeptone in 100 mL distilled water) containing 15% agar. The yeast are streaked out on the plate and incubated at 30°C. A single yeast colony is inoculated into 100 mL of YPD and incubated overnight, with shaking, at 30°C. The OD at 600 nm is determined (an absorbance at 600 nm of 1.5–1.6 is approximately equal to 80 million cells/mL).

The cells are centrifuged at 2000 rpm and pelleted, then resuspended in about 3 mL of 0.05 M EDTA (pH 7.5) and washed twice in this solution. The cells are finally resuspended in a final volume, which contains about 4×10^{10} cells/mL. The cells are then mixed with an equal volume of 2% LMT agarose in 125 mM EDTA, and 100 μL blocks are made, using gel molds. The blocks are incubated with Zymolase 5000 (Kirin Breweries), which is prepared at 2 mg/mL in 10 mM sodium phosphate buffer (pH 5.6), 50% glycerol, and incubated at 37°C overnight. The blocks are incubated in LET buffer (0.5 M EDTA, 0.01 M Tris, pH7.5) containing 7.5% 2-mercaptoethanol and incubated overnight at 37°C. This step removes the outer cell membrane and produces spheroplasts. The gel inserts are then incubated overnight in NDS solution (0.5 M EDTA, pH 9.5, 1% *N*-lauryl sarcosyl, and 1 mg/mL proteinase K). These inserts can then be stored indefinitely at 4°C without much degradation. The 100 μL blocks can be cut into ten pieces for size markers (modified method from Schwartz and Cantor [6]). Recently, a rapid method for preparation has been published that omits the zymolase step [37].

III. APPLICATIONS OF PULSED-FIELD GEL ELECTROPHORESIS

A. Cytogenetic Study of Lower Organisms

The size of the chromosomes of unicellular organisms, such as yeast and protozoa, fall into the range of size separation by PFGE; hence, the chromosomes can be directly visualized in agarose, transferred by Southern blotting, and genes can be assigned to chromosomes by hybridization procedures that are far easier than in situ hybridizations. Similarly, any chromosomal rearrangements, such as translocations or deletions, would give rise to altered patterns of chromosomal mobility, and these could be identified immediately in the gel. Many genes have been mapped in yeast [5,6,11,12]. PFGE has been used for the mapping of the variable surface glycoproteins (VSG) genes in trypanosomes [38,39]. In bacteria, such as *Escherichia coli* and *Haemophilus influenzae*, the genome is too large to be visualized directly, but after digestion with rare-cutting restriction enzymes, a number of distinct bands can be directly visualized in the agarose gel, and these have been useful for gene mapping in these organisms [40–42].

B. Linkage Analysis and Gene Mapping

The human genetic linkage map relies on the presence of polymorphic markers and family studies to determine the recombination fraction and log of the odds (LOD) scores between markers at genetic loci. PFGE makes possible the construction of a physical map between loci, which then can be compared with the genetic linkage map. A number of long-range restriction maps are already available, especially in regions of interest around loci for genetic diseases. The cystic fibrosis (CF) gene was localized to the long arm of chromosome 7 by linkage to the *met* oncogene and an arbitrary DNA sequence D7S8 [43,44]. With these flanking markers as a start, there are several groups involved in "jumping" toward the locus of the CF gene [45–47]. It has now been possible to physically map the region surrounding this gene [45]. The largest long-range restriction map reported is 12 Mb surrounding the CF locus on chromosome 7 [47]. Other regions that have been characterized include the 10 Mb surrounding the *DMD* locus [48,49] and a 3-Mb region surrounding the *HLA* locus [50,51]. The long-range restriction map of the human pseudoautosomal and adjacent sex-determining region on the Y chromosome has been determined [30,52].

Reverse genetics is the term used to identify genes from their location in the DNA, rather than from their phenotype. This technique depends on mapping genes by using a series of cloned sequences that are genetically linked to a gene of interest. *Chromosomal walking* to a gene involves a series of steps: cloning sequences, determining their orientation and their position in the walk, and then using the new clone to initiate the process again. Each step usually involves about 20 kb of DNA. Problems occur when repetitive stretches are present, because these sequences can evade cloning. From linkage studies, the regions to be cloned are often 100–2000 kb from the linked marker, and they are too large to be approached by conventional mapping techniques. A new type of cloning known as *chromosome jumping or hopping* has been developed by constructing jumping libraries. These libraries are prepared by digestion of large intact DNA with a rare restriction enzyme in the agarose blocks. The large fragments are isolated from the agarose and then circularized with a selectable marker. A second, more common, restriction enzyme is then used to cut the circular DNA molecules, and these fragments are then cloned into a suitable vector. Any recombinants that are positive for the marker will contain DNA from two regions of the genome separated out by at least the distance between the rare restriction enzyme sites. These libraries can be used to find DNA markers that are separated by hundreds of kilobases [46,53–55].

Because any chromosomal translocation will almost certainly cause a rearrangement of the rare restriction enzyme sites around the breakpoints, mapping of the chromosomal breakpoints has become an important use of

PFGE, since large regions can be analyzed. The PF fragments containing the breakpoints will be increased or decreased in length, depending on whether the translocation provided rare restriction sites closer or farther away. Patients who have genetic diseases and balanced chromosomal translocations provide good material for mapping studies and isolation of the gene involved. This has been well exemplified by the study of patients with neurofibromatosis [56].

After an anomalous band has been identified by PFGE, it is possible to isolate DNA from the gel, some of which will represent the specific fragment, and then, by further analysis, determine whether or not it represents an alteration near or within the gene of interest. In a family study, one person was a heterozygote for the sickle cell globin gene, β^S, but expressed only β^A. An anomalous PF fragment was detected. After isolation of this fragment from the gel, and further digestion with other restriction enzymes (not rare-cutters), it was shown by hybridization to the β-globin probe that the anomalous PF fragment and the β^S were on the same fragment. Thus the β^S was inactivated owing to a deletion 5′ to the Σ gene and, hence, linkage of two distant mutations was demonstrated by PFGE and subsequent analysis [57].

C. Restriction Fragment Length Polymorphism Analysis

Because of the use of restriction endonucleases that have specific recognition sites, several restriction fragment length polymorphisms (RFLPs) are expected to be detected at the pulsed-field level. The first report of a PFGE RFLP was found near the CF locus, using the rare restriction enzyme, *Not*I [54,58]. This site was not in linkage disequilibrium with the CF locus, but can be used as an additional marker in family studies. We have shown that PFGE can be used to distinguish the paternal Y pseudoautosomal region because the pulsed-field map is different in the regions adjacent to the pseudoautosomal region on the X and on the Y chromosomes [52] (Fig. 4). All previous studies relied on the presence of X-linked pseudoautosomal RFLPs to determine that the origin of the X carrying the Y material was paternal [59–63].

IV. SEX CHROMOSOME STUDIES

A. Human Sex Determination and Long-Range Mapping

We have used PFGE in an attempt to detect chromosomal breakpoints, which may have given rise to XX males and XY females, by using Y-specific and pseudoautosomal sequences to study these persons. In placental mam-

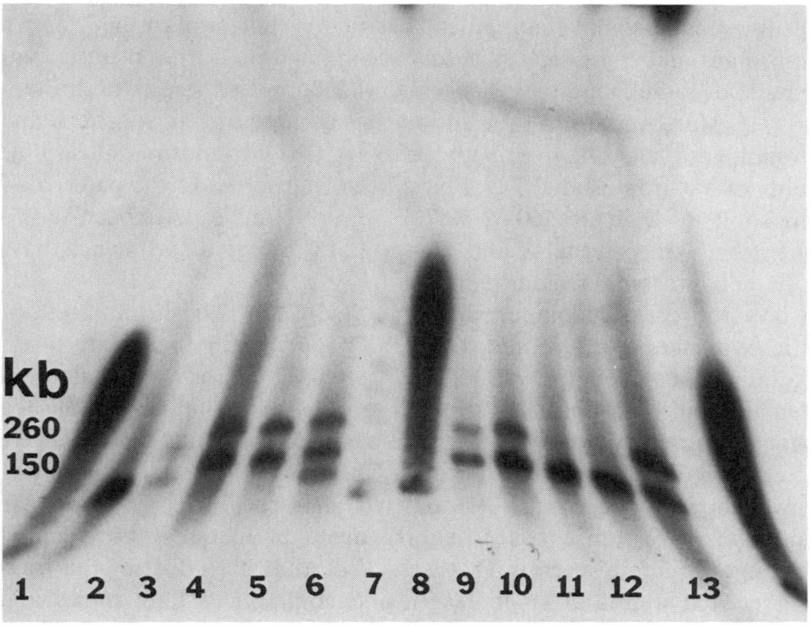

Figure 4 A Southern transfer of an OFAGE pulsed-field gel of *Sac*II-digested DNA, hybridized to the pseudoautosomal sequence (*MIC2*): The 260-kb fragment is Y-specific and the 150-kb fragment represents the X copy. Partial hybridization to the λ-ladders can be seen in lane 8 and the two outermost lanes (not numbered) and to the yeast chromosomes in lane 7. The DNA in lanes 1 and 12 are normal females, lanes 2 and 10 are normal males, lanes 3, 4, 5, 9, and 11 are XX males and, finally, lanes 6 and 13 are XY females. Note, there is no Y-specific band in the XX male in lane 11, and there is an additional 70-kb fragment in the XY female in lane 6.

mals, the presence of the Y chromosome determines maleness, and the absence thereof, femaleness. In humans, there are XX males and XY females, who are exceptions to this rule. It was proposed by Ferguson-Smith [64] that XX males arise because of an aberrant X–Y interchange during paternal meiosis, when the X and the Y chromosome pair, and that Y-specific sequences, including a gene involved in testis determination, are transferred to the short arm of the paternal X. Many studies now have confirmed this hypothesis by cytogenetic analyses showing visible translocations of Y material onto a X (presumably paternal) chromosome [65] and by high-resolution chromosome banding and in situ hybridization studies [66–69]. Molecular studies have also confirmed that Y sequences were present on the paternal X chromosome in XX males [59–63,70].

The pseudoautosomal region, identified on the short arm of the X and Y chromosomes, undergoes an obligatory recombination during meiosis and, thereby, allows exchange of genetic material between these two chromosomes [71]. Moreover, it has been shown that because of translocations and nonhomologous crossing-over between these two chromosomes, variable amounts of Y chromosomal DNA have been transferred to the paternal X chromosome in XX males [60,62,69,70]. Some XX males have been shown to inherit both the paternal X and Y copies of this region and, hence, have three pseudoautosomal regions [63].

In our study, a pseudoautosomal probe (MIC2Y) [72] and adjacent Y-specific sequences 27a [73] and 47z (DSXY5) [74] and the putative sex-determining locus, *ZFY* (pDP1007) [75], were used to study XX males and XY females and both normal males and females. A preliminary published long-range restriction map was used as a basis for this study [30]. We were able to extend the PF map on Yp using *ZFY*. This putative sex-determining locus was shown by Page et al. [75] to detect an X-specific copy on Xp. We found only two Y-specific pulsed-field fragments of 70 and 50 kb in length when using the rare restriction enzymes, *Nae*I and *Sfi*I. Most of the fragments detected were 280 kb in length and common to both males and females. Thus, it appeared that the long-range restriction map around these two loci is conserved on the X and Y chromosomes. The location of GC islands around these loci is similar. We also have shown that even though an *HTF* island was identified 3′ to *ZFY* during the chromosomal walk that resulted in the cloning of pDP1007, in genomic tissue these enzyme sites are not cut, giving rise to a much longer pulsed-field fragment. This result was obtained when using three different tissue types: fresh blood, lymphoblasts, and fibroblasts. Hence, the genomic locus is probably highly methylated. The enzymes used in this study included *BssH*II, *Nar*I, *Cla*I, *Nae*I, *Sac*II, *Sfi*I, and *Sal*I.

No translocation breakpoints have yet been detected in any of the patients studied. Our XY females had all the Y-specific fragments that were also present in normal males, including those detected with the putative sex-determining locus. As all the Y-specific fragments were present, it appears that, in the XX males, the translocation breakpoint is still centromeric to DXYS5, and other probes that are in this region will need to be studied in this group of patients. Our study did find one XX male (GM1889) without any of the Y-specific fragments detected with conventional electrophoresis or PFGE. This is one of the few typical XX males who does not have the *ZFY* copy of the *TDF* clone.

One anomaly that we found was that two XX males, GM1889 and GM2670, had an extra 150-kb *Sal*I pulsed-field fragment detected with probe 47z. Because we have detected no Y-specific material in GM1889, these anomolous fragments must arise from the long arm of the X chromosome. One XY female (GM3368) had an extra 70-kb *Sac*II fragment detected with p19B (see Fig. 4). It has been proposed that XY females also arise from translocation events with the paternal Y that has lost the *TDF* locus, resulting in an XY female. This type of translocation event has recently been reported [76]. However, because no other enzymes showed any alterations, these variants were considered to be possibly X or Y chromosomal polymorphisms.

B. Pulsed-Field Gel Mapping in the Mouse

The major question in the field of sex determination research today is whether or not the *ZFY* locus is in fact *TDF*. We are presently constructing a long-range restriction map in mice at this locus (or loci), since in mouse, there are four homologous copies to the *ZFY* locus. There are two mapping to the mouse sex determination region on the short arm of the Y chromosome [77,78]. These are termed *Zfy 1* and *Zfy 2*. It has been shown that *Zfy 2* has been deleted or lost in the sex-reversed mouse variant, *Sxr′*. This indicates that *Zfy 1* would be the crucial sequence for sex determination in the mouse. In addition, there is also an X copy and an autosomal copy on mouse chromosome 10. The human *ZFY* probe appears to hybridize more strongly to the mouse X and autosomal copies. We have cloned *Zfy 1* from an SJL/J mouse and it detects only the two Y copies of this locus [79].

To study the pulsed-field map around this region on mouse chromosomes, it is necessary to identify which pulsed-field fragments represent the different copies. It appears that after hybridization to pDP1007, only the X and autosomal copies are detected. The *Zfy 1* clone is presently being used in this study. The pulsed-field fragments appear to be larger than 800–1000 kb, as the clone hybridizes very strongly to the region of limited mobility.

Results using pDP1007, the human *ZFY* clone, show no Y-specific *Sfi*I or *Nae*I pulsed-field fragments detected in the mouse and no difference in the size of any of the bands detected between XX,*Sxr′* and XX,*Sxr* "male" mice. In addition, the pulsed-field fragments detected using *ZFY* in the mouse and the enzyme *Sac*II are a strong band at about 290 kb and a slightly weaker band at 450 kb. Because the 450-kb band is absent in humans, we predict that it represents the autosomal copy of the mouse (*Zfa*). Six bands have been detected with the enzyme *Nae*I, but all the bands were present in males and females, as well as the two types of sex-reversed mice. There is a major band at 290 kb, with less intense bands that map above and below this major band (Fig. 5). These copies are probably X and autosomal, or there is again a conservation of the long-range sites around the X and Y copies, and no Y-

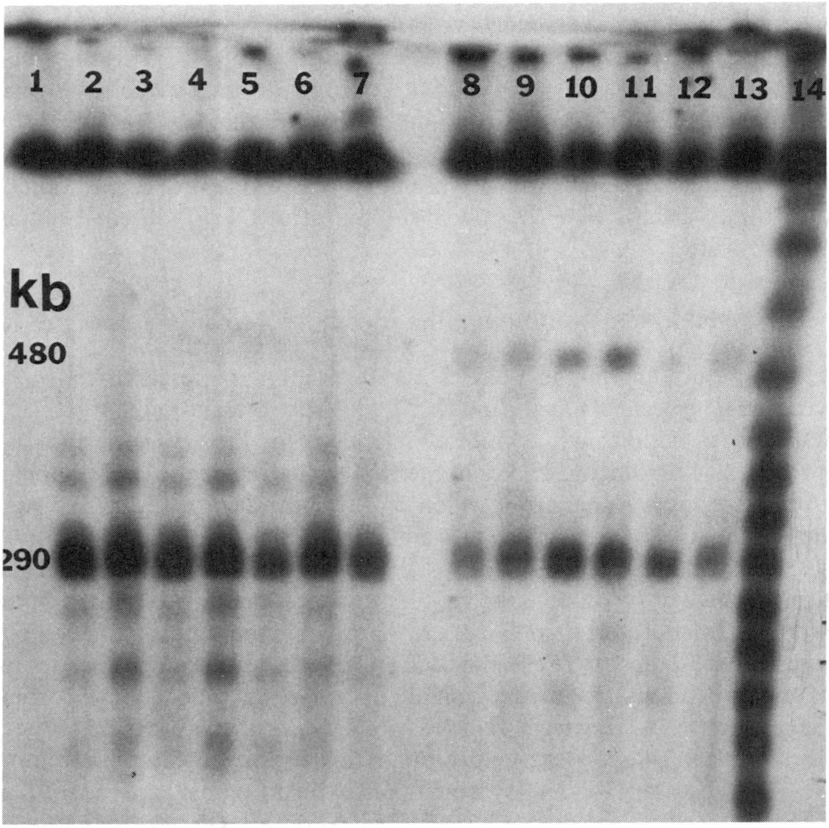

Figure 5 A Southern hybridization of human *ZFY* probe (pDP1007) to DNA of mouse digested with *Nae*I (lanes 1–7) and *Sac*II (lanes 8–13). A CHEF system was used to separate the DNA at 170 V for 40 hr with 60-sec switch times. The λ DNA ladder was hybridized to radiolabeled λ DNA and 14 distinct bands can be counted, each representing a 48.5-kb increment in size. The major fragment detected with both enzymes is 290 kb.

specific fragments are present. Comparative mapping between species is of interest to determine how the genes are conserved and the distribution of the CpG islands. We propose that, since in mice the Y copies of this locus are on the Y short arm, whereas the pseudoautosomal region in mice is on the long arm near the end, the physical separation could be why the mouse Y copies have more highly diverged from the *ZFY* copy than is the case in other species [80]. We hope to be able to determine how conserved the CpG sequences are in the regions flanking this locus. This task is more difficult in mice because of the four copies of the gene in question.

V. CONCLUSION AND SUMMARY

We have broadly discussed PFGE and many of its applications. One of the major goals remaining to be achieved is the separation of individual human chromosomes by this technique. This result will depend on whether intact, large (50-Mb) human chromosomes, containing replication forks, behave as linear molecules during electrophoresis. It has also been suggested that the size of the chromosomes may still be underestimated [24]. However, once this obstacle has been overcome, PFGE will be an even more powerful tool in genome mapping, isolation of genes, and chromosomal analysis. The problem of contamination of human chromosome-specific libraries with more than one chromosome will be easily overcome. With the wide availability of many types of apparatuses, the difficulties of PFGE may be overcome and the potential applications of this technique in molecular biology are vast. PFGE has had and will continue to have an impact in chromosome analysis.

ACKNOWLEDGMENTS

Our work has been supported by grant HD 20670.

REFERENCES

1. Tiselius, A. (1937). A new apparatus for electrophoretic analysis of colloidal mixtures. *Trans Faraday Soc. 33*:524–531.
2. Southern, E. M. (1975). Detection of specific sequences among DNA fragments separated by gel electrophoresis. *J. Mol. Biol. 98*:303–317.
3. Renwick, J. H. (1971). The mapping of human chromosomes. *Ann. Rev. Genet. 4*:81–120.
4. White, R., M. Leppert, T. Bishop, D. Barker, J. Berkowitz, C. Brown, P. Callahan, T. Holm, et al. (1985). Construction of linkage maps with DNA markers for human chromosomes. *Nature 313*:101–105.
5. Schwartz, D. C., W. Saffran, J. Welsh, R. Haas, M. Goldenberg, and C. R. Cantor (1983). New techniques for purifying large DNAs and studying their properties and packaging. *Cold Spring Harbor Symp. Quant. Biol. 47*:189–195.
6. Schwartz, D. C., and C. Cantor (1984). Separation of yeast chromosome-sized DNAs by pulsed field gradient gel electrophoresis. *Cell 37*:67–75.
7. De Gennes, P. G. (1971). Reptation of polymer chain in the presence of fixed obstacles. *J. Chem. Phys. 55*:572–579.

8. Schwartz, D. C., and M. Koval (1989). Conformational dynamics of individual DNA molecules during gel electrophoresis. *Nature 338*:520–522.
9. Smith, S. B., P. K. Aldridge, and J. B. Callis (1989). Observation of individual DNA molecules undergoing electrophoresis. *Science 243*:203–206.
10. Carle, G. F., and M. V. Olson (1984). Separation of chromosomal DNA molecules from yeast by orthogonal-field-alternation gel electrophoresis. *Nucleic Acids Res. 12*:5647–5664.
11. Carle, G. F., and M. V. Olson (1985). An electrophoretic karyotype for yeast. *Proc. Natl. Acad. Sci. USA 82*:3756–3760.
12. Carle, G. F., M. Frank, and M. V. Olson (1986). Electrophoretic separations of large DNA molecules by periodic inversion of the electric field. *Science 232*:65–68.
13. Chu, G., D. Vollrath, and R. W. Davis (1986). Electrophoretic separations of large DNA molecules by contour-clamped homogeneous electric fields. *Science 234*:1582–1585.
14. Vollrath, D., and R. W. Davis (1987). Resolution of DNA molecules greater than 5 megabases by contour-clamped homogeneous electric fields. *Nucleic Acids Res. 15*:7865–7876.
15. Clark, S. M., E. Lai, B. W. Birren, and L. Hood (1988). A novel instrument for separating large DNA molecules with pulsed homogeneous electric fields. *Science 241*:1203–1205.
16. Bancroft, I., and C. P. Wolk (1988). Pulsed homogeneous orthogonal field gel electrophoresis (PHOGE). *Nucleic Acids Res. 16*:7405–7418.
17. Southern, E. M., W. R. Brown, and R. Anand (1987). A model for the separation of large DNA molecules by crossed field gel electrophoresis. *Nucleic Acids Res. 15*:5925–5943.
18. Gemmill, R. M., J. F. Coyle-Morris, F. D. Mcpeek, L. F. Ware-Uribe, and F. Hecht (1987). Construction of long range restriction maps in human DNA using pulsed field gel electrophoresis. *Gene Anal. Tech. 4*:119–131.
19. Anand, R., J. Honeycomb, P. A. Whittaker, J. K. Elder, and E. M. Southern (1988). Clones from an 840 kb fragment containing the 5′ region of the *DMD* locus enriched by pulsed field gel electrophoresis. *Genomics 3*:177–186.
20. Smith, C. L., P. E. Warburton, A. Gaal, and C. R. Cantor (1986). Analysis of genome organization and rearrangement by pulsed field gradient electrophoresis. In *Genetic Engineering*, Vol. 8. J. K. Setlow and A. Hollaender (eds.). Plenum Press, New York, pp. 45–70.
21. Bird, A. P. (1986). CpG-rich islands and the function of DNA methylation. *Nature 321*:209–213.
22. Bird, A. P., M. Taggart, M. Frommer, O. J. Miller, and D. Macleod (1985). A fraction of the mouse genome that is derived from islands of nonmethylated, CpG-rich DNA. *Cell 40*:91–99.
23. Qiang, B., and I. Schildraut (1984). A type II restriction endonuclease with an eight nucleotide specificity from *Streptomyces fimbriatus*. *Nucleic Acids Res. 12*:4507–4516.
24. Barlow, D., and H. Lehrach (1987). Genetics by gel electrophoresis: The impact of pulsed field gel electrophoresis on mammalian genetics. *Trends Genet. 3*: 167–171.

25. Holtzwarth, G., C. B. Mckee, S. Steiger, and G. Crater (1987). Transient orientation of linear DNA molecules during pulsed field gel electrophoresis. *Nucleic Acids Res. 15*:10031–10045.
26. Cantor, C. R., P. Warburton, C. L. Smith, and A. Gaal (1986). Voltage ramp pulsed field gel electrophoresis separation of large DNA molecules. In *Electrophoresis '86*. M. Dunn (ed.). VCH Publishers, Weinheim, FRG, pp. 161–171.
27. Sor, F. (1988). A computer program allows the separation of a wide range of chromosome sizes by pulsed field gel electrophoresis. *Nucleic Acids Res. 16*: 4853–4863.
28. Lalande, M., J. Noolandi, C. Turmel, J. Rosseau, and G. W. Slater (1987). Pulsed-field electrophoresis: Application of a computer model to the separation of large DNA molecules. *Proc. Natl. Acad. Sci. USA 84*:8011–8015.
29. Birren, B. W., E. Lai, S. M. Clark, L. Hood, and M. I. Simon (1988). Optimized conditions for pulsed field gel electrophoretic separations of DNA. *Nucleic Acids Res. 16*:7563–7582.
30. Pritchard, C., P. J. Goodfellow, and P. N. Goodfellow (1987). Mapping the limits of the human pseudoautosomal region and a candidate sequence for the male-determining gene. *Nature 328*:273–275.
31. Reed, K. C., and D. A. Mann (1985). Rapid transfer of DNA from agarose gels to nylon membrane. *Nucleic Acids Res. 13*:7207–7221.
32. Church, G. M., and W. Gilbert (1984). Genomic sequencing. *Proc. Natl. Acad. Sci. USA 81*:1991–1995.
33. Feinberg, A. P., and B. Vogelstein (1984). A technique for radiolabelling DNA restriction fragments to high specific activity. *Anal. Biochem. 137*:266–267.
34. Maniatis, T., E. F. Fritsch, and J. Sambrook (1982). *Molecular cloning. A laboratory manual*. Cold Spring Harbor Laboratory, Cold Spring Harbor, NY.
35. Waterbury, P. G., and M. J. Lane (1987). Generation of lambda phage concatemers for use as pulsed field electrophoresis size markers. *Nucleic Acids Res. 15*:3930.
36. Smith, C. L., T. Matsumoto, O. Niwa, S. Kleo, J. Fan, M. Yanagida, and C. R. Cantor (1987). An electrophoretic karyotype for *Schizosaccharomyces pombe* by pulsed field gel electrophoresis. *Nucleic Acids Res. 15*:4481–4489.
37. Bellis, M., M. Pages, and G. Roizes (1987). A rapid and simple method for preparing yeast chromosomes for pulsed field gel electrophoresis. *Nucleic Acids Res. 15*:6749.
38. Johnson, P. J., and P. Borst (1986). Mapping of the VSG genes on large expression-site chromosomes of *Trypanosoma brucei* separated by pulsed-field gradient electrophoresis. *Gene* 43:213–220.
39. Bernards, A., J. M. Kooter, P. A. M. Michels, R. M. P. Moberts, and P. Borst (1986). Pulsed field gradient electrophoresis of DNA digested in agarose allows the sizing of the large duplication unit of surface antigen gene in trypanosomes. *Gene* 42:313–322.
40. Smith, C. L., and R. L. Kolodner (1988). Mapping of the *Escherichia coli* chromosomal tn5 and F insertions by pulsed field gel electrophoresis. *Genetics 119*:227–236.

41. Smith, C. L., J. G. Econome, A. Schutt, S. Klco, and C. R. Cantor (1987). A physical map of the *Escherichia coli* K12 genome. *Science 236*:1448–1453.
42. Lee, J. J., and H. O. Smith (1988). Sizing of the *Haemophilus influenzae* Rd genome by pulsed-field agarose gel electrophoresis. *J. Bacteriol. 170*:4402–4405.
43. White, R. L., S. Woodward, M. Leppert, P. O'Connell, M. Hoffman, J. Herbst, J. Lalouel, M. Dean, and G. F. Van der Woude (1985). A closely linked marker for cystic fibrosis. *Nature 318*:382–384.
44. Tsui, L.-C., M. Buchwald, D. Barker, J. C. Bramer, J. W. Schumm, H. Eiberg, J. Mohr, D. Kennedy, et al. (1985). Cystic locus defined by a genetically linked polymorphic DNA marker. *Science 230*:1054–1057.
45. Drumm, M. L., C. L. Smith, M. Dean, J. L. Cole, M. C. Iannuzzi, and F. C. Collins (1988). Physical mapping of the cystic fibrosis region by pulsed-field gel electrophoresis. *Genomics 2*:346–354.
46. Poustka, A., T. M. Pohl, D. P. Barlow, A. Frischauf, and H. Lehrach (1987). Construction and use of human chromosome jumping libraries from *Not*I digested DNA. *Nature 325*:353–355.
47. Fulton, T. R., A. M. Bowcock, D. R. Smith, L. Daneshvar, P. Green, L. L. Cavalli-Sforza, and H. Donis-Keller (1989). A 12 megabase restriction map at the cystic fibrosis locus. *Nucleic Acids Res. 17*:271–284.
48. Van Ommen, G. J. B., and J. M. H. Verkerk (1986). Restriction analysis of chromosomal DNA in the size range up to two million base pairs of pulsed field gradient electrophoresis. In *Human Genetic Diseases: A Practical Approach.* K. E. Davies (ed.). IRL, Oxford, pp. 113–133.
49. Burmeister, M., and H. Lehrach (1986). Long range restriction map around the Duchenne muscular dystrophy gene. *Nature 324*:582–585.
50. Dunham, I., C. A. Sargent, J. Trowsdale, and R. D. Campbell (1987). Molecular mapping of the human major histocompatibility complex by pulsed-field gel electrophoresis. *Proc. Natl. Acad. Sci. USA 84*:7237–7241.
51. Lawrance, S. K., C. L. Smith, R. Srivastava, C. R. Cantor, and S. M. Weissman (1987). Megabase scale mapping of the *HLA* gene complex by pulsed field gel electrophoresis. *Science 235*:1387–1390.
52. Verga, V., and R. P. Erickson (1989). An extended long range restriction map of the human sex determining region on Yp, including *ZFY*, finds marked homology on Xp and no detectable Y sequences in an XX male. *Am. J. Hum. Genet. 44*:756–765.
53. Collins, F. S., and S. M. Weissman (1984). Directional cloning of DNA fragments at a large distance from an initial probe: A circularization method. *Proc. Natl. Acad. Sci. USA 81*:6812–6816.
54. Collins, F. C., M. L. Drumm, J. L. Cole, W. K. Lockwood, G. F. Vande Woude, and M. C. Iannuzzi (1987). Construction of a general human chromosome jumping library, with application to cystic fibrosis. *Science 235*:1046–1049.
55. Iannuzzi, M. C., M. Dean, M. L. Drumm, N. Hidaka, J. L. Cole, A. Perry, C. Stewart, B. Gerrard, and F. S. Collins (1989). Isolation of additional polymorphic clones from the cystic fibrosis region, using chromosome jumping from D7S8. *Am. J. Hum. Genet. 44*:695–703.

56. Fountain, J. W., M. R. Wallace, A. M. Brereton, P. O'Connell, R. L. White, D. C. Rich, D. H. Ledbetter, et al. (1989). Physical mapping of the von Recklinghausen neurofibromatosis region on chromosome 17. *Am. J. Hum. Genet. 44*:58–67.
57. Dobkin, C. S., and M. C. Driscoll (1988). Linkage of two distant mutations in the β globin gene complex by pulsed field gel electrophoresis. *Am. J. Hum. Genet. 43*(Suppl.):A181.
58. White, R., M. Leppert, P. O'Connell, Y. Nakamura, C. Julien, S. Woodward, A. Silva, R. Wolff, M. Lathrop, and J. M. Lalouel (1986). Construction of human genetic linkage maps. I. Progress and perspectives. *Cold Spring Harbor Symp. Quant. Biol. 510*:29–38.
59. Page, D. C., and A. de la Chapelle (1984). The parental origin of X chromosome in XX males determined using restriction fragment length polymorphisms. *Am. J. Hum. Genet. 36*:565–575.
60. Page, D. C., L. G. Brown, and A. de la Chapelle (1987). Exchange of terminal portions of the X- and Y-chromosomal short arms in human XX males. *Nature 328*:437–440.
61. de la Chapelle, A. (1981). The etiology of maleness in XX men. *Nature 307*: 170–171.
62. Rouyer, F., M. C. Simmler, C. X. Johnsson, G. Vergaud, H. J. Cooke, and J. Weissenbach (1986). A gradient of sex linkage in the pseudoautosomal region of the human sex chromosomes. *Nature 319*:291–295.
63. Stalvey, J. R. D., E. J. Durbin, and R. P. Erickson (1989). Sex vesicle "entrapment": Translocation or nonhomologous recombination of misaligned Yp and Xp as alternative mechanisms for abnormal inheritance of the sex-determining region. *Am. J. Med. Genet.* 32:564–572.
64. Ferguson-Smith, M. A. (1966). X–Y chromosomal interchange in the aetiology of true hermaphroditism and of the XX Klinefelter's syndrome. *Lancet 2*:475–476.
65. Evans, H. J., K. E. Buckton, G. Spowart, and A. D. Carothers (1979). Heteromorphic X chromosomes in 46,XX males: Evidence for the involvement of X–Y interchange. *Hum. Genet. 49*:11–31.
66. Magenis, R. E., M. J. Webb, R. S. Mckean, D. Tomar, L. J. Allen, H. Kammer, D. L. Van Dyke, and E. Lovrien (1982). Translocation (X;Y) (p22.33;p11.2) in XX males: Etiology of male phenotype. *Hum. Genet. 62*:271–276.
67. Magenis, R. E., M. Casanova, M. Fellous, S. Olson, and R. Sheehy (1987). Further cytogenetic evidence for Xp–Yp translocation in XX males using in situ hybridization with Y-derived probe. *Hum. Genet. 75*:228–233.
68. Andersson, M., D. C. Page, and A. de la Chapelle (1986). Chromosomal Y-specific DNA is transferred to the short arm of the X chromosome in human XX males. *Science 233*:786–788.
69. Affara, N. A., M. A. Ferguson-Smith, J. Tolmie, K. Kwok, M. Mitchell, D. Jamieson, A. Cooke, and L. Florentin (1986). Variable transfer of Y specific sequences in XX males. *Nucleic Acids Res. 14*:5375–5387.
70. Petit, C., A. de la Chapelle, J. Levilliers, S. Castillo, B. Noel, and J. Weissenbach (1987). An abnormal X–Y interchange accounts for most but not all cases of XX maleness. *Cell 49*:595–602.

71. Burgoyne, P. S. (1982). Genetic homology and crossing over in the X and Y chromosomes of mammals. *Hum. Genet. 61*:85–90.
72. Darling, S. M., P. J. Goodfellow, B. Pym, G. S. Banting, C. Pritchard, and P. N. Goodfellow (1985). Molecular genetics of *MIC2*: A gene shared by the human X and Y chromosome. *Cold Spring Harbor Symp. Quant. Biol. 51*: 205–212.
73. Pritchard, C., P. J. Goodfellow, and P. N. Goodfellow (1987). Isolation of a sequence which maps close to the human sex-determining gene. *Nucleic Acids Res. 15*:6159–6169.
74. Geldwerth, D. A., C. Bishop, G. Guellaen, M. Koenig, G. Vergnaud, J. L. Mandel, and J. Weissenbach (1985). Extensive DNA sequence homologies between the human Y and the long arm of the X chromosome. *EMBO J. 4*:1739–1743.
75. Page, D. C., R. Mosher, E. M. Simpson, E. M. Fisher, G. Mardo, J. Pollack, B. McGillivray, A. de la Chapelle, and L. G. Brown (1987). The sex-determining region of the human Y-chromosome encodes a finger protein. *Cell 51*:1091–1104.
76. Levilliers, J., B. Quack, J. Weissenbach, and C. Petit (1989). Exchange of terminal portions of X- and Y-chromosomal short arms in human XY females. *Proc. Natl. Acad. Sci. USA 86*:2296–2300.
77. Mardon, G., R. Mosher, C. M. Disteche, Y. Nishioka, A. McLaren, and D. C. Page (1988). Duplication, deletion and polymorphism in the sex-determining region of the mouse Y chromosome. *Science 243*:78–80.
78. Nagamine, C. M., K. Chan, C. A. Kozak, and Y. Lau (1988). Chromosome mapping and expression of a putative testis-determining gene in mouse. *Science 243*:80–83.
79. Kalikin, L. M. Fujimoto, M. Witt, M. P. Verga, V., and Erickson, R. P. (1989). A genomic clone of Zfy-l from y^{DOM} mouse strain detects post-meiotic gene expression of Zfyin testes. *Biochem. and Biophys. Res. Comm. 165*:1286–1291.
80. Erickson, R. P., and V. Verga (1989). Is zinc-finger Y the sex determining gene? *Am. J. Hum. Genet. 45*: 671–674.

II

CYTOGENETIC AND LINKAGE ANALYSIS: INTRODUCTION

Cytogenetic analysis of human chromosomes has been of inestimable value in clinical diagnosis and basic research. The ability to identify chromosomes by their specific banding patterns with stains, such as Giemsa, has evolved into a sophisticated means of genetic investigation. The recognition of chromosomal abnormalities and the localization of genes to particular bands has been made possible by the technology of cytogenetics.

The power of combining cytogenetic analysis and molecular genetic techniques is demonstrated by studies of malignant melanoma, reviewed in the first chapter. Human solid tumors such as melanoma have been documented to possess chromosomal sites that display recurring alterations. The importance of chromosome 6 alterations has been demonstrated by restriction fragment length polymorphism (RFLP) analysis, pulsed-field gel electrophoresis, "linkage and jumping libraries," and microdissection of chromosome bands.

Another important topic that is discussed in this section concerns chromosome fragile sites. *Fragile sites* are breaks or gaps that appear on chromosomes prepared from cells subjected to specific treatments. Interest in fragile sites originates in their association with the most common form of mental retardation (the fragile X syndrome) and with chromosomal rearrangements in malignant disease.

Immunocytogenetics refers to procedures that reveal chromosome regions or bands with antibodies to DNA or protein components of chromosomes. The bound antibodies are visualized by secondary antibodies, fluorescence, or enzyme label. A chapter reviews the results observed for a variety of chromosomal antigens. These include DNA antigens (Z-DNA, methylated cytosine) and protein antigens (topoisomerase, RNA polymerase).

The construction of gene linkage maps for human, mouse, and other chromosomes is a major goal of current genetic research. Such maps are valuable in identifying mutant genes responsible for diseases such as cystic fibrosis, muscular dystrophy, and hemophilia. Genetic markers used to construct linkage maps have their basis in visible mutations, biochemical markers, and DNA sequence variants, particularly RFLPs.

The mouse has been a favored organism in genetic research for decades, and, therefore, a chapter concerns the assembly of linkage maps for the mouse. The availability of well-characterized strains and mutants of mice, combined with their short generation times and numerous offspring, provide an important parallel system to the human.

Genome duplication in the chromosomal maps of mice and humans is discussed next. Evidence suggests that genomes of eukaryotes have undergone multiple duplications, some of which have become rearranged with other genes. Maps of duplicated segments may have use in linkage studies and in studies of genome organization and evolution.

Sophisticated mathematical procedures have been applied to linkage analysis because of the importance and difficulty of establishing the precise order of genes on chromosomes. A chapter outlining current mathematical approaches to the problem illustrates the progress that has been made employing different strategies for ordering loci.

8

Combining Cytogenetic Analysis with Molecular Mapping of Chromosomal Alterations in Human Solid Tumors

Eckart U. Meese and Jeffrey M. Trent

University of Michigan, Ann Arbor, Michigan

I. INTRODUCTION

Over the past two decades, research on structural chromosomal changes has focused primarily on leukemias and lymphomas, with little emphasis on solid tumors [1,2]. Because solid tumors (e.g., carcinomas, sarcomas, melanoma) represent nearly 80% of human malignant neoplasms, future cytogenetic studies must increasingly include these neoplasms. As recently documented in a compendium on chromosomal rearrangements in neoplasia at the Human Gene Mapping Workshop 10 (HGM10) [3], numerous human solid tumors have been shown to contain recurring sites of chromosomal changes. These cytogenetic alterations may indicate a site in the genome that requires more detailed molecular investigation. Several recently developed techniques [e.g., pulsed-field gel electrophoresis (PFGE), chromosomal jumping, and chromosomal walking] now make it possible to narrow the gap between the chromosomal alterations and the actual genes involved in these aberrations [4–7]. In this chapter, we will present a brief overview of our research on the cytogenetic and molecular analysis of human malignant melanoma, to illustrate several different approaches that can be used to identify genes involved in the chromosomal alterations characterizing solid tumors in general. Our review will also address possible pitfalls and limitations of the described techniques.

II. CYTOGENETICS OF HUMAN MALIGNANT MELANOMA

Metastatic melanoma is currently an incurable cancer. Even the best combination of chemotherapies produces less than a 40% response rate and these responses are usually short-lived, with few long-term survivors [8]. The absolute incidence of melanoma has markedly increased over the past decade, with the Third National Cancer Institute Survey (1978) indicating an increase in the incidence of melanoma across the nation, second only to lung cancer. A choice of malignant melanoma as a model system is particularly apropos for studies at the University of Arizona. Investigators at the Arizona Cancer Center have documented a very high incidence of malignant melanoma in southern Arizona that has increased from <10 cases/100,000 in 1969 to >30 cases/100,000 at present (Fig. 1). The increasing incidence of malignant melanoma in the southwestern and southeastern portions of the United States undoubtedly reflects, in part, altered migratory patterns in the past decade [9]. Therefore, with the current rate of inmigration to the Southwest and Southeast, it is certain that the incidence of this difficult disease will continue to increase rapidly over the next decade.

Despite the aforementioned significant increase in the incidence of malignant melanoma over the past decade, surprisingly little is known about specific chromosomal abnormalities (and certainly their molecular basis) in this important disease. However, recently studies have begun to describe a

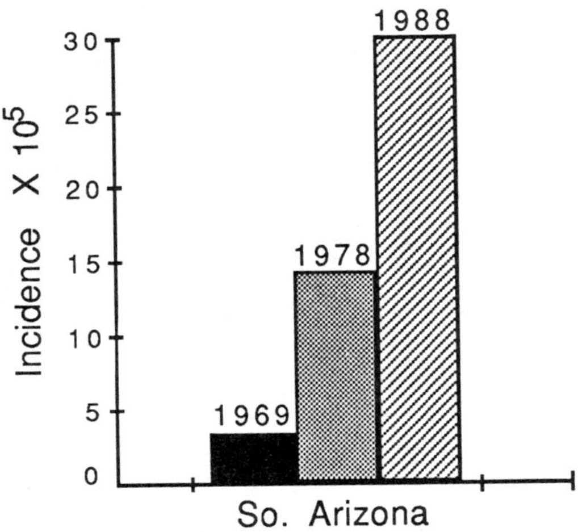

Figure 1 Incidence of malignant melanoma in southern Arizona from 1969–1988. There are approximately 27,000 new cases a year. Death rates of 50–60% are common for patients with advanced stage disease.

recurring pattern for chromosomal aberrancies in malignant melanoma. Figure 2 documents the numeric chromosomal abnormalities (gain or loss of an entire chromosome) from 28 cases of melanoma studied in our laboratory. Although, in general, loss or gain of chromosomes are relatively evenly balanced, gain of chromosome 7 and loss of chromosomes 9 and 14 are most frequently observed.

The study of banded cells from melanoma has also provided evidence for recurring (i.e., clonal) structural chromosomal abnormalities in malignant melanoma. Almost 100% of cases of metastatic melanoma demonstrate clonal abnormalities. Figure 3 provides an illustration of the decidedly nonrandom nature of structural aberrations from 40 cases of melanoma. Chromosomes 1, 6, 7, and 11 are, without question, the most frequently altered in this disorder. Of interest, although some chromosomes are altered in almost all cases (e.g., chromosome 1), other chromosomes (e.g., 20 and Y) are virtually never observed to participate in structural chromosomal rearrangements in this disease. Figure 4 illustrates a typical G-banded karyotype from the malignant cells of a patient with metastatic melanoma.

Interestingly, the gene for hereditary cutaneous malignant melanoma and its precursor lesion, dysplastic nevus, has recently been assigned to the short arm of chromosome 1(1p36) [10,11]. Recent findings also report the loss of

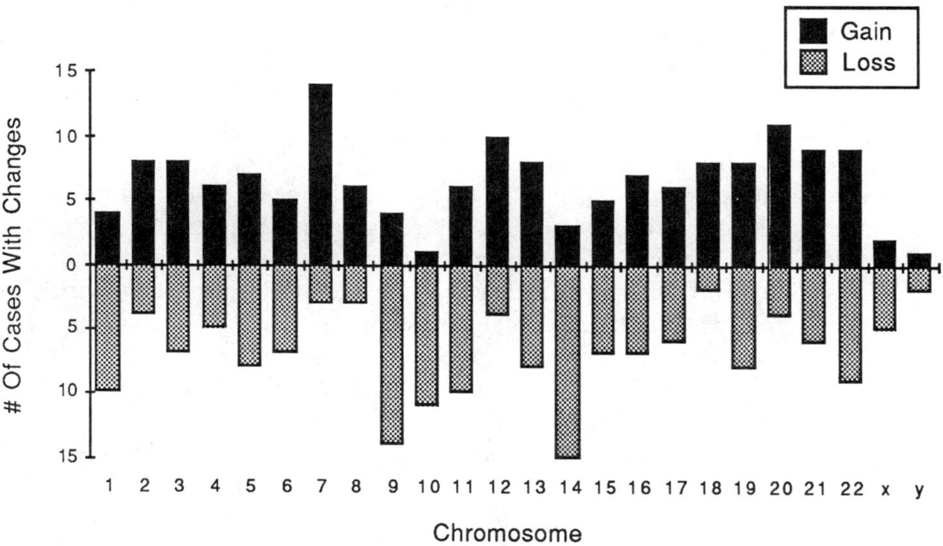

Figure 2 Numeric chromosomal abnormalities in 28 patients with malignant melanoma.

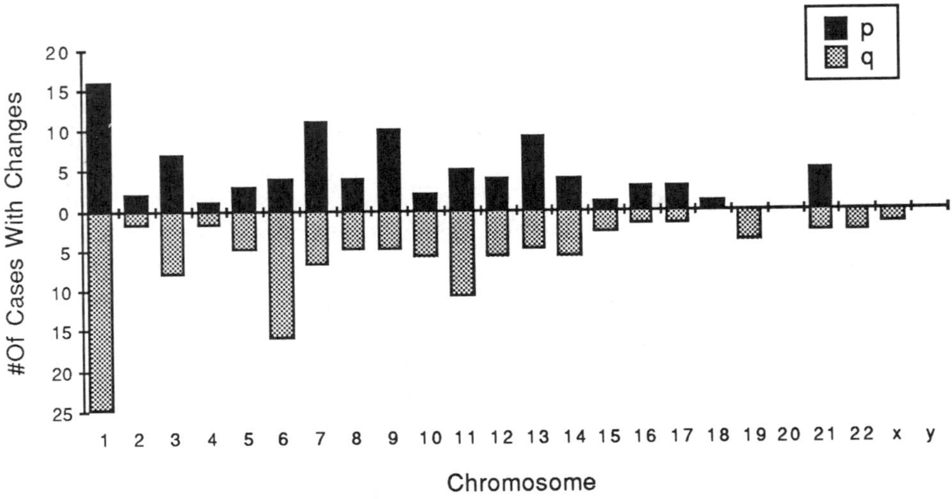

Figure 3 Structural chromosomal abnormalities in 40 patients with malignant melanoma.

alleles from the distal portion of 1p in at least a subset of melanomas [12]. However, cytogenetic changes and apparent loss of heterozygosity (LOH) for loci on chromosome 1p have been found to occur in a very large variety of cancers and, invariably, occur late in tumor progression (indicating the loss may not be directly involved in melanocyte transformation).

In contrast with the specificity of translocations that characterize many of the human hematopoietic malignancies (e.g., Ph[1] chromosome in chronic myelogenous leukemia [2]), it is much more common in solid tumors (including melanoma) for several different chromosomal alterations to be observed within a given chromosomal band region [3]. Accordingly, cytogenetic analysis of malignant melanoma is most useful now in pinpointing "regions" of specific chromosomal alterations, rather than identifying a single specific chromosomal alteration found in most patients.

After those of chromosome 1, chromosome 6 alterations represent a very frequent site of change in malignant melanoma. Work from our laboratory has recently described a nonreciprocal translocation involving chromosome 6q in melanoma (Fig. 5) [13]. Additionally, our laboratory (and many others) have documented the simple deletion of chromosome 6q as a frequent change in melanoma [14–17]. We have recently proposed that both 6q deletions and translocations may, in fact, be alternative mechanisms to ac-

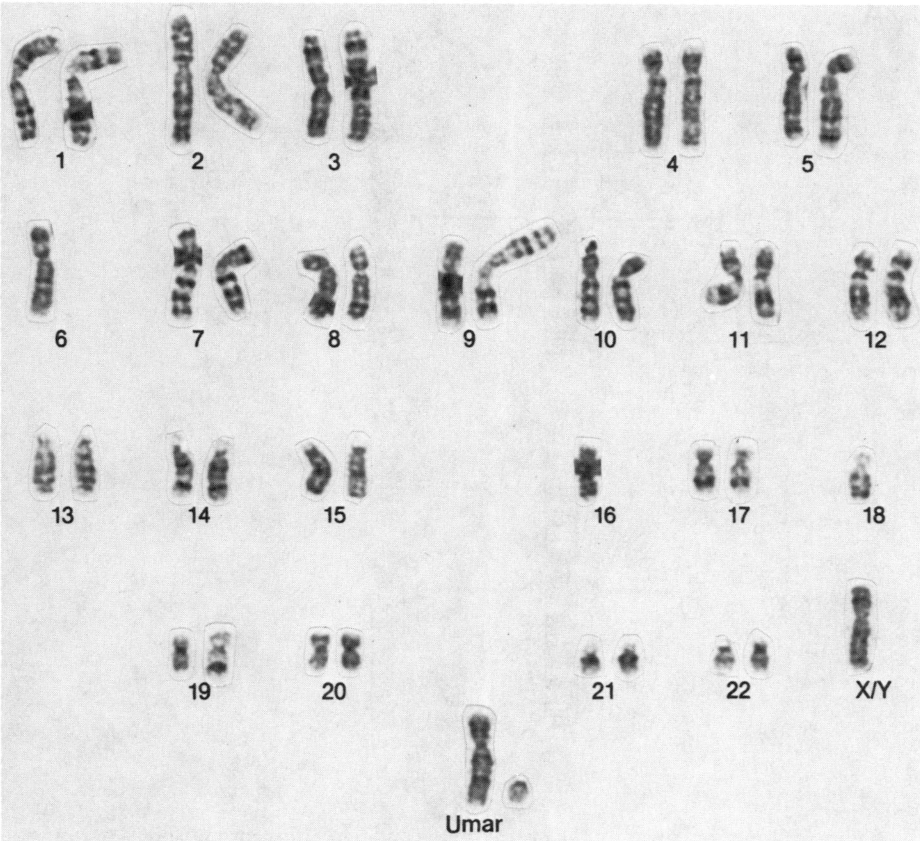

Figure 4 G-banded karyotype of the tumor cells from a patient with malignant melanoma: This tumor displayed a near-diploid modal chromosomal number of 44, with several structural and numeric chromosomal abnormalities. Clonal structural alterations included: t(1;6)(p31;p11), t(1;9)(q11;p13), t(17;19)(q21;q13). The patient's tumor cells were also monosomic for chromosomes 6 and Y. Two unidentified marker chromosome (Umars) were also observed in this tumor.

complish the loss of sequences on chromosome 6q [13]. Specifically, both the translocations (which are nonreciprocal) and the simple deletions may cause a loss of sequences along the long arm of chromosome 6, leading to the loss of genetic material and, possibly, inactivation of a tumor suppressor gene.

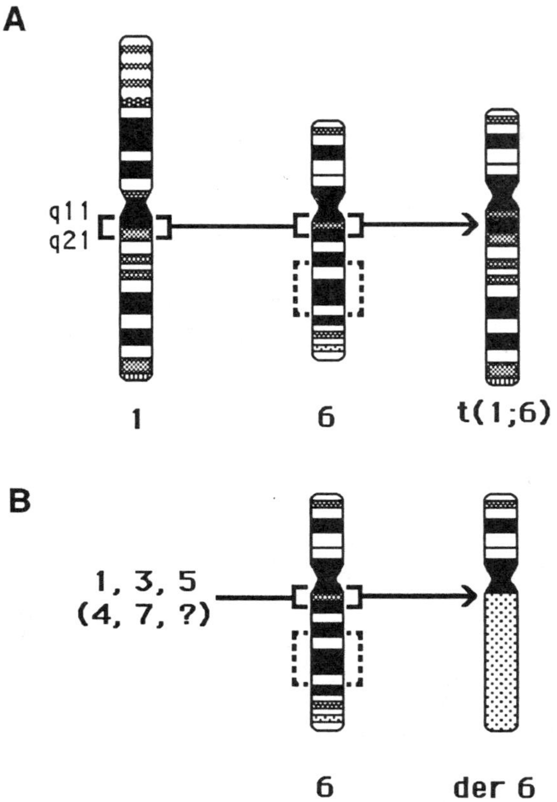

Figure 5 Idiogram diagramming breakpoints and derivative chromosomes resulting from translocations of various chromosomal regions to 6q11-21: (A) Breakpoints involving chromosome 1 and 6, with resulting derivative t (1;6). Arrows to chromosome 1: band region involved in translocation with 6q11-13 (solid brackets). (B) Chromosomes 1p, 3, 4, 5, 7, and unidentified chromosomal segments (?) were also shown to translocate to 6q11-13. Dotted brackets: region of 6q most frequently involved in simple deletions. (From Ref. 13, with permission)

Several studies have been undertaken by our laboratory in an attempt to define the importance of chromosome 6q alterations in malignant melanoma. We will briefly describe studies related to restriction fragment length polymorphism (RFLP) analysis, pulsed-field gel electrophoresis (PFGE), linkage and jumping libraries, and, finally, microdissection of banded chromosomes.

III. RESTRICTION FRAGMENT LENGTH POLYMORPHISM ANALYSIS

As previously described, chromosomal changes such as nonreciprocal translocations can lead to the loss of genetic material. However, it is increasingly clear that submicroscopic rearrangements, as well as the loss of one chromosome followed by reduplication of an apparently normal homologue, are equally likely events to generate genomic deletions [18–20]. Unfortunately, the identification of these alterations is not recognized by conventional cytogenetic analysis. However, they can be verified by an alternate procedure, restriction fragment length polymorphism (RFLP) analysis.

Normal DNA probes usually hybridize to restriction fragments of identical size, resulting in a single hybridization signal, whereas polymorphic probes hybridize to restriction fragments of different sizes, resulting in unique hybridization signals [21]. Because the different signals stem from two homologous (and usually cytogenetically indistinguishable) chromosomes, polymorphic probes enable the investigator to analyze submicroscopic deletions or rearrangements affecting every human chromosome. By means of Southern blotting, experiments are then performed by analyzing tumor DNA and comparing it with normal DNA (e.g., lymphocyte DNA) from the same patients. DNA from normal tissue is required to verify the germline heterozygosity and, subsequently, to identify the loss of heterozygosity (LOH) in tumor DNA.

Although RFLP analysis has been applied very successfully to many tumors, including melanoma, for identifying cryptic chromosomal rearrangements and deletions, this approach remains subject to significant limitations. The analysis of rearrangements in RFLP studies depends on the availability of informative polymorphic probes. Despite an increasing number of such probes, there is still an enormous gap between the size of the total human genome and the portion that can be analyzed with available RFLP probes. Additionally, the available probes are not evenly distributed along the length of each chromosome and, therefore, for some chromosomal regions, there is an abundance of probes, whereas no polymorphic probes exist for other regions.

The shortage of useful RFLP probes is also a major obstacle for additional studies of 6q in human malignant melanomas. Therefore, little information is yet available on the status of genetic loss involving chromosome 6q. The following section describes several possible methodologic strategies to overcome this problem and, thereby, to allow a more detailed study of this chromosomal region.

IV. IDENTIFICATION OF NEW RESTRICTION FRAGMENT LENGTH POLYMORPHISM PROBES

The most obvious approach to solve the aforementioned problem of a paucity of informative probes is the isolation and characterization of new probes for the target region. However, this process is very tedious and may not necessarily result in the isolation of informative polymorphic markers. As starting material, most laboratories use chromosome-sorted libraries or human–rodent hybrid cell lines that contain only one human chromosome (or a portion of a human chromosome). After the genomic DNA fragments are cloned in an appropriate vector, human clones are isolated and tested for polymorphism. If the probes detect a polymorphism and the polymorphism information content (PIC) is high enough, the probes will be regionally assigned to the target chromosome.

This regional localization can be performed by in situ hybridization or by a somatic cell hybrid panel that contains different segments of the target chromosome. Ideally, both in situ hybridization and probe localization, with the aid of a hybrid deletion-mapping panel are combined to provide unequivocal localization. After considerable effort, we have developed a chromosome 6-specific hybrid deletion-mapping panel that will assign probes to specific subregions of chromosome 6q [22]. This chromosome 6-specific panel is difficult to generate because of the relatively few functional genes mapped to chromosome 6 and the associated absence of a direct selection strategy for this chromosome. Therefore, an extensive time-consuming, labor-intensive cytogenetic screening of dozens of cell hybrids is required to select clones that retain the relevant portions of chromosome 6. An alternative approach that we have also used is to identify various translocations between 6 and other chromosomes with selectable markers. Unfortunately, few of these genetic mutant cell lines now exist.

A second approach to facilitate informative probes relates to the identification of variable-number, tandem-repeat (VNTR) markers, which greatly facilitate genetic linkage analysis [23]. The VNTRs consist of a set of tandem repeats of short sequences (up to 60 bp). Unlike RFLP probes, which depend upon alterations in restriction sites, the number of copies of the tandem repeats accounts for the difference in length of the restriction fragments. Moreover, in contrast with standard RFLP probes, VNTR markers are always highly polymorphic and, therefore, more informative than diallelic systems for the construction of linkage maps. For example, some VNTR loci have demonstrated more than 70 alleles in the general population [24]. Because of their highly polymorphic nature, VNTR markers are particularly valuable for the identification of submicroscopic rearrangements and deletions in human tumors. Also, because most patients will be heterozygous for VNTR probes, even limited amounts of tumor material can be analyzed very efficiently.

The recent description of a core sequence common to many VNTR markers provides a relatively fast approach for isolating new VNTR markers [25]. Synthetic oligonucleotides that contain the core sequence can be used to screen a cosmid library, with positive clones tested for polymorphisms using Southern blots of genomic DNAs that have been digested with frequent-cutting enzymes (e.g., *Msp*I, *Taq*I). Subsequently, new highly polymorphic markers can be genotyped and localized by linkage analysis by using previously mapped DNA markers as reference points.

To summarize this section, it is imperative that a high number of polymorphic clones are obtained to facilitate studies of a defined chromosomal region. Unfortunately, recent experiments show that some chromosomal regions (for reasons not completely understood) are underrepresented in genomic libraries and, therefore, these regions cannot now be saturated with specific probes. Consequently, it has become increasingly clear that complementary techniques are required to study rearrangements that can neither be analyzed by standard RFLP or VNTR studies nor by cytogenetic methods.

V. PULSED-FIELD GEL ELECTROPHORESIS

An area of major progress in the field of genome mapping has been made by the development of PFGE, which resolves molecules up to 9000 kb [26–29]. The tremendous impact of this technique is reflected by the rapidly increasing number of publications since its first description in 1984. Many of the initially developed PFGE systems used nonhomogeneous electric fields, causing the DNA molecules to migrate at different rates, depending on their position in the gel. This effect, in turn, led to distorted or "bent" lanes and significantly complicated the interpretation of PFGE gels. Among a variety of different PFGE systems, contour-clamped homogeneous electric field (CHEF) electrophoresis has provided excellent resolution while running DNA as a straight track [30]. Additionally, to separate these large DNA molecules, the DNA is subjected to two alternating electric fields that are arranged at an angle of 120°.

In 1987, Southern and his co-workers [31] developed a theory for the separation of DNA by PFGE that was basically confirmed by experiments from Smith [32]. According to this model, at each change in field direction the DNA molecules moved in a new direction led by one of their ends. Smith's experiments showed that by pulsing the current and changing its direction in a perpendicular fashion, the DNA molecules would advance lengthwise through the gel in a zig-zag pattern. The leading end of the DNA molecule often appeared to be condensed, suggesting that the front end of the molecule migrates slower than the back end. Figure 6 shows a simplified model of the DNA movement, without indicating the elasticity of the molecules (which alternately contract and lengthen as they move through the gel). The ends of the DNA molecules are left at different positions along the gel, depending on their overall length. As the direction of the electric field is changed, the starting point for the new movement is accordingly different for molecules of different lengths. Consequently, the overall separation of DNA molecules is a function of their difference in length. In summary and as diagrammed in Figure 6, the DNA migrates in a zig-zag fashion through the gel, resulting in a one-dimensional net migration.

In recent studies, we have used PFGE to establish a large-scale physical map within the q21–q23 region of chromosome 6 [33]. A combination of single- and double-digests with different rare-cutting restriction enzymes were used to order the corresponding restriction fragments. Figure 7 shows a typical PFGE gel used in these studies and the map that was derived from this gel. This long-range restriction map is the first step toward a physical map spanning the distal long arm of chromosome 6. With the aid of PFGE, such an extended map is, in fact, feasible. The average size of a restriction fragment generated by rare-cutting enzymes varies usually from about 1000 kb (*Not*I) to about 200 kb (*Sfi*I). Given that chromosome 6 contains about 1.5×10^8 bp and the region 6q21-qter represents about 30% of chromosome 6, only 45 *Not*I fragments or 250 *Sfi*I fragments are required to span 6q21-qter. Such a physical map would allow an extended screening for rearrangements within this region in human malignant melanoma.

Even though a large-scale map is in many ways superior to conventional RFLP analysis, some potential problems also exit for this approach and deserve further comment. First, to identify restriction fragments in a PFGE gel, the DNA is immobilized on a nylon or nitrocellulose membrane and hybridized with a radioactive or biotin-labeled probe. Therefore, an adequate number of probes are required to detect the restriction fragments spanning the selected region. Although it might appear extremely tedious to obtain sufficient probes to saturate such a large region, in contrast to the RFLP approach, the application of this strategy is not restricted to polymorphic probes. All probes that are localized within the target region can be used in this strategy. As a result, far more probes are available to saturate a given region. After identification of the large-scale restriction fragments, the position of the fragments relative to each other needs to be determined. This can be accomplished by means of "linkage libraries." This approach, which was developed by Poustka and Lehrach [34] and Collins et al. [35], will be discussed later.

A second major problem of physical mapping by PFGE is DNA methylation. The recognition sites of rare-cutting restriction enzymes contain CpG

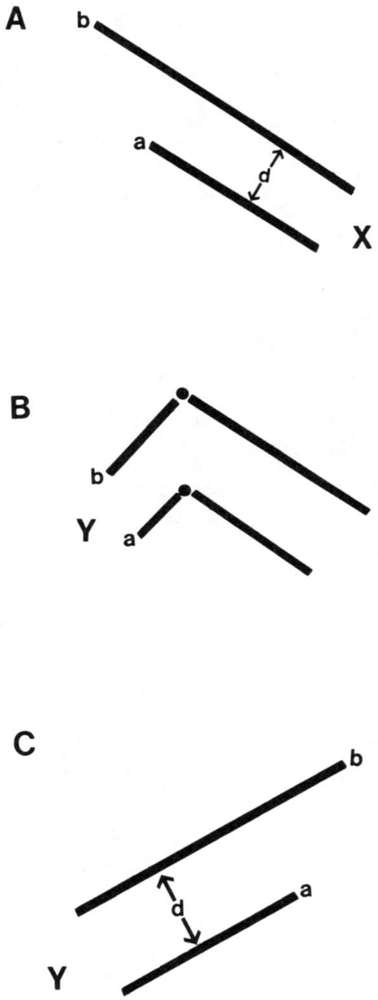

Figure 6 Simplified model of the DNA migration in electric fields generated by PFGE: (A) Two molecules a and b are aligned along the field with the ends moving in a direction X. (B) With the field reversed the molecules start to align in a new direction Y. (C) The molecules are aligned in the new direction Y. As indicated by arrows, the distance d between the molecules a and b has increased.

sequences that are very rare in mammalian DNA [36,37]. Except for *Hpa*I tiny fragments (HTF) islands that contain clusters of unmethylated CpG sequences, the CpG sequence is preferentially methylated in mammalian DNA (thereby, inhibiting the activity of the methylation-sensitive rare-cut-

(A)

(B)

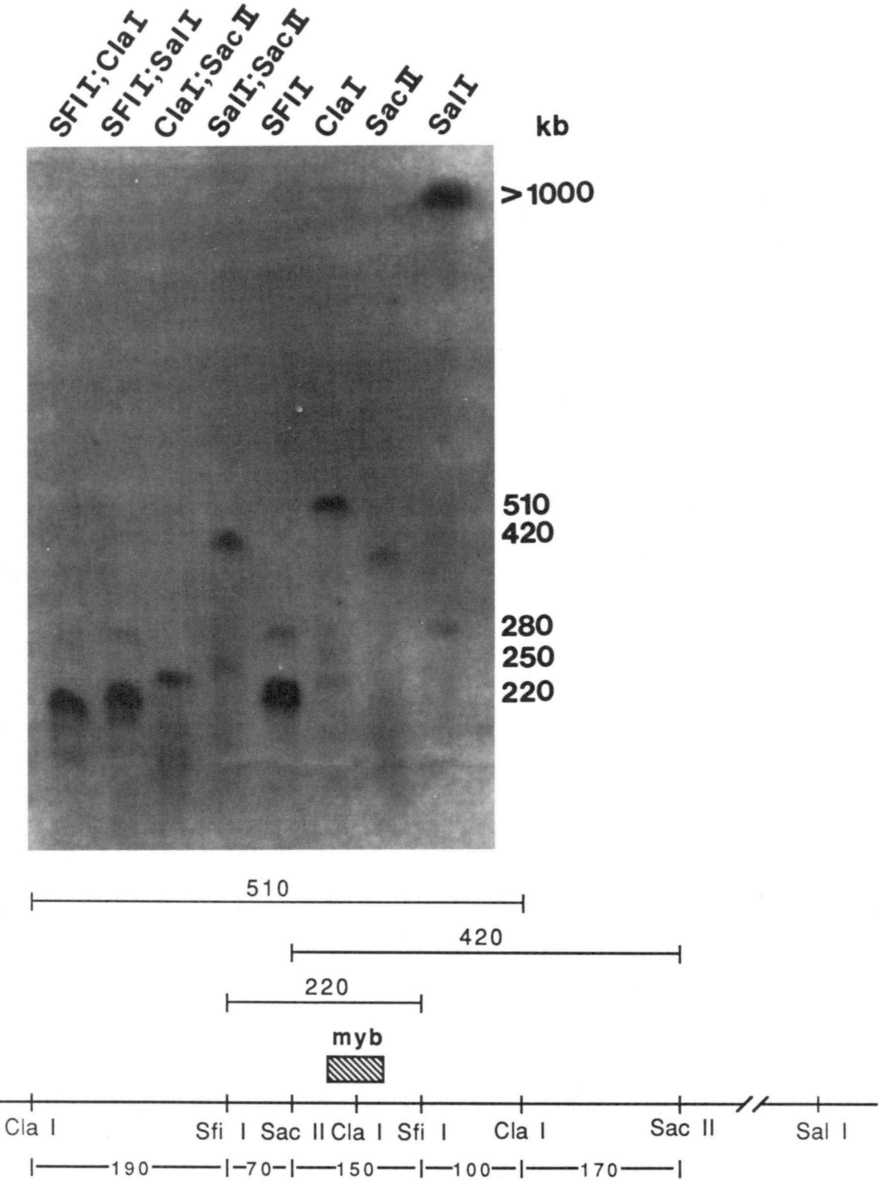

Figure 7 (A) PFGE analysis of the *MYB* gene. The DNA was cleaved with the indicated restriction enzymes to generate single and double digests. The running conditions were as follows: 120 V, 55-sec pulse time, 42-hr run time, temperature 12°–14°C, 1% agarose in 0.5% TBE (0.045 M Tris-borate, 0.001 M EDTA). (B) The long-range restriction map of the *myb* gene was derived from a combination of single and double digests. The fragments identified by single enzyme cleavage are indicated by bars. (From Ref. 33, with permission)

ting enzymes). Because the level of methylation is cell-specific as well as tissue-specific, the same enzyme is likely to yield a different restriction pattern in different tissues, even of the same individual. Hence, in contrast to RFLP studies, normal and tumor tissue from one patient is often difficult to compare. A pattern variation between tumor tissue and normal tissue may merely reflect a difference in methylation status, or it may indicate a tumor-specific rearrangement. To rule out the possibility of methylation, the control cells should stem from the same kind of tissue as the tumor cells. Additionally, some rare-cutting restriction enzymes are less methylation-sensitive than others and, consequently, they are less likely to identify a variation in the methylation status. Hence, the final proof for a tumor-specific rearrangement remains as cloning and sequencing of the target DNA.

Figure 8 demonstrates an example of a variation of the restriction pattern on a PFGE gel found between control and melanoma DNA. Melanoma DNA and normal DNA were digested with *Cla*I, separated by PFGE, and probed with c-*MYB*. Although a 510-kb signal has been found in lymphocyte DNA, melanoma DNA exclusively showed signals at 240 kb and 270 kb with *Cla*I. To determine if this variation reflected a difference in the methylation status, melanoma and normal DNA were digested using the enzyme *Sfi*I. This enzyme is known to be less methylation-sensitive than *Cla*I. With *Sfi*I, no difference between control DNA and melanoma DNA was observed, suggesting that the variation found with *Cla*I is due to the altered methylation status of a *Cla*I site within the *MYB* gene.

Because of its constant pattern in melanoma and control DNA, *Sfi*I was used for screening additional melanoma cell lines. In two cases, specific alterations of the restriction pattern were found. Interestingly, both cell lines demonstrated cytogenetic alterations of the band region 6q13. These changes in the restriction pattern have been confirmed with additional rare-cutting enzymes. The alterations appear unlikely to be due to methylation differences and lend further support for the recurring alterations of the band region 6q13-23 in malignant melanomas.

VI. LINKAGE LIBRARIES

Linkage libraries can be used to order large restriction fragments that have been separated by PFGE. *Linkage libraries* consist of genomic clones that contain the restriction site for a particular rare-cutting enzyme [38,39]. A strategy for construction of a linkage library is diagrammed in Figure 9. Genomic DNA is first digested with a frequent-cutting restriction enzyme. Following size fractionation and purification, the DNA fragments are circularized at low DNA concentrations. The low DNA concentration favors the circularization of the two ends of a single DNA molecule (intramolecular

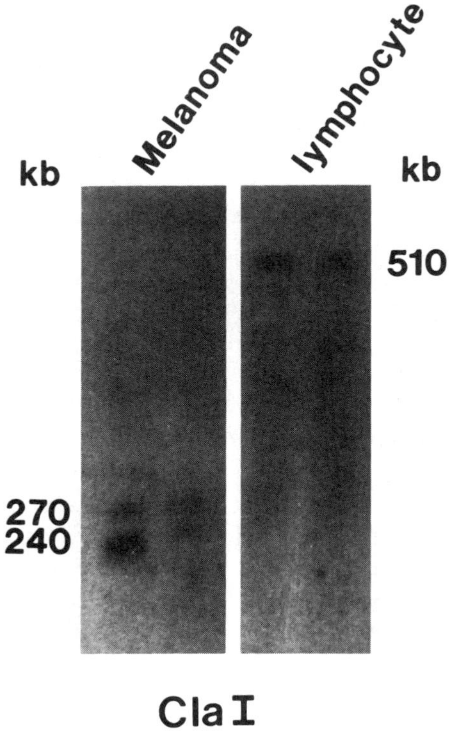

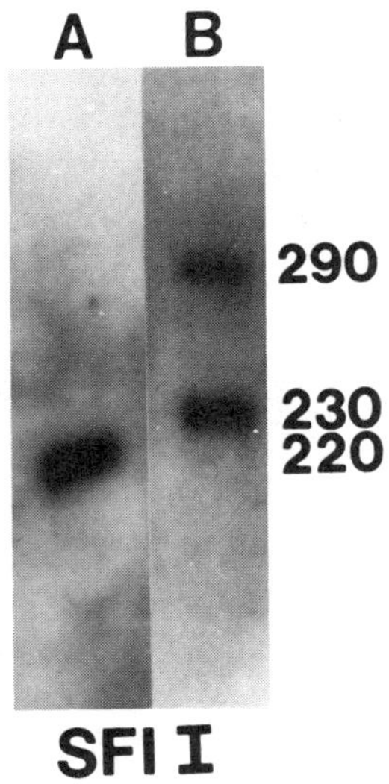

Figure 8 (A) PFGE anaylsis of the *MYB* region from human melanoma and lymphocyte DNA: The DNA was restricted by *Cla*I (running conditions as in Fig. 7). (B) PFGE anaylsis of the *MYB* region comparing DNA from two different melanoma cell lines. The DNA was cleaved by *Sfi*I (running conditions as in Fig. 7). (From Ref. 33, with permission)

ligation) over the ligation of two different molecules (intermolecular ligation). It can be advantageous to ligate a marker element into the monomer circle to enable the identification of the desired recombinant molecules in subsequent steps. In contrast with a jumping library, a linkage library can be established without a selectable marker [40]. Subsequent to the ligation, the circles are restricted with a rare-cutting enzyme and cloned into an appropriate vector system. Since only circles that contain a restriction site for a rare-cutting enzyme are linearized, this step represents a highly specific selection for linkage clones.

The linkage clones can then be used to analyze PFGE gels. Genomic DNA is digested with the same rare-cutting enzyme used for the linkage library, separated by PFGE, and probed with a linkage clone. Each linkage clone should hybridize to two DNA fragments within the gel, indicating that these two fragments are adjacent in the genome. A rapid ordering of the DNA fragments can thereby be accomplished. Theoretically, only nine linkage clones would be required to span a region of 10,000 kb, providing an average fragment size is 1000 kb. However, given a single-linkage library, it is highly unlikely that one could establish a complete physical map for a chromosomal region. Because the foregoing outlined procedure includes size selection, the linkage library will not contain all sites for a given restriction enzyme. An incomplete library may also result from a low cloning efficiency or an unequal distribution of restriction sites (CpG clusters). In addition, fragments of similar size will comigrate in a PFGE gel. These fragments cannot be distinguished by a subsequent hybridization with a linkage probe; that is, after hybridization, only one signal will be found on a PFGE blot and, consequently, identically sized (and therefore overlapping) fragments cannot be identified. These problems can be overcome by the use of several linkage libraries established with different rare-cutting enzymes. In this manner, the ordering of the restriction fragments for a particular enzyme can be confirmed and completed.

VII. MICRODISSECTION OF BANDED CHROMOSOMES

The starting material is extremely important for all of the afore-mentioned approaches for long-range physical mapping (i.e., the isolation of polymorphic probes, PFGE analysis, and the preparation of linkage libraries). The specificity of the starting material not only facilitates physical mapping, but also accounts for the completeness of a given map. Chromosome-sorted libraries or human–rodent cell hybrids are generally used to select a single chromosome or a portion of a chromosome as the starting material. Unfortunately, adventitious fragments of nonsyntenic human DNA often may be retained within flow-sorted peaks or cell hybrids. This represents

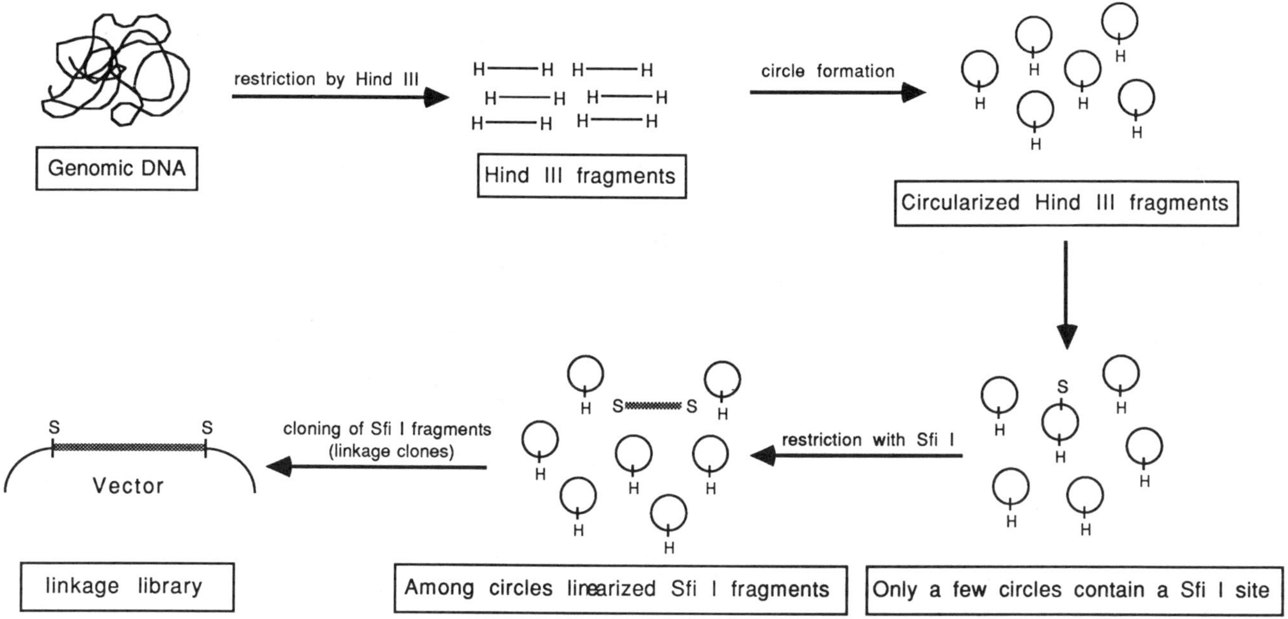

Figure 9 Preparation of an *Sfi*I linkage library: DNA is digested to completion with *Hind*III and subsequently size fractionated. Following self-ligation of the *Hind*III fragments, the circles are restricted by *Sfi*I. The linearized fragments are cloned into an appropriate vector.

a substantial problem in generating truly chromosome-specific libraries.

Therefore, a recently described technique that allows the isolation of a specifically defined chromosomal region by microdissection may be of real value [41]. By using a micromanipulator, individual bands can now be directly dissected from human metaphase chromosomes. The dissected chromosomal pieces are then treated with proteinase K and the DNA is isolated, essentially according to standard protocols (although scaled down to accommodate 1–2 pg/DNA). Following restriction digestion, the microdissected DNA is then cloned into a plasmid vector, and the inserts are amplified by the polymerase chain reaction (PCR), using primers that bind to the cloning vector (e.g., M13 pUC primer) [41].

Initial experiments with this system have documented that the approach does indeed provide very specific clones for the dissected chromosomal region. Although it remains to be seen whether microdissection is generally applicable to other chromosomal band regions, this does appear to represent an advance in the field of physical mapping. The value of these microdissection libraries will be the rapid generation of numerous new RFLP probes and linkage clones isolated for a defined chromosomal region.

VIII. CONCLUSION

The combined application of cytogenetic analysis and molecular mapping outlined in this chapter offers an increasingly important avenue to analyze tumor-specific chromosome rearrangements. Cytogenetic study remains a very important first step in pinpointing the general location of genes that may subsequently be more precisely analyzed by molecular genetic techniques.

Bridging the gap between cytogenetics (which provides the first evidence of tumor-specific rearrangements) and molecular biology (which provides definitive isolation and characterization of genes involved in chromosomal alterations) will be of major importance to cancer biologists for the foreseeable future.

REFERENCES

1. Heim, S., and F. Mitelman (1987). *Cancer Cytogenetics*. Alan R. Liss, New York.
2. Heim, S., and F. Mitelman (1989). Primary chromosome abnormalities in human neoplasia. *Adv. Cancer Res. 52*:1–43.
3. Trent, J. M., Y. Kaneko, and F. Mitelman (1989). Report of the Committee on Structural Chromosome Changes in Neoplasia, HGM10. *Cytogenet. Cell Genet. 51*:533–562.

4. Lai, E., B. W. Birren, S. M. Clark, M. I. Simon, and L. Hood (1989). Pulsed field gel electrophoresis. *Biotechniques 7*:34–41.
5. Collins, F. S., and S. M. Weissman (1984). Directional cloning of DNA fragments at a large distance from an initial probe: A circularization method. *Proc. Natl. Acad. Sci. USA 81*:6812–6816.
6. Rommens, J. M., M. C. Iannuzzi, B.-S. Kerem, et al. (1989). Identification of the cystic fibrosis gene: Chromosome walking and jumping. *Science 245*:1059–1065.
7. Poustka, A., T. M. Pohl, D. P. Barlow, et al. (1987). Construction and use of human chromosome jumping libraries from *Not*I digested DNA. *Nature 325*: 353.
8. Berdeaux, D. H., T. E. Moon, and F. L. Meyskens (1985). Clinical biologic patterns of metastatic melanoma and their effect on treatment. *Cancer Treat. Rep. 69*:397–401.
9. Schreiber, M., P. Borozo, T. Moon, and J. Mardon (1982). Solar ultraviolet radiation and skin cancer: A public education program. *Cutis 30*:516–524.
10. Bale, S. J., N. C. Dracopoli, M. A. Tucker, W. H. Clark, M. C. Fraser, B. Z. Stanger, P. Green, H. Donis-Keller, D. E. Housman, and M. H. Greene (1989). Mapping the gene for hereditary cutaneous malignant melanoma dysplastic nevus to chromosome 1p. *N. Engl. J. Med. 320*:1367–1372.
11. Friedman, R. J., E. R. Heilman, D. K. Rigel, and A. W. Kopf (1985). Dysplastic nevus: Clinical and pathologic features. *Dermatol. Clin. 3*:239–249.
12. Dracopoli, N. C., P. Harnett, S. J. Bale, et al. (1989). Loss of alleles from the distal short arm of chromosome 1 occurs late in melanoma tumor progression. *Proc. Natl. Acad. Sci. USA 86*:4614–4618.
13. Trent, J. M., F. H. Thompson, and F. L. Meyskens (1989). Identification of a recurring translocation site involving chromosome 6 in human malignant melanoma. *Cancer Res. 49*:420–423.
14. Trent, J. M., S. B. Rosenfeld, and F. L. Meyskens (1983). Chromosome 6q involvement in human malignant melanoma. *Cancer Genet. Cytogenet. 9*:177–180.
15. Becher, R., Z. Gibas, C. Karakousis, and A. A. Sandberg (1983). Nonrandom chromosome changes in malignant melanoma. *Cancer Res. 43*:5010–5016.
16. Pathak, S., H. L. Driwinga, and T. C. Hsu (1983). Involvement of chromosome 6 rearrangements in human malignant melanoma cell lines. *Cytogenet. Cell Genet. 36*:573–579.
17. Balaban, G., M. Herlyn, D. Guerry, R. Bartolo, H. Koprowski, W. H. Clark, and P. C. Nowell (1984). Cytogenetics of human malignant melanoma and premalignant lesions. *Cancer Genet. Cytogenet. 11*:429–439.
18. Lee, W.-H., R. Bookstein, F. Hong, L.-J. Young, J.-Y. Shew, and E. Y.-H. Lee (1987). Human retinoblastoma susceptibility gene: Cloning, identification, and sequence. *Science 235*:1394–1399.
19. Koufos, A., M. F. Hansen, N. G. Copeland, N. A. Jenkins, B. C. Lampkin, and W. K. Cavenee (1985). Loss of heterozygosity in three embryonal tumors suggests a common pathogenic mechanism. *Nature 316*:330–334.

20. Meese, E. U., K. D. Zang, and N. Blin (1987). Loss of heterozygosity and the origin of meningioma. *Hum. Genet.* *77*:349–351.
21. Botstein, D., R. L. White, M. Skolnick, and R. W. Davis (1980). Construction of a genetic linkage map in man using restriction fragment length polymorphisms. *Am. J. Hum. Genet.* *32*:314–331.
22. Witkowski, C. M., H. Zoghbi, I. Sawada, S. Weissman, and J. M. Trent (1989). Assignment of *Not*1 linking clones to the long arm of human chromosome 6 by a somatic cell hybrid mapping panel. 10th International Human Gene Mapping Workshop. *Cytogenet. Cell Genet.* *51*:1109.
23. Nakamura, Y., M. Leppert, P. O'Connell, R. Wolff, T. Holm, M. Culver, C. Martin, E. Fujimoto, M. Hoff, E. Kumlin, and R. White (1987). Variable number tandem repeat (VNTR) markers for human gene mapping. *Science* *237*:1616–1622.
24. Wong, Z., V. Wilson, A. J. Jeffreys, and S. L. Thein (1986). Cloning a selected fragment from a human DNA "fingerprint": Isolation of an extremely polymorphic minisatellite. *Nucleic Acids Res.* *14*:4605–4616.
25. Nakamura, Y., M. Carlson, K. Krapcho, M. Kanamori, and R. White (1988). New approach for isolation of VNTR markers. *Am. J. Hum. Genet.* *43*:854–859.
26. Schwartz, D. C., and C. R. Cantor (1984). Separation of yeast chromosome sized DNAs by pulsed field gradient gel electrophoresis. *Cell* *37*:67–75.
27. Carle, G. F., M. Frank, et al. (1986). Electrophoresis separation of large DNA molecules by periodic inversion of the electric field. *Science* *232*:65–68.
28. Carle, G. F., and M. V. Olson (1984). Separation of chromosomal DNA molecules from yeast by orthogonal field alteration gel electrophoresis. *Nucleic Acids Res.* *12*:5647–5663.
29. Anand, R. (1986). Pulsed field gel electrophoresis: A technique for fractionating large DNA molecules. *Trends Genet.* *2*:278–283.
30. Chu, G., D. Vollrath, et al. (1986). Separation of large DNA molecules by contour clamped homogeneous electric field. *Science* *234*:1582–1585.
31. Southern, E. M., R. Anand, W. R. A. Brown, and D. S. Fletcher (1987). A model for the separation of large DNA molecules by crossed field gel electrophoresis. *Nucleic Acids Res.* *15*:5925.
32. Smith, S. B., P. K. Aldridge, and J. B. Callis (1989). Observation of individual DNA molecules undergoing gel electrophoresis. *Science* *243*:203–206.
33. Meese, E., P. S. Meltzer, C. M. Witkowski, and J. M. Trent (1989). Molecular mapping of the oncogene *MYB* and rearrangements in malignant melanoma. *Genes Chromosomes Cancer* *1*:88–94.
34. Poustka, A., and H. Lehrach (1986). Jumping libraries and linking libraries: The next generation of molecular tools in mammalian genetics. *Trends Genet.* *2*:174.
35. Collins, F. S., M. L. Drumm, J. L. Cole, et al. (1987). Construction of a general human chromosome jumping library with application to cystic fibrosis. *Science* *235*:1046–1048.
36. Gardiner-Garden, M., and M. Frommer (1987). CpG islands in vertebrate genomes. *J. Mol. Biol.* *196*:261–282.

37. Bird, A. P. (1986). CpG-rich islands and the function of DNA methylation. *Nature 321*:209–213.
38. Smith, C. L., S. K. Lawrance, G. A. Gillespie, et al. (1987). Strategies for mapping and cloning macroregions of mammalian genomes. *Methods Enzymol. 151*:461–489.
39. Buiting, K., E. Passarge, and B. Horsthemke (1988). Construction of a chromosome 15-specific linking library and identification of potential gene sequences. *Genomics 3*:143–149.
40. Collins, F. S. (1988). Chromosome jumping. In *Genome Analysis a Practical Approach*. K. E. Davis (ed.). Nuffield Department of Clinical Medicine, Oxford, pp. 73–94.
41. Ludecke, H.-J., G. Senger, U. Claussen, et al. (1989). Cloning defined regions of the human genome by microdissection of banded chromosomes and enzymatic amplification. *Nature 338*:348–350.

9

The Detection of Fragile Sites on Human Chromosomes

Grant R. Sutherland

Adelaide Children's Hospital, North Adelaide, Australia

I. INTRODUCTION

Fragile sites on chromosomes are points at which the chromosome shows a nonrandom tendency to reveal a gap or break when the cells from which the chromosomes are prepared are exposed to specific conditions of tissue culture or chemical agents [1].

Fragile sites are of interest for several reasons. The most common inherited form of mental retardation, the fragile X syndrome, is associated with, if not caused by, a fragile site on the end of the long arm of the X chromosome. Fragile sites have been implicated in the genesis of chromosomal rearrangements in malignant disease, although this remains an area of controversy [2,3]. Gene mappers have used the rare fragile sites as genetic markers for linkage studies [4,5] and both rare and common fragile sites to improve the resolution of chromosomal in situ hybridization [6] (Fig. 1).

II. CLASSIFICATION OF FRAGILE SITES

Fragile sites can be divided into two major groups: those that are rare, or relatively so, and those that are common. The distinction between the rare and common fragile sites is that the most frequent of the rare fragile sites is carried by about 1 in 20 individuals in the population of Germany [7], but all individuals are probably homozygous for the common ones. Both groups

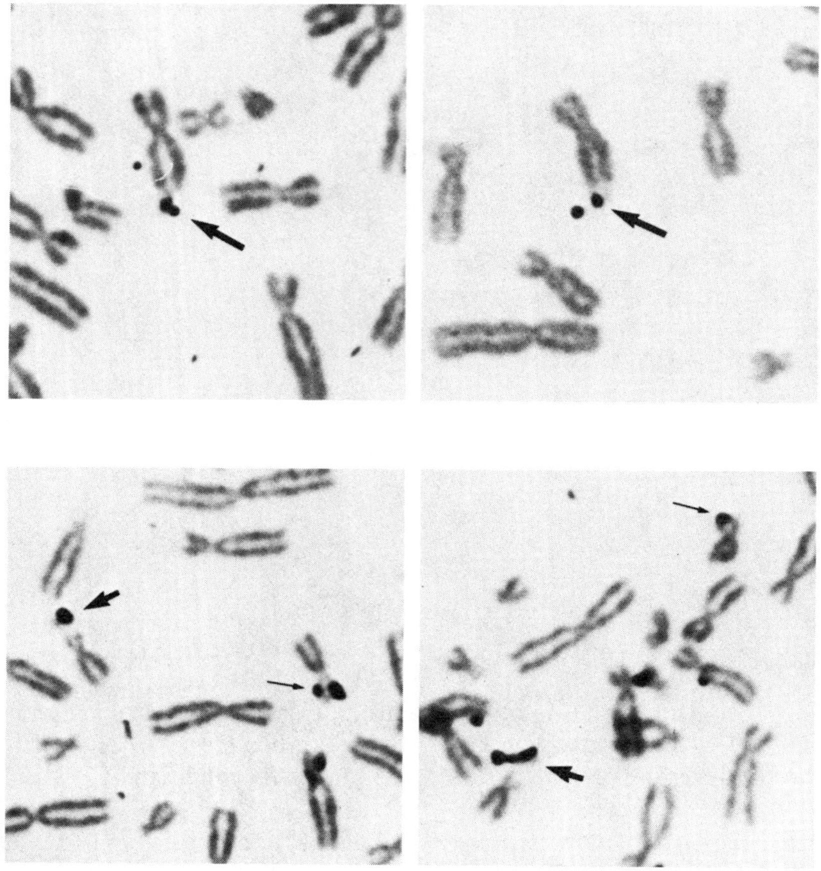

Figure 1 The use of fragile sites to increase the resolution of in situ hybridization. (a) The probe VK21C to locus *DXS296* [66] is shown to be distal to the fragile site at Xq27.3; (b) the probe *ACHF1* to locus *D16S8* [67] is distal to the rare folate-sensitive fragile site at 16q13.11 (*FRA16A*).

can be classified according to the conditions of tissue culture under which they are expressed. All known fragile sites are inducible, even though some are expressed spontaneously. The mode of induction provides a basis for classification.

Twenty-five rare fragile sites are known and, of these, 18 are folate-sensitive. Five come into a second group, the distamycin A-inducible fragile sites, and two bromodeoxyuridine (BrdU)-requiring fragile sites constitute a third group. The common fragile sites form three groups on the basis of induction by aphidicolin, 5-azacytidine, or BrdU.

A. The Rare Fragile Sites (Table 1, Fig. 2)

1. Group 1. Folate-sensitive fragile sites

The folate-sensitive fragile sites are induced in lymphocyte chromosomes by low concentrations of folic acid and thymidine in the culture medium [8]; by inhibitors of folate metabolism such as methotrexate [8,9]; by specific inhibition of thymidylate synthetase by fluorodeoxyuridine (FUdR) [10,11] or fluorodeoxycytidine (FCdR) [12]; and by high concentrations of thymidine [13,14].

2. Group 2. Distamycin A-Inducible Fragile Sites

There are two subgroups in this category:

a. This subgroup contains only fra(16)(q22) and fra(17)(p12). These are often expressed spontaneously in lymphocyte culture, but frequently require induction. Induction can be with distamycin A [15], BrdU [16], bromodeoxycytidine (BrdC) Netropsin, Hoechst 33258 [16], or diminazine aceturate (Berenil) [7]. Fra (16)(q22) is probably the most common of the rare fragile sites, at least in some populations [7]. This subgroup is distinguished from the following one, which cannot be induced by BrdU.
b. The fragile sites in this subgroup have been described only in the Japanese population. They are induced by distamycin A and the other agents that induce group 2a, except for BrdU and presumably BrdC [18–20].

3. Group 3. Bromodeoxyuridine-Requiring Fragile Sites

The fragile site at 10q25 is the second most common of the rare fragile sites, being present in 1 of 40 persons within the Australian population [21]. It can be induced with BrdU or BrdC [22,23]. There is a single report of a BrdU-requiring fragile site at 12q24.2 [24].

Table 1 The Rare Fragile Sites

Group 1	2q11.2, 2q13, 2q22.3, 6p23, 7p11.2, 8q22.3, 9p21.1, 9q32, 10q23.3, 11q13.3, 11q23.3, 12q13.1, 12q24.13, 16p13.11, 19p13, 20p11.23, 22q13, Xq27.3.
Group 2	(a) 16q22.1, 17p12.
	(b) 8q24.1, 11p15.1, 16p21.1.
Group 3	10q25.2, 12q24.2.

Source: From Ref. 37.

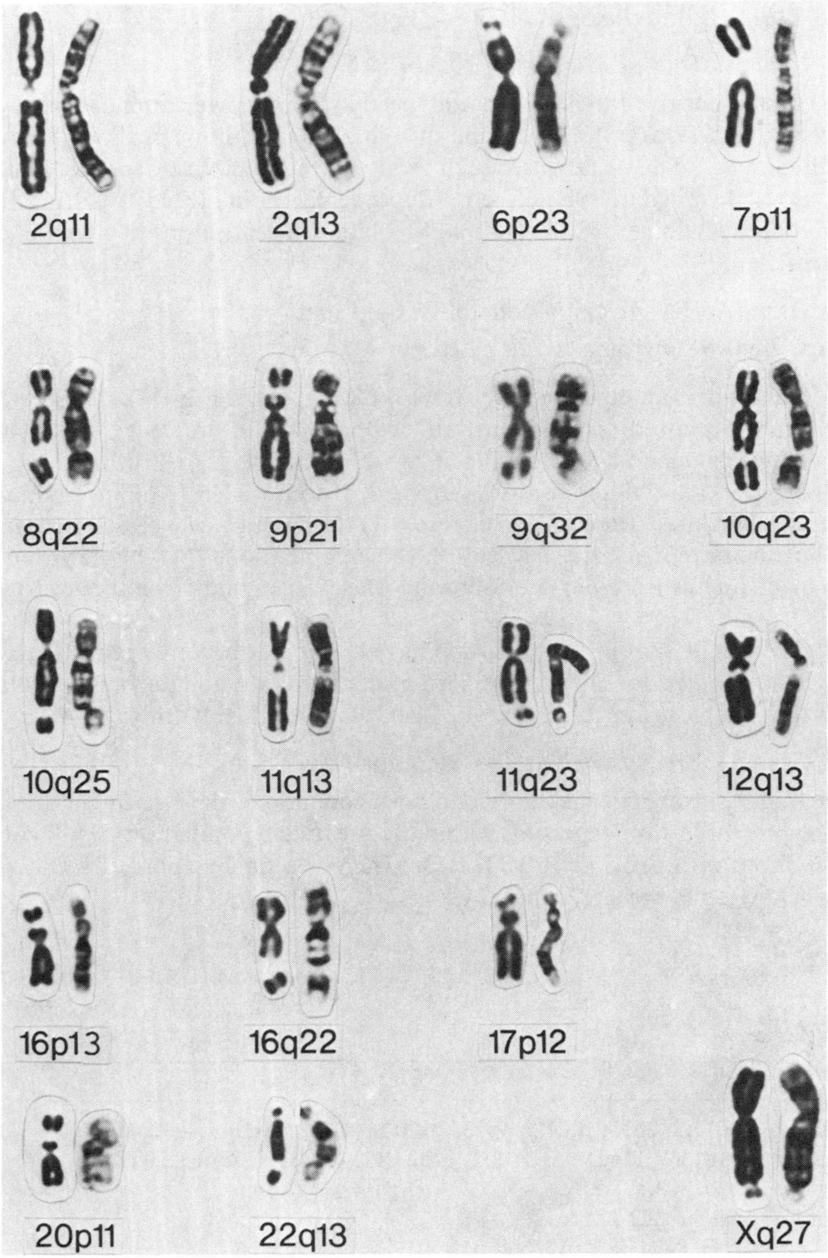

Figure 2 Eighteen of the rare fragile sites. Unbanded and G-banded chromosomes show the morphology and location of each fragile site.

B. The Common Fragile Sites

1. Group 4. Aphidicolin-Inducible Fragile Sites

The aphidicolin-inducible fragile sites are seen at low levels of expression under conditions that induce the rare folate-sensitive fragile sites [25], but can be specifically induced by aphidicolin [26]. There have been numerous lists of common fragile sites published, and, although there is considerable overlap between the lists, there is little consensus on which chromosomal bands contain one of these fragile sites [25–36]. Those listed in Table 2 have been accepted by the Human Gene Mapping Workshop [37]. This group of fragile sites has been mapped with various degrees of precision and, apparently, different fragile sites that map close to each other may be only a single site.

2. Group 5. 5-Azacytidine-Inducible Fragile Sites

These are induced in lymphocyte culture by the addition of 5-azacytidine 5–8 hr before harvest of cells for chromosomal study [34,38]. These are distinct from the noncondensation of C-band heterochromatin induced by 5-azacytidine in chromosomes 1, 7; 9, 16, and the Y.

3. Group 6. Bromodeoxyuridine-Inducible Fragile Sites

These are induced in lymphocyte culture by the addition of BrdU 4–6 hr before harvest of cells for chromosomal study [34,38].

C. Other Possible Fragile Sites

There are several other possible rare fragile sites that have been the subject of only single reports, often without adequate detail to allow one to form a conclusion on whether they fulfill the definition of a fragile site, and if they

Table 2 The Common Fragile Sites

Group 4	1p36, 1p32, 1p31, 1p31.2, 1p22, 1p21.2, 1q21, 1q25.1, 1q31, 1q44, 2p24.2, 2p16.2, 2p13, 2q21.3, 2q31, 2q32.1, 2q33, 2q37.3, 3p24.2, 3p14.2, 3q25, 3q27, 4p16.1, 4p15, 4q31.1, 5p14, 5q15, 5q21, 5q31.1, 6p25.1, 6p22.2, 6q15, 6q21, 6q26, 7p22, 7p14.2, 7p13, 7p21.2, 7q11, 7q21.2, 7q22, 7q31.2, 7q32.3, 7q36, 8q22.1, 8q24.1, 8q24.3, 9q22.1, 9q32, 10q22.1, 10q25.2, 10q26.1, 11p15.1, 11p14.2, 11p13, 11q14.2, 11q23.3, 12q21.3, 12q24, 13q13.2, 13q21.2, 13q32, 14q23, 14q24.1, 15q22, 16q22.1, 16q23.2, 17q23.1, 18q12.2, 18q21.3, 20p12.2, 22q21.2, Xp22.31, Xq22.1, Xq27.2.
Group 5	1q12, 1q42, 9q12, 19q13.
Group 6	4q12, 5p13, 5q15, 6q13, 9p21, 10q21, 13q21.

Source: From Ref. 37.

do, whether they fall into one of the foregoing groups. These reports have been summarized by Sutherland and Hecht [39]. There are many other possible common fragile sites that have been reported in some studies, but not in others.

III. NOMENCLATURE

The nomenclature for the cytogenetic description of fragile sites is given by the International System for Human Cytogenetic Nomenclature (ISCN; 1985) [40]. Briefly, a fragile site at band 10q23 in a female is designated 46,XX,fra(10)(q23); should the patient have two fragile sites on different chromosomes, they are expressed as 46,X,fra(X)(q27),fra(10)(q23). If a male had two fragile sites on the same homologue of chromosome 16, this would be expressed as 46,XY,fra(16)(p13q22).

For gene-mapping studies, nomenclature is different and is outlined in Shows et al. [41]. For example, the gene symbol for fra(X)(q27.3) is *FRAXA*.

IV. DETECTION OF FRAGILE SITES

This subject will be considered separately for each group of fragile sites and for each type of tissue. The principal methodology given is for the detection of fragile sites in phytohemagglutinin (PHA)-stimulated blood lymphocyte cultures. Standard cytogenetic technology [for details of such technologies see Ref. 42] will not be outlined in this chapter; only modifications of standard procedures required for fragile site detection will be given.

Most fragile sites have not been studied in tissues other than PHA-stimulated lymphocytes. The other tissues of main interest are amniotic fluid and chorion villus biopsies for the prenatal diagnosis of the fragile X syndrome.

A. Lymphocytes

1. The Rare Folate-Sensitive Fragile Sites (Group 1)

Four main methods are used for induction of this group of fragile sites. The simplest is given first.

a. Folic Acid-Free and Thymidine-Free Media. Sutherland's original method [8] is still the simplest reliable one for fragile site detection [12], although it is applicable to only lymphocyte cultures and is not suitable for the study of patients receiving large doses of folic acid or other folate preparations. In such cases, one of methods b or c should be used; method d has not been tested in this situation.

1. The original culture medium in this category was TC199, which contains only 0.01 mg/L of folic acid. This is not enough to inhibit the expression of folate-sensitive fragile sites. Use of TC199 often results in poor quality chromosomal preparations; other more suitable media that are commercially available include Eagle's MEM-FA (Commonwealth Serum Laboratories, Melbourne, Australia), 'M' (Gibco), and RPMI 1640 without folic acid (Irvine Scientific, Santa Ana, Calif.), which contain no folic acid or thymidine. Any other culture medium that is free of folic acid and thymidine should be suitable.
2. Buffer medium with 20 mM HEPES to initial pH of 7.5–7.6; repeat pH testing at time of addition of colchicine. If it is less than 7.3 then increase initial pH to 7.8, or increase buffer concentration up to 40 mM, or decrease cellular concentration in the culture. An adequate buffer capacity is provided by 20 mM HEPES for inoculums of 0.2 mL of whole blood in 5 mL of culture medium. Once the correct buffering capacity and initial pH for any medium have been established, it should no longer be necessary to test the pH when adding colchicine. The correct pH is vital for fragile X detection when using MEM-FA, but for medium 'M', it may not be so important [43].
3. Use less than the conventional serum supplement of 10–20%. In MEM-FA, 5% fetal bovine serum does not inhibit fragile site expression and yields cultures with a good mitotic index. Several commercial suppliers of fetal bovine serum provide assay results for folates and thymidine. If such data are available, it is advisable to select batches of serum in which these components are present in the lowest concentrations.
4. Ensure that the cultures are sterile. Microbial contamination does not always preclude conventional cytogenetic studies, but it renders cultures totally unsuitable for fragile site detection. Venous blood is preferred to capillary blood that has been collected by heel- or fingerprick, which is more likely to be microbially contaminated.
5. Allow cultures to incubate for at least 72 hr before harvest. Cells harvested after 48 hr in culture are usually adequate for conventional cytogenetics studies, but they are unsuitable for fragile site detection. Cultures harvested at 96 hr (in practice 4 days), although occasionally inferior in metaphase quality and yield, usually give slightly higher fragile site frequencies than 72-hr cultures.
6. Harvesting and staining procedures are standard.

b. Induction with Inhibitors of Dihydrofolate Reductase. This method was originally demonstrated in principle by Sutherland [8] and developed by Mattei et al. [9]. There are several essential methodologies specific for each of these inducers.

Methotrexate and Aminopterin

1. The use of medium that does not contain thymidine is important. Such media include the various modifications of Eagle's medium and RPMI 1640. Ham's F10 is unsuitable.
2. The concentration of methotrexate (MTX) to be used will need to be established for each type of complete medium, since there is a balance between the folate concentration in the medium and the concentration at which MTX becomes too toxic for adequate chromosomal preparations. Mattei et al. [9] found 10 mg/L satisfactory for TC199 with 20% human serum. Fonatsch [44] used 16 mg/L aminopterin to induce the fragile X in fibroblasts, and this should also work for lymphocytes.
3. Induction of fragile sites with MTX is maximized by adding it 12–24 hr before harvest [8]. Because MTX is toxic, it is best to reduce the time of cell exposure to a minimum. Hence, addition 16–24 hr before harvest (i.e., overnight) is preferable to a more prolonged induction.

Trimethoprim Trimethoprim has been shown [45] to induce the fragile X in lymphocyte cultures. Adding 300–600 mg/L of trimethoprim–sulfamethoxazole (Bactrim; Roche), which is equivalent to 50–100 mg/L of trimethoprim, 24 hr before harvest reliably induces fragile sites. Again a thymidine-free medium should be used, and adjustment of the concentration of this inhibitor may be required.

c. Induction with Inhibitors of Thymidylate Synthetase. This approach to fragile site induction was independently reported by Glover [10] and Tommerup et al. [11], using FUdR as an inhibitor to induce the fragile X. Jacky and Sutherland [12] used FCdR to induce several of the folate-sensitive fragile sites. FUdR is slightly more toxic than FCdR. Because thymidine can overcome the effect of the inhibitors, and folic acid can reduce their toxicities, the medium used should be thymidine-free, but contain folic acid; therefore, RPMI 1640 is suitable. Since the inhibitors are toxic, the minimal exposure consistent with fragile site induction is preferred. The optimum conditions are to add the inhibitor 16–24 hr (i.e., overnight) before harvest at 72 or 96 hr after culture initiation. The concentration of inhibitor may need to be adjusted, as small increases in concentrations of the inhibitor may cause large increases in cell toxicity. For RPMI 1640 with 5% fetal bovine serum, 0.01–0.05 mg/L of FUdR or FCdR usually proves satisfactory.

d. Induction With Excess Thymidine. Addition of excess thymidine (300 mg/L) 24 hr before harvest to lymphocytes cultured in RPMI 1640 supplemented with fetal bovine serum results in induction of this group of fragile sites [13,14]. As this concentration of thymidine has an inhibitory effect on DNA synthesis, the mitotic index of such cultures can be low. This can be improved by washing the cultured lymphocytes 7 hr before harvest twice in a folic acid- and thymidine-free medium (see foregoing Sect. IV.A.1) and continuing incubation in this medium until harvest. Usually, however, washing is not necessary, and addition of 300 mg/L thymidine 24 hr before harvest is the only departure from standard methods required for fragile site induction. The quality of chromosome preparations and the mitotic index is usually better when this method is used than when dihydrofolate reductase or thymidylate synthetase inhibitors are used.

2. The Other Rare Fragile Sites (Groups 2 and 3)

Three of these fragile sites (10q25, 16q22, 17p11) can be induced by the addition of 40 mg/L BrdU or BrdC added 16–24 hr before harvest [17]. A thymidine-free medium is preferred. If only 10q25 is of interest, then the time of addition is not so important and 10 mg/L of inducer added at culture initiation will suffice. This treatment is not satisfactory for 16q22 or 17p12, which are only minimally induced under these conditions.

If only 16q22 or 17p12 are of interest, then the method of choice for induction is 150 mg/L diminazine aceturate (Berenil) [7], added 24 hr before harvest.

There are three rare fragile sites induced by distamycin A (8q24.1, 11p15.1, and 16p12.1) that have been recorded in only the Japanese population [18,20]. These fragile sites were induced by the addition of 50 mg/L of distamycin A 24 hr before harvest. (Distamycin A is unstable in solution; therefore, a fresh aqueous solution needs to be prepared and sterilized by filtration immediately before addition to the cultures.) They were also induced by Hoechst 33258 (25 mg/L), Berenil (150 mg/L), and 4′6-diamidino-2-phenylindole (DAPI; 25 mg/L), but not by BrdU.

3. The Common Aphidicolin-Inducible Fragile Sites (Group 4)

These fragile sites will be seen at low frequencies under the conditions for induction of the rare folate-sensitive fragile sites. They are less frequently induced by high thymidine concentrations. They can be specifically induced by 0.07 mg/L of aphidicolin added to lymphocyte cultures in RPMI 1640 with 10% fetal bovine serum 24 hr before harvest [26]. Aphidicolin is insoluble in water and needs to be dissolved in dimethyl sulfoxide (DMSO) or ethanol before addition. Because aphidicolin can vary from batch to batch and its ability to induce the common fragile sites is related to the folic acid concentration in the medium used, the concentration of aphidicolin may need to be adjusted. Furthermore, it has been shown [31] that ethanol (but not DMSO) has a synergistic effect upon fragile site induction by aphidicolin.

Yunis and Soreng [28] induced this group of fragile sites by the addition of 0.1 μM (0.01 mg/L) FUdR 24 hr before harvest, followed by 2.2 mM (430 mg/L) caffeine, 6 hr before harvest, in medium MEM with 10% fetal bovine serum.

4. The Common 5-Azacytidine-Inducible Fragile Sites (Group 5)

These common fragile sites are induced in lymphocyte culture by the addition of 3–5 mg/L 5-azacytidine, 8 hr before harvest [38]. If the concentration is increased, the quality of metaphases is poor with the chromosomes becoming segmented and pulverized. If the 5-azacytidine is present for much longer than 8 hr before harvest, the fragile sites are not seen. The medium used in this study was RPMI 1640 supplemented with 5% fetal bovine serum, but the type of medium used to demonstrate these fragile sites is probably not important.

5. The Common Bromodeoxyuridine-Inducible Fragile Sites (Group 6)

The induction of these fragile sites is by the addition of 50 mg/L of BrdU 4–6 hr before harvest. If the BrdU is added for longer periods, the fragile sites are not seen. For example, none of these fragile sites are seen if BrdU is added 24 hr before harvest. Both MEM-FA and RPMI 1640 have been used to study these fragile sites, and the type of medium used appears to be unimportant [38].

6. Protocol for Fragile Site Detection in Lymphocytes

Any laboratory wishing to detect all the rare fragile sites needs to establish three lymphocyte cultures to achieve this.

a. A culture in folic acid- and thymidine-free medium to detect all the folate-sensitive fragile sites.
b. A culture in RPMI 1640 to which 40 mg/L BrdU is added 16–24 hr before harvest. This will detect the BrdU-inducible fragile sites and fra(16)(q22) and fra(17)(p11).
c. A culture in RPMI 1640 to which 150 mg/L Berenil is added 16–24 hr before harvest. This is primarily to detect the distamycin A-inducible fragile sites at 8q24.1, 11p15.1, and 16p12.1, which have been reported in only the Japanese population, but could be in other populations. This culture method will also allow detection of fragile sites at 16q22 and 17p12.
d. For any laboratory wishing to detect the common fragile sites, the methods have been given above, and a separate culture is required for each group of these fragile sites to be studied.

7. Cytogenetics

Fragile sites usually appear as a nonstaining chromosomal gap, often with some misalignment of the chromatids distal to the gap. The appearance can vary, however, from only a chromatid gap to deletion, despiralization, or duplication of the chromosomal material distal to the fragile site and other bizarre configurations (Fig. 3).

My own preference is for scoring unbanded chromosomes for fragile sites. If necessary, relevant metaphases can be subsequently banded. It is a simple technique to destain a slide (provided a Romanovsky stain has been used, rather than orcein) and then band it for chromosomal identification.

Most of the rare autosomal fragile sites are expressed in 10–40% of metaphases. Levels of expression are encountered outside this range, but only those below it present detection problems. Some obligate fragile site carriers remain refractory to cytogenetic demonstration, especially women who carry the fragile X, but such failure to detect a fragile site can also occur with autosomal fragile sites. The probability of excluding the presence of a fragile site for any number of metaphases examined can be determined from Table 3. In clinical cytogenetic laboratories in which the fragile X is being specifically sought, a standard cytogenetic analysis of 10–15 banded metaphases is also required to ensure that some other chromosomal anomaly is not present.

The common fragile sites are almost certainly present in homozygous form in all individuals, but some are much easier to demonstrate than others. For example, the sites at 3p14.2 and 16q23.1 can usually be seen in all individuals at a reasonable level of expression and, in some metaphases, both homologues will be seen to express the fragile site. Many of the other common fragile sites are seen in fewer than 5% of metaphases, and their detection may not be reproducible, even within a single laboratory [30].

The frequency of heterozygotes for the rare fragile sites in any population may vary. There have not been enough studies on the autosomal folate-sensitive sites to conclude anything other than that about 0.5–1% of individuals in populations that have been studied will have one of these [20,46,47]. The fragile X seems to have approximately the same frequency in all populations studied with about 1 : 2000–1 : 2500 children (of either sex) being mentally impaired because of the fragile X syndrome.

Of the other rare fragile sites, the BrdU-requiring fragile site at 10q25, which is present in 1 : 40 Australians [21] seems to be less frequent in the Italian [48] and Japanese populations [20]. The distamycin A-inducible fragile site at 16q22 would appear to be more common in the German population than in other populations studied [7]. The population cytogenetics of the rare fragile sites has been reviewed by Takahashi et al. [20].

B. Bone Marrow

Studies of fragile sites on chromosomes derived from bone marrow are rare. The fragile site at 10q25 has been clearly observed in bone marrow cells. The cells were from a patient with chronic myelogenous leukemia and the Philadelphia translocation. The fra(10)(q25) expressed itself spontaneously in

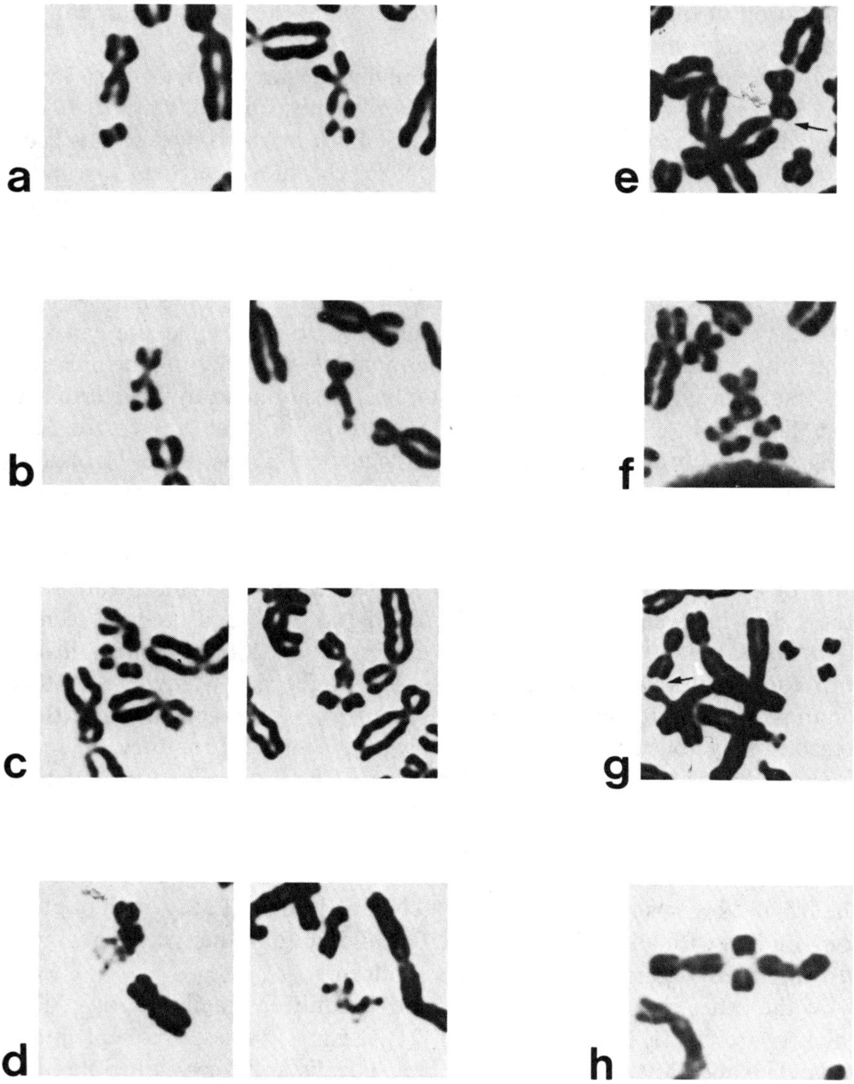

Figure 3 The distamycin A-inducible rare fragile site at 16q22 showing various forms of expression: (a) chromosome gaps; (b) a chromatid gap and a chromatid break; (c) duplication of the chromosome distal to the fragile site; (d) despiralization of the chromosome distal to the fragile site; (e) deletion at the fragile site (arrow); (f) multiradial configuration; (g) centric portion of the chromosome (arrow) separated within the metaphase from duplicated segments of chromosome distal to the fragile site; (h) quadriradial configuration.

Table 3 Percentage of Cells with a Fragile Site Excluded with 95% and 99% Confidence if Given Numbers of Cells Are Examined and not Found to Have a Fragile Site

Number of cells examined	Confidence 95%	Confidence 99%
10	26%	37%
20	14%	21%
30	10%	15%
50	6%	9%
100	3%	5%
250	2%	2%
460	1%	1%

Based on calculations by Hook [65] for the diagnosis of chromosomal mosaicism.

these bone marrow cells after a short-term incubation that did not include BrdU [49]. Haas et al. [50] observed fra(16)(q22) to be expressed spontaneously in the bone marrow of a patient with acute myeloid leukemia, but at a lower level of expression than that seen in uninduced lymphocytes.

Sutherland [51] examined 200 bone marrow metaphases from a boy with the fragile X and 60 metaphases from a girl with fra(16)(p12), but did not detect either fragile site. He used only a 2-hr culture in vitro in this study. Del Pozo and Millard [52] found the fragile X in 2% of bone marrow metaphases (compared with 38% in lymphocytes and 10% in fibroblasts) after 48-hr culture by using methotrexate synchronization with thymidine release, a method that would not have maximized fragile site expression.

Attempts to study fragile sites in bone marrow chromosomes would be experimental, and some method development would be required. The conditions for induction in lymphocyte cultures can be a guide.

C. The Fragile X and Prenatal Diagnosis

Because the rare fragile site at Xq27.3 is associated with the fragile X syndrome, its detection in cells cultured from amniotic fluid (AF) and chorionic villus samples (CVS) is important for prenatal diagnosis of the fragile X syndrome. Fragile X detection is only one aspect of prenatal diagnosis of the fragile X syndrome and is complemented by restriction fragment length polymorphism (RFLP) studies [53].

The detection of fragile sites in cultures from AF or CVS is an extension of their detection in fibroblast-type cells. Early attempts to elicit fragile site expression in fibroblasts were unsuccessful [54,55]. Fibroblasts from carriers of the fragile sites at 2q13, 10q23, 11p13, 16q22, 20p11, and Xq27.3, studied under conditions of folic acid depletion, resulted in autosomal fragile sites expressed in up to 4% of metaphases, but the fragile X was not seen [51].

The first reported demonstration of the fragile X in fibroblasts followed rigorous folate restriction [56], but this approach has not been reproducible. Other approaches have included induction by FUdR [10,11,57] and antifolates [9,44]. Gardner et al. [58] reported that increasing the L-methionine concentration of the medium by 100 mg/L enhanced fragile X expression. Following the finding that elevated levels of thymidine could induce the folate-sensitive fragile sites in lymphocytes, this approach was used [59], with reasonable success, to induce the fragile X and a number of the autosomal fragile sites in skin-derived fibroblasts. The method simply involved the addition of thymidine at a concentration of 600 mg/L to the cultures 24 hr before chromosome preparation.

Prenatal cytogenetic detection of the fragile X, to some extent, remains unsatisfactory in that both false-positive and false-negative diagnoses have been made [60–62]. More than one method of induction, FUdR, MTX, or high thymidine, should be used, and several hundred metaphases may need to be examined. Chang medium, a popular one for initiating growth in culture of AF and CVS cells, has been unsatisfactory for fragile X induction, and other media such as Ham's F10 are preferable. The method of choice is to use CVS, rather than AF. Initiate cell culture in Chang medium, subculture into this medium for 24 hr, change the medium to an induction medium, such as MEM-FA [8]. Iscove low folate medium, or RPMI 1640. The medium should be supplemented with only 5% fetal bovine serum to which either 10 mg/L MTX, 600 mg/L thymidine, or 0.01–0.05 mg/L FUdR has been added [61], and the culture harvested 24 hr later.

The recognition that there is a common fragile site at Xq27.2 that can be expressed in a very low proportion of metaphases, but that can be distinguished from the true fragile X [63], should help in the interpretation of results when only one or two metaphases expressing a fragile site on the end of the X chromosome are encountered (Fig. 4).

D. Lymphoblastoid Cell Lines

For laboratory studies of fragile sites lymphoblastoid cell lines should, in theory, be ideal. Unfortunately, they have not proved to be any more useful than fibroblasts because of the difficulties in inducing fragile sites in this type of cell. Simple folate depletion does not work [8,64]. The other induc-

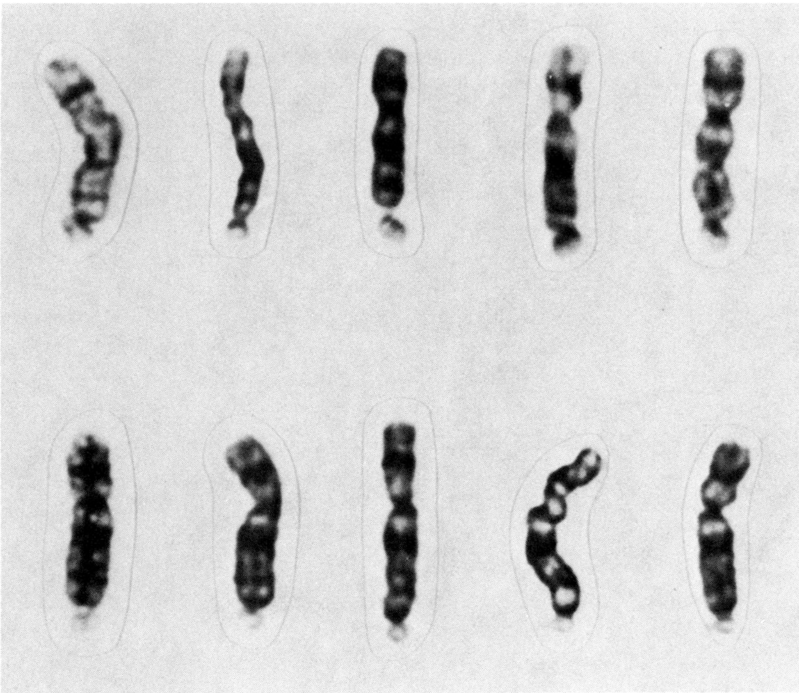

Figure 4 The rare and common fragile sites on the end of the long arm of the X chromosome. The upper row of chromosomes are expressing the common fragile site at Xq27.2, and the lower row are expressing the rare fragile site ('the fragile X') at Xq27.3.

tion methods for the folate-sensitive fragile sites can be made to work with varying degrees of success and reproducibility, but results vary from laboratory to laboratory. In general, the frequencies of expression are not as high as those obtained in lymphocytes.

The distamycin A-inducible fragile site at 16q22 seems to be virtually impossible to induce in anything other than lymphocyte cultures, but the fragile site at 16p12.1 (Fig. 5) is readily inducible by adding 25 mg/L Hoechst 33258 to cultures 24 hr before harvest [18] (see Fig. 5). The BrdU-inducible fragile site at 10q25 can also be induced in lymphoblastoid cultures, but not as readily as in lymphocytes.

ACKNOWLEDGMENTS

I thank Dr. D. F. Callen for helpful comments on an early draft of this chapter, Mrs. E. Baker for assistance in preparing the figures and for assis-

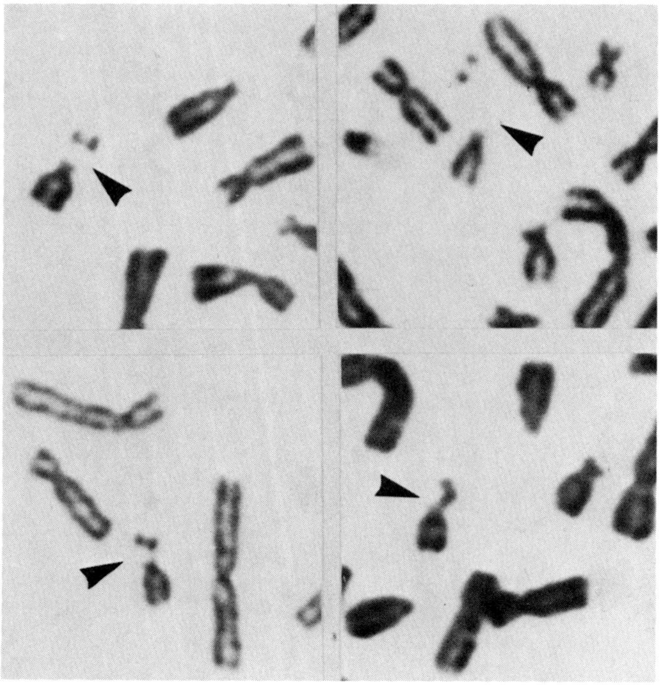

Figure 5 The distamycin A-inducible fragile site at 16p12 expressed in chromosomes from a lymphoblastoid cell line. (Cell line courtesy of Dr. T. Hori.)

tance with cytogenetic studies described. Dr. T. Hori generously provided the lymphoblastoid cell line used to prepare Figure 5.

REFERENCES

1. Berger, R., C. Bloomfield, and G. R. Sutherland (1985). Report of the committee on chromosome rearrangements in neoplasia and on fragile sites. *Cytogenet. Cell Genet. 40*:490–535.
2. Le Beau, M. M., and J. D. Rowley (1984). Heritable fragile sites in cancer. *Nature 308*:607–608.
3. Sutherland, G. R. (1988). Fragile sites and cancer breakpoints—the pessimistic view [Editorial]. *Cancer Genet. Cytogenet. 31*:5–7.
4. Mulley, J. C., C. Nicholls, and G. R. Sutherland (1983). A search for linkage in families with fragile sites. *Hum. Genet. 65*:79–81.
5. Mulley, J. C., V. J. Hyland, A. Fratini, L. J. Bates, A. K. Gedeon, and G. R. Sutherland (1989). A linkage group with *FRA16B* (the fragile site at 16q22.1). *Hum. Genet. 82:* 131–133.
6. Simmers, R. N., G. R. Sutherland, A. West, and R. I. Richards (1987). Fragile sites at 16q22 are not at the breakpoint of the chromosomal rearrangement in AMMoL. *Science 236*:92–94.

7. Schmid, M., W. Feichtinger, A. Jessberger, J. Köhler, and R. Lange (1986). The fragile site (16)(q22). I. Induction by AT-specific DNA-ligands and population frequency. *Hum. Genet. 74*:67–73.
8. Sutherland, G. R. (1979). Heritable fragile sites on human chromosomes. I. Factors affecting expression in lymphocyte culture. *Am. J. Hum. Genet. 31*: 125–135.
9. Mattei, M. G., J. F. Mattei, I. Vidal, and F. Giraud (1981). Expression in lymphocyte and fibroblast culture of the fragile X chromosome: A new technical approach. *Hum. Genet. 59*:166–169.
10. Glover, T. W. (1981). FUdR induction of the X chromosome fragile site: Evidence for the mechanism of folic acid and thymidine inhibition. *Am. J. Hum. Genet. 33*:234–242.
11. Tommerup, N., H. Poulsen, and K. Brondum-Nielsen (1981). 5-Fluoro-2-deoxyuridine induction of the fragile site on Xq28 associated with X linked mental retardation. *J. Med. Genet. 18*:374–376.
12. Jacky, P. B., and G. R. Sutherland (1983). Thymidylate synthetase inhibition and fragile site expression in lymphocytes. *Am J. Hum. Genet. 35*:1276–1283.
13. Sutherland, G. R., E. Baker, and A. Fratini (1985). Excess thymidine induces folate sensitive fragile sites. *Am. J. Med. Genet. 22*:433–443.
14. Sutherland, G. R., and E. Baker (1986). Effects of nucleotides on expression of the folate sensitive fragile sites. *Am. J. Med. Genet. 23*:409–417.
15. Schmid, M., C. Klett, and A. Niederhofer (1980). Demonstration of a heritable fragile site in human chromosome 16 with distamycin A. *Cytogenet. Cell Genet. 28*:87–94.
16. Croci, G. (1983). BrdU-sensitive fragile site on long arm of chromosome 16. *Am. J. Hum. Genet. 35*:530–533.
17. Sutherland, G. R., P. B. Jacky, and E. Baker (1984). Heritable fragile sites on human chromosomes. XI. Factors affecting expression of fragile sites at 10q25, 16q22 and 17p12. *Am. J. Hum. Genet. 36*:110–122.
18. Hori, T., E. Takahasi, and M. Murata (1988). Nature of distamycin A-inducible fragile sites. *Cancer Genet. Cytogenet. 34*:189–194.
19. Takahashi, E., Y. Kaneko, T. Ishihara, M. Minamihisamatsu, M. Murata, and T. Hori (1988). A new rare distamycin A-inducible fragile site, fra(11)(p15.1), found in two acute nonlymphocytic leukemia (ANLL) patients with t(7;11)(p15-p13;p15). *Hum. Genet. 80*:124–126.
20. Takahashi, E., T. Hori, and M. Murata (1988). Population cytogenetics of rare fragile sites in Japan. *Hum. Genet. 78*:121–126.
21. Sutherland, G. R. (1982). Heritable fragile sites on human chromosomes. IX. Population cytogenetics and segregation analysis of the BrdU requiring fragile site at 10q25. *Am. J. Hum. Genet. 34*:753–756.
22. Scheres, J. M. J. C., and T. W. J. Hustinx (1980). Heritable fragile sites and lymphocyte culture medium containing BrdU. *Am J. Hum. Genet. 32*:628–629.
23. Sutherland, G. R., E. Baker, and R. Seshadri (1980). Heritable fragile sites on human chromosomes. V. A new class of fragile site requiring BrdU for expression. *Am J. Hum. Genet. 32*:542–548.
24. Voiculescu, I., C. Hausmann, G. Wolff, and E. Back (1988). A BrdU-requiring fragile site on chromosome 12. *Hum. Genet. 78*:183–185.

25. Daniel, A., L. Ekblom, and S. Phillips (1984). Constitutive fragile sites 1p31, 3p14, 6q26 and 16q23 and their use as controls for false-negative results with the fragile X. *Am. J. Med. Genet.* *18*:483–491.
26. Glover, T. W., C. Berger, J. Coyle, and B. Echo (1984). DNA polymerase alpha inhibition by aphidicolin induces gaps and breaks at common fragile sites in human chromosomes. *Hum. Genet.* *67*:136–142.
27. Barbi, G., P. Steinbach, and W. Vogel (1984). Non-random distribution of methotrexate-induced aberrations on human chromosomes. Detection of further folic acid sensitive fragile sites. *Hum. Genet.* *68*:290–294.
28. Yunis, J. J., and A. L. Soreng (1984). Constitutive fragile sites and cancer. *Science* *226*:1199–1204.
29. Yunis, J. J., A. L. Soreng, and A. E. Bowe (1987). Fragile sites are targets of diverse mutagens and carcinogens. *Oncogene* *1*:59–69.
30. Craig-Holmes, A. P., L. C. Strong, A. Goodacre, and S. Pathak (1987). Variations in the expression of aphidicolin-induced fragile sites in human lymphocyte cultures. *Hum. Genet.* *76*:134–137.
31. Kuwano, A., and T. Kajii (1987). Synergistic effect of aphidicolin and ethanol on the induction of common fragile sites. *Hum. Genet.* *75*:75–78.
32. Tedeschi, B., B. Porfirio, P. Vernole, D. Caporossi, B. Dallapiccola, and B. Nicolette (1987). Common fragile sites: Their prevalence in subjects with constitutional and acquired chromosomal instability. *Am. J. Med. Genet.* *27*:471–482.
33. Green, R. J., D. L. Phillips, A. T. L. Chen, J. A. Reidy, and A. H. Ragab (1988). Effects of folate in culture medium on common fragile sites in lymphocyte chromosomes from normal and leukemic children. *Hum. Genet.* *81*:9–12.
34. Hecht, F., E. H. Tajara, D. Lockwood, A. A. Sandberg, and B. K. Hecht (1988). New common fragile sites. *Cancer Genet. Cytogenet.* *33*:1–9.
35. Rao, P. N., N. A. Heerema, and C. G. Palmer (1988). Fragile sites induced by FUdR, caffeine, and aphidicolin. *Hum. Genet.* *78*:21–26.
36. Fuster, C., R. Miro, C. Templado, L. Barrios, and J. Egozcue (1989). *Hum. Genet.* *81*:243–246.
37. Sutherland, G. R., and D. H. Ledbetter (1989). Report of the Committee on Cytogenetic Markers (HGM10). *Cytogenet. Cell Genet.* *51:* 452–458.
38. Sutherland, G. R., M. I. Parslow, and E. Baker (1985). New classes of common fragile sites induced by 5-azacytidine and BrdU. *Hum. Genet.* *69*:233–237.
39. Sutherland, G. R., and F. Hecht (1985). *Fragile Sites on Human Chromosomes*. Oxford University Press, New York.
40. ISCN (1985). An international system for human cytogenetic nomenclature. *Birth Defects Orig. Art. Ser.* *21*:46–47.
41. Shows, T. B., P. J. McAlpine, C. Boucheix, et al. (1987). Guidelines for human gene nomenclature. An international system for human gene nomenclature (ISGN, 1987). *Cytogenet. Cell Genet.* *46*:11–28.
42. Rooney, D. E., and B. H. Czepulkowski (eds.) (1986). Human Cytogenetics. A Practical Approach. IRL Press, Oxford.
43. Jacobs, P. A., P. A. Hunt, M. Mayer, P. Fox, J. W. Gerrard, H. G. Dunn, and D. S. Herbst (1980). X-Linked mental retardation: A study of 7 families. *Am. J. Med. Genet.* *7*:471–489.

44. Fonatsch, C. (1981). A simple method to demonstrate the fragile X chromosome in fibroblasts. *Hum. Genet. 59*:186.
45. Lejeune, J., N. Legrand, J. Lafourcade, M. O. Rethore, O. Raoul, and C. Maunoury (1982). Fragilité du chromosome X et effets de la triméthoprime. *Ann. Génét. 25*:149–151.
46. Sutherland, G. R. (1985). Heritable fragile sites on human chromosomes. XII. Population cytogenetics. *Ann. Hum. Genet. 49*:153–160.
47. Kähkonen, M., C. Tengstrom, T. Alitalo, R. Matilainen, M. Kaski, and E. Airaksinen (1989). Population cytogenetics of folate-sensitive fragile sites. II. Autosomal rare fragile sites. *Hum. Genet. 82*:3–8.
48. Sanfilippo, S., G. Neri, B. Tedeschi, N. Carolo-Stella, O. Triolo, and A. Serra (1983). Chromosomal fragile sites: Preliminary data of a population survey. *Clin. Genet. 24*:295.
49. Morgan, R., S. S. Morgan, and F. Hecht (1985). Enhanced expression of chromosome fragile site 10q25 in chronic myelogenous leukemia. *Am. J. Med. Genet. 21*:347–350.
50. Haas, O. A., E. Nacheva, and J. D. Schwarzmeier (1985). Noninvolvement of chromosome 16 in karyotype evolution of acute myeloid leukemia in a patient with a heritable fragile site on 16q22. *Cancer Genet. Cytogenet. 17*:1–12.
51. Sutherland, G. R. (1979). Heritable fragile sites on human chromosomes. II. Distribution, phenotypic effects, and cytogenetics. *Am. J. Hum. Genet. 31*: 136–148.
52. del Pozo, B. C., and P. R. Millard (1983). Demonstration of the fra(X) in lymphocytes, fibroblasts, and bone marrow in a patient with a testicular tumour. *J. Med. Genet. 20*:225–227.
53. Sutherland, G. R., and J. C. Mulley (1990). Diagnostic molecular genetics of the fragile X. *Clin. Genet. 37*:2–11.
54. Magenis, R. E., F. Hecht, and E. W. Lovrien (1970). Heritable fragile site on chromosome 16: Probable localization of haptoglobin locus in man. *Science 170*:85–87.
55. Fraccaro, M., K. Hansson, J. Lindsten, and L. Tiepolo (1971). Specific instability of the paracentric region of chromosome number 9 in a normal woman and her mongoloid son. *Ann. Génét. 14*:149–152.
56. Jacky, P. B., and F. J. Dill (1980). Expression in fibroblast culture of satellited-X chromosome associated with familial sex-linked mental retardation. *Hum. Genet. 53*:267–269.
57. Brookwell, R., A. Daniel, G. Turner, and J. Fishburn (1982). The fragile X(q27) form of X-linked mental retardation. FUdR as an inducing agent for fra(X)(q27) expression in lymphocytes, fibroblasts and amniocytes. *Am. J. Med. Genet. 13*:139–148.
58. Gardner, A. P., R. T. Howell, and A. McDermott (1982). Fragile X chromosome: Consistent demonstration of fragile site in fibroblast cultures. *Lancet 1*:101.
59. Sutherland, G. R., and E. Baker (1986). Induction of fragile sites in fibroblasts. *Am. J. Hum. Genet. 38*:573–575.
60. Tommerup, N., P. Aula, B. Gustavii, A. Heiberg, G. Holmgren, H. von Koskull, J. Leisti, M. Mikkelsen, F. Mitelman, K. B. Nielsen, P. Steinbach, S. Stengel-Rutkowski, J. Wahlström, K. Zang, and M. Zankl (1986). Second

trimester prenatal diagnosis of the fragile X. *Am. J. Med. Genet. 23*:313–324.

61. Purvis-Smith, S., S. Laing, E. Baker, and G. R. Sutherland (1988). Prenatal diagnosis of the fragile X—the Australasian experience. *Am. J. Med. Genet. 30*:337–345.
62. Shapiro, L. R., P. L. Wilmot, P. D. Murphy, and W. R. Breg (1988). Experience with multiple approaches to the prenatal diagnosis of the fragile X syndrome: Amniotic fluid, chorionic villi, fetal blood and molecular methods. *Am. J. Med. Genet. 30*:347–354.
63. Sutherland, G. R., and E. Baker (1989). The common fragile site in band q27 of the human X chromosome is not coincident with the fragile X. *Clin. Genet. 37:*167–172.
64. Jacobs, P. A., P. A. Hunt, M. Mayer, J. Wang, G. R. Boss, and R. W. Erbe (1982). Expression of the marker (X)(q28) in lymphoblastoid cell lines. *Am. J. Hum. Genet. 34*:552–557.
65. Hook, E. B. (1977). Exclusion of chromosomal mosaicism: Tables of 90%, 95% and 99% confidence limits and comments on use. *Am. J. Hum. Genet. 29*: 94–97.
66. Suthers, G. K., D. F. Callen, V. J. Hyland, H. M. Kozman, E. Baker, H. Eyre, P. S. Harper, S. H. Roberts, M. C. Hors-Cayla, K. E. Davies, M. V. Bell, and G. R. Sutherland (1989). A new DNA marker tightly linked to the fragile X locus (*FRAXA*) localises Hunter syndrome distal to *FRAXA*. *Science 264:*1298–1300.
67. Callen, D. F., V. J. Hyland, E. G. Baker, A. Fratini, A. K. Gedeon, J. C. Mulley, K. E. W. Fernandez, M. H. Breuning, and G. R. Sutherland (1989). Mapping the short arm of human chromosome 16. *Genomics 4*:348–354.

10

Immunocytogenetics: Antinuclear Antibodies as Probes for Chromosomal Structure and Function

Thomas Haaf

Stanford University School of Medicine, Stanford, California

Michael Schmid

University of Würzburg, Würzburg, Germany

I. INTRODUCTION

Only a few decades ago little information existed on the structure of the mammalian chromosome. An important milestone was the discovery that chromosomes consist of differential bands of various lengths [1]. The banding procedures reveal a level of organization immediately below that of the whole chromosome. Another breakthrough was the development of in situ hybridization as a method for chromosome mapping [2]. Banding and in situ hybridization techniques made possible impressive advances in clinical and experimental cytogenetics.

A new experimental approach to analyze chromosomal structure and function is the combining of immunological–biochemical methods with the classic cytogenetic procedures. These immunocytogenetic techniques permit a direct visualization of individual components of the chromosomes and of the cell nucleus. The basic principle of immunocytogenetics is the use of antinuclear antibodies for the in situ localization of macromolecules in chromosomes and cell nuclei. The specific binding of the antinuclear antibodies to their complementary antigens is visualized by fluorescence- or enzyme-labeled secondary antibodies directed against the primary antibodies.

Antinuclear antibodies that may be of interest to chromosome research can be divided conveniently into two categories: antibodies to acid-fast antigens and antibodies to acid-sensitive protein antigens. This is because conventional cytogenetic methods for chromosomal preparation employ an alcohol–acetic acid fixation. Immunolocalization of antigens sensitive to acid therefore requires special methods for chromosomal spreading and fixation and is technically much more complicated. This chapter deals with antibody labeling of acid-fixed and non–acid-fixed mammalian chromosomes. Although a variety of antinuclear antibodies have been successfully used for chromosome analysis, immunocytogenetics is a technique that, in our opinion, is just beginning to be established.

II. IMMUNOLOCALIZATION OF ACID-FAST ANTIGENS

Conventional acid-fixed preparations can be used for the immunolocalization of antigens that are not sensitive to acid. The acid-fast antigens are either DNA (and its conformational variants) or certain nucleotide determinant groups in single-stranded DNA. Antibodies to single-stranded DNA recognize the nucleotide bases themselves, which are hidden in the interior of the β-helix in native (double-stranded) DNA; only denaturation permits access of the antibody molecules to the antigenic determinants.

A. Nucleotide Determinant Groups

Immunocytogenetics began with the pioneering work of Erlanger, Miller, and colleagues, who first applied nucleotide-specific antibodies to metaphase chromosomal preparations [3,4]. Antibodies reacting with certain nucleotide determinant groups in single-stranded (denatured) DNA were made by immunizing animals to nucleotide–protein conjugates [5]. Immunofluorescent staining with antiadenosine and anticytosine antibodies revealed that chromosome banding patterns reflect AT- and GC-rich regions in DNA. Antiadenosine produces a consistent banding pattern, reminiscent of quinacrine staining or Giemsa G-banding; anticytosine banding is almost identical with R-banding (Fig. 1). Surprisingly enough, the quinacrine-bright heterochromatin of the human Y chromosome remains unstained by antibodies to adenosine. This indicates that differences in base pair composition alone cannot explain the phenomenon of chromosome banding; chromosomal proteins may also play an important role.

1. 5-Methylcytosine

The rare base 5-methylcytosine (MeC) is found in the DNA of plants and animals. 5-Methylcytosine itself is not incorporated directly into chromosomal DNA. Following DNA replication, symmetrically methylated strands will give rise to two hemimethylated DNA molecules. A maintenance methylase detects each methylated site in the parental DNA and adds a methyl

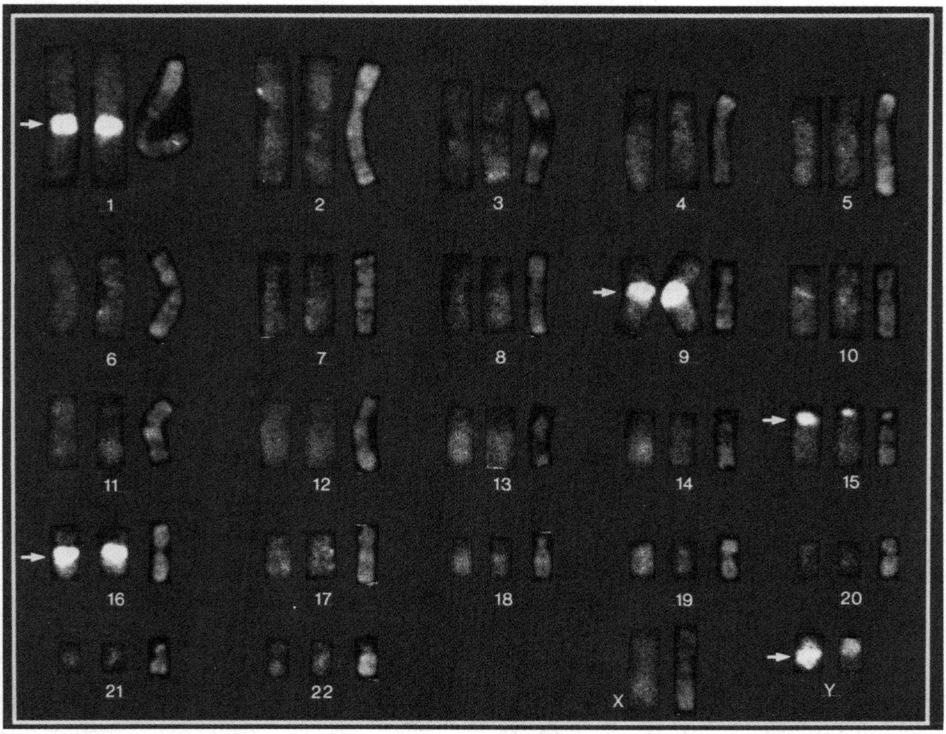

Figure 1 The 24 human chromosomes as seen by immunofluorescent banding with antibodies to cytosine (right) and 5-methylcytosine (left and center): Note the R-banding pattern produced by differential binding of anticytosine to chromosomes. The constitutive heterochromatin in the pericentromeric regions of autosomes 1, 9, 15, and 16 as well as in the distal long arm of the Y chromosome (arrows) contains high concentrations of 5-methylcytosine. (From Refs. 4 and 8)

group to the adjacent site in the newly synthesized DNA strand [6]. Antibodies to MeC can be used to detect highly methylated DNA sequences in acid-fixed and UV-irradiated chromosomes. Immunocytogenetic analyses showed that some of the constitutively heterochromatic regions of human (see Fig. 1), gorilla, mouse, and *Bovidae* are highly enriched in MeC [7–9]. The satellite DNA fractions of these species are methylated to a considerably greater extent than the euchromatic DNA sequences. Eukaryotic methylases are highly sequence-specific; methylation seems to be confined to the CpG dinucleotide [10]. Repetition of a satellite DNA sequence containing this recognition site therefore results in high concentrations of MeC.

Experimental evidence suggests that DNA methylation is involved in the regulation of gene expression. Transcriptional activity is generally associated with unmethylated DNA, whereas methylated genes are inactivated [11]. Extensive rDNA amplification is occasionally observed in homogeneously staining region (HSR)-carrying marker chromosomes of tumor cells as well as in variant Dph+ and Gph+ chromosomes of human individuals; these amplified rRNA genes are usually not transcribed. Immunocytogenetic analyses showed that anti-MeC antibodies preferentially bind to inactive rRNA genes in amplified regions; the transcriptionally active rRNA gene clusters (of similar size) exhibit only moderate antibody labeling. The inactive amplified rRNA genes appear to be hypermethylated [12,13]. Anti-MeC antibodies may have multipurpose applications for the analysis of gene or chromosome inactivation.

2. 5-Bromodeoxyuridine (BrdU)

The most commonly used technique to visualize DNA replication in metaphase chromosomes is incorporation of bromodeoxyuridine (BrdU) followed by fluorescent or fluorescence plus Giemsa (FPG) staining [14]. If added to the culture medium, the halogenated base analogue BrdU partially replaces thymidine without noticeable effects on cell viability and proliferation. Conventional staining methods for the cytological demonstration of BrdU incorporation require relatively high BrdU doses (10–100 μg/mL) and continuous exposure of the cells for at least 1 hr. Under these experimental conditions, however, BrdU itself increases the rate of sister chromatid exchanges (SCEs) and interferes with the process of chromatin condensation. The sensitivity of BrdU detection can be improved considerably by the use of monoclonal antibodies; one needs only 0.01 the BrdU concentration necessary for FPG staining. This largely precludes any artifacts on chromosomal structure caused by BrdU incorporation.

To obtain differential staining of sister chromatids, BrdU has to be substituted for two cell cycles. Anti-BrdU antibodies enable determination of the SCE frequency at very low BrdU incorporation levels [15,16]. Baseline frequencies greater than zero were measured in all cell types analyzed. Therefore, SCEs are assumed to be a normal process associated with DNA replication. Antibody labeling can also be employed for the demonstration of replication banding patterns in metaphase chromosomes [17]. One of the major advantages of the immunocytogenetic approach is that even very short BrdU pulses (5 min) can be detected cytologically. This may prove important for some fundamental studies (e.g., on asynchronous replication between homologous chromosomal regions or the detailed order of chromosome replication during S phase).

B. Conformational DNA Variants

Genomic DNA mainly exists in the classical Watson–Crick B-form. Nevertheless, conformational variants have been observed, indicating a considerable flexibility of the DNA double-helix. Some of these DNA variants are highly immunogenic, and specific antibodies have been produced. Immunocytogenetic techniques have shown that conformational variants exist in the native DNA of chromosomes and cell nuclei.

1. Z-DNA

Repetitive sequences of alternating dC and dG deoxynucleotides have a strong potential to adopt the stable Z-conformation. This left-handed Z-DNA forms a double-helix with Watson–Crick base pairs and antiparallel sugar–phosphate chains. The conformational transition between right-handed B-DNA and left-handed Z-DNA is proposed to have a regulatory role in gene transcription or chromosome condensation [18,19]. By means of specific antibodies, the distribution of Z-DNA sequences was visualized in fixed metaphase chromosomes of widely different mammalian species [20,21]. Anti-Z-DNA antibodies bind preferentially to a fraction of GC-rich constitutive heterochromatin, for example, some of the C-band regions in *Gerbillus nigeriae* (Fig. 2) and *Cebus albifrons*. In addition, the euchromatic segments of human (and *Cebus*) chromosomes exhibit a weak but consistent antibody-staining pattern resembling conventional R- and T-banding. From this, it is hypothesized that R- and T-bands are generally richer in Z-DNA sequences than G- or Q-bands. The chromosomal distribution of Z-DNA appears to be highly conserved during evolution. However, it is noteworthy that acid fixation facilitates the conversion of B-DNA to Z-DNA [22]. The fixation either unmasks naturally existing Z-DNA, previously inaccessible to antibody molecules, or induces the formation of Z-DNA in sequences capable of adopting this configuration. Although fixed chromosomes and cell nuclei react with anti-Z-DNA antibodies in a specific manner, the existence of Z-DNA in living cells has not yet been unequivocally shown.

2. Triplex DNA

A triplex DNA conformation can be formed from homologous pyrimidine · purine duplexes that have all pyrimidines on one strand and all purines on the other [23]. Monoclonal antibodies to triplex DNA react with both fixed and unfixed metaphase chromosomes of human, mouse, and other mammals [24]. The immunoreactivity of chromosomes not fixed with acid suggests that at least some DNA may occur naturally in a triplex conformation. On the other hand, acid fixation had a remarkable effect on the intensity of antibody labeling; this may be due to the formation of additional triplex

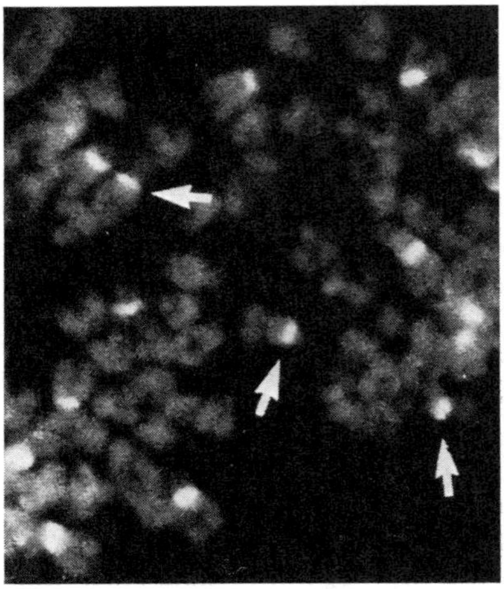

Figure 2 Metaphase spread from *Gerbillus nigeriae* after immunofluorescence staining with anti-Z-DNA antiserum: Arrows indicate some heterochromatic regions that are highly enriched in Z-DNA sequences. (From Ref. 20)

DNA complexes in sequences that do not adopt this configuration in vivo. Potential triplex DNA sequences are highly enriched in the Hoechst 33258-bright mouse heterochromatin. In acid-fixed human and murine chromosomes, the euchromatic G-bands bind more antitriplex DNA antibodies than the euchromatic R-bands. The immunobanding patterns produced by antibodies to triplex DNA and Z-DNA appear to be complementary. The triplex DNA conformation in G- and C-bands is possibly involved in the process of chromatin condensation. However, no functional role for triplex DNA or Z-DNA has yet been ascertained.

III. IMMUNOLOCALIZATION OF PROTEIN ANTIGENS

Most of the chromosomal and nuclear antigens are macromolecules that are very sensitive to acid treatment, meaning that the antigenic determinants are masked or destroyed by the conventional preparation of chromosomes. Antibody labeling of protein antigens has to be performed on reasonably intact cells, maintaining the cell membrane and cytoplasm [25]. The hypotonic treatment is designed to engorge the cell, without rupturing the cell membrane; this largely preserves the antigenic determinants. However, morphologic structure and spreading of the metaphases are distinctly worse on the average than in conventional preparations. Because cells that are not fixed with acid adhere to glass slides badly, adhesion of the cells to the slides has to be improved by technical methods. In our experience, cytocentrifugation is especially suited for preparing chromosomes without acid fixation [26]. During cytocentrifugation, the cells are sedimented from the hypotonic suspension onto vertical slides. The preparations are fixed with alcohol and acetone immediately after their removal from the cytocentrifuge to stabilize the antigenic structures and to immobilize; on the other hand, cell permeability needs to be enhanced so that the antibody molecules can enter the cells. Better antigen fixation and increased cell permeability, however, go along with a reduction of antigenicity. Therefore, antibody labeling of protein chromosomal components always constitutes a compromise between conflicting demands. As a rule, the quality of non–acid-fixed cytological preparations is barely adequate to identify the chromosomes by size and centromere index.

A. Kinetochore (Centromere)

The precision of chromosome segregation at mitosis and meiosis is a direct consequence of normal centromere/kinetochore function [27]. In mammalian cells, the centromere appears as a condensed chromosomal region, the so-called primary constriction, within which the kinetochore is found. Specific centromeric DNA sequences are believed to be responsible for centromere organization. The kinetochore is a mainly protein chromosomal component that attaches to the microtubules of the mitotic spindle. At the electron microscopic level, each kinetochore consists of a trilaminar disk, with an electron-dense outer plate, a lightly stained middle layer, and a dark inner layer, adjacent to the centromeric chromatin. The outermost plate functions as an assembly site of spindle fibers. The kinetochore is an integral part of the chromosomal scaffold [28–30].

Most patients suffering from the CREST (*c*alcinosis, *R*aynaud's disease, *e*sophageal dysmotility, *s*clerodactyly, *t*elangiectasia) syndrome of scleroderma produce high titers of circulating autoantibodies directed against the kinetochore [31]. The CREST sera contain different mixtures of antibodies reacting with several structurally related centromeric proteins [32,33]. Recently, a cDNA clone for the 80-kDa centromere protein (CENP-B), which is the major human kinetochore autoantigen, has been isolated [34]. Immunofluorescence and electron microscopic studies have shown that CREST antibodies bind specifically to the trilaminar plates of mitotic kinetochores, as well as to discrete spots in interphase nuclei corresponding to prekinetochores [35] (Fig. 3).

Since the discovery of kinetochore-specific autoantibodies [31], the antikinetochore immunofluorescent technique has found widespread use in cy-

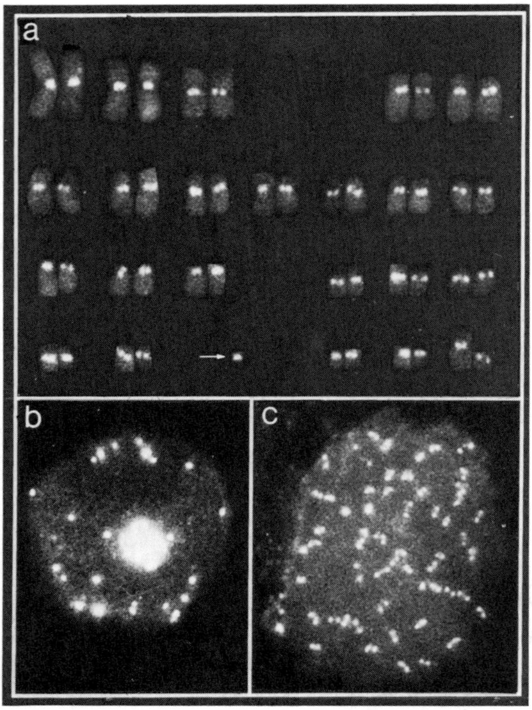

Figure 3 Immunofluorescent staining of kinetochores and prekinetochores; the chromosomes and cell nuclei are counterstained with ethidium bromide: (a) Karyotype from a human showing a minute chromosome as a constitutional change. The kinetochores of the ordinary metaphase chromosomes selectively fluoresce with a double-dot appearance. The minute chromosome (arrow) is intensely labeled by the antikinetochore antibody. (b, c) Interphase nuclei from human GLC1 tumor cells at G1 (b) and late G2 phase (c) of the cell cycle. The centromeres of G1-phase nuclei are nonrandomly positioned around the nucleolus and close to the nuclear membrane. The prekinetochores of interphase chromosomes appear as single fluorescent spots. In late G2-phase cells, all centromeres have been duplicated and show a relatively random topological distribution within the nucleus.

togenetics. The following will give a few examples. Human chromosomes exhibit considerable size differences in kinetochores, at least in terms of the antibody-binding sites [36]. The Y kinetochore is consistently smaller than the kinetochores of all other chromosomes. Smaller kinetochores are assumed to bind fewer microtubules [37], thus implying an increased risk for the generation of aneuploidy. The hypothesis that fluorescent intensity is somehow related to the activity of the kinetochore may explain the preferential loss of the human Y chromosome in vivo and in vitro. The compound kinetochores of the Indian muntjak are unique among mammals because of their large size and their unusual behavior during interphase [38]. They possibly evolved by linear fusion of smaller kinetochore units of an ancestral species. In dicentric chromosomes, somatic stability depends on centromere inactivation, which renders them ineffective for assembling kinetochore proteins and binding microtubules. Inactive centromeres either completely lose those proteins specifically recognized by CREST antibodies or conserve only moderate antigenic properties [39]. Centromere inactivation appears to be associated with premature replication of the pericentromeric heterochromatin [40]. Double minute (DM) chromosomes that persist in a variety of tumor cells consistently lack the antigenic component of the kinetochore, which is direct evidence that they do not have a centromere [41]. Despite their structural similarity to DMs, supernumerary minute chromosomes, as a constitutional change of the human karyotype [42] (see Fig. 3a), or the microchromosomes of birds and reptiles [unpublished results] possess functioning kinetochores.

Antikinetochore antibodies also provide a direct approach to the problem of internal order within the cell nucleus. Interphase nuclei of different organisms and cell types display very distinct, nonrandom centromere locations (see Fig. 3b). A universally valid principle of centromere arrangement, however, does not exist [43–45]. It is possible that specific centromere patterns in different cell types reflect specific functional and developmental capacities of these cells. In male gametogenesis of the mouse, the topological distribution of centromeres in meiotic and postmeiotic cells follows a definite pattern, according to the various stages of cellular differentiation [46,47]. There is also growing evidence indicating cell cycle-dependent patterns of centromere arrangement [38,45]. Different centromere positions relating to the cell cycle may be integral to the process of gene expression, chromatin replication, and nuclear division. Morphological duplication of kinetochores is not synchronized and occurs during the late G2-phase (see Fig. 3c). Movements and segregation of kinetochores that have been experimentally uncoupled from their chromosomal components suggest that kinetochores can even replicate independently of the genome [48].

B. Nucleolus

The nucleolus develops during telophase from the so-called nucleolar organizer regions (NORs). The NORs are those chromosomal sites that contain the tandemly arranged 18S + 28S rRNA genes. The fully formed nucleolus brings together all the components required for ribosome biogenesis, namely the fibrillar centers, the dense fibrillar component, and the granular component [49,50]. Transcription of the rRNA genes takes place in the fibrillar centers that are thought to represent the interphase counterparts of the chromosomal NORs. The primary transcripts then rapidly accumulate in the surrounding dense fibrillar component [51]. Assembly, processing, and maturation of ribosomal particles are complex processes that have been assigned to distinct nucleolar substructures [52].

Antibodies to various nucleolar proteins have successfully been used as probes in answering basic questions concerning the NOR and nucleolus. Antinucleolar antibodies of several specificities are present in human autoimmune sera, particularly from patients with progressive systemic sclerosis [53]. Indirect immunofluorescence showed that different constituents of the interphase nucleolus become differentially distributed during mitosis. Some of the nucleolar proteins remain associated with the chromosomal NORs; other proteins are transferred from the nucleolus to the surface of the condensing chromosomes or dispersed throughout the entire cytoplasm (Fig. 4).

1. Constituents of the Nucleolus Organizer Region

Antibodies that simultaneously bind the chromosomal NOR and the interphase nucleolus are directed against antigens located in the fibrillar centers of the nucleolus. At least some constituents of the fibrillar centers persist throughout mitosis associated with the rRNA genes.

RNA Polymerase I. Antibodies to RNA polymerase I (RPI) can be isolated either from the serum of patients with autoimmune diseases [54] or from animal experiments [51]. RPI is a chromatin-bound enzyme complex involved in rDNA transcription [55]. Anti-RPI antibodies allow detection of template-engaged RPI and, consequently, of transcribed rRNA genes.

Immunofluorescent staining of interphase nuclei produces a punctate nucleolar staining pattern, indicating that each nucleolus consists of an aggregate of transcriptionally active subunits [56]. Under the action of the adenosine analogue 5,6-dichloro-1-β-ribofuranosylbenzimidazole (DRB), the normally compact nucleoli are spread through the nuclear interior, forming several beaded strands. The fibrillar and the granular components of the nucleolus are segregated. The tandemly arranged granules of the experimentally induced "nucleolar necklaces" reflect the linear distribution of rRNA genes along the NORs. Each granule then represents an individual transcription unit consisting of an 18S + 28S rRNA gene and the transcriptional machinery [57]. The maximum number of RPI-positive granules in DRB-treated cells therefore corresponds to the number of transcriptionally active rRNA genes. Comparative immunocytogenetic analyses on fibroblast cells

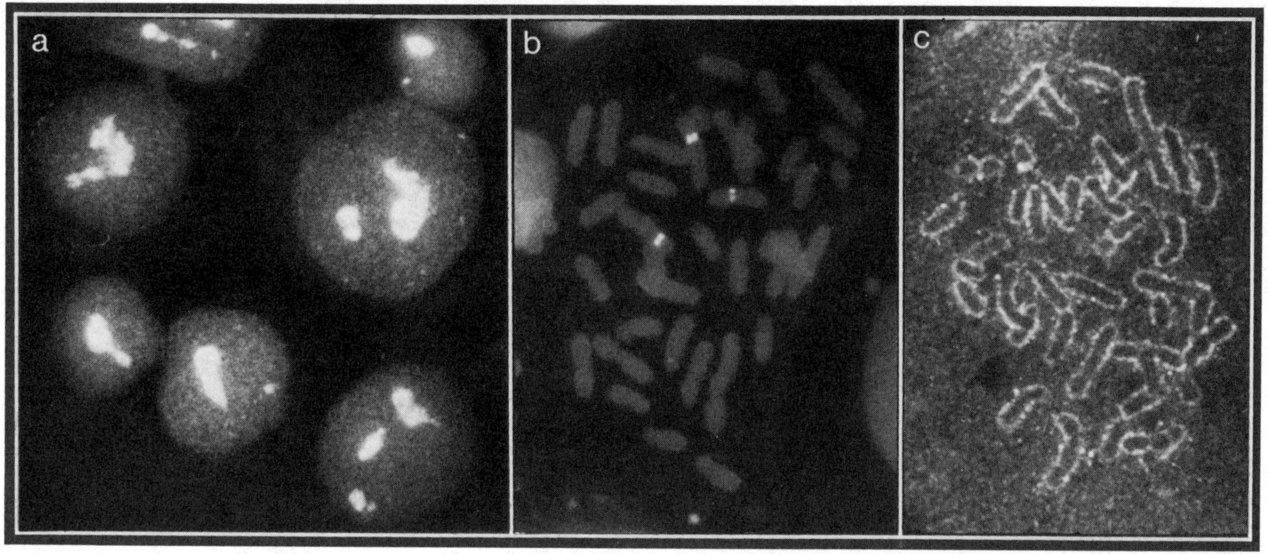

Figure 4 Immunolocalization of nucleolar antigens in interphase nuclei and metaphase spreads from a permanent fibroblast line: (a) Specific antibody labeling of nucleoli in interphase cells. This nucleolar immunofluorescence pattern is produced by antibodies of different specificities. (b) Metaphase spread stained with antibodies against RNA polymerase I and counterstained with ethidium bromide. Note the specific anti-RPI labeling of the nucleolus organizer regions in three chromosomes. (c) Metaphase spread stained with antibodies to the U3-ribonucleoprotein. The antibody-binding sites are concentrated in the perichromosomal regions; the chromosomes themselves are essentially negative.

of very different vertebrate species suggest that approximately 100 rRNA genes are transcribed during interphase [unpublished results].

The chromosomal NORs can be specifically demonstrated by labeling with anti-RPI antibodies [56]. In indirect immunofluorescence, the NORs selectively stain with a double-dot appearance (see Fig. 4b). The RPI molecules occur in the form of initiated complexes with rDNA during mitosis; when the cell enters the next interphase, the chromatin-bound RPI enzyme ensures rapid activation of the rRNA genes. Reformation of functioning nucleoli in the daughter cells seems to be dependent on the presence of an active form of RPI at the chromosomal NORs [58].

In male gametogenesis, anti-RPI labeling of nucleolar chromatin is present during the entire meiotic prophase up to the pachytene stage, is completely absent in the two meiotic metaphases, and is again demonstrable in the early spermatid nuclei [56]. The reappearance of anti-RPI labeling in the early spermatid stages indicates that the rRNA genes are reactivated in the haploid genome and remain active for a rather long period during maturation of the spermatids.

Topoisomerase I. DNA topoisomerase I is another enzyme enriched in the NORs of metaphase chromosomes, as well as in the nucleoli of interphase nuclei [59]. Topoisomerases are enzymes required for DNA decatenation (unwinding); they are possibly involved in DNA replication, transcription, repair, and recombination [60]. Human autoantibodies to topoisomerase I (Scl 70-autoantigen) are present in a subset of scleroderma sera. Anti-Scl 70 antibodies produce a bright immunofluorescent staining of nucleoli and NORs. Minor amounts of the enzyme are uniformly distributed in the entire nucleus and the metaphase chromosomes, respectively. In DRB-segregated nucleoli of interphase cells, antibody labeling is confined to the nucleolar necklaces, indicating that topoisomerase I is part of the fibrillar component of the nucleolus.

Other NOR-Associated Antigens. Apart from RPI and Scl 70, a variety of proteins are specifically associated with NORs and nucleoli. Human anti-NOR 90 autoantibodies recognize a 90-kDa protein antigen [61]. A rabbit serum localized an 80-kDa protein to the chromosomal NORs and the fibrillar centers of interphase nucleoli [62]. The phosphoprotein C23 is implicated as a silver-staining NOR protein [63]; it appears to be identical with nucleolin, a major nucleolar protein of growing eukaryotic cells [64].

Interestingly enough, there are also nonnucleolar proteins bound to the rRNA genes during mitosis. Recently, we identified two human autoimmune sera that stain the chromosomal NORs, but not the interphase nucleoli [unpublished results]. The complementary autoantigens are possible candidates for a mitotic factor attaching exclusively to the NORs of metaphase chromosomes and, thereby, blocking rRNA synthesis. The mechanism that suppresses transcriptional activity of the rRNA genes during mitosis is still unclear.

2. Constituents of the Perichromosomal Region

Human autoantibodies to U3-RNP, PM-Scl, and 7-2-RNP recognize antigens that are present not only in the interphase nucleolus, but also in the perichromosomal regions of mitotic cells [65–67]. Indirect immunofluorescence showed that substantial amounts of nucleolar material uniformly coat the metaphase chromosomes and are used in the reformation of the nucleoli of daughter cells (see Fig. 4c). Some of the nucleolar material is released into the cytoplasm, where its subsequent fate is unknown. The perichromosomal regions of mitotically dividing cells appear to comprise a variety of nucleolar constituents. From this observation, it was hypothesized that certain nucleolar proteins are involved in the condensation and decondensation of chromosomes [68].

Antibodies to U3-RNP recognize a M_r 34,000 nucleolar protein associated with the U3-ribonucleoprotein particle. Immunolocalization of this protein was in the dense fibrillar component of transcriptionally active nucleoli. The U3-RNP may be related to the early stages of rRNA processing [53,66]. The PM-Scl antigen is a complex particle consisting of at least 11 polypeptides. RNA could not as yet be identified in the antigen complex [65]. The 7-2-RNP is a ribonucleoprotein particle associated with 7-2-RNA [67]. Both PM-Scl and 7-2-RNP have been mapped to the granular component of the nucleolus. Because the granular component is the site of ribosomal assembly and packaging, this location suggests a relationship of these nucleolar autoantigens to preribosomes.

C. Chromosomal Scaffold

The chromosomal scaffold is a complex biochemical subfraction derived from chromosomes and consisting of nonhistone proteins and small quantities of DNA and RNA [69]. Indirect immunofluorescence enables the in situ localization of scaffold antigens at high resolution in chromosomes within intact cells.

1. Topoisomerase II

Among the multitude of uncharacterized scaffold proteins, two major high relative molecular mass (M_r) polypeptides, Sc 1 (170 kDa) and Sc 2 (135 kDa), have been isolated [70]. The Sc 1 was shown to be topoisomerase II, an enzyme that can unravel tangled DNA loops. Topoisomerase II appears to be placed at the bases of the radial loop domains of mitotic chromosomes; its untangling activity may be required for chromosome decondensation. An alternative explanation is that this enzyme is essential for structural functions at the chromosomal level [71,72].

2. Heterochromatin-Associated Antigens

Autoantibodies from certain patients with autoimmune disease determine a novel class of scaffold antigens specifically associated with constitutive heterochromatin of defined base pair composition [unpublished results]. The immunological affinity of antibodies for constitutive heterochromatin was demonstrated by indirect immunofluorescence of metaphase chromosomes and interphase nuclei. Several autoimmune sera preferentially label some of the heterochromatic regions in human, mouse, or sheep chromosomes. Different scaffold antigens were localized to the constitutively heterochromatic regions of human chromosomes 1, 9, 15, 16, and Y; to the paracentromeric C-band regions of mouse chromosomes (Fig. 5); and to the R–band-positive heterochromatic regions of sheep chromosomes. Minor amounts of scaffold antigens are uniformly distributed in the chromosomes of all vertebrate species. This indicates that their antigenic determinants have been highly conserved during evolution and represent an essential and ubiquitous chromosomal component. The behavior and chromosomal distribution of the scaffold antigens are not altered in interspecies somatic cell heterokaryons (see Fig. 5b, e), but apparently reflects a stable structural property germane to each chromosome type. Immunoblotting, performed with an antiserum to mouse constitutive heterochromatin, revealed a characteristic pattern of polypeptides with M_r ranging from 35 to 120 kDa. The scaffold antigens recognized by human autoantibodies may interact with heterochromatic DNA sequences at specific sites and, thus, may be related to the compactness and genetic inactivity of constitutive heterochromatin.

3. Centromere-Related Antigens

Several scaffold proteins have been shown to be components of the centromere region. The centromere proteins CENP-B (80 kDa) and CENP-C (140 kDa) recognized by CREST autoantibodies are associated with the kinetochore structure during the entire cell cycle [30,32]. The inner centromere proteins (INCENPs) determined by monoclonal antiscaffold antibodies are concentrated in the pericentromeric regions between the sister chromatids. At anaphase, as the sister chromatids separate, the INCENPs dissociate from the chromosomes completely. The inner centromere proteins are likely to play a role in pairing and separation of sister chromatids [73].

D. Synaptonemal Complex

The synaptonemal complex (SC) forms the protein axis between paired chromosomes during meiotic prophase; it is confined to meiotic cells. For-

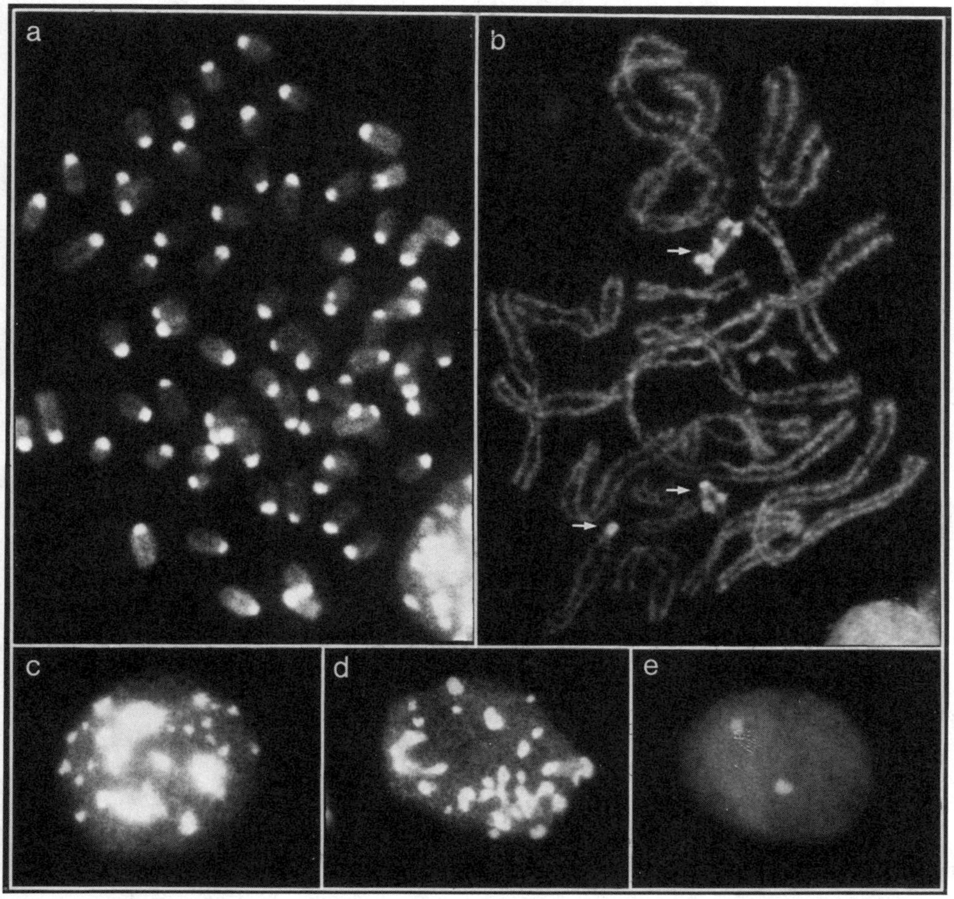

Figure 5 Chromosomes and cell nuclei labeled with human autoantibodies specific for AT-rich mouse heterochromatin: (a, b) Metaphase spreads from a spontaneously transformed mouse fibroblast (a) and a mouse–hamster (CHO) hybrid (b). Note the bright immunofluorescence of the paracentromeric heterochromatin in the acrocentric mouse chromosomes (a). The mouse–hamster heterokaryon contains one ordinary acrocentric and two marker chromosomes of murine origin (arrows); only the mouse constitutive heterochromatin is specifically labeled by human autoantibodies. The euchromatic segments of mouse chromosomes as well as all of the hamster chromosomes possess relatively few antibody-binding sites. (c–e) Interphase nuclei of mouse fibroblasts (c, d) and a mouse–hamster heterokaryon (e) as seen in indirect immunofluorescence. The mouse heterochromatin is easily identifiable by its bright staining.

mation of the SC appears to be an essential prerequisite for chromosome pairing, recombination, and homologue disjunction. At the ultrastructural level, the SC consists of two parallel lateral elements and a central element. During leptotene, single, thin axial elements appear along each chromosome. The tripartite SC structure develops in zygotene; the axes of the synapsing homologues become the lateral elements of the SC, and a central element is formed between them on the transverse filaments. During pachytene, when chromosomal pairing is complete, the SC extends from one end of the autosomal bivalent to the other. The SC exactly parallels the behavior of the chromosomal bivalent itself. The sex bivalent is exceptional in that the X and Y axes are paired only in part over an SC region of variable length. Desynapsis of bivalents occurs at diplotene. Important information on SC structure and function generally came from electron microscopic observations [74–76]. In a recent study, however, it was shown that entire SCs in meiotic prophase cells can be visualized by immunocytogenetic techniques [47,77].

1. Human Autoantigens Confined to the Synaptonemal Complex at Synapsis

Autoantibodies to SCs are spontaneously produced by patients with various autoimmune diseases. Immunofluorescent staining on surface-spread spermatocytes localized the SC autoantigen to the central element or the transverse filaments, but not to the lateral elements. In leptotene and zygotene, antigenic SC structures are not detectable. The earliest stage of meiotic prophase that shows specific antibody labeling of the SCs is pachytene. At higher magnification, the antigenic SC structure exhibits a relatively uniform longitudinal segmentation; the SC ends, which are known to attach to the nuclear envelope, are characterized by a thickened segment of variable size (Fig. 6e). This immunostaining pattern seems to reflect a structural or functional property of the SC itself. Specific antibody labeling is confined to the SC at synapsis. The unpaired homologues at leptotene and zygotene, as well as the unpaired axes of the X and Y chromosomes during pachytene (see Fig. 6a–c), are not recognized by human anti-SC antibodies. The antigenic SC component is definitely lost during diplotene. Therefore, it can be assumed that the SC autoantigen is involved in stage-specific processes such as chromosome pairing and recombination.

Cytochemical tests revealed the SC autoantigen to be a hydrophobic nonhistone protein bound to DNA. It cannot be extracted by balanced salt solutions, high ionic strength solutions, nor nonionic detergents. The nature of the SC autoantigen and its insolubility are reminiscent of the biochemical composition of the nuclear skeleton or the mitotic chromosomal scaffold,

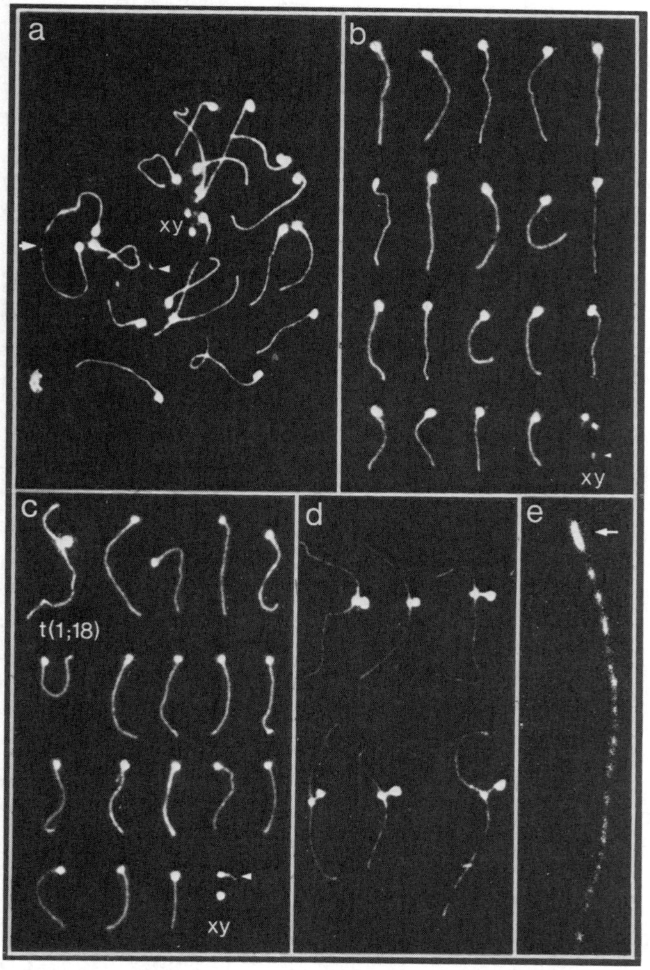

Figure 6 Double-labeling of synaptonemal complexes and kinetochores in mouse spermatocytes: (a) Pachytene spread of normal male mouse. Note the end-to-end associations of some SC long arms (arrow). The short SC fragment indicated by an arrowhead represents the synapsed regions of the XY bivalent. The kinetochores of the X and Y are not synapsed. (b) SC karyotype of normal male mouse. (c) SC karyotype of male mouse heterozygous for a Robertsonian 1/18 translocation chromosome. The SC configuration of the heteromorphic trivalent is of particular interest; the acrocentric kinetochores are always paired on the same side of the SC (*cis*-configuration). (d) Selected examples of different trivalent-SCs from a mouse interspecies hybrid (CD/NMRI). There are different lengths of the short SC side arms. (e) Antigenic SC structure as seen at high magnification. Note the longitudinal segmentation of the SC as well as the thickening of its end (arrow).

respectively. An unusual characteristic of the SC autoantigen is its species-specificity. Patients were found whose sera selectively label the SCs of human or mouse. Nevertheless, most anti-SC antisera demonstrate minor cross-reactions between different pachytene cell substrates, indicating that the SCs share some ubiquitous antigenic components [47].

The combining of anti-SC and antikinetochore antibodies provides a new method for SC karyotyping; analysis is based on SC length and kinetochore position. Figures 6a and b show a pachytene spread and SC karyotype of the normal male mouse. Immunofluorescent staining revealed 19 acrocentric SCs and a double-fluorescent spot representing the unpaired kinetochores of the sex chromosomes. The XY kinetochores are associated with a short SC fragment that corresponds to the synapsed region at the distal end of the X and Y axes. The autosomal SCs are endowed with only one large kinetochore spot; the kinetochores of the homologues are paired so closely that their fluorescent signals fuse. The synapsed kinetochores are located precisely in the SC structure, being an integral part of it.

Heterozygosity for Robertsonian (Rb) translocation chromosomes without arm homologies leads to the formation of heteromorphic trivalents in meiosis I (see Fig. 6c, d). In these trivalent-SCs, the nonhomologous kinetochores (as well as the short arms) of the two acrocentrics are paired in *cis*-configuration and juxtaposed opposite the kinetochore of the Rb chromosome; the metacentric kinetochore is part of the longitudinal SC structure. The short side arms of different trivalent-SCs appear to be somewhat different in length, probably reflecting different sizes of the synapsed nonhomologous segments of the acrocentrics (see Fig. 6d). These initial results clearly demonstrate the usefulness of an immunocytogenetic approach for SC karyotyping and the analysis of chromosomal irregularities during meiotic prophase [77].

2. Antigens Determined by Monoclonal Antibodies

Monoclonal antibodies against various SC components have been prepared by immunization of mice with whole testis protein or purified SCs. Two antibodies to the central element, SC 3 and SC 4, are not SC-specific, but also label intermediate filaments [78]. Two monoclonal antibodies to the lateral elements are assumed to be specific for meiotic prophase nuclei. Antibody II52F10, which recognizes a 30- and a 33-kDa polypeptide, reacts with paired as well as unpaired segments of zygotene, pachytene, and diplotene SCs. The lateral elements of both autosomes and sex chromosomes are bound. Another antibody, III15B8, detects an antigen that is present at the inner edges of the lateral elements and possibly functions in chromosome pairing [79–81].

E. Sex Vesicle

The sex vesicle, which consists of the condensed X and Y chromosomes, is probably the most striking structure in mammalian spermatocytes. The X and Y chromosomes form the heteropyknotic sex vesicle during early pachytene, but resume their normal structural appearance when the cell reaches diplotene. The condensed XY body is thought to prevent exchanges between the X and Y during meiotic prophase, thereby ensuring the balance for sex determination [82]. The sex vesicle may also be involved in the obligatory silencing of the transcriptional activity of the male sex chromosomes at first meiotic prophase [83].

Autoimmune sera from two patients with progressive systemic sclerosis reacted strongly with the sex vesicle of pachytene cells [84]. After antibody labeling, the sex vesicle exhibits bright and relatively uniform immunofluorescence. The remainder of the pachytene nucleus as well as all other spermatogenic cell types show only weak, nonspecific staining (Fig. 7). The high degree of condensation of the XY chromosomes (compared with the autosomal bivalents) could be mediated by a specific protein composition of the sex vesicle. Certain proteins confined to or highly enriched in the sex vesicle are probably the cytochemical target of human autoantibodies.

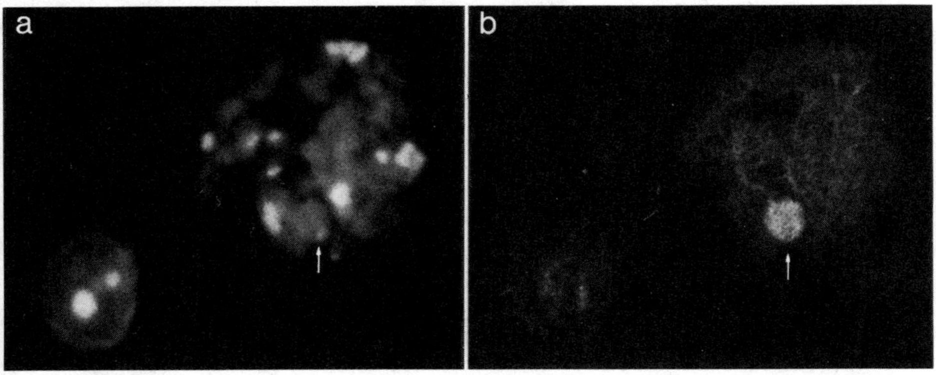

Figure 7 (a) Mouse pachytene spermatocyte and Sertoli cell stained with Hoechst 33258: The constitutive heterochromatin is characterized by its bright fluorescence. (b) The same preparation after immunofluorescent staining with human autoantibodies to the sex vesicle (arrows). Note the specific labeling of the sex vesicle.

IV. SUMMARY

Immunostaining is a simple and reliable method for the localization of structures of interest in chromosomes and cell nuclei. Antibody labeling can, for example, be used to follow the development and behavior of an antigenic component throughout the entire cell cycle and in meiosis. A number of naturally occurring or experimentally raised antibodies against chromosomal or nuclear components have been described. The antibodies used vouchsafe a high specificity and sensitivity in the demonstration of antigens. The wealth of data presented illustrate the potential power of the immunocytochemical approach in studying chromosomal structure and function. In conclusion, then, immunolocalization by use of specific antibodies provides a helpful tool in cytogenetics to supplement the information obtained by conventional chromosome banding, in situ hybridization, and electron microscopy.

ACKNOWLEDGMENTS

This work was supported by grants from the Deutsche Forschungsgemeinschaft (Ha 1374/2-1) and the Senator Kurt und Inge Schuster Stiftung.

REFERENCES

1. Caspersson, T., L. Zech, C. Johansson, and E. J. Modest (1970). Identification of human chromosomes by DNA-binding fluorescent agents. *Chromosoma 30*: 215–227.
2. Pardue, M. L., and J. B. Gall (1970). Chromosomal localization of mouse satellite DNA. *Science 168*:1356–1358.
3. Dev, V. G., D. Warburton, O. J. Miller, D. A. Miller, B. F. Erlanger, and S. M. Beiser (1972). Consistent pattern of binding of anti-adenosine antibodies to human metaphase chromosomes. *Exp. Cell Res. 74*:288–293.
4. Schreck, R. R., D. Warburton, O. J. Miller, S. M. Beiser, and B. F. Erlanger (1973). Chromosome structure as revealed by a combined chemical and immunochemical procedure. *Proc. Natl. Acad. Sci. USA 70*:804–807.
5. Erlanger, B. F., and S. M. Beiser (1964). Antibodies specific for ribonucleosides and their reaction with DNA. *Proc. Natl. Acad. Sci. USA 52*:68–74.
6. Riggs, A. D. (1975). X-Inactivation, differentiation and DNA methylation. *Cytogenet. Cell Genet. 14*:9–25.
7. Miller, O. J., W. Schnedl, J. Allen, and B. F. Erlanger (1974). 5-Methylcytosine localised in mammalian constitutive heterochromatin. *Nature 251*:636–637.
8. Schnedl, W., V. G. Dev, R. Tantravahi, D. A. Miller, B. F. Erlanger, and O. J. Miller (1975). 5-Methylcytosine in heterochromatic regions of chromosomes: Chimpanzee and gorilla compared to the human. *Chromosoma 52*:59–66.
9. Schnedl, W., B. F. Erlanger, and O. J. Miller (1976). 5-Methylcytosine in heterochromatic regions of chromosomes in *Bovidae*. *Hum. Genet. 31*:21–26.

10. Ehrlich, M., and R. Y. H. Wang (1981). 5-Methylcytosine in eukaryotic DNA. *Science 212*:1350–1357.
11. Mohandas, T., and L. J. Shapiro (1983). Factors involved in X-chromosome inactivation. In *Cytogenetics of the Mammalian X Chromosome. Part A: Basic Mechanisms of X Chromosome Behaviour*. A. A. Sandberg (ed.). Alan R. Liss, New York, pp. 271–297.
12. Tantravahi, U., R. V. Guntaka, B. F. Erlanger, and O. J. Miller (1981). Amplified ribosomal RNA genes in a rat hepatoma cell line are enriched in 5-methylcytosine. *Proc. Natl. Acad. Sci. USA 78*:489–493.
13. Tantravahi, U., W. Roy Breg, V. Wertelecki, B. F. Erlanger, and O. J. Miller (1981). Evidence for methylation of inactive human rRNA genes in amplified regions. *Hum. Genet. 56*:315–320.
14. Block, A. M. W. (1982). Sister chromatid exchange methodology. In *Sister Chromatid Exchange*. A. A. Sandberg (ed.). Alan R. Liss, New York, pp. 13–32.
15. Pinkel, D., L. H. Thompson, J. W. Gray, and M. Vanderlaan (1985). Measurement of sister chromatid exchanges at very low bromodeoxyuridine substitution levels using a monoclonal antibody in Chinese hamster ovary cells. *Cancer Res. 45*:5795–5798.
16. Tucker, J. D., M. L. Christensen, C. L. Strout, and A. V. Carrano (1986). Determination of the baseline sister chromatid exchange frequency in human and mouse peripheral lymphocytes using monoclonal antibodies and very low doses of bromodeoxyuridine. *Cytogenet. Cell Genet. 43*:38–42.
17. Vogel, W., M. Autenrieth, and G. Speit (1986). Detection of bromodeoxyuridine-incorporation in mammalian chromosomes by a bromodeoxyuridine–antibody. I. Demonstration of replication patterns. *Hum. Genet. 72*:129–132.
18. Rich, A., A. Nordheim, and A. H. J. Wang (1984). The chemistry and biology of left-handed Z-DNA. *Annu. Rev. Biochem. 53*:791–846.
19. Leng, M. (1985). Left handed Z-DNA. *Biochim. Biophys. Acta 825*:339–344.
20. Viegas-Péquignot, E., C. Derbin, F. Lemeunier, and E. Tallandier (1982). Identification of left-handed Z-DNA by indirect immunomethods in metaphasic chromosomes of a mammal, *Gerbillus nigeriae* (Gerbillidae, Rodentia). *Ann. Génét. 25*:218–222.
21. Viegas-Péquignot, E., C. Derbin, B. Malfoy, E. Tallandier, M. Leng, and B. Dutrillaux (1983). Z-DNA immunoreactivity in fixed metaphase chromosomes of primates. *Proc. Natl. Acad. Sci. USA 80*:5890–5894.
22. Robert-Nicoud, M., D. J. Arndt-Jovin, D. A. Zarling, and T. M. Jovin (1984). Immunological detection of left handed Z-DNA in isolated polytene chromosomes. Effects of ionic strength, pH, temperature and topological stress. *EMBO J. 3*:721–731.
23. Lee, J. S., M. L. Woodsworth, L. J. Latimer, and A. R. Morgan (1984). Poly(pyrimidine)·poly(purine) synthetic DNAs containing 5-methylcytosine form stable triplexes at neutral pH. *Nucleic Acids Res. 12*:6603–6614.
24. Burkholder, G. D., L. J. P. Latimer, and J. S. Lee (1988). Immunofluorescent staining of mammalian nuclei and chromosomes with a monoclonal antibody to triplex DNA. *Chromosoma 97*:185–192.

25. Stenman, S., M. Rosenqvist, and N. R. Ringertz (1975). Preparation and spread of unfixed metaphase chromosomes for immunofluorescence staining of nuclear antigens. *Exp. Cell Res. 90*:87–94.
26. Haaf, T., and M. Schmid (1988). Indirect immunofluorescence. In *Working with Animal Chromosomes*. H. Macgregor and J. Varley (eds.). John Wiley & Sons, Chichester, pp. 97–107.
27. Murray, A. W., and J. W. Szostack (1985). Chromosome segregation in mitosis and meiosis. *Annu. Rev. Cell Biol. 1*:289–315.
28. Rieder, C. L. (1982). The formation, structure, and composition of the mammalian kinetochore and kinetochore fiber. *Int. Rev. Cytol. 79*:1–58.
29. Rattner, J. B. (1987). The organization of the mammalian kinetochore: A scanning electron microscope study. *Chromosoma 95*:175–181.
30. Earnshaw, W. C., N. Halligan, C. Cooke, and N. Rothfield (1984). The kinetochore is part of the metaphase chromosome scaffold. *J. Cell Biol. 98*:352–357.
31. Moroi, Y., C. Peebles, M. J. Fritzler, J. Steigerwald, and E. M. Tan (1980). Autoantibody to centromere (kinetochore) in scleroderma sera. *Proc. Natl. Acad. Sci. USA 77*:1627–1631.
32. Earnshaw, W. C., and N. Rothfield (1985). Identification of a family of human centromere proteins using autoimmune sera from patients with scleroderma. *Chromosoma 91*:313–321.
33. Kingwell, B., and J. B. Rattner (1987). Mammalian kinetochore/centromere composition: A 50 kDa antigen is present in the mammalian kinetochore/centromere. *Chromosoma 95*:403–407.
34. Earnshaw, W. C., K. F. Sullivan, P. S. Machlin, C. A. Cooke, D. A. Kaiser, T. D. Pollard, N. F. Rothfield, and D. W. Cleveland (1987). Molecular cloning of cDNA for CENP-B, the major human centromere autoantigen. *J. Cell Biol. 104*:817–829.
35. Brenner, S., D. Pepper, M. W. Berns, E. Tan, and B. R. Brinkley (1981). Kinetochore structure, duplication, and distribution in mammalian cells: Analysis by human autoantibodies from scleroderma patients. *J. Cell Biol. 91*:95–102.
36. Cherry, L. M., and D. A. Johnston (1987). Size variation in kinetochores of human chromosomes. *Hum. Genet. 75*:155–158.
37. Mitchison, T. J., and M. W. Kirschner (1985). Properties of the kinetochore in vitro. II. Microtubule capture and ATP-dependent translocation. *J. Cell Biol. 101*:766–777.
38. Brinkley, B. R., M. M. Valdivia, A. Tousson, and S. L. Brenner (1984). Compound kinetochores of the Indian muntjak. Evolution by linear fusion of unit kinetochores. *Chromosoma 91*:1–11.
39. Peretti, D., P. Maraschio, S. Lambiase, F. Lo Curto, and O. Zuffardi (1986). Indirect immunofluorescence of inactive centromeres as indicator of centromeric function. *Hum. Genet. 73*:12–16.
40. Vig, B. K., and D. Broccoli (1988). Sequence of centromere separation: Differential replication of pericentric heterochromatin in multicentric chromosomes. *Chromosoma 96*:311–317.

41. Haaf, T., and M. Schmid (1988). Analysis of double minutes and double minute-like chromatin in human and murine tumor cells using antikinetochore antibodies. *Cancer Genet. Cytogenet. 30*:73–82.
42. Schmid, M., T. Haaf, D. Schindler, and M. Meurer (1989). Centromeric association of a microchromosome. A new category of non-random arrangement of metaphase chromosomes. *Hum. Genet. 81*:127–136.
43. Moroi, Y., A. L. Hartman, P. K. Nakane, and E. M. Tan (1981). Distribution of kinetochore (centromere) antigen in mammalian cell nuclei. *J. Cell Biol. 90*: 254–259.
44. Hadlaczky, G., M. Went, and N. R. Ringertz (1986). Direct evidence for the non-random localization of mammalian chromosomes in the interphase nucleus. *Exp. Cell Res. 167*:1–15.
45. Haaf, T., and M. Schmid (1989). Centromeric association and non-random distribution of centromeres in human tumour cells. *Hum. Genet. 81*:137–143.
46. Brinkley, B. R., S. L. Brenner, J. M. Hall, A. Tousson, R. D. Balczon, and M. M. Valdivia (1986). Arrangement of kinetochores in mouse cells during meiosis and spermiogenesis. *Chromosoma 94*:309–317.
47. Haaf, T., A. Machens, and M. Schmid (1989). Immunocytogenetics. II. Human autoantibodies to synaptonemal complexes. *Cytogenet. Cell Genet. 50*: 6–13.
48. Brinkley, B. R., R. P. Zinkowski, W. L. Mollon, F. M. Davis, M. A. Pisegna, and M. Pershouse (1988). Movement and segregation of kinetochores experimentally detached from mammalian chromosomes. *Nature 336*:251–254.
49. Schwarzacher, H. G., and F. Wachtler (1983). Nucleolus organizer regions and nucleoli. *Hum. Genet. 63*:89–99.
50. Goessens, G. (1984). Nucleolar structure. *Int. Rev. Cytol. 87*:107–158.
51. Scheer, U., and K. M. Rose (1984). Localization of RNA polymerase I in interphase cells and mitotic chromosomes by light and electron microscopic immunocytochemistry. *Proc. Natl. Acad. Sci. USA 81*:1431–1435.
52. Hadjiolov, A. A. (1985). The nucleolus and ribosome biogenesis. *Cell Biol. Monogr. 12*:1–268.
53. Reimer, G., I. Raska, E. M. Tan, and U. Scheer (1987). Human autoantibodies: probes for nucleolus structure and function. *Virchows Arch. B 54*:131–143.
54. Reimer, G., K. M. Rose, U. Scheer, and E. M. Tan (1987). Autoantibody to RNA polymerase I in scleroderma sera. *J. Clin. Invest. 79*:65–72.
55. Rose, K. M., D. A. Stetler, and S. T. Jacob (1983). RNA polymerases from higher eukaryotes. In *Enzymes of Nucleic Acid Synthesis and Modification*. S. T. Jacob (ed.). CRC Press, Boca Raton, Fla. pp. 43–74.
56. Haaf, T., G. Reimer, and M. Schmid (1988). Immunocytogenetics: Localization of transcriptionally active rRNA genes in nucleoli and nucleolus organizer regions by use of human autoantibodies to RNA polymerase I. *Cytogenet. Cell Genet. 48*:35–42.
57. Scheer, U., B. Hügle, R. Hazan, and K. M. Rose (1984). Drug-induced dispersal of transcribed rRNA genes and transcriptional products: Immunolocalization and silver staining of different nucleolar components in rat cells treated

with 5,6-dichloro-β-D-ribofuranosylbenzimidazole. *J. Cell Biol. 99*:672–679.
58. Benavente, R., K. M. Rose, G. Reimer, B. Hügle-Dörr, and U. Scheer (1987). Inhibition of nucleolar reformation after microinjection of antibodies to RNA polymerase I into mitotic cells. *J. Cell Biol. 105*:1483–1491.
59. Guldner, H. H., C. Szostecki, H. P. Vosberg, H. J. Lakomek, E. Penner, and F. A. Bautz (1986). Scl 70 autoantibodies from scleroderma patients recognize a 95 kDa protein identified as DNA topoisomerase I. *Chromosoma 94*:132–138.
60. Liu, L. F. (1983). DNA topoisomerases—enzymes that catalyse the breaking and rejoining of DNA. *CRC Crit. Rev. Biochem. 15*:1–24.
61. Rodriguez-Sanchez, J. L., C. Gelpi, C. Juarez, and J. A. Hardin (1987). Anti-NOR 90. A new autoantibody in scleroderma that recognizes a 90-kDa component of the nucleolus-organizing region of chromatin. *J. Immunol. 139*:2579–2584.
62. Courvalin, J. C., D. Hernandez-Verdun, F. Gosti-Testu, M. C. Marty, R. Maunoury, and M. Bornens (1986). A protein of M_r 80,000 is associated with the nucleolus organizer of human cell lines. *Chromosoma 94*:353–361.
63. Ochs, R., M. Lischwe, P. O'Leary, and H. Busch (1983). Localization of nucleolar phosphoproteins B23 and C23 during mitosis. *Exp. Cell Res. 146*:139–149.
64. Lapeyre, B., H. Bourbon, and F. Amalric (1987). Nucleolin, the major nucleolar protein of growing eukaryotic cells: An unusual protein structure revealed by the nucleotide sequence. *Proc. Natl. Acad. Sci. USA 84*:1472–1476.
65. Reimer, G., U. Scheer, J. M. Peters, and E. M. Tan (1986). Immunolocalization and partial characterization of a nucleolar autoantigen (PM-Scl) associated with polymyositis/scleroderma overlap syndromes. *J. Immunol. 137*:3802–3808.
66. Reimer, G., K. M. Pollard, C. A. Penning, R. L. Ochs, M. A. Lischwe, H. Busch, and E. M. Tan (1987). Monoclonal autoantibody from NZB/NZW F1 mouse and some human scleroderma sera target a M_r 34,000 nucleolar protein of the U3-ribonucleoprotein particle. *Arthritis Rheum. 30*:793–800.
67. Reimer, G., I. Raska, U. Scheer, and E. M. Tan (1988). Immunolocalization of 7-2-ribonucleoprotein in the granular component of the nucleolus. *Exp. Cell Res. 176*:117–128.
68. Luji, S. (Sze L. C.), N. Zumei, Z. Shi, G. E. Wang, and Y. Yang (1987). Involvement of a nucleolar component, perichromonucleolin, in the condensation and decondensation of chromosomes. *Proc. Natl. Acad. Sci. USA 84*: 7953–7956.
69. Earnshaw, W. C. (1988). Mitotic chromosome structure. *BioEssays 9*:147–150.
70. Lewis, C. D., and U. K. Laemmli (1982). High order metaphase chromosome structure: Evidence for metalloprotein interactions. *Cell 29*:171–181.
71. Earnshaw, W. C., B. Halligan, C. A. Cooke, M. M. S. Heck, and L. F. Liu (1985). Topoisomerase II is a structural component of mitotic chromosome scaffolds. *J. Cell Biol. 100*:1706–1715.
72. Earnshaw, W. C., and M. M. S. Heck (1985). Localization of topoisomerase II in mitotic chromosomes. *J. Cell Biol. 100*:1716–1725.

73. Cooke, C. A., M. M. S. Heck, and W. C. Earnshaw (1987). The inner centromere protein (INCENP) antigens: Movement from inner centromere to midbody during mitosis. *J. Cell Biol. 105*:2053–2067.
74. Gillies, C. B. (1975). Synaptonemal complex and chromosome structure. *Annu. Rev. Genet. 9*:91–109.
75. Moses, M. J. (1977). The synaptonemal complex and meiosis. In *ICN-UCLA Symposia on Molecular and Cellular Biology. Vol. VII. Molecular Human Cytogenetics*. R. S. Sparkes, D. E. Comings, and C. F. Fox (eds.). Academic Press, New York, pp. 101–125.
76. Solari, A. J. (1980). Synaptonemal complexes and associated structures in microspread human spermatocytes. *Chromosoma 81*:315–337.
77. Haaf, T., H. Winking, and M. Schmid (1989). Immunocytogenetics. III. Analysis of trivalent and multivalent configurations in mouse pachytene spermatocytes by human autoantibodies to synaptonemal complexes and kinetochores. *Cytogenet. Cell Genet. 50*:14–22.
78. Dresser, M. E. (1987). The synaptonemal complex and meiosis: An immunocytochemical approach. In *Meiosis*. P. B. Moens (ed.). Academic Press, New York, pp. 245–274.
79. Heyting, C., P. B. Moens, W. van Raamsdonk, A. J. J. Dietrich, A. C. G. Vink, and E. J. W. Redeker (1987). Identification of two major components of the lateral elements of synaptonemal complexes. *Eur. J. Cell Biol. 43*:148–154.
80. Heyting, C., R. J. Dettmers, A. J. J. Dietrich, E. J. W. Redeker, and A. C. G. Vink (1988). Two major components of synaptonemal complexes are specific for meiotic prophase nuclei. *Chromosoma 96*:325–332.
81. Moens, P. B., C. Heyting, A. J. J. Dietrich, W. van Raamsdonk, and Q. Chen (1987). Synaptonemal complex antigen location and conservation. *J. Cell Biol. 105*:93–103.
82. Solari, A. J. (1974). The behaviour of the XY pair in mammals. *Int. Rev. Cytol. 38*:273–317.
83. Lifschytz, E., and D. L. Lindsley (1972). The role of X-chromosome inactivation during spermatogenesis. *Proc. Natl. Acad. Sci. USA 69*:182–186.
84. Haaf, T., A. G. Groscurth, A. Machens, and M. Schmid (1990). Human autoantibodies to spermatogenic antigens and Sertoli cells. *J. Histochem. Cytochem. 38*:129–134.

11

Molecular Linkage Genetics of the Mouse

Peter D'Eustachio

New York University Medical Center, New York, New York

I. INTRODUCTION

Constructing precise, detailed linkage maps of mammalian chromosomes is worthwhile on two levels. At a practical level, it provides information essential for activities such as the accurate prenatal diagnosis of inherited disorders, and it can provide the starting point for the identification of mutant genes, the properties of which are known only at the level of the whole organism. The recent cloning of the human cystic fibrosis gene is a spectacular example of a physical assault, guided almost entirely by genetic linkage data, on a gene [1,2]. At a basic level, such maps are beginning to yield insight into the biology of the recombination process itself as it takes place in mammals.

This review will center on the construction of linkage maps of mouse chromosomes that are based on DNA markers, although parallels in the human will be noted. This focus partly reflects my own interests and expertise, but also reflects the properties of the mouse as an object of genetic analysis. Mice have short generation times and numerous offspring. They tolerate inbreeding well, perhaps because they tend to breed in small, closed populations in the wild [3–5], and large numbers of highly inbred, well-characterized strains are available. Hundreds of mutations with visible effects have been described and characterized genetically [6,7]. The phenotypes of many of these mutations parallel those of human hereditary diseases, accordingly, their analysis can provide physiological and genetic data useful to the clinician.

II. FROM LINKAGE MAPS TO DNA MAPS

How detailed and accurate a map is needed? Molecular experiments have yielded evidence of interactions in *cis* among DNA sequence elements spanning distances of up to a few thousand kilobases (kb), the size of a large multigene family and its associated regulatory elements. Techniques for the physical analysis of DNA fragments allow molecules of up to several thousand kilobases to be visualized and manipulated [8]. The physical size of the haploid mouse genome is 2.7×10^6 kb [9], and estimates of its genetic size made by counting chiasmata in meiotic figures range from 1600 to 2000 centimorgans (cM) [10, pp. 290–300], for an average of 1350–1700 kb of DNA per centimorgan. If the amount of DNA per unit of the genetic map is even roughly constant, then, methods exist to bridge a physical map and a 1-cM resolution genetic map of a mammalian chromosome.

For bridging to occur, however, the average density of markers in the genetic map must be high. If a linkage map consists of randomly distributed markers at a mean frequency, μ, of one per centimorgan, then a fraction of $e^{-\mu}$ or 36.7% of all 1-cM intervals is expected to be unmarked, and the odds of having at least one marker in every 1-cM interval of a 50-cM chromosome are approximately $1 : 10^{10}$ [11]. Even a 10-cM segment has only 1% odds of having a marker in every 1-cM interval. A mean frequency of six markers per centimorgan is needed to give 95% odds of marking every centimorgan in a 10-cM segment, and a mean of 11 markers/cM is needed to give 95% odds of at least one marker falling in every 1-cM interval of a 50-cM chromosome.

Construction of linkage maps is not a new activity. The first such map, of six X-chromosome genes in *Drosophila*, was published in 1913 [12], and the first mouse linkage, of the pink-eye dilution (*p*) and albino (*c*) genes on what is now known to be chromosome 7, was published 2 years later [13]. The major factor limiting the improvement of mammalian linkage maps over the following 65 years was the shortage of useful genetic markers. The hundreds of known visible mutations in the mouse constitute the largest such collection for any mammal, but are insufficient by 2 orders of magnitude for construction of a bridging map. Furthermore, the deleterious effects of many of these mutations, and the complex interactions of their phenotypes with one another, make it impractical to study the inheritance of more than a few such markers at a time [14,15]. The various codominant biochemical and serological markers of laboratory mice catalogued over the past 30 years [16] are phenotypically neutral under laboratory conditions and mostly do not interact with one another. Their number, however, is still insufficient to yield a bridging map. The problem is compounded by the tendency of these markers to cluster, rather than to distribute randomly over the genome [17].

The manifesto of Botstein et al. [18] called attention to the usefulness of DNA sequence variants as heritable markers for the construction of mammalian linkage maps. The widespread use of these markers has had its most visible effects on the construction of human linkage maps [19]. The effects of DNA markers on genetic analysis in the mouse have been more subtle, but nevertheless profound. We will first discuss the nature of DNA sequence variation, emphasizing the differences between the patterns of variation seen in laboratory mice and in humans. We will then discuss the use of DNA markers to assemble multilocus linkage maps of mouse chromosomes, and will end by discussing the conclusions that can be drawn from these maps concerning phenomena such as the effects of interference and sex on the recombination process in mammals.

III. CHARACTERIZATION OF RESTRICTION FRAGMENT LENGTH POLYMORPHISMS

A priori, DNA sequence variation between individuals of a species could be due either to point substitution or to DNA rearrangements such as insertions or deletions. Variations within functional genes are largely of the former kind, and initial surveys of polymorphisms associated with anonymous DNA fragments—those cloned without regard to function and, therefore, likely to have no function at all—likewise revealed variations in restriction fragment patterns that could be explained by point substitution [20,21]. More recent work has revealed the existence of numerous islands of short, tandemly repeated DNA sequences in the human genome. The number of copies of the tandem-repeat sequence in any one island typically varies from individual to individual. Such islands have been found near both functional genes and anonymous DNA sequences. The enormous degree of polymorphism associated with such islands has made them the markers of choice for many kinds of human gene-mapping work, including the identification of forensic material [22–25]. The relative frequencies of insertion–deletion and point substitution polymorphisms in the human genome remains a point of investigation, but of more than 1890 polymorphisms now catalogued, the majority appear to be point substitutions [19,26].

This classification is only an estimate, however, as the structural basis of most DNA variants has been deduced from Southern blotting analysis and not from direct comparison of their sequences. That is, the observation that a given probe reacts with either a single restriction fragment of 10 kb or with two, of 4 and 6 kb, is most readily explained by the hypothesis that the first allele differs from the second by a point substitution that destroyed a restriction cleavage site, but only sequence analysis of that region of both alleles can prove it. Despite this caveat, there is no obvious experimental or theoretical reason to doubt the overall accuracy of the classification of kinds and frequencies of human DNA polymorphism.

The number of characterized DNA markers for the mouse is smaller, and published lists of these markers have not classified them according to the nature of the DNA variation associated with each [27,28]. The set of DNA markers for mouse chromosome 12 appears to be typical, however, for laboratory mice. The set consists of ten anonymous single-copy sequences isolated because they adjoin mouse-specific reiterated DNA sequences, as well as probes for *Apob* (apolipoprotein B), *Odc* (ornithine decarboxylase), the *Nmyc* and *Fos* protooncogenes, the *Aat* (α_1-antitrypsin) gene complex, *Spnb* (β-spectrin), *Crip* (cysteine-rich intestinal protein), *Ck-3* (creatine kinase brain form), the *Igh-V* and *Igh-C* (immunoglobulin heavy-chain variable and constant region) gene complexes, and ten proviral integration sites. The proviral integration sites are of necessity insertion–deletion variants. Of the remaining 20 markers, all but one, *Nmyc*, are associated with DNA restriction fragment variants in surveys of laboratory strains of mice, and most of these cannot be accounted for by point substitutions [29–41; P. D'Eustachio, unpublished].

Both the large proportion of variants, and the complexity of the variations seen at each, distinguish this collection of mouse DNA markers from those for human chromosomes. There is no evidence that different mutational mechanisms have operated in the evolution of rodent and primate genomes, or that they have operated at significantly different rates. These patterns of genetic variation thus suggest the need for a reexamination of the possible origins of laboratory strains of mice. Laboratory strains of mice are thought to be derived from wild populations of *Mus domesticus*. Many instances of non-*M. domesticus* information fixed in these strains have been described, however, and the amount of strain-to-strain variation exceeds that found in wild *M. domesticus* populations [42,43]. A survey of published typing data for inbred and wild mice suggested that fixation of non-*M. domesticus* alleles in one or more inbred strains had occurred at about 20% of autosomal genes. Furthermore, computer simulations of the inbreeding process suggest that the median length of the chromosomal segments introgressed as the result of an interspecies cross would be 17 cM. These simulations accurately predicted the distributions of progenitor strain segment lengths for chromosomes 1, 9, and 12 in recombinant inbred strains of mice. Together, these results support the model that initial hybridization among a few highly divergent stocks or species, followed by segregation and fixation of the resulting "patchwork" genomes, is a plausible and sufficient explanation for the genetic diversity of present-day laboratory strains of mice [44].

The excess of DNA polymorphisms among laboratory strains of mice is in contrast with the pattern of polymorphism found in wild mice. *Mus minutoides*, *Arvicanthus niloticus*, and *Praomys* spp. trapped in East Africa have been surveyed for polymorphism associated with four of the chromosome 12 markers and a sample of transcriptionally active genes from other chromosomes. The sample of mice is large (over 100 each of *M. minutoides* and *A. niloticus*, and approximately 20 *Praomys*). Nevertheless, the RFLPs found have been limited: typically, no more than three alleles of a marker have been found in any one species, and these can often be accounted for, at the level of resolution allowed by genomic Southern blotting, by point substitutions [Y. Cole, C. Jolly, and P. D'Eustachio, unpublished].

The complex origin of laboratory strains of mice leads to a small, but important, point of nomenclature concerning DNA sequence variations found among them. Because these strains cannot be treated as individual members of a population of a good single species, restriction fragment variants studied only among laboratory strains cannot be called polymorphisms. A variant allele is conventionally considered to be polymorphic when it occurs at a frequency of 1% or more in a natural population of a species. Without a natural population or a reliable model of one to survey, polymorphism has no meaning.

IV. ASSEMBLY AND ANALYSIS OF LINKAGE MAPS

The assembly of a linkage map from recombination frequencies measured among informative genetic markers in a cross proceeds identically whether these markers are visible phenotypes such as coat color, biochemical phenotypes such as the electrophoretic mobility of an enzyme, or DNA fragment variants. Likewise, the assembly of a map is independent of the eukaryotic species from which these data are collected, at least conceptually. Nevertheless, the ability to follow arbitrarily many different informative marker genes simultaneously in a cross, by use of DNA probes, has changed the practical design of experiments to search for linkage and to build maps, and provides a way to study the biology of the recombination process itself in mammalian systems. Accordingly, it is useful to summarize some aspects of strategies for building linkage maps.

Assembling a map proceeds in two independent stages. First, the data are used to demonstrate linkage among a group of markers. Second, contingent on the hypothesis of linkage, optimal map positions for the markers are determined. That is, the same data, recombination fractions, are used sequentially to test the two distinct hypotheses.

A. Detecting Linkage

Linkage between a test gene and a marker of interest is demonstrated experimentally by measuring a recombination fraction between them that is significantly smaller than the value of 0.5 expected for unlinked genes. Carter and Falconer [14] coined the term *swept radius* to mean the greatest genetic distance that could separate two markers and still allow a high probability of observing such a significant deviation among a given number of progeny. They calculated swept radii for various numbers of progeny and intermarker spacings by using binomial probabilities, and suggested, based on these calculations, optimal strategies for searching for linkage between a test marker of interest and genetic landmarks, defined mostly by dominant visible mutations fixed in laboratory stocks of mice.

The binomial statistics used by Carter and Falconer test the odds of obtaining the observed data set on the hypothesis of linkage. Bayesian statistics allows the more rigorous, and experimentally more relevant, test of the odds that the markers are linked, given the data set [38,45]. Application of this more stringent test to the computation of swept radii has two major consequences (Table 1). Swept radii are sharply reduced, and it is possible to calculate numbers of recombinant progeny that allow the hypothesis of linkage to be explicitly excluded. Thus, if 20 progeny from a backcross are typed for inheritance of alleles at two markers and two or fewer are recombinant, the probability that the two markers are linked (i.e., lie within 50 cM of one another) is greater than 0.95. If eight or more are recombinant, the probability that the two genes are linked is less than 0.05. If three to seven progeny are recombinant, no conclusion is justified on linkage between the two markers at a high confidence level.

The distance swept by a marker in a collection of 50–100 progeny is thus small, and the swept distance increases slowly with increases in numbers of progeny (see Table 1). DNA markers alleviate this problem in two ways. First, because they are codominant, backcrosses can always be performed, and intercrosses, yielding two independently informative chromosomes per progeny individual, are often possible. Second, the enormous number of DNA variants known, even among laboratory strains [27,28], allows progeny from single crosses to be screened for multiple, closely spaced markers of every chromosome, so that the small distance swept per marker need not limit a search for linkage [46]. Indeed, strains of mice have been constructed that harbor endogenous ecotropic proviral DNA sequences in their germline genomes at multiple well-mapped sites. With progeny from a cross between a strain carrying a distinctive allele of a gene of interest and such a proviral testing stock, a substantial fraction of the genome can be swept for linkage in a single Southern blotting experiment [47]. The recent development of techniques that also allow nonecotropic proviral DNA sequences to be used as markers raises the possibility of following hundreds of markers, widely scattered over the mouse genome, in single crosses, with simple, efficient assays [41,48].

Table 1 Distances Swept by Codominant Markers in Backcrosses

			Bayesian linkage				Bayesian unlinkage
N	p_r	m_r	R	R/N	SE	m_r	R
10	0.18	0.19					4
20	0.28	0.31	2	0.10	0.07	0.10	8
30	0.32	0.37	5	0.17	0.07	0.17	12
40	0.34	0.42	8	0.20	0.06	0.21	16
50	0.36	0.45	12	0.24	0.06	0.26	21
60	0.37	0.48	15	0.25	0.06	0.27	25
70	0.38	0.50	19	0.27	0.05	0.30	29
80	0.39	0.52	23	0.29	0.05	0.33	34
90	0.39	0.53	27	0.30	0.05	0.34	38
100	0.40	0.55	31	0.31	0.05	0.36	43
200	0.43	0.64	72	0.36	0.03	0.45	88
300	0.44	0.70	116	0.39	0.03	0.51	134
400	0.45	0.74	160	0.40	0.02	0.55	181
500	0.46	0.77	201	0.40	0.02	0.55	228
600	0.46	0.79	251	0.42	0.02	0.60	276
700	0.46	0.82	297	0.42	0.02	0.63	323
800	0.46	0.83	343	0.43	0.02	0.64	371
900	0.47	0.84	389	0.43	0.02	0.66	419
1000	0.47	0.86	436	0.44	0.02	0.67	467
1500	0.47	0.91	670	0.45	0.01	0.72	707
2000	0.48	0.94	906	0.45	0.01	0.75	949

Recombination fractions (p_r) and map distances swept (m_r), calculated according to Carter and Falconer [14], are shown for various number, N, of informative progeny chromosomes. The Bayesian approach of Silver and Buckler [45], as modified by Blank et al. [38], was used to calculate the maximum number of recombinant chromosomes, R, consistent with the conclusion at $p \geq 0.95$ of linkage for each value of N. Recombination fractions (R/N), their standard errors, and values of m_r corresponding to each recombination fraction are shown. Finally, the Bayesian approach [38] was used to calculate the minimum value of R that supports the conclusion at $p \geq 0.95$ that the test and marker loci are not linked (Bayesian unlinkage).

B. Ordering Linked Markers

Given a group of markers shown to be linked to one another, the recombination fractions measured among them can then be used to construct a linkage map for the markers. Sturtevant [12] assembled his linkage map of the *Drosophila* X chromosome making three assumptions: the chromosome is linear; each gene occupies a single point on the line; and crossover events are rare, randomly spaced, and independent of one another. Therefore, the best order for a group of linked markers is the one that minimizes the number of multiple crossover events needed to account for the observed data. Such a best map should be nearly additive: if three markers are in the order A–B–C, the pairwise recombination fraction measured between A and C should equal, within experimental error, the sum of the fractions observed between A and B and between B and C. As before, the reasoning is independent of the assays used to define markers and has remained the starting point for thinking about the construction of linkage maps. It has two crucial limitations when applied to the large collections of markers made possible for the first time in mammals by the use of DNA probes: assessing alternative orders of markers systematically becomes unwieldy as the number of markers grows; and the assumption that recombination events are independent and random precludes the investigation of interference [19,49].

The first limitation has been addressed by dealing with the data hierarchically in the algorithms now in use for assembly of linkage maps. Best maps are determined for groups of three or four markers for which large amounts of data are available. These groups are then assembled into larger maps, and markers for which only limited data are available can be interpolated into the growing map at a late stage [38,50–52]. In this way, many possible orders of a group of markers need never be considered in searching for the best one. Furthermore, these algorithms all yield quantitative measures of the fit of the map to the data.

This quantitation is often crucial. If, for example, one wants to use map information to identify markers on each side of a chromosomal segment of interest to use as starting points for a physical walk through the segment, one would want a way of measuring the odds that the positions assigned to the markers are correct.

Our analysis of mouse chromosome 12 provides an example of such a map-building exercise [38]. Map-building starts from a pair of markers, adds additional loci one at a time, and calculates the spacings among the loci that maximize the odds of obtaining all of the observed recombination fractions between markers considered two at a time. By calculating the region of the linkage map within which each marker resides at any desired confidence level (empirically, a 99% confidence level appears reliable), given the placement of the other markers in the map, it is also possible to measure the probability that a map arrived at in this way represents the correct order of the markers. To construct a framework linkage map of the chromosome, inheritance of restriction fragment variants associated with four anonymous DNA markers (*D12Nyu1*, *2*, *3*, and *4*), the *Fos* protooncogene, the *Mtv-9* viral integration site, and the α_1-antitrypsin (*Aat-1*) and immunoglobulin heavy chain (*Igh*) gene families in the mouse was followed in a backcross experiment between the inbred laboratory strains C57BL/6J and SWR/J. Two groups of markers studied in recombinant inbred strains of mice could be interpolated into the map: *Es-25* [53], *D12Nyu10*, *D12Nyu7*, and *Apob* [35] form a cluster proximal to *D12Nyu2*; and *Ly-18*, *Ah*, and *D12Nyu5* form a cluster between *D12Nyu2* and *D12Nyu1*. Together, these data establish an unambiguously ordered linkage group including *Igh* and *Aat-1* that spans most of chromosome 12 [38]. Markers such as *Fv-4* [54] and *Spnb* [37], studied in more limited backcross experiments, can be interpolated accurately into this map, as can the gene defined by the *iv* mutation [46].

More recent work has expanded this map to include the transcriptionally active ornithine decarboxylase (*Odc*) gene, 2-cM proximal to *D12Nyu2* [40]; the *Igh-V* complex, spanning a 0.5-cM region distal to *Igh-C*; and the creatine kinase–brain form (*Ck-3*) structural gene, 2-cM proximal to *Igh-C* [P. D'Eustachio, unpublished]. In addition, preliminary results suggest that the *Akt* protooncogene and the *Crip* (cysteine-rich intestinal protein [32]) map within a few tenths of a centimorgan of *Igh-C*. The map now spans 58 cM, from *Odc* to *Igh-V*, and most markers are localized, at 95% confidence, to within a few centimorgans. In some regions, notably near the middle of the chromosome, markers as close as 1–2 cM apart are unambiguously ordered.

Similar results have been achieved for several human chromosomes [19] and for large portions of several mouse chromosomes, including numbers of 1 [55], 2 [56,57], 9 [58], 11 [59], 19 [57], and the X chromosome [60].

V. BIOLOGY OF MULTILOCUS LINKAGE MAPS

The use of extensive maps based on DNA markers as a framework to localize biologically interesting genes has been discussed in the foregoing. The maps are also intrinsically interesting because of the insights they are starting to provide into the organization of genetic information in mammals and its reorganization at meiosis.

A. Sex

Sex differences in recombination fractions measured between pairs of markers are common in the mouse, but the extent and even the direction of the differences vary with the markers examined [10, pp. 223–239; 61,62]. Only for chromosome 12 have linkage maps been built separately with data from female and male meioses. Here, the male map was significantly shorter overall. Most human linkage maps likewise show reduced male meiotic lengths, although the extent of the reduction varies from chromosome to chromosome [19]. Indeed, human chromosome 14q, homologous to the distal half of mouse chromosome 12, shows a slightly longer male than female meiotic length [63]. Despite the obvious gaps in the data, it is attractive to speculate that in mice, as in humans, male meiotic maps will be shorter overall than female ones. This effect seems unlikely to be uniform, however, and multilocus map building will provide a powerful tool, not only for identifying overall trends, but also for pinpointing anomalous regions.

B. Segregation Distortion

Significant deviations from 1 : 1 transmission of alleles over extended regions of mouse chromosomes 2, 4, 10, 12, and X have been observed in crosses between laboratory strains of mice and *M. spretus*. These deviations do not appear to result from differential lethality, and other regions of the same and other chromosomes, mapped in the same crosses, are free from such effects [56,64–66]. They suggest the existence of several unlinked genes, alleles of each of which are associated with a segregation distortion phenotype.

C. Interference

Analysis of interference in mammals requires a deviation from Sturtevant's strategy, because that strategy assumes there is none. That is, it assumes that each recombination event occurs randomly and independently of all other events. Thus, the odds of there being two events in a given interval are the square of the odds of there being a single event. This assumption allows the conclusions that recombination fractions are an accurate measure of distances between markers and that the optimal linkage map is the one that requires the fewest multiple crossover events to explain observed data sets. It is often a good approximation to reality, as demonstrated by 75 years of successful map building in diverse eukaryotes. Indeed, for human genetic data, because most progeny in a cross are informatively polymorphic for only a few of the markers being assembled into a linkage map, optimal use of all available data to build a linkage map requires that an assumption be made in advance about the extent of interference, and the assumption usually made is that there is none [19,52].

Because existence of interference is well known in other eukaryotic systems, it is clearly of interest to determine whether or not it exists and what its effects are quantitatively in mammals [67].

To use a map built on these assumptions to test for possible nonrandomness in the distribution of crossover events, however, is a circular argument. Other map-building procedures must be used to allow the nature and extent of interference to be examined. In the mouse, the extensive DNA sequence variation observed, even between laboratory strains of mice, makes it practical to generate hundreds of progeny simultaneously informative for numerous markers of a single chromosome of interest. The use of only recombination fractions measured between pairs of markers in such a cross to deduce the order of the markers, as discussed previously [38], yields a map independent of assumptions about the interactions among recombination events. It is then legitimate to tally numbers of crossover events observed per chromosome among the progeny of the cross, and search for deviation from the distribution of events per chromosome expected on the hypothesis that events are random relative to one another. The Poisson distribution can be used to predict the expected distribution of events, and a χ^2 calculation can be used to test for significance any deviation from expectation [11].

In the linkage group on mouse chromosome 12 that extends from *Odc* to *Igh-V*, for example, among 317 progeny chromosomes, 143 were nonrecombinant, 161 had undergone a single recombination event, and 13 had undergone two. None had undergone three or more. Thus, there were 188 events among 317 chromosomes. If the events were randomly distributed, 175.1 zero-event chromosomes, 103.9 one-event chromosomes, 30.8 two-event chromosomes, and 7.1 chromosomes with three or more events would be expected. The difference between the two distributions is significant (χ^2 equals 56.2 with 2 degrees of freedom, $p < 0.005$) [38; P. D'Eustachio, unpublished data]. Qualitatively, the proportions of nonrecombinant and multiply recombinant chromosomes are significantly smaller than expected in the absence of chiasma interference.

This result is not peculiar to the one cross or to the one chromosome. The same linkage group, examined in a backcross between laboratory mice and individuals of a *M. spretus* stock, showed similarly strong positive chiasma interference [65], and other linkage groups studied in the progeny of this interspecies cross also showed chiasma interference [55]. Indeed, it is plausible to speculate that strong positive chiasma interference is a universal feature of the recombination process in mice. Figure 1 shows a plot of deviations from Poisson expectation found in a total of 12 multilocus linkage analyses carried out, over the past 5 years, with various combinations of mouse strains and species, for various chromosomes, and in various laboratories. The plot suggests that the odds of finding interference are nearly a linear function of the size of the data set: the larger the total number of recombination events found, the stronger the statistical evidence for interference. Thus, although neither multilocus analysis nor the use of DNA markers is necessary a priori to detect it, the use of both greatly facilitates the process at a practical level.

The growing set of multipoint linkage maps also makes it worthwhile to reexamine the question of chromatid interference in mammalian systems. That is, in those meiotic tetrads that do undergo multiple recombination

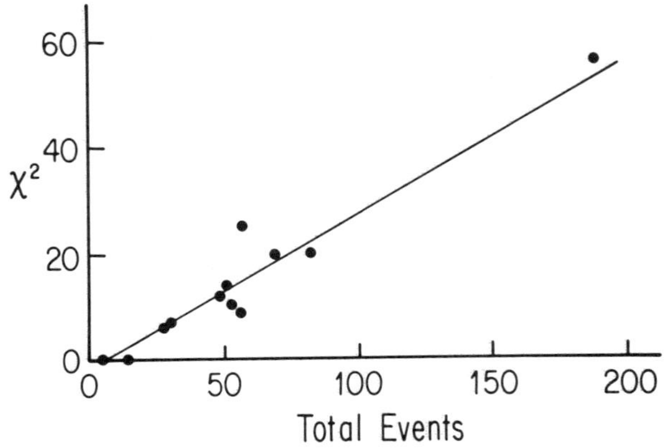

Figure 1 Positive chiasma interference in the mouse. Distributions of recombination events in 12 backcross data sets, including linkage groups on chromosomes 2, 4, 6, 11, 12, 19, and X [38,56,57,59,60,65,73–75], were analyzed for deviation from the distribution expected on the hypothesis of no interference as described in the text for chromosome 12. For each set, the calculated value of χ^2 is plotted as a function of the total number of recombination events in the data set.

events, are the pairs of chromatids involved in each event random relative to one another? If all four products of the meiosis could be recovered, randomness would yield a 1 : 2 : 1 ratio of two-stranded : three-stranded : four-stranded double recombination events. Chromatid interference would cause deviation from this ratio. Analysis of ordered tetrads in fungi and of markers of attached-X chromosomes in *Drosophila* reveals no evidence of positive chromatid interference. There is either no effect or a marginal excess of two-stranded double events [68,69].

The progeny in a breeding experiment represent a random sample of meiotic products, so a direct search for chromatid interference is impossible. Carter [70], however, calculated the effect that such interference would have on the relationship between recombination fractions and overall map distances across linkage groups in the mouse (Fig. 2). The data available to him in 1954, shown by the open circles in the figure, suggested that in mice, as in *Drosophila* and fungi, there was little or no chromatid interference. However, this analysis was based on small amounts of data and limited linkage groups. Interpolation of data from mouse chromosomes 11 and 12 into Carter's graphs suggests that positive chromatid interference may be taking place. As the shape of the curves makes clear, the critical test of this notion will come as multilocus maps of linkage groups significantly longer than 50 cM overall are analyzed in this way.

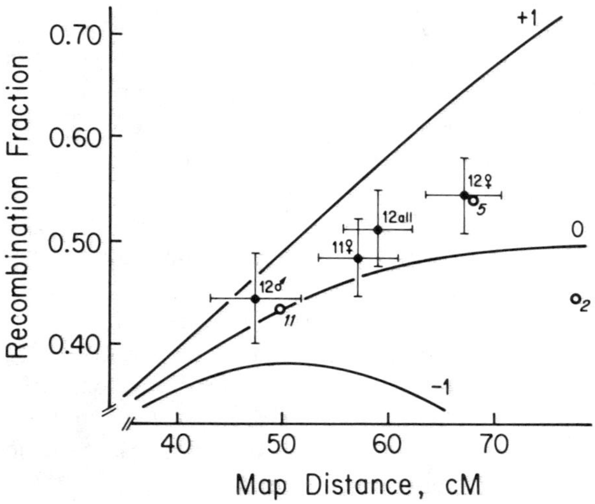

Figure 2 Chromatid interference in the mouse. Recombination percentages observed between the end markers of linkage groups are plotted as a function of the map distance between them. The three curves, redrawn from Carter [70], show the relationships expected if chromatid interference is complete and positive (+1), complete and negative (−1), or absent (0). Recombination percentages between end markers are the pairwise recombination fractions measured between these markers in a backcross. Map distances are obtained by summing the distances measured between intervening markers. (Thus, in the map A–B–C–D, the recombination fraction would be the observed percentage recombination between A and D, whereas the map distance would be the sum of the recombination percentages observed between A and B, B and C, and C and D.) Carter's [70] results for linkage groups on chromosomes 2, 5, and 11 are shown with open circles. Recent data for chromosomes 12 (from heterozygous females, males, and combined [38]) and for chromosome 11 (female only [59]) are shown, ± standard errors.

D. Comparison of Maps

The genetic distance between two markers is often not a constant, but rather, varies according to the genetic backgrounds of the mice in which it is being measured [62]. Analysis of data from multilocus crosses is useful in studying such differences, as it is for studying the aforementioned differences owing to sex. As in that case, available results are insufficient to support many general conclusions, but they do suggest hypotheses and indicate directions for further investigation.

First, such differences are real and can affect large regions of a chromosome. For example, the map of chromosome 12 constructed from analysis of an interspecies backcross is significantly shorter overall than one constructed with data from a cross between two laboratory strains of mice [65]. Similar results have been obtained for a segment of chromosome 5 [71]. Analysis of a group of markers spanning the central third of mouse chromosome 2 yielded maps that differed significantly in overall length when crosses involving different combinations of laboratory mouse strains were compared [64].

Second, such differences affect gene spacing, but not gene order. A possible exception to this hypothesis is the proximal region of chromosome 17, the organization of which in *M. spretus* and t-haplotype mouse stocks differs from that seen in laboratory mice and wild *M. domesticus* and *M. musculus* [72; L. Silver, personal communication]. Otherwise, however, no such differences in gene order have been observed.

These results provide an appropriate end for this review, because they illustrate both how the use of DNA markers has changed our view of mammalian linkage maps and, at the same time, they indicate how much remains unknown. DNA markers have made a large fraction of the genome, perhaps all of it, readily accessible to genetic analysis. The increased density and uniformity of genetic maps that have resulted is already beginning to affect our thinking about basic biological processes, such as recombination itself, and to lead to experiments to manipulate these processes in mammalian systems. At an applied level, the abundance of markers has already led to maps that are starting points for the physical analysis of extended regions of mouse and human chromosomes, an activity that will increase rapidly in importance in the next few years.

ACKNOWLEDGMENTS

Work from my laboratory discussed in this review was supported by grants from the NIH and the National Foundation–March of Dimes.

REFERENCES

1. Rommens, J. M., M. C. Iannuzzi, B.-S. Kerem, M. L. Drumm, G. Melmer, M. Dean, R. Rozmahel, J. L. Cole, D. Kennedy, N. Hidaka, M. Zsiga, M. Buchwald, J. R. Riordan, L.-C. Tsui, and F. S. Collins (1989). Identification of the cystic fibrosis gene: Chromosome jumping and walking. *Science 245*:1059–1065.
2. Kerem, B.-S., J. M. Rommens, J. A. Buchanan, D. Markiewicz, T. K. Cox, A. Chakravarti, M. Buchwald, and L.-C. Tsui (1989). Identification of the cystic fibrosis gene: Genetic analysis. *Science 245*:1073–1080.

3. Selander, R. (1970). Behavior and genetic variations in natural populations. *Am. Zool. 10*:53–66.
4. Klein, J. (1975). *Biology of the Mouse Histocompatibility-2 Complex*. Springer-Verlag, New York, pp. 291–292.
5. Berry, R. J. (1986). Genetic processes in wild mouse populations. Past myth and present knowledge. *Curr. Top. Microbiol. Immunol. 127*:86–94.
6. Green, M. (1989). Catalog of mutant genes and polymorphic loci. In *Genetic Variants and Strains of the Laboratory Mouse*, 2nd ed. M. F. Lyon and A. G. Searle (eds.). Oxford University Press, New York, pp. 12–403.
7. Davisson, M. T., T. H. Roderick, A. L. Hillyard, and D. P. Doolittle (1989). The locus map of the mouse. *Mouse News Lett. 84*:15–23.
8. Smith, C. L., S. K. Lawrence, G. A. Gillespie, C. R. Cantor, S. M. Weissman, and F. S. Collins (1987). Strategies for mapping and cloning macroregions of mammalian chromosomes. *Methods Enzymol. 151*:461–489.
9. Laird, C. D. (1971). Chromatid structure: Relationship between DNA content and nucleotide sequence diversity. *Chromosoma 32*:378–406.
10. Robinson, R. (1972). *Gene Mapping in Laboratory Mammals*, Part B. Plenum Press, New York.
11. Snedecor, G. W., and W. G. Cochran (1967). *Statistical Methods*, 6th ed. Iowa State University Press, Ames, pp. 223–226.
12. Sturtevant, A. H. (1913). The linear arrangement of six sex-linked factors in *Drosophila*, as shown by their mode of association. *J. Exp. Zool. 14*:43–59.
13. Haldane, J. B. S., A. D. Sprunt, and N. M. Haldane (1915). Reduplication in mice. *J. Genet. 5*:133–135.
14. Carter, T. C., and D. S. Falconer (1951). Stocks for detecting linkage in the mouse, and the theory of their design. *J. Genet. 50*:307–323.
15. Green, E. L. (1981). *Genetics and Probability in Animal Breeding Experiments*. Oxford university Press, New York, pp. 109–113.
16. Roderick, T. H., and J. H. Guidi (1989). Strain distribution of polymorphic variants. In *Genetic Variants and Strains of the Laboratory Mouse*, 2nd ed. M. F. Lyon and A. G. Searle (eds.). Oxford University Press, New York, pp. 663–772.
17. Lyon, M. F. (1976). Distribution of crossing-over in mouse chromosomes. *Genet. Res. 28*:291–299.
18. Botstein, D., R. White, M. Skolnick, and R. W. Davis (1980). Construction of a genetic map in man using restriction fragment length polymorphisms. *Am. J. Hum. Genet. 32*:314–331.
19. White, R., and J.-M. Lalouel (1988). Sets of linked genetic markers for human chromosomes. *Annu. Rev. Genet. 22*:259–279.
20. Nei, M. (1987). *Molecular Evolutionary Genetics*. Columbia University Press, New York, pp. 64–110.
21. Barker, D., M. Schafer, and R. White (1984). Restriction sites containing CpG show a higher frequency of polymorphism in human DNA. *Cell 36*:131–138.
22. Jeffreys, A. J., V. Wilson, and S. L. Thein (1985). Hypervariable “minisatellite” regions in human DNA. *Nature 314*:67–73.

23. Nakamura, Y., M. Leppert, P. O'Connell, R. Wolff, T. Holm, M. Culver, C. Martin, E. Fujimoto, M. Hoff, E. Kumlin, and R. White (1987). Variable number of tandem repeat (VNTR) markers for human gene mapping. *Science 235*:1616–1622.
24. King, M.-C. (1989). Genetic testing of identity and relationship. *Am. J. Hum. Genet. 44*:179–181.
25. Lander, E. (1989). DNA fingerprinting on trial. *Nature 339*:501–505.
26. Kidd, K. K., A. M. Bowcock, J. Schmidtke, R. K. Track, F. Ricciuti, G. Hutchings, A. Bale, P. Pearson, and H. F. Willard (1989). Report of the DNA committee and catalogs of cloned and mapped genes and DNA polymorphisms. 10th International Workshop on Human Gene Mapping. *Cytogenet. Cell Genet. 51*:622–947.
27. Elliott, R. (1989). DNA restriction fragment variants. *Mouse News Lett. 83*: 126–148.
28. Eppig, J. T. (1989). List of mouse DNA clones and probes. *Mouse News Lett. 83*:73–116.
29. Blatt, C., K. Mileham, M. Haas, M. N. Nesbitt, M. E. Harper, and M. I. Simon (1983). Chromosomal mapping of the mink cell focus-inducing and xenotropic *env* gene family in the mouse. *Proc. Natl. Acad. Sci. USA 80*:6298–6302.
30. D'Eustachio, P. (1984). A genetic map of mouse chromosome 12 composed of polymorphic DNA fragments. *J. Exp. Med. 160*:827–838.
31. Wejman, J. C., B. A. Taylor, N. A. Jenkins, and N. G. Copeland (1984). Endogenous xenotropic murine leukemia virus-related sequences map to chromosomal regions encoding mouse lymphocyte antigens. *J. Virol. 50*:237–247.
32. Birkenmeier, E. H., and J. I. Gordon (1986). Developmental regulation of a gene that encodes a cysteine-rich intestinal protein and maps near the murine immunoglobulin heavy chain locus. *Proc. Natl. Acad. Sci. USA 83*:2516–2520.
33. Rossomando, A., and D. Meruelo (1986). Viral sequences are associated with many histocompatibility genes. *Immunogenetics 23*:233–245.
34. Peters, G., M. Placzek, S. Brookes, C. Kozak, R. Smith, and C. Dickson (1986). Characterization, chromosome assignment, and segregation analysis of endogenous proviral units of mouse mammary tumor virus. *J. Virol. 59*:535–544.
35. Lusis, A. J., B. A. Taylor, D. Quon, S. Zollman, and R. C. LeBoeuf (1987). Genetic factors controlling structure and expression of apolipoproteins B and E in mice. *J. Biol. Chem. 262*:7594–7602.
36. Bauer, S. R., L. A. D'Hoostelaere, and K. Huppi (1988). Restriction fragment length polymorphism near the IgH locus on mouse chromosome 12. *Nucleic Acids Res. 16*:8200.
37. Birkenmeier, C. S., E. C. McFarland-Starr, and J. E. Barker (1988). Chromosomal location of three spectrin genes: Relationship to the inherited hemolytic anemias of mouse and man. *Proc. Natl. Acad. Sci. USA 85*:8121–8125.
38. Blank, R. D., G. R. Campbell, A. Calabro, and P. D'Eustachio (1988). A linkage map of mouse chromosome 12: Localization of *Igh* and effects of sex and interference on recombination. *Genetics 120*:1073–1083.

39. Campbell, G. R., K. Zimmerman, R. D. Blank, F. W. Alt, and P. D'Eustachio (1989). Chromosomal location of N-*myc* and L-*myc* genes in the mouse. *Oncogene Res. 4*:47–54.
40. Villani, V., P. Coffino, and P. D'Eustachio (1989). Linkage genetics of mouse ornithine decarboxylase (*Odc*). *Genomics 5*:636–638.
41. Frankel, W. N., J. P. Stoye, B. A. Taylor, and J. M. Coffin (1990). A linkage map of endogenous murine leukemia proviruses. *Genetics 124*:221–236.
42. Rice, M. C., and S. J. O'Brien (1980). Genetic variance of laboratory outbred Swiss mice. *Nature 283*:157–161.
43. Fitch, W. M., and W. R. Atchley (1985). Evolution in inbred strains of mice appears rapid. *Science 228*:1169–1175.
44. Blank, R. D., G. R. Campbell, and P. D'Eustachio (1986). Possible derivation of the laboratory mouse genome from multiple wild *Mus* species. *Genetics 114*: 1257–1269.
45. Silver, J., and C. E. Buckler (1986). Statistical considerations for linkage analysis using recombinant inbred strains and backcrosses. *Proc. Natl. Acad. Sci. USA 83*:1423–1427.
46. Brueckner, M., P. D'Eustachio, and A. L. Horwich (1989). Linkage mapping of a mouse gene, *iv*, that controls left–right asymmetry of the heart and viscera. *Proc. Natl. Acad. Sci. USA 86*:5035–5038.
47. Taylor, B. A., and L. Rowe (1989). A mouse linkage testing stock possessing multiple copies of the endogenous ecotropic murine leukemia virus genome. *Genomics 5*:221–232.
48. Stoye, J. P., and J. M. Coffin (1988). Polymorphism of murine endogenous proviruses revealed by using virus class-specific oligonucleotide probes. *J. Virol. 62*:168–175.
49. Goldgar, D. E., P. R. Fain, and W. J. Kimberling (1989). Chiasma-based models of multilocus recombination: Increased power for exclusion mapping and gene ordering. *Genomics 5*:283–290.
50. Lathrop, G. M., J.-M. Lalouel, C. Julier, and J. Ott (1984). Strategies for multilocus linkage analysis in humans. *Proc. Natl. Acad. Sci. USA 81*:3443–3446.
51. Lathrop, M., Y. Nakamura, P. Cartwright, P. O'Connell, M. Leppert, C. Jones, H. Takeishi, T. Bragg, J.-M. Lalouel, and R. White (1988). A primary genetic map for markers of human chromosome 10. *Genomics 2*:157–164.
52. Lander, E. S., P. Green, J. Abrahamson, A. Barlow, M. J. Daly, S. E. Lincoln, and L. Newburg (1987). MAPMAKER: An interactive computer package for constructing primary genetic linkage maps of experimental and natural populations. *Genomics 1*:174–181.
53. von Deimling, O., and B. A. Taylor (1987). Esterase-25 (*Es-25*): Identification and characterization of a new kidney arylesterase of the house mouse genetically linked to *Ly-18* on chromosome 12. *Biochem. Genet. 25*:639–646.
54. Ikeda, H., H. Sato, and T. Okada (1981). Mapping of the mouse *Fv-4* gene controlling resistance to murine leukemia viruses. *Int. J. Cancer 28*:237–240.
55. Seldin, M. F., and G. D. Kruh (1989). Mapping of *Abll* within a conserved linkage group on distal mouse chromosome 1 syntenic with human chromosome 1 using an interspecific cross. *Genomics 4*:221–223.

56. Siracusa, L. D., A. M. Buchberg, N. G. Copeland, and N. A. Jenkins (1989). Recombinant inbred strain and interspecific backcross analysis of molecular markers flanking the murine *agouti* coat color locus. *Genetics 122*:669–679.
57. Glaser, T., K. E. Matthews, J. W. Hudson, P. Seth, D. E. Housman, and M. M. Crerar (1989). Localization of the muscle, liver and brain glycogen phosphorylase genes on linkage maps of mouse chromosomes 19, 12, and 2, respectively. *Genomics 5*:510–521.
58. Kingsley, D. M., N. A. Jenkins, and N. G. Copeland (1989). A molecular genetic linkage map of mouse chromosome 9 with regional localizations for the *Gsta*, *T3g*, *Ets-1* and *Ldlr* loci. *Genetics 123*:165–172.
59. Buchberg, A. M., E. Brownell, S. Nagata, N. A. Jenkins, and N. G. Copeland (1989). A comprehensive genetic map of murine chromosome 11 reveals extensive linkage conservation between mouse and human. *Genetics 122*:153–161.
60. Mullins, L. J., S. G. Grant, D. A. Stephenson, and V. M. Chapman (1988). Multilocus molecular mapping of the mouse X chromosome. *Genomics 3*:187–194.
61. Dunn, L. C., and D. Bennett (1967). Sex differences in recombination of linked genes in animals. *Genet. Res. 9*:211–220.
62. Davisson, M. T., T. H. Roderick, and D. P. Doolittle (1989). Recombination percentages and chromosomal assignments. In *Genetic Variants and Strains of the Laboratory Mouse*, 2nd ed. M. F. Lyon and A. G. Searle (eds.). Oxford University Press, New York, pp. 432–505.
63. Nakamura, Y., M. Lathrop, P. O'Connell, M. Leppert, M. I. Kamboh, J.-M. Lalouel, and R. White (1989). Frequent recombination is observed in the distal end of the long arm of chromosome 14. *Genomics 4*:76–81.
64. Carlson, G. A., P. A. Goodman, M. Lovett, B. A. Taylor, S. T. Marshall, M. Peterson-Torchia, D. Westaway, and S. B. Prusiner (1988). Genetics and polymorphism of the mouse prion gene complex: Control of scrapie incubation time. *Mol. Cell Biol. 8*:5528–5540.
65. Seldin, M. F., T. A. Howard, and P. D'Eustachio (1989). Comparison of linkage maps of mouse chromosome 12 derived from laboratory strain intraspecific and *Mus spretus* interspecific backcrosses. *Genomics 5*:24–28.
66. Biddle, F. G. (1987). Segregation distortion of X-linked marker genes in interspecific crosses between *Mus musculus* and *Mus spretus*. *Genome 29*:389–392.
67. Morton, N. E., C. J. MacLean, and R. Lew (1985). Tests of hypotheses on recombination frequencies. *Genet. Res. 45*:279–286.
68. Strickberger, M. W. (1985). *Genetics*, 3rd ed. Macmillan, New York, pp. 327–329.
69. Fincham, J. R. S., P. R. Day, and A. Radford (1979). *Fungal Genetics*. University of California Press, Berkeley, pp. 100–105.
70. Carter, T. C. (1954). A search for chromatid interference in the male house mouse. *Z. Vererbungsl. 86*:210–223.
71. Cheng, S. V., G. R. Martin, J. H. Nadeau, J. L. Haines, M. Bucan, C. A. Kozak, M. E. MacDonald, J. L. Lockyer, F. D. Ledley, S. L. C. Woo, H. Lehrach, C. Gilliam, and J. F. Gusella (1989). Synteny on mouse chromosome

5 of homologs for human DNA loci linked to the Huntington disease gene. *Genomics 4*:419–426.
72. Schimenti, J., and L. M. Silver (1986). Amplification and rearrangement of DNA sequences during the evolutionary divergence of t haplotypes and wild-type forms of mouse chromosome 17. *Curr. Top. Microbiol. Immunol. 127*: 247–252.
73. Nadeau, J. H., F. G. Berger, K. A. Kelley, P. M. Pitha, C. L. Sidman, and N. Worrall (1986). Rearrangement of genes located on homologous chromosomal segments in mouse and man: The location of genes for alpha- and beta-interferon, alpha-1 acid glycoprotein-1 and -2, and aminolevulinate dehydratase on mouse chromosome 4. *Genetics 104*:1239–1255.
74. Dandoy, F., J. De Maeyer-Guignard, and E. De Maeyer (1989). Linkage analysis of the murine *mos* proto-oncogene on chromosome 4. *Genomics 4*:546–551.
75. Bucan, M., T. Yang-Feng, A. M. Colberg-Poley, D. J. Wolgemuth, J.-L. Guenet, U. Francke, and H. Lehrach (1986). Genetic and cytogenetic localisation of the homeo box containing genes on mouse chromosome 6 and human chromosome 7. *EMBO J. 5*:2899–2905.

12

Genome Duplication and Comparative Gene Mapping

Joseph H. Nadeau

Jackson Laboratory, Bar Harbor, Maine

I. INTRODUCTION

Genome duplication is one of the three principal ways by which genes duplicate. Its characteristic feature is that unlinked duplicated genes are part of duplicated chromosomal segments. In this chapter, evidence for duplicated chromosomal segments in genome maps for mouse and man is reviewed. Approximately 60% of unlinked duplicated genes appear to be part of these duplicated chromosomal segments. The genome appears to be composed of comparable numbers of short segments that are not interrupted by unrelated genes and long segments that are always interrupted. Analysis of simulated maps of a random gene distribution suggests that half of the map could result from coincidental combinations of duplicated genes, with the remaining half presumably reflecting ancestral arrangements. Together these results suggest that short segments are ancestral, whereas long segments are coincidental. Similar to traditional comparative gene maps, these maps of duplicated chromosomal segments within the genome of a single species can be used in mapping gene families and in analyzing genome organization and evolution.

II. BACKGROUND

Comparative gene mapping, the analysis of the chromosomal location of homologous genes in different species, is not only a powerful aid in gene mapping but also a valuable approach for studying genome organization and evolution [1–5]. Comparative mapping has been used to predict linkage of an unmapped locus in one species, given the chromosomal location of its homologue in another species. Examples are numerous [6–10]. Comparative mapping has also been used to identify candidate mouse models for human genetic disorders. An excellent example is the apparent homology between Grieg cephalopolysyndactyly and the extra toes (*Xt*) mutation in the mouse [11,12]. Comparative mapping has been used to identify the chromosomal location of the murine counterpart of human disease loci by mapping murine homologues of marker loci flanking the disease locus [13–15]. Finally, lengths of conserved segments have been used to analyze the chromosomal distribution of rearrangement breakpoints and to estimate the rate of chromosomal rearrangement during mammalian evolution [16]. Recent estimates, for example, indicate approximately one disruption per million years since divergence of lineages leading to mouse and man [2,17]. With comparative maps becoming better defined, these applications become increasingly valuable.

Analysis of the chromosomal distribution of unlinked duplicated genes within a species is an analogous and equally valuable alternative to traditional comparative gene mapping. Many genes are members of families composed of both linked and unlinked duplicates [18–25]. Although much is known about rules governing the size and composition of families of tandemly duplicated genes [18–32], much less is known about rules governing the chromosomal distribution of unlinked duplicates [33–37].

There is considerable evidence that the eukaryotic genome has undergone multiple duplications. This evidence is primarily based on the mass of nuclear DNA per cell and the presence of unlinked duplicated genes [21,22]. An important consequence of genome duplication is that genes are duplicated as part of duplicated chromosomes. If the rate of subsequent chromosomal rearrangement has been modest, then maps showing the chromosomal distribution of unlinked duplicated genes not only provide evidence for genome duplication, but can also be used for comparative gene mapping. This idea is not new, however; it originated with Haldane [38], was formally developed by Ohno [21,22], and elaborated by Comings [39], Renwick [40], Lundin [41] and others [42,43]. Surprisingly, however, genome duplication has not been widely used as a means for studying genome organization and evolution, and its uses in gene mapping are largely unexploited. Molecular techniques now provide increasingly rigorous criteria for identifying duplicated genes [44]. In addition, the exponential rate at which genes are being added to the genome maps for mouse and man [2,45] leads to increasingly rigorous data that can be used for intraspecific comparative gene mapping. In this chapter, I evaluate evidence for duplicated chromosomal segments in genome maps for mouse and human, propose criteria for distinguishing ancestral and derived segments, and discuss applications of these maps in mapping gene families and in analyzing genome organization and evolution.

III. THE ORIGIN OF DUPLICATED GENES AND DUPLICATED CHROMOSOMAL SEGMENTS

A. Origins of Duplicated Genes

The three principal means by which genes duplicate are tandem duplication, retrotransposition, and genome duplication. Tandem duplication produces adjacent gene duplicates through unequal crossing-over [19,20,25–27,32]. This process can lead to considerable expansion and contraction of gene family size. Subsequent chromosomal rearrangements either increase the distance between duplicates or distribute duplicates to other chromosomes, without regard to the chromosomal location of other members of the gene family. These rearrangements are expected to be rare, however, given the original proximity of the duplicates and the very low fixation probability for reciprocal translocations [46–52]. Consequently, tandem duplicates tend to remain physically linked.

Retrotransposition produces gene duplicates that lack introns and that move to new chromosomal locations through an RNA intermediate [53–58]. The chromosomal location of these processed pseudogenes is thought to be independent of the location of other members of the gene family from which they originated.

Duplication of the genome involves groups of genes that are simultaneously duplicated [21,22]. The characteristic feature of genomic duplication is that unlinked gene duplicates tend to be located on duplicated chromosomal segments. Genomic duplication has probably been rare during amphibian, reptilian, and mammalian evolution, because of difficulties in sex determination in individuals with duplicated sex chromosomes [21,22]. The most recent genome duplication is believed to have occurred approximately 300 million years ago, long before divergence of lineages leading to mouse and man [21,22].

The chromosomal location of gene duplicates and the presence of introns within duplicates readily distinguish these three alternative origins. Tandem duplicates are adjacent and have introns; processed pseudogenes are located in unrelated chromosomal locations and lack introns; duplicated genes, resulting from genome duplication, are part of duplicated chromosomal segments and have introns.

B. Evolution After Genome Duplication

Two events usually follow genome duplication (Fig. 1). The first involves resolution of tetraploidy (or tetrasomy) through chromosome rearrange-

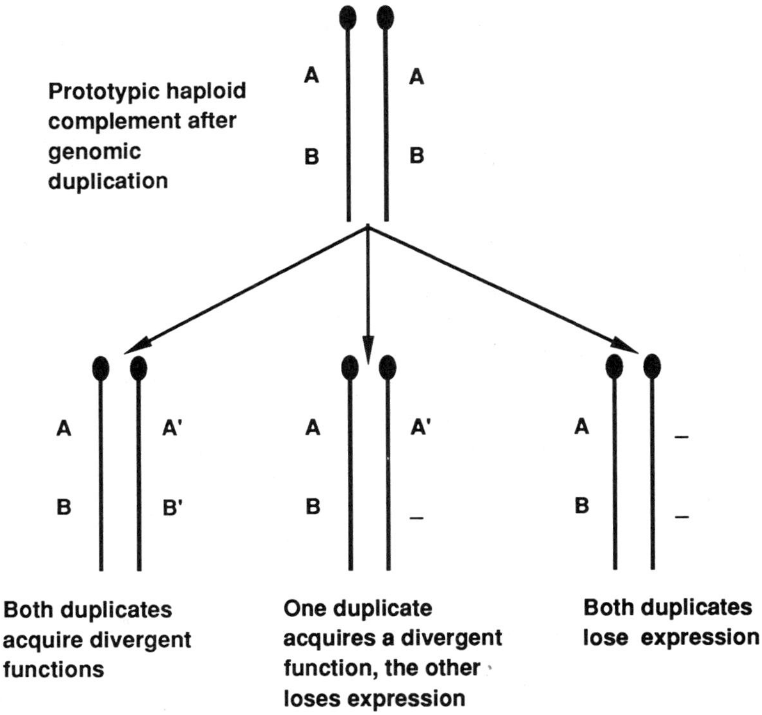

Figure 1 A model for genome duplication and subsequent genome evolution: A and B designate independent genes with unrelated functions, prime (′) designates a paralogous gene with a divergent function, and – designates a gene that is no longer expressed (i.e., a pseudogene).

ments that restore diploidy (or disomy) [21,22]. This event produces two bivalents from a single quadrivalent. The second event involves functional divergence, resulting either in duplicated genes with related functions or in loss of gene expression of one of the duplicated genes. It is assumed that one gene with the ancestral function will be retained because the gene has an essential function both before and after genome duplication. Pseudogenes with introns are expected to be part of duplicated chromosomal segments.

C. Composition of Duplicated Chromosomal Segments

There are three alternative arrangements of genes in these duplicated chromosomal segments. The first occurs when one gene duplicate acquires a divergent function. The resulting pair of chromosomes is marked by pairs of genetically and functionally related genes. The second arrangement occurs when one duplicate acquires a divergent function and the other loses expression and thereby becomes an unprocessed pseudogene. The resulting chromosomes are marked by one pair of related genes and by a gene and its unprocessed pseudogene duplicate. The third arrangement occurs when one member of each pair of duplicates loses expression. The resulting chromosomes are marked by a gene and its unprocessed duplicate. Segments marked by pseudogenes are expected to be more common than those marked by functionally divergent genes, because most gene duplicates lose expression [19,20,33,34].

D. Factors That Obscure Evidence for Genome Duplication

Several factors obscure evidence for duplicated chromosomal segments. First, physical loss of the gene obviously removes evidence for duplication. Second, sequence divergence reduces the ability to identify gene duplicates. If a gene became nonfunctional shortly after duplication, it is very doubtful that its functional homologue could be identified because of the high rate of nucleotide substitutions in pseudogenes [19,20,59–62]. Third, chromosomal rearrangements disrupt ancestral linkages and syntenies. These rearrangements, which reduce the size and increase the number of duplicated segments, make it appear as if there were more genomic duplications than there really were.

E. Definitions: Paralogous and Orthologous Chromosomal Segments

It is important to distinguish between conserved linkages and syntenies within and between species. The terms paralogous linkage and paralogous syntenies are therefore proposed. *Paralogous linkage* refers to conserved gene order and distances for genes composing a pair of duplicated segments, and *paralogous synteny* refers to conserved chromosomal assignments, regardless of gene order or distance. The corresponding relationships between species are defined as *orthologous linkages* and *orthologous syntenies*. These definitions are consistent with those proposed by Lundin [41], but not with those proposed by Fitch [63].

IV. EVIDENCE FOR DUPLICATED CHROMOSOMAL SEGMENTS IN GENOME MAPS OF MOUSE AND MAN

A. Criteria Used for Identifying Duplicated Genes

Genome maps for mouse and man were used for identifying and characterizing duplicated chromosomal segments. Criteria proposed by the Committee for Comparative Gene Mapping [44] were used to identify duplicated genes.

Gene sequence and protein function were the principal criteria used. Application of these criteria, for example, led to inclusion of genes, such as the isozyme-encoding genes malate dehydrogenase-1 and malate dehydrogenase-2, because of their similar enzymatic properties. Subsequent DNA sequence analysis supports the conclusion that these are gene duplicates [64]. Although it is possible that genes such as these arose through convergent evolution, gene duplication is the more parsimonious explanation. These criteria excluded genes such as dilute (*d*), leaden (*ln*), and pink-eyed dilution (*p*), which were used previously in comparable analyses [40,41]. Similarities between genes such as these are based on phenotypic analogies, rather than genetic homologies.

Closely linked genes with related functions were counted as a single marker because they probably arose by tandem duplication, rather than by genome duplication [18–27] and, therefore, do not provide independent evidence. Genes within the β-globin complex are an example of tandem duplicates counted as a single marker. An effort was made to include only functional genes and unprocessed pseudogenes. Genes, the status of which as a functional gene is uncertain, were excluded (e.g., *Pomc-1* and *Pomc-2* in the mouse and *TCP1L1* and *TCP1L2* in man). Pseudogenes, such as *Hba-ps3* in the mouse and *PGK1P1* and *PGK1P2* in man, were excluded if they were known or suspected to have originated through retrotransposition [53–58]. The autosomal phosphoglycerate kinase locus in both mouse and man was similarly excluded because, although it is a functional (expressed) gene, it lacks introns and is thought to have arisen through an RNA intermediate [65]. Pseudogenes of this type do not provide information concerning genome duplication because they arise as a duplication of a single gene, rather than as part of a duplicated chromosomal segment. Proposed paralogous genes in the July 1988 genome maps for mouse and man are listed in the Appendix.

B. Evidence for Mouse

Gene duplicates in the mouse are located on 18 of the 19 autosomes and on the X chromosome (Chr) (Fig. 2). There were 40 gene families, with 28 families composed of two members, 10 of three members, and 2 of four members. The total number of unlinked duplicated genes is 94, and the average number of members in these 40 gene families is 2.4. Sixty percent (56/94) of the unlinked duplicated genes in the mouse are located in 12 chromosomal segments composed of at least two pairs of gene duplicates (Table 1). *Ldh-2* and *Kras-2*, for example, are located on Chr 6, and *Ldh-1* and *Hras-1* are located on Chr 7. Eight genes (*Apoa-1, Apoa-2, Apob, Apoe, Atpa-2, Atpa-3, Hras-1*, and *Mylf*) are part of more than one duplicated segment. *Hras-1* on Chr 7, for example, is part of duplicated segments involving Chrs 6 and 7 in one set and Chrs 3 and 7 in another. Chromosome rearrangements are not responsible for these overlapping segments because rearrangements change the chromosomal location, but not the number of duplicated genes.

C. Evidence for Man

The genome map for man provides comparable evidence for paralogous segments. There are 52 gene families with 34 composed of two genes, 14 composed of three genes, and 4 of four genes (see Table 1). The total number of gene duplicates is 126, and the average number of members in these 52 families is 2.4. These duplicates are located on each of the 22 autosomes and on the X chromosome. Sixty-four percent (81/126) of the unlinked duplicated genes occur in 19 chromosomal segments composed of at least two pairs of duplicated genes (see Table 1). Twelve genes (*ALDH3, ALDOC, APOA1, APOA2, APOB, APOE, HOX1, HOX2, HOX4, HRAS2, KRAS2*, and *NRAS*) are part of more than one duplicated segment. Overlapping segments together with the average number of members of dispersed gene families is compelling evidence for multiple duplications in the progenitor species of lineages leading to mouse and man.

V. CHARACTERISTICS OF PARALOGOUS SEGMENTS

A. Lengths of Paralogous Segments

The length of chromosome segments that have remained intact during evolution can provide considerable insight into genome organization and evolution [16]. These lengths represent the recombination distance between chromosomal rearrangements that have disrupted linkage. Therefore, segment lengths can be used to determine whether or not rearrangement breakpoints are randomly distributed within the genome (i.e., Are there too many long segments? Are there too many short segments?) [16]. Segment lengths can also be used to estimate the number of rearrangements that have occurred since divergence of lineages leading to species such as mouse and man [16]. It should be possible to use lengths of paralogous segments in comparable ways.

Ancestral paralogous segments are expected to be short, whereas derived (or coincidental) segments are expected to be long. To test these expectations, recombination frequency, which is a reasonably close approximation of physical distance for most intervals, was used for estimating segment length. Data for the mouse were obtained from published summaries [66]. Estimates of segment lengths were available for 17 of the 24

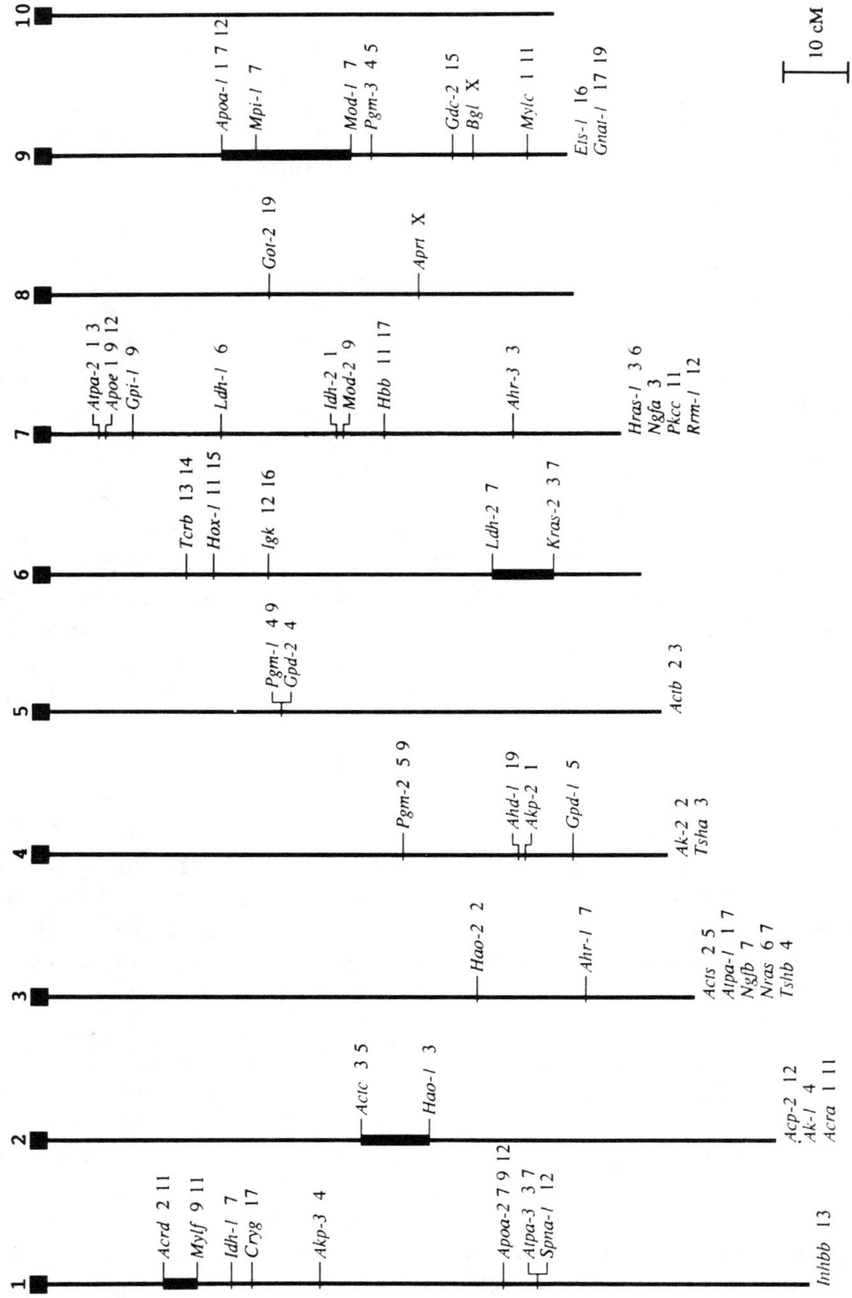
1
Acrd 2 11
Mylf 9 11
Idh-1 7
Cryg 17
Akp-3 4
Apoa-2 7 9 12
Atpa-3 3 7
Spna-1 12
Inhbb 13
2
Actc 3 5
Hao-1 3
Acp-2 12
Ak-1 4
Acra 1 11
3
Hao-2 2
Ahr-1 7
Acts 2 5
Atpa-1 1 7
Ngfb 7
Nras 6 7
Tshb 4
4
Pgm-2 5 9
Ahd-1 19
Akp-2 1
Gpd-1 5
Ak-2 2
Tsha 3
5
Pgm-1 4 9
Gpd-2 4
Actb 2 3
6
Tcrb 13 14
Hox-1 11 15
Igk 12 16
Ldh-2 7
Kras-2 3 7
7
Atpa-2 1 3
Apoe 1 9 12
Gpi-1 9
Ldh-1 6
Idh-2 1
Mod-2 9
Hbb 11 17
Ahr-3 3
Hras-1 3 6
Ngfa 3
Pkcc 11
Rrm-1 12
8
Got-2 19
Aprt X
9
Apoa-1 1 7 12
Mpi-1 7
Mod-1 7
Pgm-3 4 5
Gdc-2 15
Bgl X
Mylc 1 11
Ets-1 16
Gnat-1 17 19
10
10 cM

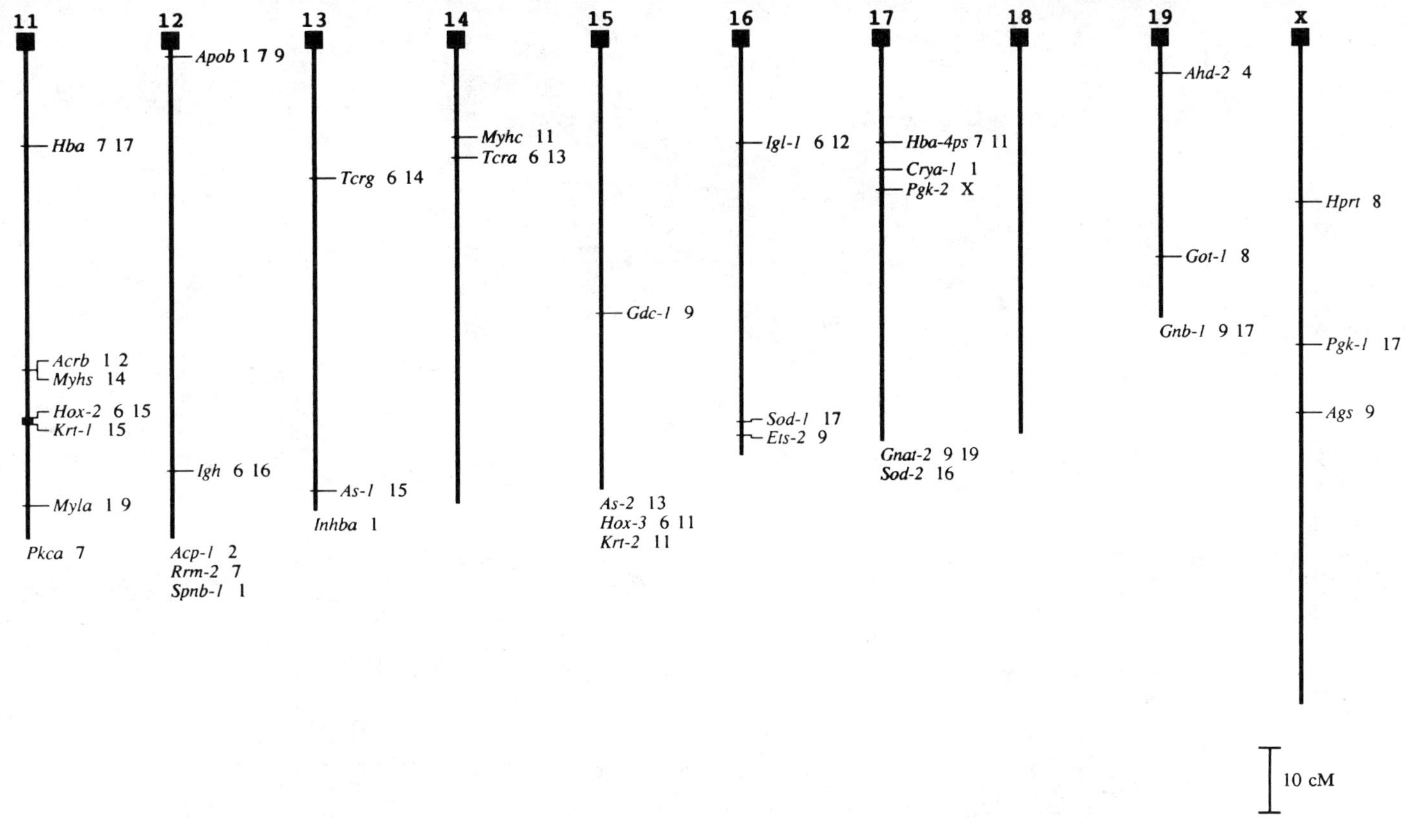

Figure 2 Chromosomal distribution of unlinked duplicated genes in the mouse. Tandem duplicates (e.g., genes within the *Hox-2* family on mouse Chr 11) were counted as a single marker because they do not provide independent evidence for duplication, diploidization, and divergence. Segments are highlighted that are marked by at least two pairs of unlinked paralogous genes and are not interrupted by unrelated genes.

Table 1 Putative Duplicated Chromosomal Segments in Mouse and in Man

Markers	Chromosome	Interrupted	Recombination frequency (%)
Mouse			
**Acrd–Mylf*	1	No	5
Acrb–Myla	11	Yes	25
Apoa-2–Atpa-3-*Idh-1*	1	Yes	45
Apoe–Atpa-2-*Idh-2*	7	Yes	35
****Apoa-2–****Spna-1*	1	No	3
Apob–*Spnb*	12	?	28
Mylf–Apoa-2	1	Yes	47
*Mylc–**Apoa-1***	9	Yes	45
**Hao-1–Actc*	2	No	12
Hao-2–Acts	3	?	NK
Ahr-1–Atpa-1–Ngfb–Nras	3	?	NK
*Ahr-3–**Atpa-2**–Ngfa–**Hras-1***	7	Yes	61
Pgm-2–Gpd-1	4	Yes	25
**Pgm-1–Gpd-2*	5	No	<8
**Ldh-2–Kras-2*	6	No	9
*Ldh-1–**Hras-1***	7	?	NK
Apoe–Mod-2–Gpi-1	7	Yes	35
Apoa-1–Mod-1–Mpi-1	9	Yes	19
Apoe–Rrm-1	7	?	NK
Apob–Rrm-2	12	?	NK
Hbb–Pkcc	7	?	NK
Hba–Pkca	11	?	NK
**Hox-2–Krt-1*	11	No	<1
**Hox-3–Krt-2*	15	No	3
Man			
*ACADM–ENO1–**NRAS***	1p13-p36	Yes	
*ACADS–ENO2–**KRAS2***	12qter-p12	Yes	
ACAD–?–HRAS2	X	?	
*ACP1–**APOB–RRM2***	2p23-p25	?	
*ACP2–**APOA1–RRM1***	11p11-qter	?	
ACTA–LAMB2	1p21-qter	?	
ACTB–LAMB1	7pter-q31	?	
AK2–SPTA1	1p34-q25	?	
AK1–SPTAN1	9q33-q34	No	
ALDHI–ALDOB	9q	?	
ALDH3–ALDOC	17	?	
ALDH2–CYK4	12	?	
ALDH3–*CYK13*	17	?	
ALDOA–PKCB–TK2	16	?	

Table 1 (Continued)

Markers	Chromosome	Interrupted	Recombination frequency (%)
ALDOC–*PKCA*–*TK1*	17	?	
ALPP–*CRYG1*–***HOX4***	2q33-q37	No	
ALPPL–*CRYB1*–***HOX2***	17	?	
APOA2*–*ATP1A1	1	?	
APOE–*ATP1A3*	19q13	No	
APOA2*–*NRAS	1p22-q23	Yes	
APOA1–*HRAS1*	11p15-qter	Yes	
APOB–*TGFA*	2p13-p24	No	
APOE–*TGFB*	19p13	No	
APOA2–*UGP1*–*MYCL*	1p32-q23	?	
APOB–*UGP2*–*MYCN*	2p24-p23	No	
COL1A2–*ERBB*–***HOX1***–*MYH5*	7p21-q22	?	
COL1A1–*ERBA*–***HOX2***–*MYH1*	17pter-q22	?	
FUCA1–*PGM1*	1p34-p22	?	
FUCA2–*PGM3*	6q12	?	
GNAT2–*SRC2*	1	?	
GNAS–*SRC1*	20q12-q13	?	
GOT1–*LIPA*	10q25	?	
GOT2–*LIPB*	16q12-q22	?	
HOX4–*IFNB3*–*INHBB*–*MDH1*	2p23-qter	?	
HOX1–*IFNB2*–*INHBA*–*MDH2*	7p21-qter	?	
HOX1–*NKNA*	7p21-q22	?	
HOX3–*NKNB*	12q21-q12	?	
HRAS1–*LDHA*	11p14-p15	No	
KRAS2–*LDHB*	12p12	No	

Genes, the location of which on the linkage map has been established, were evaluated to determine whether or not evidence is available that the duplicated segment they mark is interrupted by unrelated genes. The segment marked by *Apoa-1* and *Mylc* on Chr 9, for example, appears to be duplicated by the segment marked by *Apoa-2* and *Mylf* on Chr 1. The segment on Chr 9, however, is interrupted by four unrelated markers (see Fig. 2). Genes that are part of more than one duplicated segment are printed in bold. NK indicates that the recombination frequency is not known. An asterisk indicates putative ancestral segments (see text) and ? indicates that information about whether unrelated genes interrupt a segment is not available.

segments (see Table 1). The average recombination frequency between loci that bound duplicated segments was 23.9% (SD, 18.5%) and ranged from 1% for *Hox-2* and *Krt-1* on Chr 11, to 61% for *Ahr-3*–*Atpa-2*–*Ngfa*–*Hras-1* on Chr 7. The distribution of segment lengths was clearly bimodal, however, with the mean of one mode being 5.9% (SD, 3.9%; N = 7) and the mean of the other being 36.5% (SD, 12.9%; N = 10). One explanation for the bimodal distribution is that short segments are ancestral and long segments derived.

The genome map of paralogous genes provides information concerning whether or not unrelated genes occur between genes marking duplicated segments and, thereby, provides additional evidence for which segments are more likely to be ancestral and which derived. For example, *Hox-2* and *Krt-1* are located between *Acrb* and *Myla* on Chr 11 (see Fig. 2) and, therefore, represent unrelated genes interrupting a putative paralogous segment. Of the 16 informative segments, seven show no evidence for interruption (see Table 1). More importantly, there was a perfect correlation between interruption and segment length (L): none of the segments in the lower mode ($L \leq 12\%$) were interrupted, whereas all segments in the higher mode ($L > 12\%$) were interrupted. This analysis, therefore, provides strong evidence that short segments are ancestral, whereas long segments are derived.

The average length of these segments is biased in two ways: first, because genes used to identify each segment are not located at the bounds of the segment, segment length is underestimated; second, exclusion of segments marked by a single locus and of segments that remain to be identified truncate the distribution of segment length [16]. Nadeau and Taylor [16] proposed methods that can be used to address these biases. If we accept the average recombination length of ancestral paralogous segments (those in the lower mode) to be 5.9% (SD, 3.9%; see foregoing), then the estimate of the average true length of these segments is 17.6% (SD, 11.8%), and the estimate of the length of all ancestral paralogous segments in the genome is $7.4\% \pm 1.7\%$.

B. Orthologous Segments: Conservation of Paralogous Segments in Mouse and Man

Ancestral paralogous segments and old coincidental (or derived) arrangements should also be orthologous segments. To test this prediction, the chromosomal locations of paralogous genes that have been mapped in both mouse and man were compared [2,16,17,44,67]. Synteny was conserved for 8 of 19 informative segments and disrupted for the other 11 (Table 2). Because the most recent genome duplication is believed to have occurred 300 million years (myr) ago [21] and, therefore, long before the divergence of lineages leading to mouse and man, only 23% (70 myr/300 myr) of the ancestral paralogous segments are expected to be conserved. Although data are few, conservation of 42% of segments is at least roughly consistent with expectations.

Table 2 Conservation of Duplicated Chromosomal Segments Between Mouse and Man

Duplicated segment	Chromosomal location	Nature of synteny
In Mouse	**In man**	
Kras-2-Ldh-2	12p, 12p	Conserved
Hras-1-Ldh-1	11p, 11p	Conserved
Hox-2-Krt-1	12q, 12	Conserved
Hox-3-Krt-2	17q, 17	Conserved
Idh-1-Atpa-3	2, 1	Disrupted
Idh-2-Atpa-2	15, ?	?
In Man	**In mouse**	
ACP1-APOB-RRM2	12, 12, 12	Conserved
ACP2-APOA1-RRM1	2, 9, 7	Disrupted
APOA2-NRAS	1, 3	Disrupted
APOA1-HRAS1	9, 7	Disrupted
HOX1-COL1A2-ERBB-MYH5	6, 16, 11, ?	Disrupted
HOX4-COL1A1-ERBA-MYH1	12, 11, 11, 11	Partially conserved
ALPP-CRYG-HOX4	1, 1, 12	Partially conserved
APLP-CRYB-HOX2	4, ?, 11	Disrupted
APOA2-NRAS	1, 3	Disrupted
APOA1-HRAS1	9, 7	Disrupted
COL1A2-ERBB-HOX1-MYH5	?, 11, 6, ?	Disrupted
COL1A1-ERBA-HOX2-MHY1	?, 11, 11, 11	Conserved
HOX4-IFNB3-INHBB-MDH1	12, 4, 1, ?	Disrupted
HOX1-IFNB2-INHBA-MDH2	6, 4, 13, 5	Disrupted

? Indicates that the chromosomal location of the orthologous gene in the other species is not known. Most segments in Table 1 had to be excluded from this listing because the necessary gene-mapping data were not available. Genes and their chromosomal location are listed in the same order.

VI. COMPARISON OF EMPIRICAL AND "RANDOM" MAPS

An important, but often overlooked, problem in comparative gene mapping is the role of coincidence in producing the distribution of genes in the map. It is obvious that chromosomal rearrangements could, by chance, bring together pairs of paralogous genes and produce combinations that appear ancestral, but are, in fact, more recently derived. This problem is especially important when the time since duplication is considerable, as for genomic duplication.

To evaluate the contribution of coincidence to the observed distribution of unlinked gene duplicates in the mouse, "random maps" were analyzed. These maps were made in the following way: genes in each family were assigned a random number, ranging in value from 1 to 20 and corresponding to one of the 19 autosomes or the X chromosome. The X chromosome was included in the analysis because at the time of the most recent genome duplication, the XY system of sex determination was not established [21,22]. These random numbers were obtained using a nonlinear, additive, feedback random number generator. The probability of sampling each of the 20 numbers was weighted by the recombination length of each chromosome [68]. For example, 7.20%, rather than 5%, of the random numbers were 1; 6.88% of the numbers were 2, and so on. This process was used to produce 20 random maps, each of which was analyzed in the same manner as the empirical map.

Empirical and random maps differed in at least two significant ways (Table 3). First, approximately twice as many paralogous segments were observed (N = 13) than expected (N = 6.75). Second, significantly more genes were included in paralogous segments (N = 56) than were expected (N = 25.60; goodness-of-fit: $\chi^2 = 51.15$; $p < 0.001$). These results suggest that half of the putative paralogous segments are ancestral and half are derived. Although human data were not analyzed in a comparable way, there is no reason to suppose that similar results would not be obtained.

These results are extremely important because they suggest that it is possible to distinguish not only the origin of particular segments, but also the relative number of the two kinds of segments: approximately half of the unlinked duplicated genes appear to be part of short, ancestral segments,

Table 3 Patterns of Unlinked Gene Duplicates in the Mouse Genome and Comparison with Averages for 20 Replicates of a Random Distribution

Parameter	Observed	Simulation
No. duplicated segments with 2 genes	12	6.10 (1.33; 3–8)
No. duplicated segments with 3 genes	0	0.10 (0.31; 0–1)
No. duplicated segments with 4 genes	1	0
No. triplicated segments with 2 genes	0	0.55 (0.69; 0–2)
No. overlapping segments	4	1.45 (1.39; 0–4)
No. genes in duplicated segments	56	25.60 (5.04; 15–37)

The analyses are based on the March, 1988 map. The standard deviation and range among the 20 replicates are given in parentheses.

and the other half part of longer, coincidental segments. The putative ancestral segments in the mouse are identified in Table 1.

VII. GENE MAPPING APPLICATIONS OF MAPS OF PARALOGOUS GENES

As with comparative gene maps for different species, maps of paralogous genes can be used to expedite gene mapping studies: given the chromosomal location of one member of a gene family, the chromosomal location of other, unmapped members of the family can be predicted. This approach has recently been used successfully. Many, if not all, members of the type I family of keratin genes are closely linked to the *Hox-2* complex on mouse Chr 11 [69]. We predicted, and have subsequently verified, linkage of the type II keratin gene family to the *Hox-3* complex on Chr 15 [J. G. Compton and J. H. Nadeau, in preparation]. Because 60% of unlinked duplicated genes are part of paralogous segments, maps of paralogous genes (see Fig. 2) should be an increasingly valuable aid in mapping gene families, the members of which are located on more than one chromosome.

Another important application of maps of paralogous genes involves identifying putative paralogous mutations and disease loci. For example, the mouse engrailed-1 (*En-1*) gene is closely linked to the dominant hemimelia (*Dh*) mutation on mouse Chr 1, whereas the engrailed-2 (*En-2*) gene is closely linked to the hemimelic extra toes (*Hx*) mutation on Chr 5 [70]. Because *En-1* and *En-2* mark paralogous chromosomal segments, Martin et al. [70] predict that *Dh* and *Hx*, which have similar phenotypic effects, will be found to be mutations in paralogous genes. Nadeau et al. [69] provide another example involving mutations affecting skin and hair development that are closely linked to the families of keratin genes and homeobox genes on mouse Chrs 11 and 15. These predictions can obviously be used to guide experiments to molecularly identify and characterize the genes involved in these mutations because, for example, molecular clones for one locus could be used to identify clones for the paralogous locus.

VIII. OTHER FEATURES OF GENOME DUPLICATION

A. Chromosomal Distribution of Unprocessed Pseudogenes

Pseudogenes resulting from genome duplication and subsequent loss of duplicated gene function are expected to have introns. The only known unprocessed pseudogene that is not linked to its progenitor locus is the pseudo α-globin locus *Hba-ps4*, which is located on mouse Chr 17 [53–56,71,72], but there is no evidence that *Hba-ps4* is part of a paralogous segment (see Fig. 2). In addition, there is no evidence for an orthologous gene in man, as would be expected because genome duplication occurred before divergence of lineages leading to mouse and human. Given the rate of nucleotide substitution in the average gene in mouse and man [62], sufficient homology should remain for the human homologue to be identified, if it existed. Two explanations could account for the anomalous chromosomal location of *Hba-ps4*: The chromosomal segment marked by the progenitor locus *Hba* duplicated after divergence of lineages leading to mouse and man, and other paralogous genes marking this segment remain to be identified or, alternatively, *Hba-ps4* arose from an unprocessed transcript [see Ref. 58 for preliminary evidence for the latter explanation].

All other pseudogenes in mouse and man that are not linked to their progenitor locus are either known to be processed or information is not available concerning their origin. The reason for the paucity of evidence for unprocessed pseudogenes is that most duplicates rapidly lose function, accumulate nucleotide substitutions, and, consequently, lose homology to the paralogous gene [19,20,59–62]. As a result, one of the more important predictions of the genome duplication hypothesis, namely that unprocessed pseudogenes should be part of paralogous segments, is now untestable.

B. Comparison of Genetic and Cytogenetic Evidence for Paralogous Segments

Comings [39] compared the banding patterns of human chromosomes and identified six pairs of putatively paralogous chromosomes: Chr 1 and 2, 4 and 5, 7 and 8, 11 and 12, 14 and 15, and 16 and 17. More recent surveys provide additional evidence [73]. The human map of paralogous segments supports Comings' argument for three of the six pairs of chromosomes; Chrs 1 and 2, 11 and 12, and 16 and 17 have two or more pairs of genes that mark paralogous segments (see Table 1). Because these paralogous segments must be long to be detected by cytogenetic analyses, it is significant that three genes mark two of these duplicated segments.

IX. COMPARISON OF INTERSPECIFIC AND INTRASPECIFIC COMPARATIVE GENE MAPPING

There are two factors that make traditional comparative gene mapping so feasible. The first is that genome duplication (i.e., divergence of lineages leading to mouse and man) is relatively recent, about 70 million years ago. The importance of recent divergence is that orthologous genes are readily identified and that the number of chromosomal rearrangements disrupting synteny or linkage has been modest. The second factor is that speciation represents a single duplication of the genome, with one duplicate now represented in the mouse, the other in man. As a result, all conserved linkages and chromosomal rearrangements can be traced to a single duplication event.

Genome duplications that produced paralogous segments have neither of these advantages. It has been estimated that the most recent genome duplication occurred 300 million years ago [21,22]. This long evolutionary time has provided considerable opportunity for sequence divergence. Identification of paralogous genes is consequently difficult. In addition, it is likely that several independent genome duplications occurred. With multiple duplications, paralogous segments sometimes overlap (see Table 1). As a result, it is difficult to attribute segments to a given duplication and, consequently, to characterize the rates of disruption of duplicated segments. It is important that these two factors be remembered when evaluating maps of duplicated chromosomal segments.

Despite these limitations, however, analysis of duplicated chromosomal segments represents a new and potentially powerful way to analyze genome organization and evolution. This new approach can be called intraspecific comparative gene mapping.

ACKNOWLEDGMENTS

I thank Drs. Eva Eicher, Janan Eppig, Mark Erhart, Jim Stone, and Ben Taylor for many helpful comments on a draft of this paper, and the students and assistants in my laboratory for many enlightening and enjoyable discussions of this problem. This work was supported by NIH grants GM32461, GM39414, and GM42970.

REFERENCES

1. Lalley, P. A., J. D. Minna, and U. Francke (1978). Conservation of autosomal synteny groups in mouse and man. *Nature 274*:160–162.
2. Nadeau, J. H. (1989). Maps of linkage and synteny homologies between mouse and man. *Trends Genet. 5*:82–86.
3. O'Brien, S. J., H. N. Seuanez, and J. E. Womack (1985). On the evolution of genome organization in mammals. In *Molecular Evolutionary Genetics*. R. J. MacIntyre (ed.). Plenum Press, New York, pp. 519–589.
4. O'Brien, S. J., H. Seuanez, and J. E. Womack (1988). Mammalian genome organization: An evolutionary view. *Annu. Rev. Genet. 22*:323–351.
5. Ohno, S. (1973). Ancient linkage groups and frozen accidents. *Nature 244*:259–262.
6. Buchberg, A. M., E. Brownell, S. Nagata, N. A. Jenkins, and N. G. Copeland (1989). A comprehensive genetic map of murine chromosome 11 reveals exten-

sive linkage conservation between mouse and human. *Genetics 122*:153–161.
7. Crosby, J. L., S. J. Phillips, and J. H. Nadeau (1989). The cardiac actin locus (*Actc-1*) is not on mouse chromosome 17 but is linked to beta 2-microglobulin on chromosome 2. *Genomics 5*:19–23.
8. Kingsley, D. M., N. A. Jenkins, and N. G. Copeland (1989). A molecular genetic linkage map of mouse chromosome 9 with regional localizations for the *Gsta, T3g, Ets-1*, and *Ldlr* loci. *Genetics 123*:165–172.
9. Siracusa, L. D., A. M. Buchberg, N. G. Copeland, and N. A. Jenkins (1989). Recombinant inbred strain and interspecific backcross analysis of molecular markers flanking the murine agouti coat color locus. *Genetics 122*:669–679.
10. Stone, J. C., J. L. Crosby, C. A. Kozak, A. R. Schievella, R. Bernards, and J. H. Nadeau (1989). The murine retinoblastoma homolog maps to chromosome 14 near *Es-10*. *Genomics 5*:70–74.
11. Brueton, L., S. M. Huson, R. M. Winter, and R. Williamson (1988). Chromosomal localisation of a developmental gene in man: Direct DNA analysis demonstrates that Grieg cephalopolysyndactyly maps to 7p13. *Am. J. Hum. Genet. 31*:799–804.
12. Winter, R. M., and S. M. Huson (1988). Grieg cephalopolysyndactyly syndrome: A possible mouse homologue (*Xt*-extra toes). *Am. J. Hum. Genet. 31*: 793–798.
13. Cheng, S. V., J. H. Nadeau, R. E. Tanzi, P. C. Watkins, J. Jagadesh, B. A. Taylor, J. L. Haines, N. Sacchi, and J. F. Gusella (1988). Comparative mapping of DNA markers from the familial Alzheimer disease and Down syndrome regions of human chromosome 21 to mouse chromosomes 16 and 17. *Proc. Natl. Acad. Sci. USA 85*:6032–6036.
14. Cheng, S. V., G. R. Martin, J. H. Nadeau, J. L. Haines, M. Bucan, C. A. Kozak, M. E. MacDonald, J. L. Lockyer, F. D. Ledley, S. L. C. Woo, H. Lehrach, T. C. Gilliam, and J. F. Gusella (1989). Synteny of mouse chromosome 5 homologs for human DNA loci linked to the Huntington disease gene. *Genomics 4*:419–426.
15. Coyle, J. J., M. L. Oster-Granite, R. H. Reeves, and J. D. Gearhart (1988). Down syndrome, Alzheimer disease and trisomy 16 in the mouse. *Trends Neurosci. 11*:390–394.
16. Nadeau, J. H., and B. A. Taylor (1984). Lengths of chromosomal segments conserved since divergence of man and mouse. *Proc. Natl. Acad. Sci. USA 81*: 814–818.
17. Nadeau, J. H., and A. H. Reiner (1989). Linkage and synteny homologies in mouse and man. In *Genetic Variants and Strains of the Laboratory Mouse*, 2nd ed. M. F. Lyon and A. G. Searle (eds.). Oxford University Press, Oxford, pp. 506–536.
18. Hood, L., J. H. Campbell, and S. C. R. Elgin (1975). The organization, expression, and evolution of antibody genes and other multigene families. *Annu. Rev. Genet. 9*:305–353.
19. Li, W.-H. (1982). Evolution of duplicate genes and pseudogenes. In *Evolution of Genes and Proteins*. M. Nei and R. K. Koehn (eds.). Sinauer Associates, Sunderland, Mass. pp. 14–37.

20. Li, W.-H. (1982). Evolutionary change of duplicated genes. *Isozymes: Curr. Top. Biomed. Med. Res. 6*:55–92.
21. Ohno, S. (1970). *Evolution by Gene Duplication*. Springer-Verlag, New York.
22. Ohno, S., U. Wolf, and N. B. Atkin (1968). Evolution from fish to mammals by gene duplication. *Hereditas 59*:169–187.
23. Ohta, T. (1980). Evolution and variation of multigene families. *Lect. Notes Biomath. 37*:1–131.
24. Ohta, T. (1983). On the evolution of multigene families. *Theor. Popul. Biol. 23*: 216–240.
25. Ohta, T. (1988). Multigene and supergene families. *Oxford Surv. Evol. Biol. 5*: 41–65.
26. Arnheim, N. (1983). Concerted evolution of multigene families. In *Evolution of Genes and Proteins*. M. Nei and R. K. Koehn (eds.). Sinaur Associates, Sunderland, Mass. pp. 38–61.
27. Brown, D. D., P. C. Wensink, and E. Jordan (1972). A comparison of the ribosomal DNAs of *Xenopus laevis* and *Xenopus mulleri*: The evolution of tandem genes. *J. Mol. Biol. 63*:57–73.
28. Long, E.O., and I. B. Dawid (1980). Repeated genes in eukaryotes. *Annu. Rev. Biochem. 49*:721–748.
29. Maniatis, T., E. F. Fritsch, J. Lauer, and R. M. Lawn (1980). The molecular genetics of human hemoglobins. *Annu. Rev. Genet. 14*:145–178.
30. Smith, G. P. (1974). Unequal crossover and the evolution of multigene families. *Proc. Cold Spring Harbor Symp. Quant. Biol. 38*:507–513.
31. Smith, G. P. (1976). Evolution of repeated DNA sequences by unequal crossover. *Science 191*:528–535.
32. Zimmer, E. A., S. L. Martin, S. M. Beverly, Y. W. Kan, and A. C. Wilson (1980). Rapid duplication and loss of genes coding for the alpha chains of hemoglobin. *Proc. Natl. Acad. Sci. USA 7*:2158–2162.
33. Bailey, G. S., R. T. M. Poulter, and P. A. Stockwell (1978). Gene duplication in tetraploid fish: Model for gene silencing at unlinked duplicated loci. *Proc. Natl. Acad. Sci. USA 75*:5575–5579.
34. Ferris, S. D., and G. S. Whitt (1977). Loss of duplicate gene expression after polyploidisation. *Nature 265*:258–260.
35. Johnson, K. R., J. E. Wright, and B. May (1987). Linkage relationships reflecting ancestral tetraploidy in salmonid fish. *Genetics 116*:579–591.
36. Ohta, T., and G. A. Dover (1983). Population genetics of multigene families that are dispersed into two or more chromosomes. *Proc. Natl. Acad. Sci. USA 80*:4079–4083.
37. Takahata, N., and T. Maruyama (1979). Polymorphism and loss of duplicate gene expression: A theoretical study with application to tetraploid fish. *Proc. Natl. Acad. Sci. USA 76*:4521–4525.
38. Haldane, J. B. S. (1927). The comparative genetics of colour in rodents and carnivora. *Biol. Rev. Camb. Philos. Soc. 2*:199–212.
39. Comings, D. E. (1972). Evidence of ancient tetraploidy and conservation of linkage groups in mammalian chromosomes. *Nature 238*:455–457.

40. Renwick, J. H. (1972). Comments on the chromosome maps of mouse and vertebrates. *Bull. Eur. Soc. Hum. Genet.* Nov:108–124.
41. Lundin, L.-G. (1979). Evolutionary conservation of large chromosomal segments reflected in mammalian gene maps. *Clin. Genet. 16*:72–81.
42. Morizot, D. C. (1987). Comparative gene mapping: Evidence for chromosome duplications during vertebrate evolution. *Cytogenet. Cell Genet. 46*:664.
43. Tolan, D. R., J. Niclas, B. D. Bruce, and R. V. Lebo (1987). Evolutionary implications of the aldolase-A, -B, -C and -pseudogene chromosome locations. *Am. J. Hum. Genet. 41*:907–924.
44. Lalley, P. A., S. J. O'Brien, N. Creau-Goldberg, M. T. Davisson, T. H. Roderick, G. Echard, J. E. Womack, J. M. Graves, D. P. Doolittle, and J. N. Guidi (1987). Report of the Committee on Comparative Mapping. *Cytogenet. Cell Genet. 46*:367–389.
45. Eicher, E. M. (1981). Foundation for the future: Formal genetics of the mouse. In *Mammalian Genetics and Cancer: The Jackson Laboratory Fiftieth Anniversary Symposium*. E. S. Russell (ed.). Alan R. Liss, New York, pp. 7–49.
46. Bengtsson, B. O., and W. F. Bodmer (1976). On the increase of chromosome mutations under random mating. *Theor. Popul. Biol. 9*:260–281.
47. Hedrick, P. W. (1981). The establishment of chromosomal variants. *Evolution 35*:322–332.
48. Hedrick, P. W., and D. A. Levin (1984). Kin-founding and the fixation of chromosomal variants. *Am. Nat. 124*:789–797.
49. Lande, R. (1979). Effective gene sizes during long-term evolution estimated from rates of chromosomal rearrangement. *Evolution 33*:234–251.
50. Lande, R. (1985). The fixation of chromosomal rearrangements in a subdivided population with local extinction and colonization. *Heredity 54*:323–332.
51. Slatkin, M. (1981). Fixation probabilities and fixation times in a subdivided population. *Evolution 35*:477–488.
52. Wright, S. (1941). On the probability of fixation of reciprocal translocations. *Am. Nat. 75*:513–522.
53. Leder, A., D. Swan, F. Ruddle, P. D'Eustachio, and P. Leder (1981). Dispersion of alpha-globin genes of the mouse to three different chromosomes. *Nature 293*:196–200.
54. Lueders, K., A. Leder, P. Leder, and E. Kuff (1982). Association between a transposed alpha-globin pseudogene and retrovirus-like elements in the BALB/c mouse genome. *Nature 295*:426–428.
55. Nishioka, Y., A. Leder, and P. Leder (1980). Unusual alpha-globin-like gene that has cleanly lost both globin intervening sequences. *Proc. Natl. Acad. Sci. USA 77*:2806–2809.
56. Popp, R. A., P. A. Lalley, J. B. Whitney, and W. F. Anderson (1981). Mouse alpha-globin genes and alpha-globin-like pseudogenes are not syntenic. *Proc. Natl. Acad. Sci. USA 78*:6362–6366.
57. Sharp, P. A. (1983). Conversion of RNA to DNA in mammals: *Alu*-like elements and pseudogenes. *Nature 301*:471–472.
58. Vanin, E. F. (1985). Processed pseudogenes: Characteristics and evolution. *Annu. Rev. Genet. 19*:253–272.

59. Gojobori, T., W.-H. Li, and D. Graur (1982). Patterns of nucleotide substitution in pseudogenes and functional genes. *J. Mol. Evol. 18*:360–369.
60. Jukes, T. H., and M. Kimura (1984). Evolutionary constraints and the neutral theory. *J. Mol. Evol. 21*:90–92.
61. Li, W.-H., C.-I. Wu, and C.-C. Wu (1984). Nonrandomness of point mutation as reflected in nucleotide substitutions in pseudogenes and its evolutionary implications. *J. Mol. Evol. 21*:58–71.
62. Miyata, T., and H. Hayashida (1981). Extraordinarily high evolutionary rate of pseudogenes: Evidence for the presence of selective pressure against changes between synonomous codons. *Prod. Natl. Acad. Sci. USA 78*:5739–5743.
63. Fitch, W. M. (1977). Molecular evolutionary clocks. In *Molecular Evolution*. F. J. Ayala (ed.). Sinauer Associates, Sunderland, Mass, pp. 160–178.
64. Setoyama, C., T. Joh, T. Tsuzuki, and K. Shimada (1988). Structural organization of the mouse cytosolic malate dehydrogenase gene: Comparison with that of the mouse mitochondrial malate dehydrogenase gene. *J. Mol. Biol. 202*:355–364.
65. McCarrey, J. R., and K. Thomas (1987). Human testis-specific PGK gene lacks introns and possesses characteristics of a processed gene. *Nature 326*:501–504.
66. Davisson, M. T., and T. H. Roderick (1989). Recombination frequencies. In *Genetic Variants and Strains of the Laboratory Mouse*, 2nd ed. M. F. Lyon and A. G. Searle (eds.). Oxford University Press, Oxford, pp. 432–505.
67. Searle, A. G., J. Peters, M. F. Lyon, J. G. Hall, E. P. Evans, J. H. Edwards, and V. J. Buckle (1989). Chromosome maps of man and mouse. IV. *Ann. Hum. Genet. 53*:89–140.
68. Committee on Standardized Genetic Nomenclature for Mice (1972). Standard karyotype of the mouse, *Mus musculus. J. Hered. 63*:69–72.
69. Nadeau, J. H., F. G. Berger, D. R. Cox, J. L. Crosby, M. T. Davisson, D. Ferrara, E. Fuchs, C. Hart, L. Hunihan, P. A. Lalley, S. H. Langley, G. R. Martin, L. Nichols, S. J. Phillips, T. H. Roderick, D. R. Roop, F. H. Ruddle, L. C. Skow, and J. G. Compton (1989). A family of type I keratin genes and the homeo box-2 gene complex are closely linked to the *rex* locus on mouse Chromosome 11. *Genomics 5*:454–462.
70. Martin, G. R., M. Richman, S. Reinsch, J. H. Nadeau, and A. Joyner (1989). Mapping of the two mouse engrailed-like genes: Close linkage of *En-1* to dominant hemimelia (*Dh*) on chromosome 1 and of *En-2* to hemimelic extra-toes (*Hx*) on chromosome 5. *Genomics 6*:309–315.
71. D'Eustachio, P. D., B. Fein, J. Michaelson, and B. A. Taylor (1984). The alpha-globin pseudogene on mouse chromosome 17 is closely linked to *H-2*. *J. Exp. Med. 159*:958–963.
72. Fox, H. S., L. M. Silver, and G. R. Martin (1984). An alpha globin pseudogene is located within the mouse *t* complex. *Immunogenetics 19*:125–130.
73. Sawyer, J. R., and J. C. Hozier (1986). High resolution of mouse chromosome: Banding conservation between man and mouse. *Science 232*:1632–1635.
74. Cohen, I. H., D. A. Consiglio, U. Francke, K. K. Kidd, R. L. Miller, A. J. Pakstis, C. W. Partridge, S. R. Perretta, F. H. Ruddle, R. K. Track, T. B. Shows, and P. J. McAlpine (1987). Human gene map. *Genet. Maps 3*:518–533.

75. McKusick, V. A. (1987). The human gene map. *Genet. Maps 3*:534–580.
76. Davisson, M. T., and T. H. Roderick (1987). Locus map of the mouse (*Mus musculus*). *Genet. Maps 3*:430–451.
77. Kozak, C. A. (1987). Mouse linkage map of retroviral and cancer-related genes. *Genet. Maps 3*:452–455.
78. Peters, J. (1988). Gene list. *Mouse News Lett. 80*:10–90.

APPENDIX GENE FAMILIES WHOSE MEMBERS ARE LOCATED ON MORE THAN ONE CHROMOSOME

Gene symbol	Gene name	Chromosome
Mouse		
Acp-1	Acid phosphatase-1	12
Acp-2	Acid phosphatase-2	2
Acra	Acetylcholine receptor, α	2
Acrb	Acetylcholine receptor, β	11
Acrd	Acetylcholine receptor, δ	1
Actb	Actin, cytoplasmic, β	5
Actc	Actin, cardiac, α	2
Acts	Actin, cardiac, α	3
Ags	Galactosidase, α	X
Bgl	Galactosidase, β	9
Ahd-1	Aldehyde dehydrogenase-1	4
Ahd-2	Aldehyde dehydrogenase-2	19
Ahr-1	Aldehyde reductase-1	3
Ahr-3	Aldehyde reductase-3	7
Ak-1	Adenylate kinase-1	2
Ak-2	Adenylate kinase-2	4
Akp-2	Alkaline phosphatase-2	4
Akp-3	Alkaline phosphatase-3	1
Apoa-1	Apolipoprotein-1	9
Apoa-2	Apolipoprotein-2	1
Apob	Apolipoprotein-B	12
Apoe	Apolipoprotein-E	7
Aprt	Adenine phosphoribosyl transferase	8
Hprt	Hypoxanthine phosphoribosyl transferase	X
As-1	Arylsulfatase B	13
As-2	Arylsulfatase A	15
Atpa-1	Na, K-ATPase-1	3
Atpa-2	Na, K-ATPase-2	7
Atpa-3	Na, K-ATPase-3	1
Crya-1	Crystallin, α_1	17
Cryg	Crystallin, γ	1

Gene symbol	Gene name	Chromosome
Ets-1	E26 avian leukemia oncogene-1	9
Ets-2	E26 avian leukemia oncogene-2	16
Gdc-1	Glycerol phosphate dehydrogenase-1	15
Gdc-2	Glycerol phosphate dehydrogenase-2	9
Gnat-1	Guanine nucleotide-binding protein, α_1	9
Gnat-2	Guanine nucleotide-binding protein, α_2	17
Gnb-1	Guanine nucleotide-binding protein, β_1	19
Got-1	Glutamate oxaloacetate transaminase-1	19
Got-2	Glutamate oxaloacetate transaminase-2	8
Gpd-1	Glucose phosphate dehydrogenase-1	4
Gpd-2	Glucose phosphate dehydrogenase-2	5
Gpi-1	Glucose phosphate isomerase	7
Mpi-1	Mannose phosphate isomerase	9
Hao-1	Hydroxyacid oxidase-1	2
Hao-2	Hydroxyacid oxidase-2	3
Hba	Hemoglobin, α	11
Hba-4ps	Hemoglobin, α, pseudogene	17
Hbb	Hemoglobin, β	7
Hox-1	Homeo box-1	6
Hox-2	Homeo box-2	11
Hox-3	Homeo box-3	15
Hras-1	Harvey rat sarcoma oncogene-1	7
Kras-2	Kirsten rat sarcoma oncogene-2	6
Nras	Neuroblastoma oncogene	3
Idh-1	Isocitrate dehydrogenase-1	1
Idh-2	Isocitrate dehydrogenase-2	7
Igh	Immunoglobulin, heavy chain	12
Igk	Immunoglobulin, kappa chain	6
Igl	Immunoglobulin, light chain	16
Inhba	Inhibin β_A	13
Inhbb	Inhibin β_B	1
Krt-1	Keratin, type I	11
Krt-2	Keratin, type II	15
Ldh-1	Lactate dehydrogenase-1	7
Ldh-2	Lactate dehydrogenase-2	6
Mod-1	Malic enzyme-1	9
Mod-2	Malic enzyme-2	7
Myhc	Myosin heavy chain, cardiac	14
Myhs	Myosin heavy chain, skeletal	11
Myla	Myosin light chain, atria	11
Mylc	Myosin light chain, ventricles	9
Mylf	Myosin light chain, fast skeletal	1
Ngfa	Nerve growth factor, α	7

Gene symbol	Gene name	Chromosome
Ngfb	Nerve growth factor, β	3
Pgk-1	Phosphoglycerate kinase-1	X
Pgk-2	Phosphoglycerate kinase-2	17
Pgm-1	Phosphoglucomutase-1	5
Pgm-2	Phosphoglucomutase-2	4
Pgm-3	Phosphoglucomutase-3	9
Pkca	Protein kinase C, α	11
Pkcc	Protein kinase C, γ	7
Rrm-1	Ribonucleotide reductase M1	7
Rrm-2	Ribonucleotide reductase M2	12
Sod-1	Superoxide dismutase-1	16
Sod-2	Superoxide dismutase-2	17
Spna-1	Spectrin, α	1
Spnb	Spectrin, β	12
Tcra	T-cell receptor, α	14
Tcrb	T-cell receptor, β	6
Tcrg	T-cell receptor, γ	13
Tsha	Thyrotropin, α	3
Tshb	Thyrotropin, β	4
Man		
ACAD	Acyl-coenzyme A dehydrogenase	X
ACADM	Acyl-coenzyme A dehydrogenase C-4	1p31
ACADS	Acyl-coenzyme A dehydrogenase C-2	12q22-qter
ACO1	Aconitase 1	9p22-q32
ACO2	Aconitase 2	22q11.2-q13.31
ACP1	Acid phosphatase 1	2p25 or 2p23
ACP2	Acid phosphatase 2	11p11
ACTA	Actin, α, skeletal	1p21-qter
ACTB	Actin, β	7pter-q22
ACTC	Actin, α, cardiac	15q11-q26
ADRAR	Adrenergic receptor, α	10q23-q25
ADRBR	Adrenergic receptor, β	5q31-q32
AK1	Adenylate kinase 1	9q34.1-q34.2
AK2	Adenylate kinase 2	1p34
AK3	Adenylate kinase 3	9p24-p13
ALDH1	Aldehyde dehydrogenase 1	9q
ALDH2	Aldehyde dehydrogenase 2	12
ALDH3	Aldehyde dehydrogenase 3	17
ALDOA	Aldolase A	16q22-q24
ALDOB	Aldolase B	9q21.3-q2.1
ALDOC	Aldolase C	17
ALPP	Placental alkaline phosphatase	2q37-q37

Gene symbol	Gene name	Chromosome
ALPPL	Placental alkaline phosphatase-like	17
APOA1	Apolipoprotein AI	11q23-qter
APOA2	Apolipoprotein AII	1q21-q23
APOB	Apolipoprotein B	2p24-p23
APOE	Apolipoprotein E	19q12-q13
APRT	Adenine phosphoribosyl transferase	16q24
HPRT	Hypoxanthine phosphoribosyl transferase	Xq26
ARAF1	Murine sarcoma 3611 oncogene 1	Xp21-q11
ARAF2	Murine sarcoma 3611 oncogene 2	7p14-q21
ARSA	Arylsulfatase A	22q13.31-qter
ARSB	Arylsulfatase B	5p11-q13
ATP1A1	Na, K-ATPase, α_1	1p21-cen
ATP1A3	Na, K-ATPase, α_3	19
BCT1	Branched-chain aminotransferase 1	12pter-q12
BCT2	Branched-chain aminotransferase 2	19
CGA	Chorionic gonadotropin, α	6p21.1-23
CGB	Chorionic gonadotropin, β	19q13.3
CKBB	Creatine kinase, brain	14q32
CKMM	Creatine kinase, muscle	19q13
COL1A1	Collagen, type 1, α_1	17q21.3-q22
COL1A2	Collagen, type 1, α_2	7q21-q22.1
CRYA1	Crystallin, α_1	21
CRYB1	Crystallin, β_1	17cen-q21
CRYB2	Crystallin, β_2	22q11.2-q12.2
CRYG1	Crystallin, γ	2q33-q35
CTSB	Cathepsin B	8p22
CTSD	Cathepsin D	11p15
CTSH	Cathepsin H	15q24-q25
CYK4	Cytokeratin 4	12
CYK13	Cytokeratin 13	17
DIA1	Diaphorase 1	22q13.31-qter
DIA4	Diaphorase 2	16q12-q22
ENO1	Enolase 1	1pter-p36.13
ENO2	Enolase 2	12p11-qter
ERBA	Avian erythroblastic leukemia oncogene A	17q11-q21
ERBB	Avian erythroblastic leukemia oncogene B	7p14-p12
ETS1	Avian erythroblastosis E26 oncogene	11q23.3
ETS2	Avian erythroblastosis E26 oncogene	21q22.3
FUCA1	Fucosidase, α_1	1p34
FUCA2	Fucosidase, α_2	6
GAA	Glucosidase, α	17q23
GBA	Glucosidase, β	1q21

Gene symbol	Gene name	Chromosome
GLA	Galactosidase, α	Xq21-q22
GLB1	Galactosidase, β_1	3p21-cen
GLB2	Galactosidase, β_2	22
GNAT1	Guanine nucleotide-binding protein, α_1	3
GNAT2	Guanine nucleotide-binding protein, α_2	1
GNAS	Guanine nucleotide-binding protein, α_S	20
GOT1	Glutamate oxaloacetate transaminase 1	10q25.3
GOT2	Glutamate oxaloacetate transaminase 2	16p12-q22
GST2	Glutathione *S*-transferase 2	6
GST3	Glutathione *S*-transferase 3	11q13-qter
HBA	Hemoglobin, α	16p13
HBB	Hemoglobin, β	11p15.5
HEXA	Hexosaminidase A	15q22-q25.1
HEXB	Hexosaminidase B	5q13
HOX1	Homeo box 1	7p21-p14
HOX2	Homeo box 2	17q11-q22
HOX3	Homeo box 3	12q12-q13
HOX4	Homeo box 4	2q31-q37
HRAS1	Harvey rat sarcoma oncogene 1	11p15.5
HRAS2	Harvey rat sarcoma oncogene 2	X
KRAS2	Kirsten rat sarcoma oncogene 2	12p12.1
NRAS	Neuroblastoma oncogene	1p22
IDH1	Isocitrate dehydrogenase 1	2q32-qter
IDH2	Isocitrate dehydrogenase 2	15q21-qter
IFNA	Interferon, α	9p22-p13
IFNB2	Interferon, β_2	7
IFNB3	Interferon, β_3	2p23-qter
IGH	Immunoglobulin, heavy chain	14q32.3
IGK	Immunoglobulin, κ-chain	2p12
IGL	Immunoglobulin, λ-chain	22q11.1-q11.2
INHBA	Inhibin, β_A	7p15-p13
INHBB	Inhibin, β_B	2cen-q13
LAMB1	Laminin B1	7q22
LAMB2	Laminin B2	1q25-q31
LDHA	Lactate dehydrogenase A	11pter-p13
LDHB	Lactate dehydrogenase B	12p12.2-p12.1
LIPA	Lipase A	10
LIPB	Lipase B	16
MANA	Mannosidase, α	15q11-qter
MANB	Mannosidase, β	19p13.2-q13.2
MDH1	Malate dehydrogenase 1	2p23
MDH2	Malate dehydrogenase 2	7p13-q22

Gene symbol	Gene name	Chromosome
MYC	Myelocytomatosis oncogene	8q24
MYCL	Myelocytomatosis oncogene, lung	1p32
MYCN	Myelocytomatosis oncogene, neuroblastoma	2p24-p23
MYH1	Myosin heavy chain, skeletal 1	17pter-p11
MYH5	Myosin heavy chain, skeletal 5	7cen-q11.2
NKNA	Neurokinin A	7q21-22
NKNB	Neurokinin B	12q13-21
PCCA	Propionyl CoA carboxylase, α	13
PCCB	Propionyl CoA carboxylase, β	3q13.3-q22
PFK3	Phosphofructokinase, platelet	10pter-p11.1
PFKL	Phosphofructokinase, liver	21q22.3
PFKM	Phosphofructokinase, muscle	1cen-q32
PGK1	Phosphoglycerate kinase 1	Xq13
PGK2	Phosphoglycerate kinase 2	19
PGM1	Phosphoglucomutase 1	1p22.1
PGM2	Phosphoglucomutase 2	4p14-q12
PGM3	Phosphoglucomutase 3	6q12
PKCA	Protein kinase C, α	17q22-q24
PKCB	Protein kinase C, β	16p12-p11.1
PKCG	Protein kinase C, γ	19q13.2-q13.4
POLA	DNA polymerase, α	Xp22.1-p21.3
POLB	DNA polymerase, β	8q24
RAF1	Murine leukemia oncogene 1	3p25-p24
RAF2	Murine leukemia oncogene 2	4
RRM1	Ribonucleotide reductase M1	11p15
RRM2	Ribonucleotide reductase M2	2p25-p24
SOD1	Superoxide dismutase 1	21q22.1
SOD2	Superoxide dismutase 2	6q21
SPTA1	Spectrin, α, erythrocytic 1	1q22-q25
SPTAN1	Spectrin, α, nonerythrocytic 1	9q33-q34
SPTB	Spectrin, β, erythrocytic	14
SRC1	Avian sarcoma oncogene 1	20q12-q13
SRC2	Avian sarcoma oncogene 2	1p36-p34
TCRA	T-cell receptor, α	14q11.2
TCRB	T-cell receptor, β	7q32
TCRG	T-cell receptor, γ	7p15
TGFA	Transforming growth factor, α	2p13
TGFB	Transforming growth factor, β	19q13.1-q13.3
TK1	Thymidine kinase 1	17q21-q22
TK2	Thymidine kinase 2	16
TUBA1	Tubulin A1	2q

Gene symbol	Gene name	Chromosome
TUBB	Tubulin B	6pter-p21
UGP1	UDP-glucose pyrophosphorylase 1	1q21-q22
UGP2	UDP-glucose pyrophosphorylase 2	2
YES1	Yamaguchi sarcoma oncogene 1	18q21.3
YES2	Yamaguchi sarcoma oncogene 2	6

References for chromosomal assignments can be found in Cohen et al. [74] and McKusick [75], for man, and in Davisson and Roderick [76], Kozak [77], and Peters [78], for mouse.

13

Human Linkage Analysis: Strategies for Locus Ordering

Daniel E. Weeks*

College of Physicians & Surgeons of Columbia University and New York State Psychiatric Institute, New York, New York

I. INTRODUCTION

Linkage analysis in human genetics is a conceptually rich discipline involving such diverse mathematical areas as hypothesis testing, parameter estimation, numerical optimization, probability theory, and combinatorics. Even though research in the area of linkage analysis may require skill in these areas, the underlying model is quite simple, dating back to Mendel's experiments with pea plants in the late 1800s. These experiments led Mendel to three main postulates, which, in modern terminology, are:

1. Each trait of an individual is determined by two "factors" or "genes." One of these genes is inherited from the father, the other from the mother. Before Mendel's time, the traits of the parents were thought to blend in their children.
2. Law of segregation: "A reproductive cell produced by an individual bears, for each character, one and only one of the two genes which the individual carries." [1, p. 7].
3. Law of independent assortment: Members of different gene pairs segregate to gametes independently of one another.

The law of independent assortment is violated for two traits when the genes determining these traits lie so close together on the same chromosome that they tend to be inherited together. Thus, it may be possible to determine

---

**Current affiliation:* University of Pittsburgh, Pennsylvania

if two or more traits are linked by following the traits as they segregate down through a pedigree. Although there are physical mapping techniques that do not require family data such as somatic cell hybrids, pulse-field gel electrophoresis, and mapping by the polymerase chain reaction on single sperm [2–5], linkage analysis based on nonindependent assortment requires pedigree data. In addition, pedigree data are often necessary for determining the mode of inheritance, for risk estimates, and for paternity testing [6].

Statistical tests for linkage can be either nonparametric or parametric. Because most of the nonparametric methods focus on the problem of detecting linkage to a disease trait, they concentrate on the question: Do pairs of affected relatives share marker alleles identical-by-descent more often than expected? This can often be answered essentially by counting marker similar and dissimilar relatives and applying a χ^2 test. In contrast, the parametric methods focus on estimating the recombination fraction Θ between two loci by likelihood theory. If $P(X|\Theta = \Theta_1)$ is the likelihood of the pedigrees X at recombination fraction Θ_1, then the LOD score at Θ_1 is

$$\log_{10}\left[\frac{P(X|\Theta = \Theta_1)}{P(X|\Theta = 0.5)}\right]$$

which is the log of the odds in favor of linkage. Two loci are statistically proved to be linked when the LOD score exceeds 3.0 and unlinked (at the given Θ_1) when the LOD score is less than -2.0 [7].

In Section II, we first define the locus ordering problem, giving a four-locus example based on simulated data, and then discuss two general algorithms for ordering problems. Section III gives several examples of the heuristic methods that have actually been used to create chromosome maps, and Section IV.A gives a description, in chronological order, of several preliminary ordering methods that rely only on pairwise information. Section IV.B discusses two evaluations of several pairwise methods. Since ordering can also be carried out while considering all the loci jointly, Section IV.C discusses several intriguing approaches to efficient and rapid multilocus analysis.

II. THE LOCUS ORDERING PROBLEM

With the discovery of restriction fragment length polymorphisms (RFLPs), it became possible to generate enough genetic markers to make a linkage map of the human genome. Once many markers are shown to be on one chromosome, we then face the problem of ordering these syntenic loci. The ideal ordering approach would be to compute a multipoint likelihood for each of the N!/2 orders of N loci, and then to rank the orders, choosing as the best order the one with the largest likelihood. However, "for even moder-

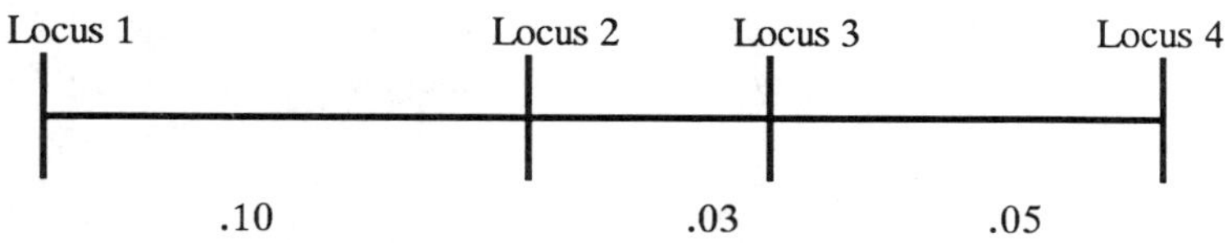

Figure 1 True recombination fractions for each of the three intervals in the simulated data.

ate size N it is not feasible to perform all N!/2 possible multipoint linkage analyses, and geneticists are forced to use ad hoc methods to reduce the number of orders considered" [8, p. 236]. For example, even if it ever became possible to evaluate a 12-point likelihood in 1 sec, it would still take about 7.6 years to evaluate all the likelihoods for each of the 239,500,800 possible orders of 12 loci.

Let us illustrate the locus ordering problem by ordering four codominant linked loci, which were simulated in 29 large three-generation pedigrees. This sample of 380 people is similar in size to the CEPH (Centre D'Étude du Polymorphisme Humain) reference panel [9], often used to create chromosome maps [10–14]. The simulations were conducted under the assumptions of equal female and male recombination fractions, Hardy-Weinberg and linkage equilibrium, and no chiasma interference. Figure 1 exhibits the recombination fractions used in the simulation; we then analyzed the resulting data, calculating pairwise LOD scores and corresponding recombination fraction estimates (Table 1). Although the pairwise LOD scores also indicate order, the best-supported order was found by the statistically sound, but more time-consuming method of ranking each of the 12 possible orders by its four-point loglikelihood (Table 2). Note that the true order, 1–2–3–4, is

Table 1 LOD Scores and Maximum Likelihood Estimates of Recombination Fractions Between Four Simulated Loci[a]

	Locus 1	Locus 2	Locus 3	Locus 4
1	–	.0881 ±.0346	.1670 ± .0408	.1947 ± .0483
2	10.13	–	.0272 ± .0190	.0651 ± .0260
3	7.95	13.77	–	.0550 ± .0240
4	4.61	13.51	16.27	–

[a]Pairwise LOD scores at the maximum likelihood estimates are displayed below the diagonal. Estimated recombination fractions and their standard errors are displayed above the diagonal.

Table 2 The 12 Possible Orders of the Four Simulated Loci Ranked by Their Multipoint Loglikelihoods, with the Corresponding Recombination Fraction Estimates

Order	Loglikelihood	θ_{AB}	θ_{BC}	θ_{CD}
4-3-2-1	-58.72	.05358	.03065	.09806
4-2-3-1	-57.16	.05968	.01848	.1206
3-2-4-1	-55.57	.02586	.04726	.1316
3-4-2-1	-55.11	.04407	.05432	.1007
2-3-4-1	-54.64	.02095	.05382	.1407
2-4-3-1	-54.15	.04649	.04349	.1262
4-3-1-2	-49.59	.04571	.1025	.05329
4-1-2-3	-48.64	.1059	.06993	.02305
4-1-3-2	-47.28	.1046	.08269	.01713
3-4-1-2	-46.89	.04359	.1219	.05528
4-2-1-3	-46.81	.05782	.06246	.09332
3-1-4-2	-44.39	.09501	.09100	.04151

the best-supported order, with odds of 1 : 36 against the next best-supported order 1–3–2–4.

The locus ordering problem falls into the category of combinatorial minimization problems, in which there are a large number of discrete states (i.e., orders), each of which can be assigned a numerical value (here, a multipoint loglikelihood) by a cost or objective function. The goal of the optimization process is to find the state taking on the minimum (or maximum) cost. Combinatorial optimization problems are often very difficult to solve, because the number of states is typically so large that it is not possible to evaluate the cost function for every state. Since the states are discrete, there is no notion of a "downhill direction" that proves useful in continuous optimization. Thus, heuristic methods may provide the only feasible approach to such problems. As Wedd [15] points out, the problem of ordering loci belongs to a class of mathematical problems that are not yet solvable by any polynomially bounded algorithm, despite extensive study. In other words: "it is impossible to write an algorithm or a computer program which can be guaranteed to find the best ordering of the loci on a chromosome in a reasonable time" [15, p. 334].

One of the classic combinatorial problems is the traveling salesman problem, for which the salesman's goal is to minimize the cost of visiting once each of the cities on his itinerary [16–18]. Since the locus ordering problem is similar to the traveling salesman problem, we can benefit from examining heuristic algorithms developed for the latter problem. One class of these approaches consists of the iterative improvement algorithms [19]. Another class, containing the iterative construction algorithms, takes advantage of the concepts of distance and order inherent to the problem of locus ordering.

The first approach may be outlined as follows:

Order Improvement Algorithm

1. Start with an arbitrary order of *all* the loci
2. Create another tentative order by permutation and evaluate its likelihood or similar measure of "cost"
3. If the tentative order is better, let it replace the old order
4. Loop

Unfortunately, this process can easily get stuck in a local minimum, especially if the rearrangement rules call for monotonic downhill movement at each step.

The other approach may be outlined as follows:

Order Construction Algorithm

1. Start with a suborder
2. Select a locus to add
3. Insert the locus into the suborder
4. Loop until done

Typically a set of well-spaced highly informative "anchor" loci is chosen as the starting suborder, forming a framework onto which the remaining loci can be iteratively added. However, in a large number of construction algorithms developed for the traveling salesman problem [16–18], the selection/insertion criteria require that the symmetric matrix of distances $[d_{ij}]$ satisfy the triangle inequality; that is, $d_{ij} + d_{jk} \geq d_{ik}$ for all $1 \leq i,j,k \leq N$, where N is the number of cities (or loci) and d_{ij} is the distance between city i and city j. Unfortunately, neither the matrix of LOD scores nor of distances (in centimorgans) need satisfy this inequality. For example, in Table 1, $d_{12} + d_{23} = 8.90 + 2.72 = 11.62 < d_{13} = 17.37$, where d_{ij} are genetic distances in centimorgans.

The two algorithms described above may be implemented using the "divide-and-conquer" strategy [19], in which the original problem is broken down into smaller subproblems, which are typically simpler to solve. Once the subproblems are solved, the overall solution is found by piecing together all the solutions of the subproblems. This strategy often performs well if the original problem lends itself to natural partitioning. Because groups of closely linked loci, as identified by pairwise linkage analyses, form natural partitions, the problem of locus ordering is often attacked by the divide-and-conquer strategy. In this approach, after first ordering loci within their subgroups, one considers only those orders of all the loci consistent with the best-supported suborders.

III. PRACTICAL EXAMPLES

Let us briefly describe some of the heuristic methods actually used to order several loci. Drayna and White [20] ordered 21 markers on the X chromosome by first using the physical order of six loci as a framework. Then four-point mapping was performed on quartets of loci to find the best-supported order consistent with the physical data and with the results of the pairwise linkage analyses and previous four-point analyses of overlapping sets of loci. In those cases for which genetic evidence for order was limited, five-point analyses were performed.

Markers may be regionally localized and ordered by creating numerous mutant cell lines with partial chromosomal deletions by irradiation [21]. If two markers are absent from one of the cell lines that is missing only one chromosomal segment, it is likely that they are both on the missing segment. Since pedigree analysis indicated that one of their eight loci, p7H4, was well-separated from the other loci, Leach et al. [22] relied on deletion data from 53 irradiated cell lines to locate p7H4 distal to the HLA loci. Four of the remaining seven loci were then lumped into one HLA haplotype on the basis of tight linkage, leaving only four loci to be ordered. To order these, they calculated all 12 multipoint likelihoods and found that two orders were much more likely than the other 10 orders. One of the two best-supported orders was then eliminated by regional localization using the irradiated cell lines. Note that if they had tried to calculate multipoint likelihoods for all possible orders of eight loci, they would have had to consider 20,160 orders!

Leppert et al. [23] ordered seven markers on chromosome 13q by first using pairwise analyses to reduce the number of loci considered in the multipoint analyses. Two loci were removed from consideration until later in the analysis because one showed only very loose linkage to all of the other loci and the other was very tightly linked to one of the remaining five markers. The two-point data also identified three clusters of closely linked loci, further reducing the complexity of the ordering problem to that of ordering the three clusters. This use of clusters was checked by performing three-point tests to "verify that any member of a cluster was very unlikely to be located between members of other clusters" [23, p. 429]. Allowing for all possible permutations within and between clusters, only 12 five-point tests were carried out instead of 60 five-point tests or 2520 seven-point tests. After the best supported five-point map was found, the remaining two loci were added to the map by using location scores [24] and several six-point tests. Location scores can be used to add one new locus to a well-established map. The genetic locations of the old markers are fixed while calculating the odds in favor of the new marker being in a particular location ω versus the new marker being unlinked to the established map. The location score $l(\omega)$ is simply twice the natural logarithm of these odds. Thus the highest point on a graph of $l(\omega)$ identifies the best supported location for the new marker.

Donis-Keller et al. [10] constructed a linkage map of the human genome containing 403 polymorphic markers. Their strategy for ordering these loci involved first calculating the recombination fraction and LOD score between each of the 81,003 pairs of loci, using these results to divide the loci up into linkage groups. Each group was then assigned to a chromosome by the physical localization of at least one of its members. Two iterative strategies with similar results were next employed for ordering the loci on the more densely covered chromosomes.

In the first strategy, they chose the two loci with the highest number of informative meioses. Then the next most informative locus was added to the map in a certain order if that order was 100-fold more likely than the other possible orders. At each step involving N loci, N orders are considered. When no more loci could be added in this manner, the criterion was changed to "add the locus with the smallest number of permissible orders."

The second approach used as its starting point "a set of informative loci separated at intervals of roughly 10–20 centimorgans" [10, p. 324] as determined by two-point linkage analyses. Multipoint analysis of each possible order of this set was used to find the best supported order (usually by an odds ratio of 10,000 : 1). This order was then used as a framework into which the remaining loci were added (if favored by 100 : 1 odds). When many loci were assigned to the same interval in the framework, they were ordered by likelihood analysis for all possible orders within that interval. Those loci that could be placed in more than one interval were added to all the favored intervals and multipoint mapping on a region of the map was performed to find the most likely order.

O'Connell et al. [25] ordered 16 loci into a continuous genetic map of chromosome 12. Starting out with 11 loci, the analysis of two loci was deferred, leaving nine loci to order. By using pairwise recombination estimates, an initial trial map was constructed by use of Lalouel's [26] nonparametric multidimensional scaling approach. The trial map contained three clusters of closely linked loci, which were used to reduce the number of orders from $9!/2 = 181{,}440$ orders to a more manageable number (i.e., 72 analyses). The validity of each cluster was evaluated by determining if the best supported placement of each locus not assigned to a given cluster was actually outside the cluster. The order of the loci within each cluster was evaluated with only those members of the cluster. For two of the clusters, one order was much more likely than the other orders. Fixing the order within each of these two clusters and permuting the locus orders in the remaining third cluster, a total of 72 nine-point analyses were performed, resulting in a best-supported nine-locus map. The PRB cluster of four closely linked loci was ordered by four-point analysis, and the two best-supported orders were entered into each interval in the nine-locus map by 13-locus analysis in which only the recombination fractions involving the four new loci were allowed to vary. The remaining three loci were added to the map by the method of location scores [24].

Lathrop et al. [11] incorporated several heuristic approaches often used to order loci into a computer program, named GMS (for gene mapping system), which relies on the divide-and-conquer strategy. Their algorithm commences by constructing a preliminary trial map from two-point LOD scores by a variety of methods including Lalouel's nonparametric multidimensional scaling approach [26], seriation [27,28], and Thompson's [29] minimization of crossover counts. The loci in this trial map are divided into subgroups based on multilocus analysis; the loci are then ordered within each subgroup, and next the subgroups are oriented and ordered relative to each other (Fig. 2). If the resulting map differs from the initial trial map, the process is repeated using the new map as the trial map. Once this iterative procedure converges, the final order is then subjected to a variety of permutations to verify that it is the best-supported order overall. The GMS algorithm has been employed successfully in many mapping problems by its developers, including genetic maps of chromosome 10 [11], chromosome 17 [13], chromosome 15 [12], and chromosome 1 [14].

IV. THEORETICAL METHODS

In the foregoing practical examples, the first step in ordering a set of linked loci is to choose a starting order or suborder based on the pairwise LOD score curves. Because it can often be difficult to discern overall patterns by peering at a large matrix of pairwise recombination fractions, a variety of preliminary methods for ordering loci have been proposed that rely on only pairwise information [8,15,26,27,30–34]. These methods are discussed in chronologic order in Section IV.A, whereas Section IV.B discusses two evaluations of several of the pairwise methods. Alternative approaches to reducing the computational burden of ordering loci aim at improving or developing multipoint algorithms which simultaneously consider all the loci [29,35–39]. These are discussed in Section IV.C.

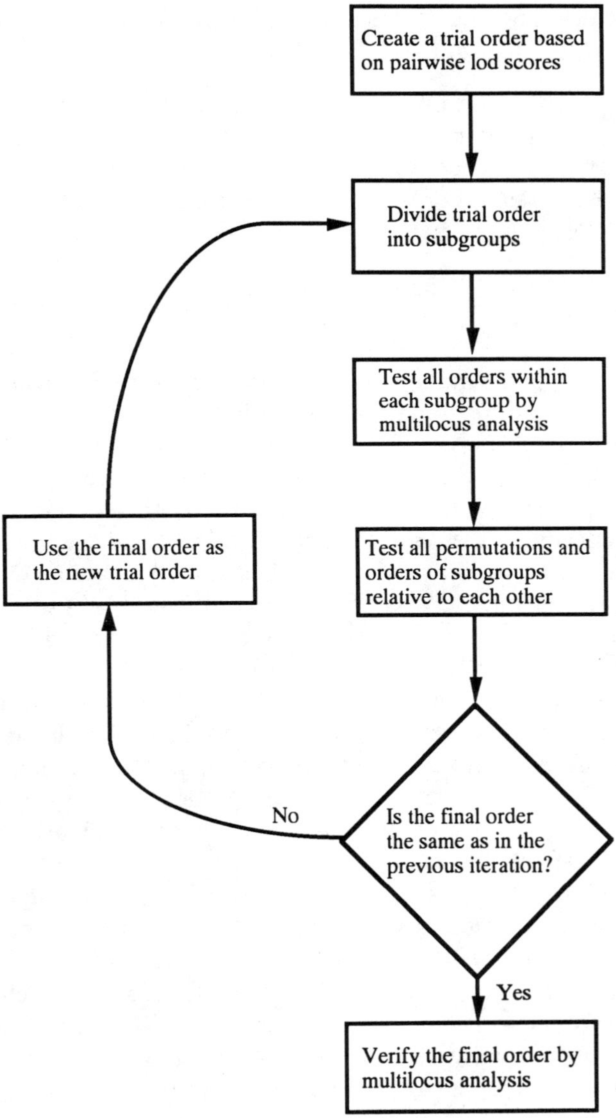

Figure 2 The gene mapping system (GMS) algorithm of Lathrop et al. [11].

A. Preliminary Ordering Methods Based on Pairwise Data

1. Sturt

Sturt [30] ranked orders by a Bayesian approach which depends on the concepts of posterior and prior probabilities. Suppose we wish to order three loci A, B, and C based on pedigree linkage data X. Then the posterior probability P(order ABC|X) of the order ABC given the data X depends on the prior probability P(order ABC) in the following manner:

$$P(\text{order ABC}|X) = \frac{P(X|\text{order ABC})\, P(\text{order ABC})}{P(X)}$$

Note that P(X|order ABC) is the multipoint likelihood of the pedigree data. In the absence of any prior information, the P(order ABC) may be taken to be $1/3$ since there are three possible orders [6, p. 183–184]. However, Sturt [30] defines the prior probabilities by conditioning on the distances between the marker loci. To recover the desired posterior P(order ABC|X), these distances must be removed by integrating the product of the likelihood and the prior over all possible values of the recombination fractions under each possible order. "In effect we are trying out every possible position for each locus given the gene order and the length of the chromosome" [30, p. 256]. Unfortunately, because Sturt's method requires integration, it can be difficult to apply, especially if one tried to integrate the multipoint likelihood. Therefore, Sturt substituted the product of pairwise likelihoods for the multipoint likelihood, reasoning that if each pairwise loglikelihood came from an independent data set, then the multipoint likelihood would simply be the product of the appropriate pairwise likelihoods. However, usually this is not true because markers are all run on the same families in most mapping studies.

Since Sturt's method requires integration for each order under consideration, it was necessary to carry out 60 separate integrations to order five loci on chromosome 1. Such exhaustive ranking of all possible orders will rapidly become impossible as the number of loci increases. In addition, such integration would be too time-consuming if true multipoint likelihoods are used instead of the product of pairwise likelihoods. Although these computational difficulties may be overcome by fast computers, Sturt's method will always suffer from the fact that it is crucially dependent on the definition of the prior probability. As the prior probability is difficult to define, Sturt's method is seldom used.

2. Lalouel: Nonparametric Multidimensional Scaling

Lalouel's [26] method for ordering loci is an implementation of multidimensional scaling [40–43], which seeks to optimally construct a map from a matrix of pairwise distances. Since defining a mapping function which relates pairwise recombination estimates $\hat{\theta}_{ij}$ to map distances requires essentially unverifiable assumptions about interference, Lalouel's nonparametric algorithm avoids mapping functions by focusing on order relations. The algorithm strives to find a set of locus coordinates x_i on a line such that

$$\begin{cases} \text{whenever } \hat{\theta}_{ij} \le \hat{\theta}_{kl} \\ \text{then as much as feasible } |x_i - x_j| \le |x_k - x_l| \end{cases}$$

Leaving out some details, the algorithm essentially consists of the following steps, which are demonstrated using the simulated data displayed in Table 1:

1. Sort the observed recombination fractions $\hat{\theta}_{ij}$ in ascending order, as shown in the top line of Table 3.
2. Starting with an arbitrary set of coordinates x_i for the loci, compute the corresponding distances $|x_i - x_j|$. Sort these distances by size in the same order as the $\hat{\theta}_{ij}$, labeling them as $\hat{d}_{ij}$. Note that no mapping function is required because the permuted distances $\hat{d}_{ij}$ are **not** transformed $\hat{\theta}_{ij}$, but rather are derived from the coordinates x_i. The recombination estimates influence the final solution only through the numerical ordering of their values $\hat{\theta}_{ij}$ [26]. For Table 3, we arbitrarily assumed that the loci were in the order 4–1–2–3 with coordinates (0.00, 0.10, 0.17, 0.21). The second line in Table 3 displays the $|x_i - x_j|$, whereas the sorted $\hat{d}_{ij}$ are displayed in the final line.
3. Find the new locus coordinates x_i which minimize

$$\frac{1}{4} \sum_{ij} e_{ij}(|x_i - x_j| - \hat{d}_{ij})^2$$

where the e_{ij} are the weights.

The final order and the distances between the loci may be easily inferred from the locus coordinates x_i. In our example, if we simply set the $e_{ij} = 1$,

Table 3 Application of Multidimensional Scaling to the Simulated Data

Locus Pairs	(2,3)	(3,4)	(2,4)	(1,2)	(1,3)	(1,4)
Sorted $\hat{\theta}_{ij}$	.0272	.0550	.0651	.0881	.1670	.1947
$\lvert x_i - x_j \rvert$	0.04	0.21	0.17	0.07	0.11	0.10
Sorted $\hat{d}_{ij}$	0.04	0.07	0.10	0.11	0.17	0.21

then the final coordinates are (0.00, 0.12, 0.16, 0.22), which implies that the algorithm converged to the true order 1–2–3–4. Lalouel successfully applied his method to *D. melanogaster* chromosome 3 recombination data [26] and to human chromosome 1 irradiation data [31].

3. Rao and Colleagues: Sum of All Pairwise LOD Scores

To discuss the ordering method of Rao et al. [31], it is necessary to briefly define the smallest region of overlap (SRO) and the likely region of overlap (LRO) methods: If a collection of "abnormal" chromosomes with duplications, deletions, or other rearrangements can be assayed for a locus, then the SRO method assigns that locus to the smallest region consistent with all of the chromosomes. But an SRO cannot be found if some of the inclusive segments are inconsistent with each other. Williams et al. [44] created the LRO method to statistically combine all the data on which chromosomal segments include the locus, even if the data are inconsistent.

Rao et al. [31] developed an approximate maximum likelihood method for mapping loci, based on both the interpolated sum of *all* pairwise LOD scores and a male chiasma map. A chiasma map gives the locations of crossovers (chiasmata) in relation to banding patterns and other cytological markers visible under the microscope. In Rao's method, the best locus locations are taken to be those that maximize the sum of the pairwise LOD scores while satisfying the constraints induced by the chiasma map. Although this method appeared to work quite well ordering 13 chromosome 1 markers [31], the best map placed several markers outside of their SRO or LRO. Seeking to avoid the problem of loci mapping outside their SROs or LROs, Keats et al. [45] extended the method of Rao et al. [31] to convert SROs and LROs into equivalent LOD scores as a function of the size of the region of overlap. However, a few years later Sherman et al. [46] decided that the chiasma map was too inaccurate to warrant inclusion in a mapping program, so they used the method of Rao et al. [31] on pairwise recombination data only. None of these three papers [31,45,46] discuss the methods' convergence performance or sensitivity to initial conditions. Although including information from chiasma maps and somatic cell hybrids is commendable, a map based only on pairwise LOD scores is no longer an acceptable endpoint; rather, most maps today are based on multipoint data.

4. Wedd

Wedd [15] noted that if one were ordering loci interactively, one would not consider many orders because they are "obviously" impossible. Instead, one would consider only those orders that are "close" to the current order. In an effort to imitate this intuitive approach, he developed an iterative algorithm which, given a locus order, tries to replace it with a better "neighboring" order. This process is repeated until no order with a larger maximum likelihood can be found. Wedd discovered that the definition of "neighboring" order was quite important. In fact, only exchanging pairs of adjacent loci did not work well, so he recommends, in addition, exchanging and/or reversing adjacent groups of closely linked loci.

5. Buetow and Chakravarti: Seriation

The seriation algorithm of Buetow and Chakravarti [27] is based on the pairwise recombination fractions and employs a conceptually simple order construction algorithm:

1. Pick a locus i.
2. Place the locus nearest to locus i to the right of locus i. Now we have two ordered loci.
3. *Selection step:* From the remaining unordered loci, pick the locus closest to locus i.
4. *Insertion step:* Place the new locus on the left end of the ordered loci if it is closer to the leftmost locus than the rightmost locus; otherwise, add it to the right end of the cluster of ordered loci.
5. Repeat the selection and insertion steps 3 and 4 until all the loci are ordered.

Buetow and Chakravarti provide rules for dealing with ties when the new locus is equidistant from both the leftmost and rightmost loci (step 4). The foregoing algorithm is repeated for each of the N loci, resulting in N orders, which will frequently not all be identical. These orders are then combined into a consensus order, by either "averaging" the orders or by choosing the order with the lowest continuity index, where the continuity index compares the observed recombination fraction between two loci to the ordinal distance between their locations in a given order [27, p. 183].

Although the seriation algorithm performs well [28,47,48], it requires *all* pairwise LOD scores. If some pairwise LOD scores are missing, as is frequently the case in published data, then the seriation algorithm can be applied to only those subgroups of loci for which all pairwise LOD scores are available. Hazout et al. [49] point out that another potential weakness of the seriation algorithm is the restriction of the insertion step to the addition of a new locus onto either *end* of the suborder; the existing suborder cannot be modified by information provided by the new locus. In addition, seriation does not consider the relative amount of information in the recombination fractions, which is reflected by the magnitude of the LOD score and by the standard errors of the recombination fractions. However, Buetow and Chakravarti [27] point out that treating all observations equally should not pose much of a problem if adequate sample sizes are used to estimate recombination fractions.

6. Weeks and Lange: Four Ranking Criteria and Simulated Annealing

Weeks and Lange [8] investigated two ranking criteria for preliminary locus ordering based on pairwise LOD scores and recombination estimates $\hat{\theta}_{ij}$. The first criterion is simply the sum of adjacent LODs (SAL) under a given order. For example, if we use the simulated data in Table 1, the SAL for the order 1–2–3–4 is 10.13 + 13.77 + 16.27 = 40.17. Note that SAL criterion requires that *all* pairwise LOD scores be available.

The definition of the second criterion is slightly more technical: Let $\hat{d}_{ij}$ represent the corresponding map distance between loci i and j. If we let d_i be the theoretical map distance from some arbitrary origin and p_i be the theoretical map distance between the ith and $i + 1^{st}$ loci (i.e., $p_i = d_{i+1} - d_i$), then loglikelihood of the $\hat{d}_{ij}$ is minus half the sum of

$$\sum \frac{\left[\hat{d}_{ij} - \sum_{\min(i,j)}^{\max(i,j)-1} p_k\right]^2}{\mathrm{Var}(\hat{d}_{ij})}$$

where the sum is taken over all pairs of loci with observed recombination estimates. Weeks and Lange [8] denote the minimum of this sum by LLS, for Lalouel's least squares criterion. Each possible locus order has a corresponding LLS value, which may be used to rank the orders relative to each other. The LLS has an advantage over some of the other ranking criteria in that not all pairwise recombination fractions need be available.

When the number N of loci is small, it is possible to evaluate the SAL or LLS for every order and then rank the orders by these numbers. When N becomes larger, exhaustive ranking is no longer feasible, and then Weeks and Lange [8] use a simulated annealing algorithm [19,50], which is a method for combinatorial optimization that imitates the formation of an ordered crystal out of a slowly cooled liquid. Suppose the liquid consists of a collection of a large number of molecules (or loci), which may take on an inconceivably large number of different configurations, each with a measurable "energy." This liquid will often crystallize into an ordered state of minimum energy when subjected to the simulated annealing algorithm (Fig. 3), which may be outlined as follows:

1. Suppose a current configuration C has energy E_C.
2. Randomly pick a new configuration N from a neighborhood of the current configuration and calculate its energy E_N.
3. Depending on chance and the relative magnitudes of E_N and E_C, move to this new configuration or remain in the original configuration as

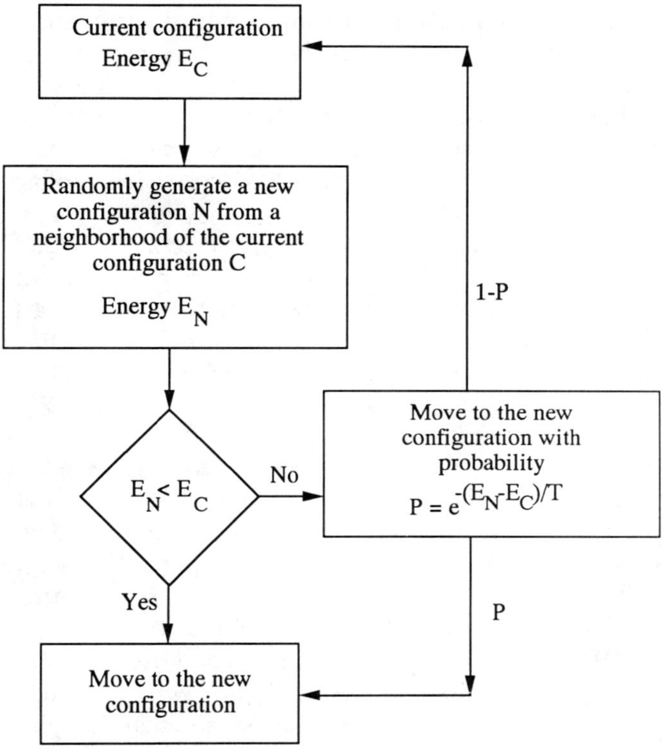

Figure 3 A flow diagram of the simulated annealing algorithm at a fixed temperature T. (From Ref. 52)

indicated in Fig. 3. Note that since the algorithm does not always move strictly downhill, it may avoid becoming trapped in a local minimum.

4. Loop, gradually lowering the "temperature" T, which controls movement from one configuration to another, until convergence.

Strict order improvement algorithms often get stuck in local minimums, especially if they always move monotonically downhill. Local minimums may be avoided by starting the optimization method in many different randomly chosen points, hoping that at least one of these many optimizations attains the global maximum. The simulated annealing algorithm usually avoids this problem because, at high temperatures, "we are in essence simulating many random starting solutions in a controlled, simultaneous fashion" [51, p. 221]. Then, as the temperature is lowered, the simulated annealing algorithm becomes more and more likely to accept only those configurations that are an improvement over the current configuration.

Simulated annealing, in conjunction with easily computed energy measurements (such as LLS or SAL), appears to be a reliable method for generating a list of the best candidate orders, without sampling all possible orders. When the number of loci becomes large, this ability becomes more and more advantageous. For instance, in a nine-locus example, simulated annealing produced a complete list of the 25 best LLS orders while sampling fewer than 47% of all possible orders [8].

In addition to the LLS and SAL criteria, we briefly present two other criteria: the SAD (sum of adjacent distances) and SARF (sum of adjacent recombination fractions) criteria [52]. SAD was suggested to us by Michael Boehnke [personal communication]. Assuming the order 1–2–3–4 and using the simulated data in Table 1, the SARF criterion is 0.0881 + 0.0272 + 0.0550 = 0.1703, whereas the SAD criterion is 0.0890 + 0.0272 + 0.0552 = 0.1754 if we use Kosambi's [53] mapping function. We applied all four criteria [52] to some chromosome 13 data involving six linked loci on 380 people in 29 pedigrees (Table 4); these data were submitted to the CEPH database by Leppert et al. [23]. As needed for the criteria, we computed all 15 pairwise LOD scores and estimated recombination fractions (Table 5). Also, we computed all 360 six-point LOD scores to find the best-supported order. To apply the SAD method, we used Kosambi's [53] map function $d = M(\Theta) = 1/4 \ln[(1 + 2\Theta)/(1 - 2\Theta]$ to transform the nonadditive recombination fractions $\hat{\Theta}_{ij}$ into additive map distances $\hat{d}_{ij}$. A problem arises when $\hat{\Theta} = 0.5$, since then $M(0.5) = \infty$. Therefore, if $\hat{\Theta}$ were greater than 0.49, we arbitrarily set $M(\hat{\Theta})$ equal to 5.18. Alternately, an upper bound on $M(\hat{\Theta})$ might be derived from the sum of known map distances on the chromosome of interest. Note that the SARF ranking criterion corresponds closely to the criterion employed by Buetow and Chakravarti [27]. Even though the SAD and SARF methods are computationally much quicker than the LLS method, they suffer from two main drawbacks: (1) They do not take into account the standard errors of the recombination fractions. (2) They require *all* pairwise recombination fractions. If one or more pairwise recombination fractions are 0.50 because of a lack of informative families, this could have

Table 4 Names of the Loci Analyzed on Chromosome 13q

Probe name	PHU10	P7F12	P7D2	P9D11	P1E8	P9A7
HGM8 name	D13S6	D13S1	D13S10	D13S2	D13S4	D13S3
Number	1	2	3	4	5	6

Source: Ref. 52.

Table 5 Maximum Likelihood Estimates of Recombination Fractions and LOD Scores for Linkage Between Six Loci on Chromosome 13q[a]

Loci	1	2	4	5	6	3
1	-	9.44	4.78	6.63	.365	1.38
2	.057±.032	-	4.42	1.83	0	5.82
4	.183±.048	.239±.045	-	19.54	.067	1.12
5	.173±.041	.288±.054	.021±.014	-	.889	1.74
6	.378±.074	.50	.462±.068	.354±.053	-	.048
3	.177±.093	.098±.042	.144±.095	.228±.065	.445±.116	-

[a]Estimated recombination fractions and their standard errors are displayed on the subdiagonal. On the superdiagonal, LOD scores at the maximum likelihood estimates are displayed.
Source: Ref. 52.

an adverse effect on the SAD and SARF rankings. In such a case, SAD and SARF should not be blindly applied. Fortunately, most mapping projects are usually performed on many families, so that this situation seldom occurs. For example, the $\hat{\theta} = 0.50$ between loci 2 and 6 in the chromosome 13q data set is not due to uninformative families.

Table 6 displays the 25 best-supported orders and their rankings by the various criteria, including six-point LOD scores. In evaluating these results, it must be remembered that the goal is ferreting out the best-supported order, rather than creating a ranking that correlates perfectly with the LOD score throughout its whole range. Each of the four ranking methods contain the best two supported orders in its top 25 orders (or 6.9% of all 360 possible orders). The SAD method performs the worst, because loci 2 and 6 are adjacent in several of the best-supported orders. These two loci have a pairwise recombination fraction $\hat{\theta} = 0.50$, which is transformed to a large distance (5.18), drastically affecting the SAD rank whenever loci 2 and 6 are contiguous.

Lange and Weeks [54] present an algorithm for using preliminary ranking procedures such as SAL and LLS. Let N be the number of loci to be ordered. After the 40 best candidate orders have been identified by simulated annealing, these 40 orders are ranked by multipoint LOD scores to find the best N-point order A. Then order A is permuted by inverting a single block of contiguous loci, generating $\binom{N}{2}$ new orders. These orders are ranked by multipoint LOD scores. If the best order B is still order A, then we choose order A (equals B) as the best-supported order; otherwise, the

Table 6 Results of Exhaustive Maximum Likelihood (ML), LLS, SAL, SARF, and SAD Ranking of all 360 orders of Six Loci on Chromosome 13q

		Rankings				
Order	lod score	ML[§]	LLS	SAL	SARF	SAD
1.2.3.4.5.6.	44.54540	1	13	6	1	1
3.2.1.4.5.6.	44.47844	2	5	3	3	2
1.2.3.5.4.6.	44.11774	3	14	10	14	41
3.2.1.5.4.6.	43.99253	4	8	1	8	29
2.3.1.4.5.6.	43.67308	5	3	33	11	7
2.1.3.4.5.6.	43.49877	6	1	35	4	4
1.3.2.4.5.6.	43.48905	7	23	46	16	10
3.1.2.4.5.6.	43.40129	8	16	16	13	9
2.3.1.5.4.6.	43.16381	9	7	30	25	55
5.4.1.2.3.6.	43.12917	10	11	4	7	19
1.3.2.5.4.6.	43.11433	11	22	78	61	83
5.4.3.2.1.6.	43.07994	12	18	12	2	3
4.5.3.2.1.6.	43.03864	13	17	5	5	5
4.5.1.2.3.6.	43.02309	14	12	2	6	17
2.1.3.5.4.6.	43.01587	15	2	42	30	59
3.1.2.5.4.6.	42.98697	16	15	39	47	69
5.4.1.3.2.6.	42.25816	17	9	53	41	245
4.5.1.3.2.6.	42.11914	18	10	31	38	244
4.5.2.3.1.6.	42.09644	19	26	77	34	24
5.4.3.1.2.6.	42.09388	20	4	54	18	242
5.4.2.3.1.6.	42.07645	21	28	52	22	15
5.4.2.1.3.6.	42.07395	22	20	24	29	42
4.5.2.1.3.6.	42.05792	23	19	41	43	57
4.5.3.1.2.6.	41.99000	24	6	44	42	246
3.2.4.5.1.6.	39.91472	25	112	8	21	14

[a]ML, ranking by multipoint loglikelihood.
Source: Ref. 52.

permutation process is repeated with the new best order B. When Lange and Weeks [54] applied this to simulated ten-point data using SAL and LLS, the true order was recovered on the first iteration after calculating only 116 multipoint LOD scores!

This strategy for using the preliminary ranking procedures has been criticized for arbitrarily considering only the 40 best-ranked orders. As the number of loci increases, it seems clear that the number of top orders considered should increase. But how fast? It would be interesting to investigate the effects of making the number of top orders considered dependent on the magnitude of the difference in loglikelihoods between the best-supported and worst-supported of these top-ranked orders. That is, we would like to sample enough orders so that the relative odds between our best-supported and worst-supported orders is greater than, say, 1000 : 1. However, this approach may not work well if the list of top orders includes some very poorly supported orders. For example, the chromosome 13q order ranked sixth by the SAD criterion is 10^{15} times less likely than the order ranked first.

Another possible rule for choosing the number of top orders considered might be this: Take the 20 top-ranked orders and compute their multipoint likelihoods. Regress the value of the ranking criterion for each order on the corresponding multipoint likelihood and, from the regression curve, find the ranking criterion value V_{1000} corresponding to a decrease in the likelihood of 1000. Then continue to rank additional orders until the lowest-ranked order has a criterion value less than V_{1000}.

7. Wilson: Product of Adjacent Recombination Fractions

Wilson proposed that the optimal order of a group of linked loci is the "order with the minimum value of the product of the adjoining pairwise recombination rates" [32, p. 75]. This follows from two common assumptions: (1) the rate of recombination in a subinterval is less than or equal to the recombination fraction in the larger interval; and (2) absence of interference. The first assumption is true on biological grounds, whereas the second assumption is invoked by most multipoint mapping programs [24,55]. Kammerer and MacCluer [47] found that Wilson's ranking procedure was more powerful in ordering loci than Lalouel's [26] nonparametric multidimensional scaling method. However, as might be expected, the power of ordering methods depends on the spacing of the loci and their alleleic frequencies; Wilson's procedure performs better when the loci are equally spaced.

8. Morton and Andrews: Multiple Pairwise Linkage Analysis

Morton and Andrews [34] defined a good order as one that has a high value of the sum of measures between adjacent pairs of loci, where the "measure" is one of three criteria: $1-2\Theta$, Z, or Z', where Θ is the estimated recombination rate, Z is the maximum LOD score, and Z' is the maximum LOD score divided by the equivalent number of meioses contributing to the LOD score. The equivalent number of meioses is the number of fully informative meioses that would be necessary to yield a maximum LOD score and an estimate of Θ identical with the observed ones [56,57]. When the measure is Z, then the sum of adjacent measures (SAM) criterion is identical with the SAL (sum of adjacent LODs) criterion of Weeks and Lange [8]. These measures are used in the most recent version of a two-step ordering program called MAP [31,34,45,46], which apparently uses only one of these measures at a time. MAP first generates a trial order and then attempts to iteratively improve the order:

1. Generate a trial order: The initial order is produced by sorting the pair of loci according to the measure and then, starting with the pair of loci with the highest measure, successively choosing the locus pair with the next highest measure that does not create a circular map. The order may be deduced from the pairs chosen because the pairs specify adjacent loci. "For example, the ordered string 34, 13, 23, 24, 14, 12 generates the map 1342 by omission of the 23, 14, and 12 elements" [34, p. 265]. Any loci omitted by this process are added to the map by inserting them into each interval and choosing the map with the highest measure.
2. Improve the map: Given a trial order of N loci, the algorithm evaluates the SAMs of the N − 1 orders generated by permuting two adjacent loci in the trial order. If any of these permutations has a better SAM than the trial order, it becomes the new trial order, and the process is repeated until an order is found that is better than all its N − 1 permuted orders. Then the orders (among the N − 1 permuted orders) whose SAMs differ by no more than a set amount from the best order's SAM are compared with the best order and the loci spanning the discordances are permuted to generate many other orders. If any of these permuted orders are better than the current best order, the whole process is repeated again; otherwise, the final order is printed out with odds against certain alternate orders.

The MAP program of Morton and Andrews [34] was used to order 23 loci on chromosome 1 on the basis of published two-point linkage data [58]. Although MAP generated an order that was very similar to orders derived from multipoint data, the resulting map was about 30% shorter than the multipoint maps. This is probably because the two-point method is less sensitive to double recombinants [58].

9. Falk: Minimum Distance Map

Falk [33] wrote a program utilizing simulated annealing [50] for preliminary ordering of loci, called MDMAP, in which the optimal order is defined as the one with the minimum sum of adjacent recombination fractions. She called this a "minimum distance map," although the recombination fractions were not converted into distances. Falk [33] found that although this method performed well ordering relatively loosely linked markers, it did not fare so well when ordering groups of closely linked loci. Weeks [52] independently investigated the same method, which he called SARF, for "sum of adjacent recombination fractions" (see above). The SARF ranking criterion suffers from the fact that it does not take into account the relative amount of information in each of the pairwise recombination fractions, as pointed out earlier.

B. Evaluation of Preliminary Methods for Locus Ordering

1. Kammerer and MacCluer: Evaluation of Three Methods by Simulation

With use of simulated data on 50 and 100 pedigrees, Kammerer and MacCluer [47] empirically evaluated three ordering methods: seriation [27], the product of recombination frequencies [32], and two nonmetric multidimensional scaling methods, including Lalouel's program GENMAP [26]. They found that "as the distance between locus pairs increased, GENMAP was more likely to misorder loci than were the other locus-ordering methods" [47, p. 969]. In addition, they found that "the multidimensional scaling method was less powerful than either seriation or the product of recombination frequencies. The latter two methods were approximately equally powerful" [47, p. 964]. As might be expected, all the methods were more accurate with larger numbers of pedigrees and larger distances between adjacent pairs of loci. As there must be at least one recombinant between each pair of adjacent loci, it is not surprising that it is quite difficult to order closely linked loci. In fact, two groups [59,60] found that, for ordering, the optimum spacing between loci is about 20 cM.

2. Olson and Boehnke: Evaluation of Eight Methods by Simulation

Olson and Boehnke used a simulation approach to compare eight preliminary ordering procedures [61,62]. Five of these have already been discussed: SARF (sum of adjacent recombination fractions, [33,52]), PARF (product of adjacent recombination fractions, [32]), LLS (Lalouel's least squares, [8]), SER (seriation, [27,28]), and SAL (sum of adjacent LODs, [8]). The PDR (probability of double recombinants [63]) scheme, which seeks to minimize the probability of double recombinants, is new, as are the following two different forms of SAL:

1. The sum of adjacent LODs weighted by the product of the polymorphism information contents (SALPIC). For a codominant marker, the *polymorphism information content* (PIC) is defined as the probability that the marker genotype will identify which gamete was passed on to a given child from the heterozygous parent in a Dd $\times$ dd mating, where D is the disease allele causing a rare dominant disease [64].
2. The sum of adjacent LODs weighted by the equivalent number of meioses (SALEQ). The *equivalent number of meioses* is the number of fully informative meioses that would be necessary to yield a maximum LOD

score and an estimate of Θ identical with the observed ones [56,57]. SALEQ is similar to the "normalized" LOD score measure of Morton and Andrews [34].

Olson and Boehnke [62] proposed these two new forms of SAL after observing that SAL is very sensitive to how informative the loci are. They argue as follows: Suppose we have three loci in the order 1–2–3 for which loci 1 and 3 are much more informative than locus 2. If, as expected, the maximum LOD score Z_{13} between loci 1 and 3 is much larger than the maximum LOD scores Z_{12} and Z_{23}, then SAL will favor those orders that place 1 and 3 next to each other. Both SALEQ and SALPIC attempt to remedy this defect of SAL by weighting each pairwise LOD score by a measure of informativeness of the two loci contributing to the LOD score.

The most difficult part of a comparison of ordering methods is defining an appropriate measure of power. Whereas Kammerer and MacCluer [47] defined *power* as percentage of correct orders found in 50 replications, this alone would not be a suitable measure of power for the methods proposed by Weeks and Lange [8] or Olson and Boehnke [61,62], since their methods (except SER) generate a ranking of possible orders, rather than converging to one specific order. For such methods, it is not clear if ranking the true order first, while obtaining no similar order in the list of top orders, is better than missing the true order, while including many orders very similar to the true order. In response to this problem, Olson and Boehnke [61,62] measured performance in terms of "the mean rank of the true order, the mean number of pairwise permutations from the best order to the true order, and percent of replicates in which the best order equals the true order" [61, p. A155]. Because SER does not generate a ranking of orders, the mean rank criterion cannot be used to evaluate the SER method.

Olson and Boehnke [62] found that although most of the methods performed comparably, PDR, SER, and SAL were the worst. As expected, they [62] found that SAL lost power when highly informative markers were at the ends of the map. However, if the highly informative markers were adjacent to each other, SAL significantly outperformed all the other methods. In contrast, the LLS method was best when the highly informative markers were at the ends of the map. This suggests to us that one possible approach would be to use both the SAL and the LLS methods together, bearing in mind their relative strengths. However, the LLS method requires substantial amounts of computer time, so Olson and Boehnke [62] recommend SALEQ and SARF as the methods of choice because, on average, theses methods performed well under a variety of conditions. Olson and Boehnke [62] also explored how many candidate orders should be ranked to include the true order with high probability.

C. Improved Algorithms for Joint Analysis of Multipoint Data

Although ordering procedures that are based on pairwise information may show promise, it is well known that pairwise data contain less information than joint multipoint data [59]. Indeed, "for the sizes of sample available in human medical genetics, pairwise analysis may be a statistically inefficient and inconclusive procedure. Three-way data provide very much more information, particularly with regard to the problem of ordering loci" [65, p. 31]. However, multipoint analysis of linkage data may be orders of magnitude more time-consuming than pairwise analysis. There are three avenues of improvement that have been explored:

1. Substitute a different algorithm for the traditional Elston–Stewart [66] algorithm for calculating the pedigree likelihood.
2. Use more efficient routines to maximize the loglikelihood: even if the pedigree likelihoods can be calculated very efficiently, the speed of pedigree analyses depends crucially on the maximization algorithm employed. Ideally, the optimization algorithm should find an extremum quickly and reliably with a minimum of function evaluations. Pedigree analysis may be open to improvement by judicious selection and implementation of an optimization algorithm.
3. Utilize novel joint measures of support other than the time-consuming multipoint likelihood.

1. Background: The Elston–Stewart Algorithm

The use of likelihood theory to detect linkage and to order loci requires algorithms for computing the likelihood of a pedigree and then maximizing it. The Elston–Stewart [66] algorithm is most commonly employed to calculate the likelihood of a pedigree as follows: Consider a pedigree with m members where $\mathbf{x} = (x_1, \ldots, x_m)$ is the vector of observed phenotypes for the pedigree (i.e., x_i is the phenotype of individual i). If G_i is the set of genotypes g_i compatible with person i's phenotype, then the likelihood of the pedigree can be written as a sum of products,

$$P(x_1, \ldots, x_m) = \sum_{g_1 \in G_1} \cdots \sum_{g_m \in G_m} \prod_{\substack{i \\ \text{observed}}} P(x_i|g_i) \prod_{\substack{k \\ \text{founders}}} P(g_k) \prod_{\{i_1,i_2,i_3\}} P(g_{i_1}|g_{i_2}, g_{i_3}) \qquad \text{(Eq. 1)}$$

where $\{i_1,i_2,i_3\}$ is an arbitrary offspring–parent triple and "i observed" refers to those individuals with observed phenotypes [67,68]. The probability $P(x_i|g_i)$ of individual i having phenotype x_i given his genotype g_i is known as the *penetrance*. Since a *founder* k is an individual whose parents are not members of the pedigree, the prior probability $P(g_k)$ is a function of the population gene frequencies. The transmission probability $P(g_{i_1}|g_{i_2}, g_{i_3})$ is the probability that parents with genotypes g_{i_2} and g_{i_3} produce a child with genotype g_{i_1}. Note that equation (1) may be evaluated most easily by writing it as an iterated sum. The Elston–Stewart algorithm can be viewed as a method for choosing an order for performing this iterated sum to minimize the number of additions and multiplications. It is clearly advantageous to eliminate as many superfluous genotypes as possible in each genotype G_i before computing the likelihood [54,68,69].

2. Lander and Green

Lander and Green [38], Green [39], and Lander et al. [37] present a method for rapid construction of multilocus linkage maps that is directed toward CEPH-type pedigrees. Their approach is based on a novel view of the phenotype and genotype vectors. In the traditional Elston–Stewart algorithm [66] each element of these vectors corresponds to a person, whereas in the Lander–Green algorithm each element of these vectors corresponds to a locus. In other words, if there are N loci, the vector of phenotypes is written as $\mathbf{x}_L = (x_{L_1}, x_{L_2}, \ldots, x_{L_N})$, where x_{L_i} contains the phenotypes at locus i for each of the m people in the pedigree. Likewise, let $\mathbf{g}_L = (g_{L_1}, \ldots, g_{L_N})$ be a vector of multilocus ordered genotypes for the pedigree. An ordered genotype is one in which, for each locus, the maternally inherited allele appears first, followed by the paternally inherited allele. If L_i is the set of ordered genotypes at locus i compatible with the set of phenotypes, then, following Lander and Green [38], the likelihood of the pedigree can be written as a sum of products:

$$P(x_{L_1}, x_{L_2}, \ldots, x_{L_N})$$

$$= \sum_{g_{L_1} \in L_1} \cdots \sum_{g_{L_N} \in L_N} \left[\prod_{i=1}^{N} P(x_{L_i}|g_{L_i}) \right] P(g_{L_N}|g_{L_{N-1}}, g_{L_{N-2}}, \ldots, g_{L_1}) \ldots P(g_{L_2}|g_{L_1}) P(g_{L_1})$$

If we assume no interference and that the genotypes are ordered, then

$$P(g_{L_K}|g_{L_{K-1}}, g_{L_{K-2}}, \ldots, g_{L_1}) = P(g_{L_K}|g_{L_{K-1}})$$

To compute $P(\mathbf{x}_L)$ efficiently as a telescoping sum, let

$$A(g_{L_1}) = P(g_{L_1})$$

and, for $1 < i < N$, let

$$A(g_{L_{i+1}}) = \sum_{g_{L_i} \in L_i} P(g_{L_{i+1}}|g_{L_i}) P(x_{L_i}|g_{L_i}) A(g_{L_i})$$

Then

$$P(\mathbf{x}_L) = \sum_{g_{L_N} \in L_N} P(x_{L_N} | g_{L_N}) A(g_{L_N}).$$

Note that to evaluate each $A(g_{L_{i+1}})$ for $1 < i < N$ requires $2|L_i|$ multiplications and $|L_i| - 1$ additions. If everyone is typed and every locus is codominant, then each L_i has at most 2^m elements and at worst the evaluation of $P(\mathbf{x}_L)$ requires $2^m N(3 \cdot 2^m - 1) - 2^{2m}$ operations. Note that the operation count scales linearly with the number of loci N and scales exponentially with the number of pedigree members m. This is quite different from the traditional Elston–Stewart algorithm, for which the operation count scales linearly with m and exponentially with N. However, Green [39] notes that "if there are missing individuals, as is the case in many disease pedigrees, then the amount of calculation in fact grows exponentially in the number of loci because the number of genotypes that must be tried in the missing individuals grows exponentially." Thus, although the Lander–Green algorithm can be very useful for multilocus mapping problems in which nearly everyone is typed at codominant loci, it is restricted to small pedigrees. The Lander–Green algorithm has been implemented in a computer program called MAPMAKER, which is designed for making linkage maps based on two- and three-generation nuclear families [38].

3. Lathrop and Colleagues: Factorization

In some special cases, a multipoint likelihood may factor into a product of terms involving disjoint sets of contiguous loci. In such cases, the factorized likelihood may be rapidly maximized by optimizing each "simple" term separately, whereas the unfactorized likelihood may be too complex and time-consuming to maximize or even evaluate. Lathrop et al. [35] developed two very specific factorization rules that apply only to codominant loci segregating in simple "CEPH-type" families consisting of children, parents, and grandparents, in which at least the parents must be typed at the locus of interest. For example, consider a set of n loci with a presumptive order 1 through n. Then, factorization rule 1 reads: "Suppose the ith locus is fully informative for linkage in one family and that, if the grandparents are typed, neither grandparental mating is an intercross at this locus. Then, the likelihood for this family can be factorized into two terms, the first involving recombination in the segments delimited by the loci 1 . . i, and the second involving recombination in the segments delimited by i . . n." [p. 42–43]. Lathrop et al. [35] present three transformation rules that modify genotypes and recombination rates to meet the requirements for application of the factorization rules. This approach, although very useful for CEPH-type families, is limited by its specificity.

4. Edwards

Edwards [36] describes how to determine locus order in the specialized case in which haplotypes may be inferred at codominant loci. He suggests using upper- and lower-case letters to distinguish phase at each of the loci and provides a computer program for analyzing such data. He argues that the advantages of rapid and simple analysis warrant the exclusion of those relatively few cases in which phase cannot be determined. Such an approach is also employed by Lander et al. [37] in their MAPMAKER program, which allows one to perform very rapid initial analyses by excluding phase-unknown data.

5. Thompson: Minimize Obligatory Crossover Count

Thompson [29] developed a method for ordering loci that considers loci jointly, while avoiding the burden of computing a multipoint likelihood for each order considered. She states that: "For the problem of ordering multiple markers, it seems that more may be lost by failure to take account of loci jointly than by failing to extract all the statistical information. That is, a simple scoring method based on joint data may prove to produce better results than a full likelihood analysis of pairwise data" [29, p. 94]. Her method is based on minimizing the obligatory crossover count when the data indicate grandparental origin of the received gametes, if known. This is the criterion one would intuitively use if one were trying to order loci by hand on phase-known data [6, p. 175–176]. The locus order with the minimum obligatory crossover count is found by a branch-and-bound search, which has an advantage over simulated annealing in that it is guaranteed to find the global optimum. No matter which method is employed, numerous informative gametes are always necessary for accurate locus ordering because "loci cannot be ordered until there is at least a recombinant distinguishing the members of every adjacent pair" [29, p. 98]. Thompson successfully applies her method to simulated data and mathematically proves that the true order will asymptotically have the minimum crossover count as the number of informative gametes goes to infinity. However, if loci are not equally informative, then minimum crossover criterion may not work well because it counts all observed recombinants equally.

V. SUMMARY

The problem of ordering several syntenic loci is not trivial, for it belongs to a class of mathematical problems that are not yet solvable by any polynomially bounded algorithm, despite extensive study. This is because as the number of loci increases, the number of possible orders increases exponentially. Such an exponential increase prohibits evaluation of multipoint likelihoods for *all* orders, except when the number of loci is small. This combinatorial complexity has forced linkage analysts to turn to heuristic algorithms, which fall into two broad categories: order improvement and order construction. In addition to employing such algorithms, it is often helpful to divide the overall ordering problem into smaller problems delineated by groups of closely linked loci. The heuristic approaches have been formalized into certain computer packages, which, after extensive computation, generate the "best" multipoint map for a given set of linkage data.

In multilocus linkage analysis, there are two types of information: pairwise, in which only two loci are considered at a time; and multipoint, in which more than two loci are considered. Similarly, the methods for ordering loci may be split into two categories, depending on whether or not they rely mainly on pairwise or multipoint information.

The pairwise LOD score curves often form the basis for choosing a starting order or suborder to begin a locus-ordering algorithm. Because it can often be difficult to discern overall patterns by peering at a large matrix of pairwise recombination fractions, a variety of preliminary methods for ordering loci have been proposed that rely on only the information contained in the pairwise LOD score curves. The cornerstone of many of these methods is the replacement of multipoint likelihoods with rapidly computable measures based on pairwise information. For example, a partial list of criteria that have been used includes: sum of adjacent distances, sum of adjacent recombination fractions, product of adjacent recombination fractions, and the sum of adjacent LODs. The ideal measure should take into account the relative amount of information in each pairwise LOD score curve. For example, a LOD score curve based on two informative families should influence the final order less than a LOD score curve based on ten informative families.

Although ordering procedures based on pairwise information show promise, it is well known that pairwise data contain less information than joint multipoint data [59]. In fact, an order based on multipoint information is more credible and trustworthy than an order based only on pairwise information. However, multipoint analysis of linkage data may be orders of magnitude more time-consuming than pairwise analysis. There are three avenues of improvement that have been discussed: (1) substitute a different algorithm for the traditional Elston–Stewart [66] algorithm for calculating the pedigree likelihood; (2) use more efficient routines to maximize the loglikelihood; and (3) utilize novel joint measures of support, other than the time-consuming multipoint likelihood.

There has been little research comparing different ordering algorithms. Although Weeks and Lange [8] have compared two methods, Kammerer and MacCluer [47] have compared three methods, and Olson and Boehnke

[61,62] have compared eight methods, much more research is needed. As the final goal is "correct" ordering, the methods should be evaluated as they would be used in actual practice. For example, if we were using seriation to order our loci, we would not simply accept the order generated by seriation as the final and best order. Rather, we would compute multipoint likelihoods for that order and for many permutations of it, until we were satisfied that, in fact, it was the best order. To compare ordering methods, they must be completely automated and well-defined, for, as Morton and Andrews [34, p. 264] pointed out, a nonautomated method "virtually precludes comparison of algorithms, since art rather than science gives the final map."

Although there seem to be a plethora of different strategies for ordering loci, we hope the final result will be accurate and extensive genetic linkage maps, which can be very useful in several areas [70,71].

Before marker maps existed, disease genes were tested against randomly selected markers, hoping by chance to find a linkage. Today, it is much more efficient to use linkage maps to strategically choose sets of markers that span each chromosome [71]. Indeed, it is now possible to purchase "chromosome kits," each of which contains a set of highly informative markers optimally spaced for detecting linkage to a given chromosome.

Well-defined genetic maps of markers can make it much easier to localize genes causing genetic diseases. For example, testing whether a locus maps to an interval between two flanking markers is much more powerful than testing whether the locus links to a single isolated marker [72,73]. Lander and Botstein found that, with markers every 52 cM, this interval mapping approach requires "about 40% fewer families to detect linkage to a dominant trait and about 60% [fewer] for a recessive trait" than needed by the single-marker approach [73, p. 7357].

More accurate and dense linkage maps permit more accurate genetic counseling, because such counseling often depends on the exact location of the disease gene relative to the markers in use. In addition, families are more likely to be informative for at least one marker if there are many markers available in the region of interest. For example, in Duchenne muscular dystrophy, use of several linked and localized markers has led to marked increases in diagnostic accuracy [74,75].

Linkage maps will help in detecting genetic heterogeneity because linked markers provide so much more information than solitary markers. Lander and Botstein [72,73] proposed a method for detecting genetic heterogeneity called "simultaneous search," which simultaneously analyzes a set of marker loci segregating on several families, attempting to find a set of markers for which, in each family, one of the markers is closely linked to the trait.

Ultimately, high-resolution maps enhance the chances of moving from the genetic map to the physical map and, finally, to cloning the disease gene. For instance, consider the successful story of cystic fibrosis, for which the disease gene was localized to increasingly smaller and smaller intervals by using linkage information and, eventually, cloned after combining data on linkage, linkage disequilibrium, and physical DNA clones [76–78].

ACKNOWLEDGMENTS

I thank Suzanne Leal, Thomas Lehner, Jurg Ott, and Joe Terwilliger for their helpful comments. Since portions of this chapter were taken from my dissertation [52], I also would like to thank my doctoral committee members: Rita M. Cantor, Robert W. Haile, Elliot M. Landaw, Roderick J. A. Little, and Kenneth Lange (Chair). Support by the Muscular Dystrophy Association, the W. M. Keck Foundation, and the New York State Psychiatric Institute is gratefully acknowledged.

REFERENCES

1. Jacquard, A. (1974). *The Genetic Structure of Populations.* Springer-Verlag, New York.
2. Watson, J. D., N. H. Hopkins, J. W. Roberts, J. A. Steitz, and A. M. Weiner (1987). *Molecular Biology of the Gene.* Benjamin/Cummings Publishing. Menlo Park, California.
3. Cox, D. W., M. A. Walter, S. E. Coulson, K. H. Grzeschik, H. F. Willard, R. Hill, and P. Boyd (1987). Regional localization of alpha 1-antichymotrypsin (AACT) to 14q32.1 and its proximity to alpha 1-antitrypsin (PI) by pulsed field gel electrophoresis. *Human Gene Mapping 9 (1987): Ninth International Workshop on Human Gene Mapping. Cytogent. Cell Genet. 46*:600.
4. Li, H., U. B. Gyllensten, X. Cui, R. K. Saiki, H. A. Erlich, and N. Arnheim (1988). Amplification and analysis of DNA sequences in single human sperm and diploid cells. *Nature 335*:414–417.
5. Boehnke, M., N. Arnheim, H. Li, and F. S. Collins (1989). Fine-structure genetic mapping of human chromosomes using the polymerase chain reaction on single sperm: Experimental design considerations. *Am. J. Hum. Genet. 45*:21–32.
6. Ott, J. (1985). *Analysis of Human Genetic Linkage.* Johns Hopkins University Press, Baltimore.
7. Morton, N. E. (1955). Sequential tests for the detection of linkage. *Am. J. Hum. Genet. 7*:277–318.
8. Weeks, D. E., and K. Lange (1987). Preliminary ranking procedures for multilocus ordering. *Genomics 1*:236–242.

9. Dausset, J. (1986). Le centre d'étude du polymorphisme humain. *Presse Méd.* *15*:1801–1802.
10. Donis-Keller, H., P. Green, C. Helms, S. Cartinhour, B. Weiffenbach, K. Stephens, T. P. Keith, D. W. Botstein, G. Akots, K. S. Rediker, T. Gravius, V. A. Brown, M. B. Rising, C. Parker, J. A. Powers, D. E. Watt, E. R. Kauffman, A. Bricker, P. Phipps, H. Muller-Kahle, T. R. Fulton, S. Ng, J. W. Schumm, J. C. Braman, R. G. Knowlton, D. F. Barker, S. M. Crooks, S. E. Lincoln, M. J. Daly, and J. Abrahamson (1987). A genetic linkage map of the human genome. *Cell 51*:319–337.
11. Lathrop, M., Y. Nakamura, P. Cartwright, P. O'Connell, M. Leppert, C. Jones, H. Tateishi, T. Bragg, J. M. Lalouel, and R. White (1988). A primary genetic map of markers for human chromosome 10. *Genomics 2*:157–164.
12. Nakamura, Y., M. Lathrop, P. O'Connell, M. Leppert, J. M. Lalouel, and R. White (1988). A mapped set of DNA markers for human chromosome 15. *Genomics 3*:342–346.
13. Nakamura, Y., M. Lathrop, P. O'Connell, M. Leppert, D. Barker, E. Wright, M. Skolnick, S. Kondoleon, M. Litt, J. M. Lalouel, and R. White (1988). A mapped set of DNA markers for human chromosome 17. *Genomics 2*:302–309.
14. O'Connell, P., G. M. Lathrop, Y. Nakamura, M. L. Leppert, R. H. Ardinger, J. L. Murray, J. M. Lalouel, and R. White (1989). Twenty-eight loci form a continuous linkage map of markers for human chromosome 1. *Genomics 4*:12–20.
15. Wedd, N. (1984). A computer program for constructing a maximum-likelihood map from linkage data and its application to human chromosome 1. *Ann. Hum. Genet. 48*:333–345.
16. Foulds, L. R. (1984). *Combinatorial Optimization for Undergraduates.* Springer-Verlag, New York.
17. Papadimitriou, C. H., and K. Steiglitz (1982). *Combinatorial Optimization: Algorithms and Complexity.* Prentice-Hall, Englewood Cliffs, N.J.
18. Rosenkrantz, D. J., R. E. Stearns, and P. M. Lewis II (1977). An analysis of several heuristics for the traveling salesman problem. *SIAM J. Comput. 6*:563–581.
19. Kirkpatrick, S., C. D. Gelatt, Jr., and M. P. Vecchi (1983). Optimization by simulated annealing. *Science 220*:671–680.
20. Drayna, D., and R. White (1985). The genetic linkage map of the human X chromosome. *Science 230*:753–758.
21. Kavathas, P., F. H. Bach, and R. DeMars (1980). Gamma ray-induced loss of expression of HLA and glyoxalase I alleles in lymphoblastoid cells. *Proc. Natl. Acad. Sci. USA 77*:4251–4255.
22. Leach, R., R. DeMars, S. Hasstedt, and R. White (1986). Construction of a map of the short arm of human chromosome 6. *Proc. Natl. Acad. Sci. USA* *83*:3909–3913.
23. Leppert, M., W. Cavenee, P. Callahan, T. Holm, P. O'Connell, K. Thompson, G. M. Lathrop, J. M. Lalouel, and R. White (1986). A primary genetic map of chromosome 13q. *Am. J. Hum. Genet. 39*:425–437.

24. Lathrop, G. M., J. M. Lalouel, C. Julier, and J. Ott (1984). Strategies for multilocus linkage analysis in humans. *Proc. Natl. Acad. Sci. USA 81*:3443-3446.
25. O'Connell, P., G. M. Lathrop, M. Law, M. Leppert, Y. Nakamura, M. Hoff, E. Kumlin, W. Thomas, T. Elsner, L. Ballard, P. Goodman, E. Azen, J. E. Sadler, G. Y. Cai, J. M. Lalouel, and R. White (1987). A primary genetic linkage map for human chromosome 12. *Genomics 1*:93–102.
26. Lalouel, J. M. (1977). Linkage mapping from pair-wise recombination data. *Heredity 38*:61–77.
27. Buetow, K. H., and A. Chakravarti (1987). Multipoint gene mapping using seriation I. General methods. *Am. J. Hum. Genet. 41*:180–188.
28. Buetow, K. H., and A. Chakravarti (1987). Multipoint gene mapping using seriation II. Analysis of simulated and empirical data. *Am. J. Hum. Genet. 41*: 189–201.
29. Thompson, E. A. (1987). Crossover counts and likelihood in multipoint linkage analysis. *IMA J. Math. Appl. Med. Biol. 4*:93–108.
30. Sturt, E. (1975). The use of LOD scores for the determination of the order of loci on a chromosome. *Ann. Hum. Genet. Lond. 39*:255–260.
31. Rao, D. C., B. J. Keats, J. M. Lalouel, N. E. Morton, and S. Yee (1979). A maximum likelihood map of chromosome 1. *Am. J. Hum. Genet. 31*:680–696.
32. Wilson, S. R. (1988). A major simplification in the preliminary ordering of linked loci. *Genet. Epidemiol. 5*:75–80.
33. Falk, C. T. (1989). A simple scheme for preliminary ordering of multiple loci: Application to 45 CF families. In *Multipoint Mapping and Linkage Based upon Affected Pedigree Members: Genetic Analysis Workshop 6*. R. C. Elston, M. A. Spence, S. E. Hodge, and J. W. MacCluer (eds.). Alan R. Liss, New York, pp. 17–22.
34. Morton, N. E., and V. Andrews (1989). MAP, an expert system for multiple pairwise linkage analysis. *Ann. Hum. Genet. 53*:263–269.
35. Lathrop, G. M., J. M. Lalouel, and R. L. White (1986). Construction of human linkage maps: Likelihood calculations for multilocus analysis. *Genet. Epidemiol. 3*:39–52.
36. Edwards, J. H. (1987). The locus ordering problem. *Ann. Hum. Genet. 51*: 251–258.
37. Lander, E. S., P. Green, J. Abrahamson, A. Barlow, M. Daly, S. Lincoln, and L. Newburg (1987). Mapmaker: An interactive computer package for constructing primary genetic linkage maps of experimental and natural populations. *Genomics 1*:174–181.
38. Lander, E. S., and P. Green (1987). Construction of multilocus genetic linkage maps in humans. *Proc. Natl. Acad. Sci. USA 84*:2363–2367.
39. Green, P. (1988). Rapid maximum likelihood estimation. (Draft).
40. Guttman, L. A. (1968). A general nonmetric technique for finding the smallest coordinate space for a configuration of points. *Psychometrika 33*:469–506.

41. Schiffman, S. S., M. L. Reynolds, and F. W. Young (1981). *Introduction to Multidimensional Scaling*. Academic Press, New York.
42. Coxon, A. P. M. (1982). *The User's Guide to Multidimensional Scaling*. Heinemann Educational Books, Exeter, N.H.
43. Young, F. W. (1987). *Multidimensional Scaling: History, Theory, and Applications*. R. M. Hamer (ed.). Lawrence Erlbaum Associates, Hillsdale, N.J.
44. Williams, W. R., N. E. Morton, R. Lew, and S. Yee (1979). The likely region of overlap (LRO) method for physical assignment of loci. *Hum. Genet. 47*:297–304.
45. Keats, B. J. B., N. E. Morton, and D. C. Rao (1981). Reduction of physical assignments to a standard LOD table: Chromosome 1. *Hum. Genet. 56*:353–359.
46. Sherman, S. L., J. King, E. B. Robson, and S. Yee (1984). A revised map of chromosome 1. *Ann. Hum. Genet. 48*:243–251.
47. Kammerer, C. M., and J. W. MacCluer (1988). Empirical power of three preliminary methods for ordering loci. *Am. J. Hum. Genet. 43*:964–970.
48. Buetow, K. H., D. Y. Nishimura, P. Green, Y. Nakamura, M. Schull, J. L. Weber, and J. C. Murray (1989). A multipoint genetic map and new RFLPs for human chromosome 1. *Human Gene Mapping 10 (1989): Tenth International Workshop on Human Gene Mapping. Cytogenet. Cell Genet. 51*:972
49. Hazout, S., M. C. Babron, and F. Clerget-Darpoux (1989). Pairwise permutation algorithm for ordering loci. In *Multipoint Mapping and Linkage Based upon Affected Pedigree Members: Genetic Analysis Workshop 6*. R. C. Elston, M. A. Spence, S. E. Hodge, and J. W. MacCluer (eds.). Alan R. Liss, New York, pp. 35–39.
50. Press, W. H., B. P. Flannery, S. A. Teukolsky, and W. T. Vetterling (1986). *Numerical Recipes*. Cambridge University Press, New York.
51. Golden, B. L., and W. R. Stewart (1985). Empirical analysis of heuristics. In *The Traveling Salesman Problem: A Guided Tour of Combinatorial Optimization*. E. L. Lawler, J. K. Lenstra, A. H. G. R. Kan, and D. B. Shmoys (eds.). John Wiley & Sons, New York.
52. Weeks, D. E. (1988). New mathematical methods for human gene mapping [Dissertation]. University of California, Los Angeles.
53. Kosambi, D. D. (1944). The estimation of map distances from recombination values. *Ann. Eugen. 12*:172–175.
54. Lange, K., and D. E. Weeks (1989). Efficient computation of LOD scores: Genotype elimination, genotype redefinition, and hybrid maximum likelihood algorithms. *Ann. Hum. Genet. 53*:67–83.
55. Lange, K., D. Weeks, and M. Boehnke (1988). Programs for pedigree analysis: MENDEL, FISHER, and dGENE. *Genet. Epidemiol. 5*:471–472.
56. Edwards, J. H. (1971). The analysis of X-linkage. *Ann. Hum. Genet. 34*:229–250.
57. Edwards, J. H. (1976). The interpretation of LODs in linkage analysis. In *Human Gene Mapping 3*. D. Bergsma (ed.). Birth Defects: Original Article Series Vol. XII, No. 7, The National Foundation–March of Dimes, S. Karger, Basel.

58. Bruns, G. A. P., and S. L. Sherman (1989). Report of the Committee on the Genetic Constitution of Chromosome 1. In *Human Gene Mapping 10 (1989): Tenth International Workshop on Human Gene Mapping. Cytogenet. Cell Genet. 51*:67–90.
59. Bishop, D. T. (1985). The information content of phase-known matings for ordering genetic loci. *Genet. Epidemiol. 2*:349–361.
60. Ott, J., and G. M. Lathrop (1987). Goodness-of-fit tests for locus order in three-point mapping. *Genet. Epidemiol. 4*:51–57.
61. Olson, J. M., and M. Boehnke (1989). Comparison of preliminary multilocus ordering schemes. *Am. J. Hum. Genet. 45 (Suppl)*:A155.
62. Olson, J. M., and M. Boehnke (1990). Monte Carlo comparison of preliminary methods for ordering multiple genetic loci. *Am. J. Hum. Genet. 47*:470–482.
63. Knapp, M., M. Neugebauer, R. Fimmers, S. A. Seuchter, and M. P. Baur (1989). Preliminary ordering of multipoint linkage data. In *Multipoint Mapping and Linkage Based upon Affected Pedigree Members: Genetic Analysis Workshop 6*. R. C. Elston, M. A. Spence, S. E. Hodge, and J. W. MacCluer (eds.). Alan R. Liss, New York, pp. 41–46.
64. Botstein, D., R. L. White, M. Skolnick, and R. W. Davis (1980). Construction of a genetic linkage map in man using restriction fragment length polymorphisms. *Am. J. Hum. Genet. 32*:314–331.
65. Thompson, E. A. (1984). Information gain in joint linkage analysis. *IMA J. Math. Appl. Med. Biol. 1*:31–49.
66. Elston, R. C., and J. Stewart (1971). A general model for the genetic analysis of pedigree data. *Hum. Heredity 21*:523–542.
67. Ott, J. (1974). Estimation of the recombination fraction in human pedigrees: Efficient computation of the likelihood for human linkage studies. *Am. J. Genet. 26*:588–597.
68. Lange, K., and M. Boehnke (1983). Extensions to pedigree analysis. V. Optimal calculation of mendelian likelihoods. *Hum. Heredity 33*:291–301.
69. Lange, K., and T. M. Goradia (1987). An algorithm for automatic genotype elimination. *Am. J. Hum. Genet. 40*:250–256.
70. Spence, M. A. (1987). Genetic linkage: Sampling issues and multipoint mapping. *J. Psychiat. Res. 21*:631–637.
71. White, R., and J. M. Lalouel (1988). Sets of linked genetic markers for human chromosomes. *Annu. Rev. Genet. 22*:259–279.
72. Lander, E. S., and D. Botstein (1986). Mapping complex genetic traits in humans: New methods using a complete RFLP linkage map. *Cold Spring Harbor Symp. Quant. Biol. 51*:49–62.
73. Lander, E. S., and D. Botstein (1986). Strategies for studying heterogeneous genetic traits in humans by using a linkage map of restriction fragment length polymorphisms. *Proc. Natl. Acad. Sci. USA 83*:7353–7357.
74. Hejtmancik, J. F., S. G. Harris, C. C. Tsao, P. A. Ward, and C. T. Caskey (1986). Carrier diagnosis of Duchenne muscular dystrophy using restriction fragment length polymorphisms. *Neurology 36*:1553–1562.
75. Caskey, C. T., P. Ward, and F. Hejtmancik (1988). DMD carrier detection and prenatal diagnosis via recombinant DNA methods. In *Advances in Neurology,*

Vol. 48: Molecular Genetics of Neurological and Neuromuscular Disease. S. DiDonato et al. (eds.). Raven Press, New York.

76. Kerem, B., J. M. Rommens, J. A. Buchanan, D. Markiewicz, T. K. Cox, A. Chakravarti, M. Buchwald, and L. Tsui (1989). Identification of the cystic fibrosis gene: Genetic analysis. *Science 245*:1073–1080.
77. Riordan, J. R., J. M. Rommens, B. Kerem, N. Alon, R. Rozmahel, Z. Grzelczak, J. Zielenski, S. Lok, N. Plavsic, J. Chou, M. L. Drumm, M. C. Iannuzzi, F. S. Collins, and L. Tsui (1989). Identification of the cystic fibrosis gene: Cloning and characterization of complementary DNA. *Science 245*:1066–1073.
78. Rommens, J. M., M. C. Iannuzzi, B. Kerem, M. L. Drumm, G. Melmer, M. Dean, R. Rozmahel, J. L. Cole, D. Kennedy, H. Hidaka, M. Zsiga, M. Buchwald, J. R. Riordan, L. Tsui, and F. S. Collins (1989). Identification of the cystic fibrosis gene: Chromosome walking and jumping. *Science 245*:1059–1065.

III

CELLULAR TECHNIQUES: INTRODUCTION

Advanced methods to investigate chromosomes at the cellular level have recently been developed. These procedures have been introduced for studies of both chromosome structure and chromosome function. Especially fruitful has been the application of molecular biology approaches to research in cell and developmental biology.

Among cell biologists, there is a growing appreciation of the significance of the three-dimensional (3-D) structure of the cell. This appreciation is similar to the recognition by biochemists that the 3-D structure of enzymes, solved by x-ray crystallography, determines enzyme function. Understanding the 3-D arrangement of chromosomes in the intact cell is important because chromosome function is related to organization. DNA replication, gene transcription, and oncogene activation may all be involved with higher-order nuclear structure. Investigation of 3-D cellular structure has been made possible by advanced methodology in electron and light microscopy and in image analysis and computer graphics. As Chapter 14 shows, the 3-D organization of chromosomes can now be rigorously characterized, and this will allow further detailed studies of the influence of chromosome spatial organization on function.

The production of transgenic animals demonstrates the power of utilizing molecular techniques, such as those discussed in Part I, in problems of developmental biology. Transgenic animals are created by the incorporation of foreign genes, usually by the microinjection of purified DNA into the cells of embryos. Chapter 15 reviews studies concerned with questions such as the expression in transgenic mice of introduced immunoglobulin, globin, and viral genes. Other new procedures are also mentioned that deal with gamete and embryo micromanipulation. A topic receiving attention is the use of microsurgery of eggs to improve in vitro fertilization methods.

Testing the genotoxic effects of environmental mutagens is a major biomedical concern. The recognition that exposure to chemical mutagens is related to cancer has made environmental mutagenesis an active area of research. A chapter, therefore, is concerned with protocols employing mammalian cells to evaluate genotoxic agents. The advantages of human cells and other mammalian cells for identifying chemicals that induce chromosomal aberrations (clastogens) are outlined.

The karyotypic analysis of solid tumors is the subject of the chapter that follows. Methods are described to improve culture conditions of primary solid tumor cells and, thereby, obtain a greater number of metaphase cells for karyotyping. The cytogenetic and molecular characterization of chromosomal abnormalities found with solid tumors, such as lung cancer, lags behind that for hematologic malignancies, particularly leukemia.

Research on the structure and function of mitotic chromosomes requires the availability of workable amounts of mitotic cells. Colchicine is a plant alkaloid that has been widely employed to arrest dividing cells in mitosis by preventing the formation of spindle microtubules. The concluding chapter in this section presents a practical guide for the use of colchicine and its derivative demecolcine (Colcemid).

14

Arrangement of Chromosomes in Human Mitotic Cells: Three-Dimensional Computer Reconstructions

Kenneth W. Adolph

University of Minnesota Medical School, Minneapolis, Minnesota

I. INTRODUCTION

During mitosis, the condensation of chromosomes and their arrangement for distribution to daughter cells produces the highest level of chromosome organization in the cell division cycle. Understanding chromosome structural changes during cell division is one of the fundamental challenges of modern biological research.

Because chromosome function is related to organization, it is particularly important to determine the three-dimensional (3-D) arrangement of chromosomes. For example, a common phenomenon may be the exchange of genetic material between chromosomes, and such exchanges could have a role in oncogene activation and other similar events. In addition, the higher-order structure of the interphase nucleus has a major role in DNA replication and gene transcription, and nuclear genome structure is likely to be related to the spatial ordering of mitotic chromosomes.

Production of the 3-D reconstructions required three stages. Thin sections through the distribution of chromosomes were first prepared and electron micrographs taken. Serial sections were obtained from metaphase, anaphase, and telophase cells. The micrographs were then digitized, contours of the chromosome boundaries were computed, and the contours in different sections were aligned to give "wire frame" 3-D images. Solid surfaces of the chromosomes were then reconstructed by applying advanced computer graphics software.

With this approach, the spatial arrangement of HeLa chromosomes during the basic stages of cell division were determined. The locations and configurations of chromosomes were shown by the reconstructions to depend upon their sizes. During metaphase, chromosomes with the longest arms are distributed around the perimeter of the metaphase plate. Each chromosome arm is directed outward. A location near the center of the plate is observed for the smaller chromosomes. The organization of chromosomes as a plate is also found in the anaphase arrangement of chromosomes and the structure of the newly formed telophase nucleus. These results are probably general features of chromosome packing in mitotic cells.

II. MATERIALS AND METHODS

A. Preparation of Samples for Electron Microscopy

HeLa cells were grown as monolayers in minimum essential medium (MEM) with 5% fetal bovine serum added. Mechanical detachment ("mitotic shake-off") was used to obtain mitotic cells. Samples were prepared for electron microscopy (EM) by standard procedures. Cells in warm growth medium were fixed with glutaraldehyde (0.8%) and then further treated with osmium tetroxide (1%) [1–3]. Quetol 651 was used as the embedding medium [A. Abad and R. Kuehn, personal communication]. The embedding medium was cured as a thin layer and small squares were mounted for sectioning.

B. Thin Sectioning and Electron Microscopy

A Reichert Ultracut E microtome with a diamond knife was employed for serial sectioning of the Quetol-embedded cells. The section thickness was about 100 nm. Single-slot grids, covered with formvar–carbon films, were used to collect the ribbons of sections. The sections were stained with uranyl acetate in water and, usually, with lead citrate. A Hitachi H600 electron microscope, operated at 50 kV and at a magnification of 5000×, was utilized to record the images. The EM film was DuPont Cronalar EM-7.

C. Image Processing

Electron micrographs on a light box were digitized using a Dage MTI68 Newvicon video camera coupled to a Model 75 image processor (International Imaging Systems, Milpitas, Calif.) and a Masscomp MC535 minicomputer. Grey-level images of 512 × 512 × 8 bits/pixel were obtained. The registration capability of the image-processing system was used to align each section, as it was displayed on a video monitor, with the preceding section that was already digitized. A radiometric transform was applied to each digitized image to enhance contrast. The image was then thresholded to isolate the chromosomes. Each thresholded image was subject to a rectangular 3 × 3-median filter operation [4]. This procedure removes single pixel noise and, without dilation or erosion, slightly smoothes the object boundaries. A boundary tracking or "contouring" program was used to extract the chromosome contours. These were stored as a 3-D database in a compact chain-coded format including the x, y, z coordinates of the contours [5].

D. Three-Dimensional Computer Graphics

To reconstruct chromosomes and nuclei, Wavefront 3-D modeling and rendering software (Wavefront Technologies, Santa Barbara, Calif.) running on an IRIS 2400T graphics workstation (Silicon Graphics, Mountain View, Calif.) was used. After the 3-D database of contours was input, the entire set of contours could be viewed interactively on the IRIS workstation at any chosen angle and scale. In this way, the alignment of sections could be slightly corrected and the contours edited to rectify artifacts. Contours in different sections were then connected with a polygonal mesh, using the Wavefront software, for representation of chromosomes and nuclei as solid surfaces. For the actual reconstructions, two methods were utilized. The first procedure of 3-D representation generated a surface through "extruding" each section as a thick slice to abut the following section. The second procedure created a more complete and realistic model by connecting vertices of neighboring contours with the polygonal mesh. Individual metaphase chromosomes and telophase nuclei were reconstructed with the second method; the metaphase plate and anaphase structure were reconstructed using the thick-slice method. The colors and reflective properties of the surfaces were chosen simply for viewing purposes. A Dunn Instruments Multicolor 35-mm film recorder was utilized to capture the reconstructed images.

III. RESULTS

A. Electron Microscopy of Thin Sections

An electron micrograph selected from serial sections through a metaphase HeLa cell is shown in Figure 1. The section is close to a middle of the plate. The micrograph indicates that the larger chromosomes are distributed around the perimeter of the plate; chromosomes near the center of the plate are generally shorter. In addition, the larger chromosomes are not extended

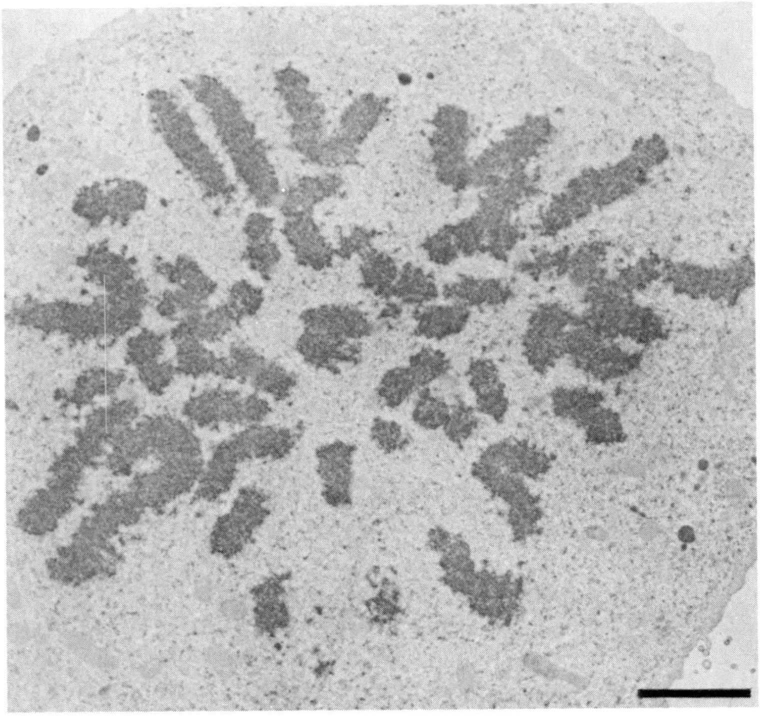

Figure 1 Electron micrograph from a consecutive series of thin sections through a metaphase HeLa cell: The thin section intersects the face of the metaphase plate; a section near the middle of the plate is shown. The micrograph indicates that the larger chromosomes are situated around the perimeter of the plate and the chromosome arms are radially arranged. The magnification bar represents 2.0 μm.

X-shaped structures. Each chromosome arm is V-shaped because the arms are folded at the centromeres.

Figure 2 contains electron micrographs that display in greater detail the configuration of an individual metaphase chromosome. The micrographs reveal the unusual bent shape of metaphase chromosomes within the living cell. These 100-nm thick sections provide the necessary resolution for reconstructing individual chromosomes with 3-D computer graphics.

Consecutive sections cut into the side of a metaphase plate, about 90° to those in Figure 1, further show the larger chromosomes to be arranged around the periphery with the smaller chromosomes toward the center (Fig. 3A). The shorter chromosomes may have some preferential orientation normal to the plate.

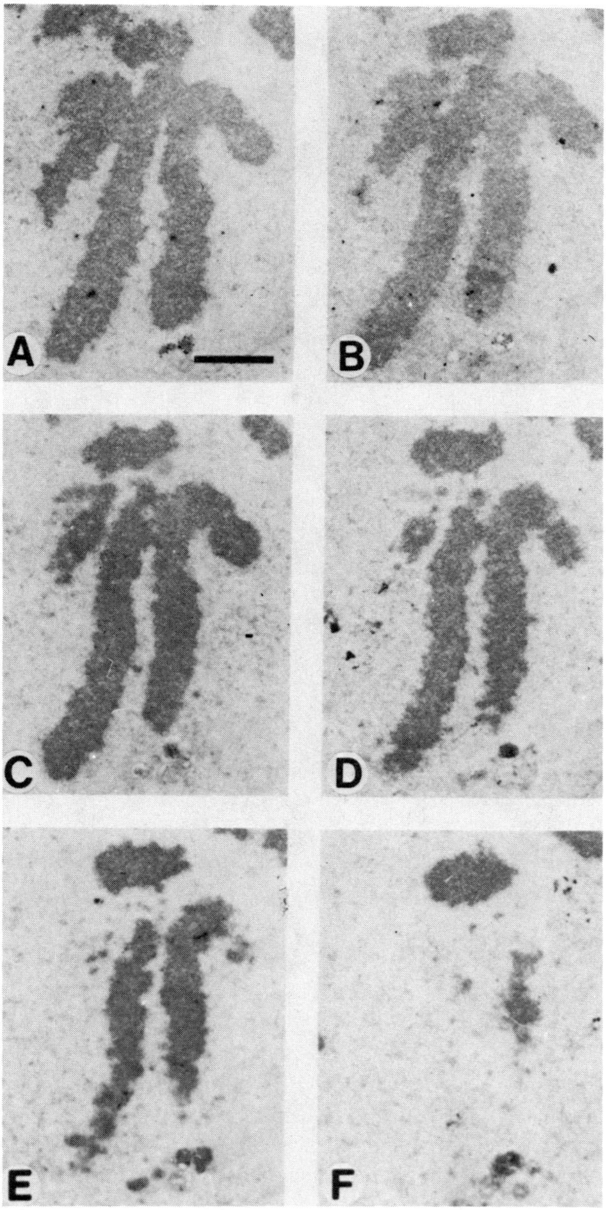

Figure 2 Serial sections through a particular HeLa metaphase chromosome: The 100-nm thick sections were prepared as described in Materials and Methods. The example shown is one of the large chromosomes located near the perimeter of the metaphase plate. The chromatid arms are bent at the centromere region so that the telomeres are directed radially outward. Bar, 1.0 μm.

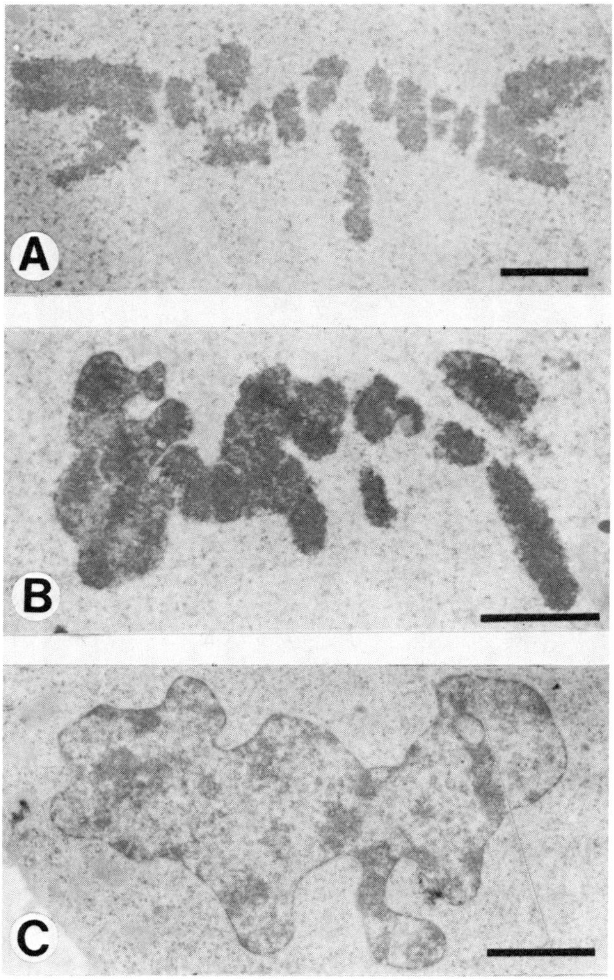

Figure 3 Characteristic electron micrographs of serially sectioned metaphase (A), anaphase (B), and telophase (C) cells: The section in each case intersects the edge of the platelike chromosome structure. The micrographs illustrate the morphological changes that accompany cell division. Individual metaphase chromosomes can be distinguished in (A), but in (C) the nuclear envelope completely surrounds the mass of decondensed telophase chromosomes. Bar, 2.0 μm.

Determining changes in chromosome organization during the cell division cycle was another major purpose of this investigation. Electron micrographs of thin sections through anaphase cells demonstrate that chromosomes are closely packed and often in extensive contact (see Fig. 3B). The beginning of nuclear envelope formation around the chromosomes gives a sharp definition to the chromosome boundaries. Because daughter chromosomes are separating on the mitotic spindle, the anaphase structure has a characteristic rounded appearance with trailing chromosome arms.

In late telophase the nuclear envelope has coalesced to form a continuous structure around the decondensing chromosomes (see Fig. 3C). The electron micrograph shows that the nucleus is not spherical, but has a disklike structural appearance. This shape, which is determined by the plate of metaphase and anaphase chromosomes, has therefore been preserved through each stage of mitosis. The telophase nucleus, about 4-μm thick and 14 μm in diameter, is not perfectly flat, but has undulating faces and an irregular margin.

B. Image Processing: Contouring and Alignment of Chromosomes

Stacked sections for metaphase and telophase cells are included in Figures 4 and 5. Distortion of sections was not a significant problem, and the contours are well registered and progress smoothly between sections. The greatest challenge was resolving individual metaphase chromosomes, because these chromosomes are densely packed. But, as the aligned sections illustrate, the procedure is capable of resolving structural details of individual chromosomes and nuclei.

C. Three-Dimensional Reconstructions

To obtain accurate and directly perceivable representations of chromosome changes during cell division, solid models were computed. The 3-D reconstructions will also allow precise measurements of the dimensions, volumes, shapes, and positions of chromosomes through mitosis. And the results could serve as the basis of studies involving the location of chromosome regions, DNA sequences, and so forth.

For metaphase cells, the aim was to determine the structure of individual chromosomes as well as their spatial position and orientation. Anaphase chromosomes are in contact, dispersing, and being coated by the nuclear membrane, so that reconstructing the overall structure of the chromosome cluster was the goal for this stage of mitosis. The nuclear membrane completely surrounds the entire disk of chromosomes in telophase cells; the reconstructions were thus carried out to relate the telophase protonucleus to the metaphase and anaphase structures.

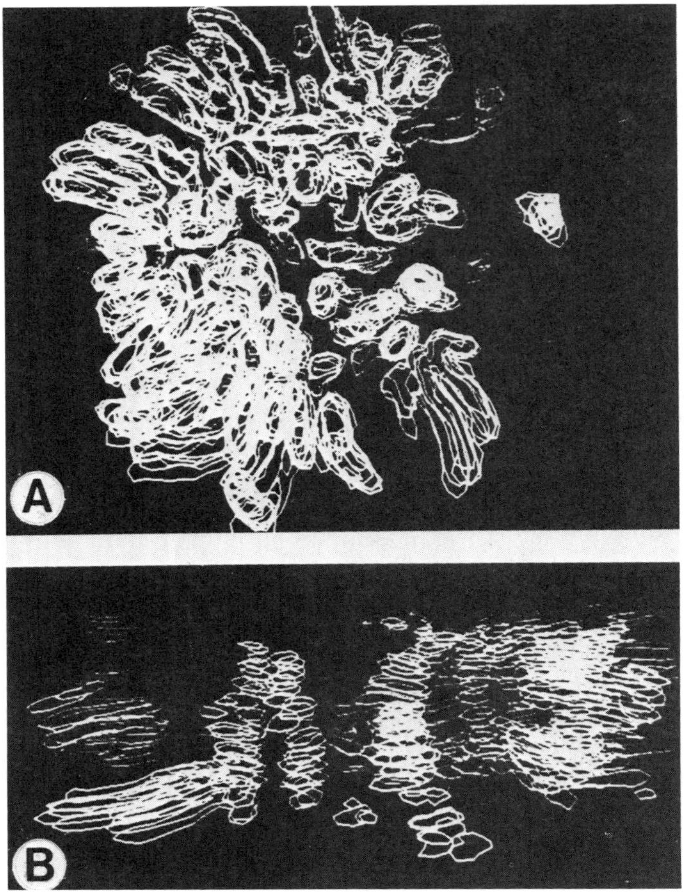

Figure 4 Computer images of digitized and contoured electron micrographs of a serially sectioned metaphase plate: The contours of individual chromosomes in each section were generated from the digitized images by utilizing a boundary-tracking program. Two different views of the stacked sections, rotated almost 90° about a horizontal axis, are shown in (A) and (B). The larger chromosomes at the periphery of the metaphase plate are most clearly recognizable.

Figures 6–8 show representative 3-D reconstructions of chromosomes and nuclei. Chromosomes assembled to form the metaphase plate have unique spatial configurations, as Figure 6 demonstrates. The large chromosomes do not exist as extended structures, but are arranged such that the telomeres are all in the same radial direction. The longer arm of each chromatid has its

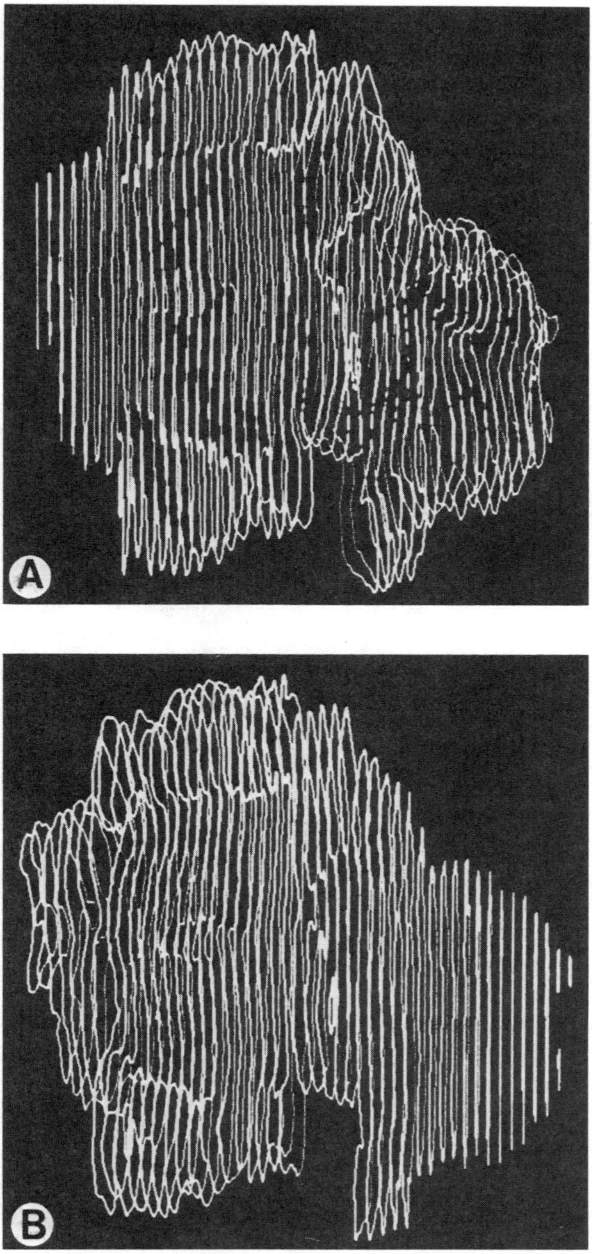

Figure 5 Aligned contours of a telophase nucleus: The two views are of the same nucleus rotated about a vertical axis. The computer images display every third section of the set of consecutive sections. The flattened and irregular surfaces of the telophase nucleus can be discerned.

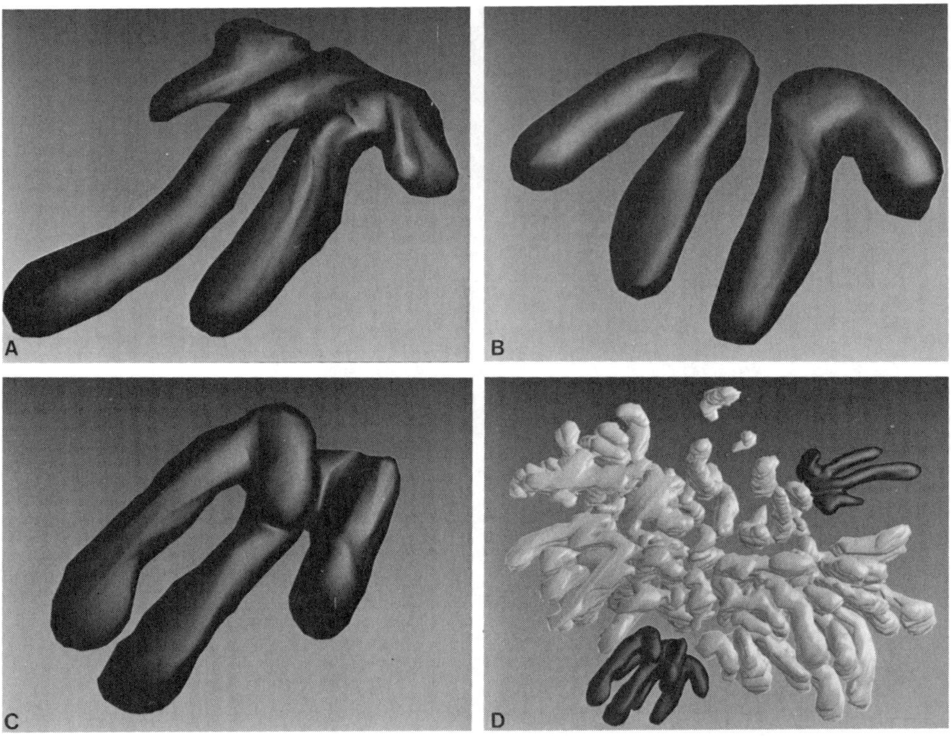

Figure 6 Three-dimensional computer reconstructions of individual metaphase chromosomes (A–C) and the metaphase plate containing the chromosomes (D): These large chromosomes are located at the periphery of the plate and have their arms bent outward at the centromeres. In (D), chromosome C partly covers chromosome B. (Copyright 1988 Kenneth W. Adolph)

centromere proximal and its telomere distal to the center of the metaphase plate. The shorter arms are folded outward, in the same direction as the longer arms, with their telomeres also distal to the center of the plate. Each of the chromosomes in panels A, B, and C display these structural features. (The examples were chosen because the relative lengths of the two arms are different in each.) This unexpected morphologic arrangement is probably a consequence of the dense packing of chromosomes in the metaphase plate and the flexibility of the chromosomes.

Two other observations on the reconstructed chromosomes in Figure 6 are worth noting. In (B) and (C), the sister chromatids of the chromosomes are beginning to separate with the approach of anaphase. And the contraction

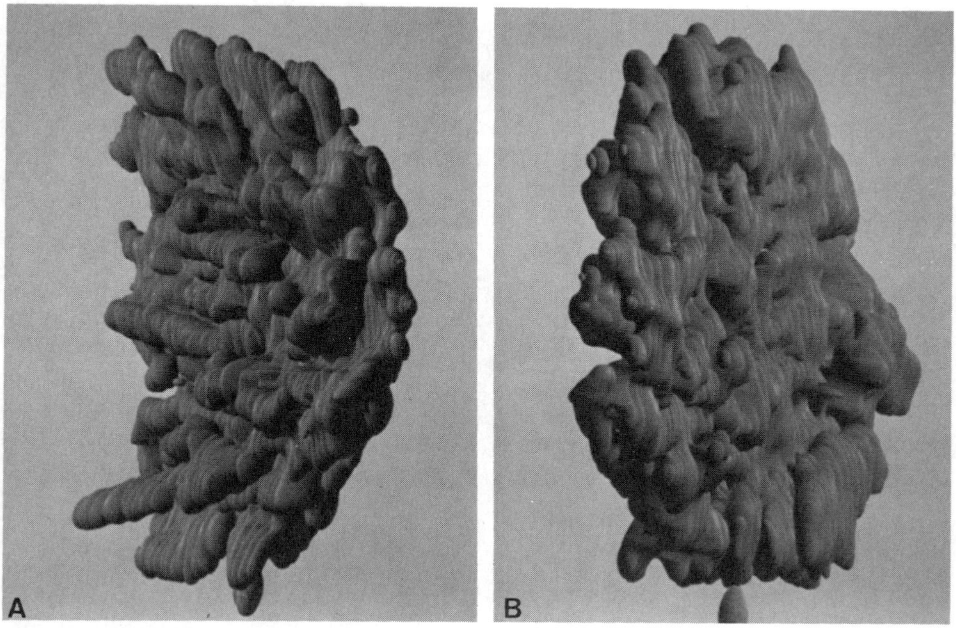

Figure 7 The anaphase arrangement of chromosomes as shown by 3-D reconstructions: The chromosome cluster is disk shaped, and the side seen in (A), away from the direction of movement on the mitotic spindle, is characterized by chromosomes that project from the anaphase configuration. The leading side (B) is flat with fewer structural features. The anaphase disk in A is rotated, from edge-on, 55° counterclockwise about a vertical axis, whereas the rotation in B is 305°. (Copyright 1988 Kenneth W. Adolph)

and lengthening of the two chromatids of A, B, and C are not precisely synchronous through metaphase: the chromatids are seen to have different lengths.

The positions of the three chromosomes in the metaphase plate are shown in Figure 6D. In the panel, the reconstructed chromosomes have been imported into a thick-slice representation of the plate. (Chromosome B is partly hidden, since it is beneath C.) The chromosomes have the peripheral location and radial orientation characteristic of large chromosomes. The length of the chromosome arms is thus observed to be the primary factor that determines the position and configuration of chromosomes in the plate.

The arrangement of anaphase chromosomes is substantially different from that in metaphase because sister chromatids have separated and the

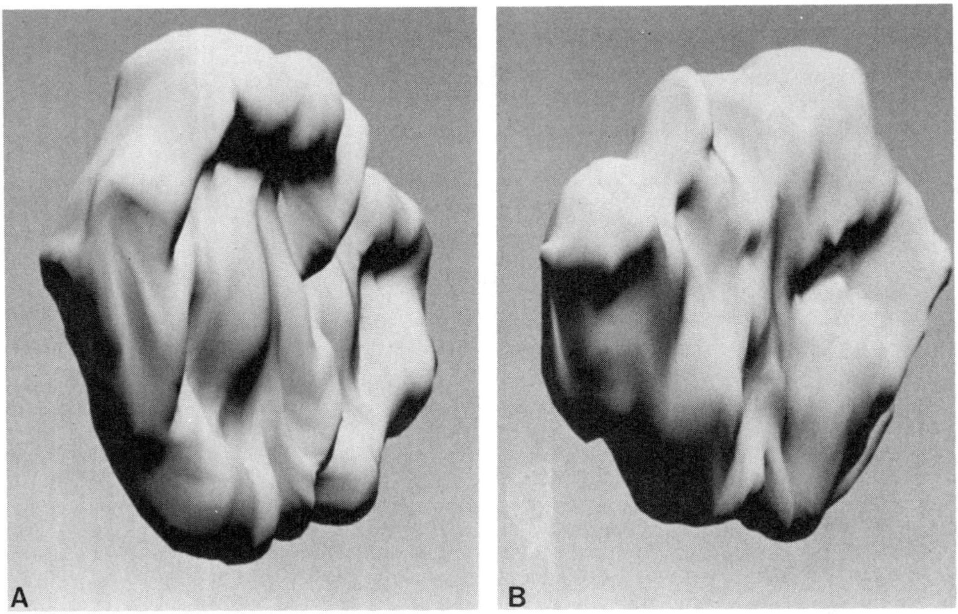

Figure 8 Reconstructions of a telophase nucleus: The nucleus is a disk structure with an irregular margin and indentations and bulges on its faces. The stage of cell division of the nucleus is precisely known, since electron micrographs show that the cell is still connected to its sister at the midbody. In A, the nucleus is rotated around a vertical axis 120° counterclockwise from edge on. In B, a further 180° rotation, to 300°, shows the other side of the same nucleus. (Copyright 1988 Kenneth W. Adolph)

chromosomes are associated with the mitotic spindle. A more amorphous distribution of chromosomes is observed in anaphase, with chromosomes touching and serving as templates for coating by the nuclear membrane. The 3-D reconstructions included in Figure 7 reveal that the grouping of chromosomes retains the plate or disk arrangement established in metaphase. The two sides of the structure have different appearances. The side closest to the center of the cell (see Fig. 7A) is characterized by chromosomes projecting from the disk. The other side (see Fig. 7B), facing the direction of movement on the spindle, is flatter with few projecting chromosomes. Even though the metaphase distribution of chromosomes has been deformed, the relative positions of chromosomes should be unchanged.

The telophase nucleus is not a simple spherical structure, but has a complex external morphology (Fig. 8). The boundary of the nucleus at this stage of cell division has an appearance like that of the metaphase and anaphase structures. The disklike telophase structure has similar overall dimensions, and the bulges and furrows reflect the positions of anaphase chromosomes. The reconstructions in panels A and B are of the same nucleus rotated 180°, and illustrate the typical features of the two sides. The irregular nuclear perimeter and faces are characteristic. Telophase nuclei are also distinguished by variability, even between sister nuclei, since they are dynamic structures that have started the process of rounding-up into the spherical shape of mature nuclei.

IV. DISCUSSION

This investigation was concerned with determining the configuration of chromosomes and the mode of packing in metaphase, and the structural changes in anaphase and telophase. The reconstructions were completed by the computation of solid models representing the chromosomes and nuclei. This volume structuring will facilitate further studies of the configuration, substructure, volume, and identity of chromosomes.

A major benefit of using thin sections for the reconstructions is that the in vivo distribution of chromosomes is preserved. This is because living, intact cells are fixed for electron microscopy. The high resolution of EM will be an advantage for additional experiments concerned with the locations of specific genes and proteins. The computer procedure employed for image processing avoids manual tracing of the chromosome boundaries, and thus qualitative decisions are minimized during the reconstruction process.

Attaining the maximum density of chromosome packing would account for the arrangement of the larger chromosomes around the perimeter of the metaphase plate and their outward orientation. The longer, unwieldy chromosome arms would disrupt the packing less by protruding into the more available volume at the periphery. Similar considerations will probably govern chromosome organization in other types of cells that have different numbers, sizes, and shapes of chromosomes. Other factors that would influence the spatial arrangement of chromosomes are the possible existence of specific DNA connections between chromosomes and the requirement for correct attachment to kinetochore microtubules.

The structure of the metaphase plate continues to be mirrored in the organization of chromosomes in anaphase and telophase. A platelike structure remains present through anaphase as the separated chromosomes move toward the spindle poles. More surprising is the morphologic appearance of the telophase nucleus. The nucleus is shaped as a somewhat irregular disk following nuclear envelope formation around the chromosomes.

The 3-D reconstructions should be considered in the light of recent results concerning eukaryotic chromosome structure and function. Research advances have been particularly noteworthy for the DNA and protein components of chromosomes. Active areas include the role of DNA-binding proteins [6], topological properties of DNA [7], and the analysis of gene sequences [8]. Studies of the cell biology of meiosis [9–11], X and Y chromosomes [12,13], and mitosis [14] are also yielding exciting data. Moreover, fundamental information continues to be provided by special eukaryotic systems such as polytene chromosomes [15] and yeast [16]. The 3-D reconstructions of chromosomes in other mammalian cells have not yet been reported. Mitotic cells of grasses have been reconstructed from electron micrographs [17], and *Drosophila melanogaster* polytene chromosomes have been investigated by optical sectioning microscopy [18].

The reconstructions show that the arrangements of chromosomes in human mitotic cells are complex, yet they possess a high degree of organization. This information is important because it should have fundamental implications for understanding chromosome function.

V. SUMMARY

Three-dimensional (3-D) computer reconstruction techniques were applied to determine the arrangement of chromosomes in mitotic HeLa cells. Metaphase, anaphase, and telophase cells were fixed for electron microscopy and consecutive thin sections were obtained. This procedure preserved the distributions of chromosomes in intact cells without distortion. Electron micrographs of the serial sections revealed that the positions and orientations of metaphase chromosomes depend on their sizes and shapes. The larger chromosomes are located around the periphery of the metaphase plate with the chromosome arms directed radially outward. The 3-D reconstructions produced by image processing and computer graphics extended these observations. Solid models of chromosomes were computed after electron micrographs were digitized, chromosome boundaries were contoured, and sections were aligned. The reconstructions established the peripheral distribution and radial orientation of the larger chromosomes. The arms of these chromosomes (chromosomes 1, 2, 3 . . .) are bent back, making the chromatids V-shaped, with the telomeres pointing outward. The smaller chromosomes (22, 21, 20 . . .) are found closer to the center of the plate. Changes in the arrangements of chromosomes in anaphase and telophase were also determined by the 3-D reconstruction procedure. The anaphase configuration of chromosomes is a disk-shaped structure, one side of which, trailing the direction of movement on the spindle, is characterized by protruding chromosome arms. The other side is relatively flat and featureless. The reconstructed telophase nucleus also has a disklike structure, with an uneven perimeter and faces with bulging and concave regions. Telophase nuclear morphology thus continues to reflect the arrangement of chromosomes in metaphase and anaphase.

ACKNOWLEDGMENTS

The reconstructions shown in Figures 6, 7, and 8 were carried out by Dr. C. K. Knox of the Biomedical Image Processing Laboratory, Department of Cell Biology and Neuroanatomy. The expertise and advice of A. Abad and R. Kuehn in serial sectioning and electron microscopy greatly benefited this investigation. A grant from the National Institutes of Health made available the facilities of the Biomedical Image Processing Laboratory. Financial support for this study was provided by a grant from the Graduate School of the University of Minnesota.

REFERENCES

1. Adolph, K. W. (1988). Arrangement of chromatin fibers in metaphase chromosomes. In *Chromosomes and Chromatin*. K. W. Adolph (ed.). CRC Press, Boca Raton, Fla., pp. 3–27.
2. Adolph, K. W. (1981). A serial sectioning study of the structure of human mitotic chromosomes. *Eur. J. Cell Biol. 24*:146–153.
3. Adolph, K. W., L. R. Kreisman, and R. L. Kuehn (1986). Assembly of chromatin fibers into metaphase chromosomes analyzed by transmission electron microscopy and scanning electron microscopy. *Biophys. J. 49*:221–231.
4. Rosenfeld, A., and C. K. Avinash (1982). *Digital Picture Processing*. Academic Press, New York, pp. 1–435.
5. Castleman, K. R. (1979). *Digital Image Processing*. Prentice-Hall, New York, pp. 1–429.
6. Goodwin, G. H., R. H. Nicolas, P. N. Cockerill, S. Zavou, and C. A. Wright (1985). The effect of salt extraction on the structure of transcriptionally active genes; evidence for a DNAse I-sensitive structure which could be dependent on chromatin structure at levels higher than the 30-nm fiber. *Nucleic Acids Res. 13*:3561–3579.
7. White, J. H., N. R. Cozzarelli, and W. R. Bauer (1988). Helical repeat and linking number of surface-wrapped DNA. *Science 241*:323–327.
8. Schmid, C. W., and W. R. Jelinek (1982). The *Alu* family of dispersed repetitive sequences. *Science 216*:1065–1070.
9. Hotta, Y., and H. Stern (1984). The organization of DNA segments undergoing repair synthesis during pachytene. *Chromosoma 89*:127–137.
10. Busby, S., and A. H. Bakken (1980). Transcription in developing sea urchins: Electron microscopic analysis of cleavage, gastrula and prism stages. *Chromosoma 79*:85–104.
11. Risley, M. S., S. Einheber, and D. A. Bumcrot (1986). Changes in DNA topology during spermatogenesis. *Chromosoma 94*:217–227.

12. Gartler, S. A., and A. D. Riggs (1983). Mammalian X-chromosome inactivation. *Annu. Rev. Genet. 17*:155–190.
13. Hennig, W. (1985). Y chromosome function and spermatogenesis in *Drosophila hydei. Adv. Genet. 23*:179–234.
14. Balczon, R. D., and B. R. Brinkley (1987). Tubulin interaction with kinetochore proteins: Analysis by in vitro assembly and chemical cross-linking. *J. Cell Biol. 105*:855–862.
15. Jamrich, M. (1986). Polytene chromosomes. In *Chromosome Structure and Function*. M. S. Risley (ed.). Van Nostrand Reinhold, New York, pp. 221–242.
16. Zakian, V. A., H. M. Blanton, and L. Wetzel (1986). Distribution of telomere associated sequences in yeast. In *Extrachromosomal Elements in Lower Eukaryotes*. R. B. Wicker, A. Hinnebusch, I. C. Gunsalus, A. M. Lambowitz, and A. Hollaender (eds.). Plenum Press, New York, pp. 493–498.
17. Heslop-Harrison, J. S., J. B. Smith, and M. D. Bennett (1988). The absence of the somatic association of centromeres of homologous chromosomes in grass mitotic chromosomes. *Chromosoma 96*:119–131.
18. Hochstrasser, M., D. Mathog, Y. Gruenbaum, H. Saumweber, and J. W. Sedat (1986). Spatial organization of chromosomes in the salivary gland nuclei of *Drosophila melanogaster. J. Cell Biol. 102*:112–123.

15

Production of Transgenic Animals and Other New Techniques of Gamete and Embryo Micromanipulation

Jon W. Gordon

Mt. Sinai School of Medicine, New York, New York

I. INTRODUCTION

In recent decades, advances in several areas of research have made possible new and powerful approaches to the study of mammalian fertilization and development. In vitro culture techniques have developed to the point at which it is now possible to fertilize oocytes from a large number of mammalian species, including humans, in the petri dish, reimplant the zygotes, and obtain normal live births. Moreover, fertilized eggs of many species can be cultured in vitro to the blastocyst stage, then reimplanted without significant loss of viability. Preimplantation embryos can also be grown under conditions that allow establishment of permanent cell lines, known as embryonic stem (ES) cells. Although these cells can be cultured in vitro for apparently indefinite periods and manipulated as a cultured cell line, they can also be reinserted into embryos and induced to contribute to normal cell lineages of an adult animal.

A second important area of technological development has been micromanipulation. Equipment has been invented that permits very controlled modifications of single mammalian cells. The oocyte zona pellucida can be manipulated to increase the efficiency of fertilization, and sperm can be microinjected under the zona or even directly into eggs. Blastomeres can be sampled from early embryos without compromising development. Pronuclei of fertilized eggs can be removed and replaced, and nuclei of more advanced cells can be transferred into eggs and early embryos. Finally, it is possible to insert genes directly into the pronuclei of fertilized eggs by micromanipulation.

Another important advance has been the arrival of recombinant DNA technology. The ability to clone genes and to transfer them into cells under conditions that allow their expression can be applied to problems of early mammalian development. Thus, it is possible to transfer genes into the embryo, integrate them into the host cell genome, and follow their expression throughout development.

These advances, originally developed as unrelated research programs, have now coalesced to create an entirely new investigative approach to developmental genetics. The result has been an explosion of valuable information relevant to the problem of gene regulation during mammalian ontogeny, a problem that had previously stood as among the most baffling, yet intriguing, in all the biological sciences. Understandably, these changes have profound implications not only for clarifying important fundamental mechanisms of development, but also for applied sciences in agriculture and medicine. Powerful forms of genetic engineering, once considered in the realm of science fiction, are now near realities. As we stand at the threshold of this frontier in the biomedical sciences and contemplate our newfound power, some of us are alarmed, some excited, but all who appreciate the implications of this historic moment are awestruck.

The purpose of this chapter is to review these new advances in mammalian gamete and embryo micromanipulation. I will first consider the area of gene transfer into embryos, because this important research illustrates how recombinant DNA technology, in conjunction with embryo culture and manipulation capabilities, has provided insights into fundamental mechanisms of developmental gene regulation and opened the door to a new form of genetic engineering. Next, the area of gamete micromanipulation will be reviewed. This field has had the most immediate and direct impact on clinical disorders, most notably, for the problem of male infertility. Finally, pronucleus removal and embryo biopsy will be discussed. This field, in which the micromanipulation technologies required are quite sophisticated, awaits further small improvements before it can have practical use in reproductive medicine.

The sequence in which these subjects are to be described is taken because principles of development elucidated by gene transfer experiments have, in some instances, revealed both limitations and new potentials in the other areas. Therefore, the dependence of applied sciences and basic findings is best underscored by this approach. In each section, however, the potential application of the findings will be emphasized. Where it is considered relevant, the ethical implications of these new technologies will also be discussed.

II. GENE TRANSFER INTO THE EMBRYO

A. Current Approaches to Gene Transfer

The phenomenon of differential gene function, manifest throughout the development of all multicellular eukaryotes, including mammals, has been among the most difficult to understand in all of biology. Given that all cells of the organism are genetic clones of the fertilized egg, how is it possible that through intercellular communication, these genetically identical entities succeed in differentially activating portions of their genomes? This question is critical, not only to understanding development, but also to understanding diseases relating to faulty execution of the genetic program. Such diseases are among the most severe and include birth defects, cancer, autoimmune disorder and, perhaps, aging processes as well. It has long been recognized that the process of establishing specific patterns of gene expression in differentiated cells can be divided into two phases: determination and differentiation. Determination is a process by which cells become committed to a specific pathway of differentiation. Thus, at the moment of determination, the decision has been made to execute a specific genetic program and to exclude all others that might characterize other differentiated cell types. The phenomenon of differentiation is one whereby the designated program of gene expression is established.

Prior to the recent developments in embryo culture, micromanipulation, and recombinant DNA technology, the processes of determination and differentiation could be investigated only in poikilothermic species by skin grafting. For example, organ primordia in the developing amphibian could be grafted from one site to another (e.g., the presumptive eye grafted to the flank), to determine if transplantation altered the genetic program. If the process of determination had not occurred at the time of grafting, a structure appropriate for the recipient graft site would develop. If, on the other hand, determination had occurred before transplantation, an abnormally situated structure characteristic of the donor site would develop [e.g., Ref. 1].

Experiments of this kind could not be extended to mammals until embryo culture techniques improved. Then, in 1961, Tarkowski [2] showed that genetically distinct mouse embryos at the eight-cell stage could be aggregated in vitro to form single composite embryos. Similar findings were also reported by Mintz, who also subsequently demonstrated that after transfer of such embryos to the uteri of pseudopregnant recipients, adult genetic mosaics could be derived [3]. This technology of chimeric or "allophenic" mouse production thereby allowed one genotype to be "grafted" to another, and the genetic differences between the two embryos could be used as markers to follow cells through development. The technique was subsequently exploited to study many phenomena such as the genetic basis for immune tolerance, susceptibility to malignancy, or allocation of stem cells to populate adult organs [reviewed in Ref. 4]. However, this technology fell short of the ideal experiment, which was to isolate a specific gene and determine the factors that enabled it to undergo a specific pattern of developmental regulation.

In 1973, Graham and Van der Eb demonstrated that an individual gene's transfer and expression could be selected for in cultured cells exposed to DNA [5]. This led to systems in which selection for expression of donor genetic material, inserted into cells by calcium phosphate coprecipitation, could detect rare, successful gene transfer events [6–8]. These DNA-mediated gene transfer methods were subsequently exploited to study various aspects of gene regulation, and they approached the goal of understanding the factors that regulate specific genes. However, these systems fell short of the ideal experiments, in which donor genes could be studied throughout mammalian development from the one-cell stage to the adult.

Then in 1980, Gordon et al. [9] published a protocol for transferring individual cloned genes into developing mice. This technique, which entailed microinjection of genes into the pronuclei of fertilized eggs, was based on previous demonstrations that introduction of DNA into cultured cell nuclei by microinjection was associated with a high rate of foreign gene integration and expression [10,11]. In this first report [9], pronuclear stage zygotes were microinjected with a recombinant plasmid containing the simian virus 40 (SV40) early promoter and the herpesvirus thymidine kinase gene. When newborn mice were evaluated by Southern blot hybridization, two animals were shown to unequivocally carry DNA sequences homologous to the microinjected material. Subsequently, it was demonstrated that such material was integrated into the host genome and transmitted through the germline to succeeding generations of animals [12–14]. Gordon and Ruddle called such animals *transgenic* [12], a term that is now used to describe organisms carrying transferred genes in a variety of animal and plant systems. The protocol for producing transgenic animals, which was subsequently described in detail by Gordon and Ruddle [15], is shown in Figure 1.

In the next few years, two other methods for introducing cloned genetic material into the mammalian germline were developed. The first was based upon the initial observation by Jaenisch [16] that Moloney murine leukemia virus (MoMuLV) could infect mouse embryos and insert a proviral DNA copy of its RNA genome into the chromosomal material of the mouse.

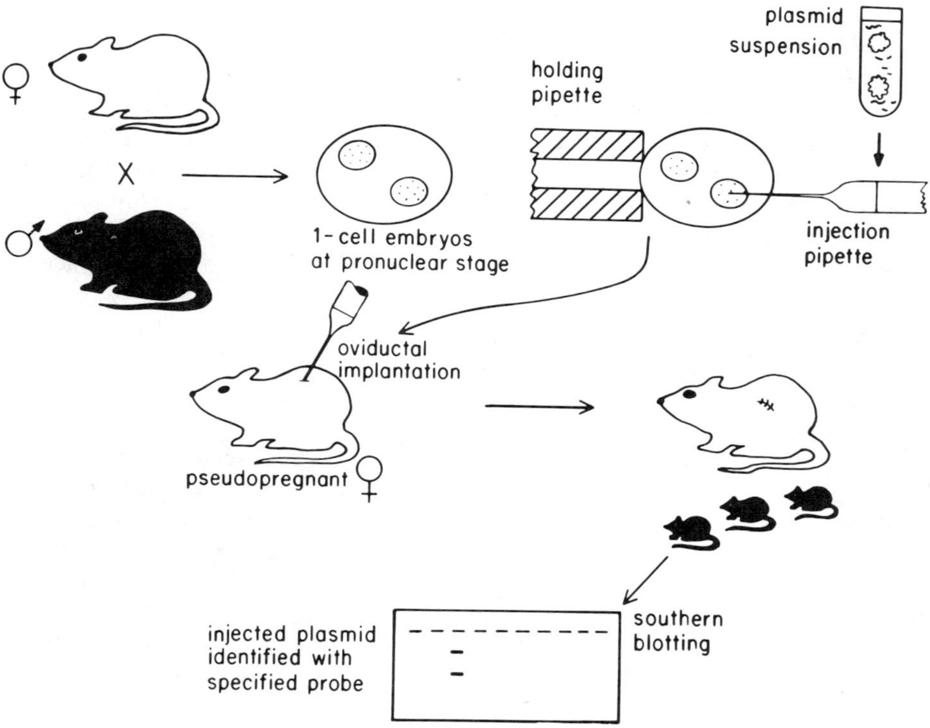

Figure 1 Protocol for production of transgenic mice by pronuclear microinjection [9]: Female mice are superovulated and mated to males (upper left). On the day after mating, one-celled embryos at the pronuclear stage are recovered and microinjected with DNA (upper right). After embryo transfer (center) and development to term, animals are tested by Southern hybridization for integration of the injected DNA.

Although this system was at first limited to insertion of retroviral genomes into mice, subsequent development of cell lines that produce retroviral packaging proteins led to the ability to substitute heterologous genes for large regions of the retroviral genome, package such recombinant retroviruses, and use the infectious characteristics of the recombinants to transfer the new genes into recipient cells [e.g., Ref. 17]. This system not only set the stage for somatic gene therapy strategies [18], it also allowed the insertion of recombinant DNA into early embryos [19,20]. The use of recombinant retroviruses for production of transgenic mice has the limitations that the size of the donor molecule transferred is restricted by the need to package the recombinant retrovirus into an infectious particle, and by the fact that recombinant retroviruses generally poorly express their heterologous genes [20]. Nonetheless, this system has many of the valuable attributes of the microinjection approach and is diagrammed in Figure 2.

A third method of introducing new genes into mammals is based upon the development of ES cell lines. In 1975 Mintz and Illmensee [21] and Papaiannou et al. [22] showed that teratocarcinoma cells, when microinjected into the blastocyst cavities of mouse embryos, could become "domesticated" and could participate in the formation of normal tissues in a chimeric mouse. Teratocarcinoma cells could be used for insertion of foreign genes [23] and, hence, could theoretically serve as vectors for transfer of new traits into the mouse [24]. Unfortunately, teratocarcinoma cells only rarely contributed to the germ cell populations of chimeric mice; therefore, it was difficult to establish pedigrees of transgenic animals by breeding. Subsequently, cell lines derived from preimplantation embryos maintained on feeder layers, and termed embryonic stem cells [25,26], were demonstrated to have all the advantageous characteristics of teratocarcinoma cells, but could also frequently contribute to the germline. Genetic deletions introduced into ES cells could be propagated through the germline to establish pedigrees of animals with experimentally induced genetic deletions [27,28]. Moreover, the ES cell approach has a uniquely powerful attribute—it can be used to mutate genes by directed integration. Previous work with fibroblasts had demonstrated that genes transferred into cultured mammalian cells can rarely integrate into those cells by homologous recombination [29]. When

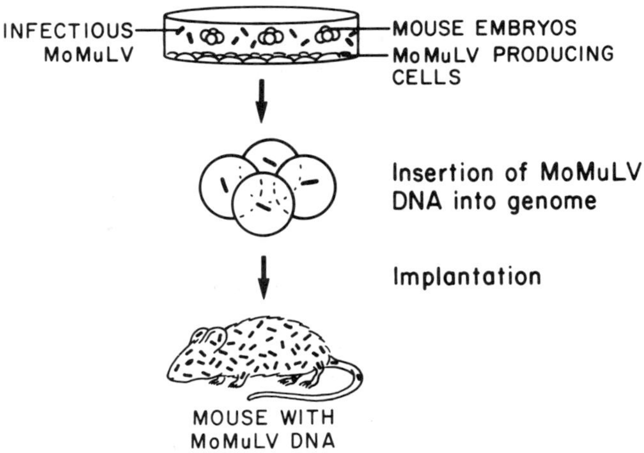

Figure 2 Retrovirus-mediated gene transfer into embryos [16]: Cleaving embryos are exposed to cell cultures, which produce retroviruses (top), or to concentrated viral stocks. After embryo transfer, newborn animals carry retrovirus proviral DNA in their genomes.

fragments of cloned genes are linked to irrelevant DNA sequences and inserted, the cloned fragments can integrate by homologous recombination, carry with them the linked irrelevant material, and cause insertional destruction of a host locus. It appears that subjecting ES cells to such manipulations reduces the likelihood that they will contribute to the germ cell population of the chimeric mouse they subsequently colonize. However, the feasibility of creating directed mutations by this approach has been clearly demonstrated [30]. The ES cell system has not been as widely used as microinjection because ES cells must be carefully monitored for chromosomal instabilities that render germline colonization less likely. In addition, production of mice by this method requires sophisticated cell culture technology as well as expertise in mouse embryo manipulation. However, the unique attributes of this system assure that it will assume an increasingly important role in transgenic technology. The ES cell methodology is illustrated in Figure 3.

Recently, a highly simplified approach to germline gene transfer has been reported [31]. With this approach, sperm are simply incubated with recombinant DNA for 30 min–2 hr and then used for in vitro fertilization. It was reported that this technique results in production of transgenic mice in about 30% of newborn animals, a percentage similar to that reported for microinjection [31]. Several questions remain to be answered before the feasibility of this protocol is established. First, the method must be repeated in other laboratories. Second, aberrations in the structure of the donor DNA in mice produced by this approach must be both explained and eliminated. In the existing published report [31], a recombinant molecule pSV2CAT, which contains the SV40 early promoter and the bacterial chloramphenicol acetyl transferase gene cloned into a plasmid, was used for sperm exposure. However, evaluation of 5 of the approximately 80 animals produced by this procedure showed extensive structural modification of the pSV2CAT molecule in most, perhaps all, of the animals. It is presently unclear why such alterations of structure took place. However, rearrangement of donor DNA during the gene-transfer process could preclude expression, thereby limiting the value of the methodology.

Transgenic mice have proved to be the most powerful tool ever devised for studying gene regulation. Two features of the system make it especially useful. First, expression of genes enables specific regulatory changes to be introduced, and second, random integration of donor DNA into the mouse genome can disrupt and mutate host genes, creating what are termed "insertional mutations." In the next sections, applications of transgenic technology to fundamental problems in mammalian developmental genetics will be outlined. Where possible, emphasis will be placed on applications of gene insertion to agriculture and medicine. It will not be possible here to review

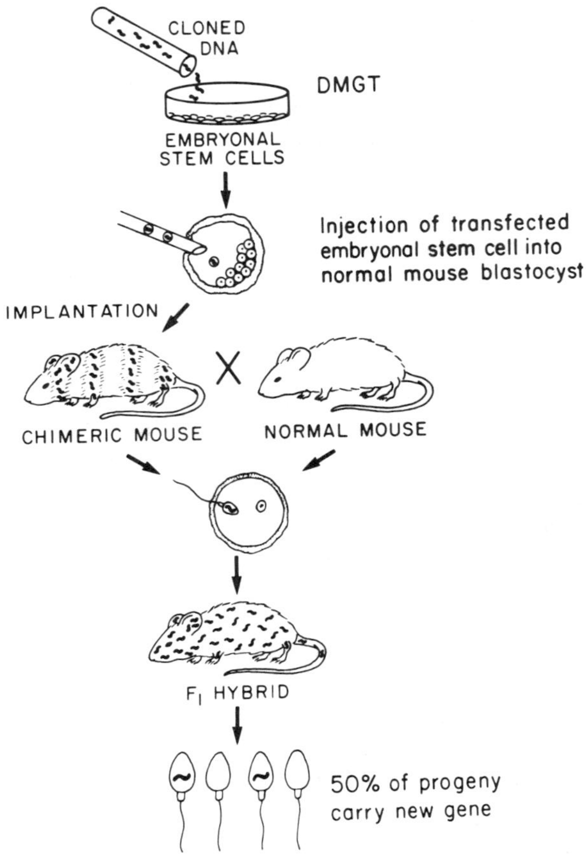

Figure 3 Embryonal stem cell-mediated gene transfer: Cultured ES cells are subjected to DNA-mediated gene transfer (DMGT, top). Transformed cells are microinjected into the blastocyst cavity of a normal embryo, and the implanted blastocyst gives rise to a genetic chimera (middle left). The chimeras, usually males, then transmit the transferred gene through their sperm to an F_1 animal. In the F_1, 50% of gametes will carry the new gene (bottom).

all major refinements provided by transgenic animals to our current concepts of mammalian development. Examples of such contributions to our basic understanding of gene regulation, immunology, and oncology will be presented. In addition, the use of these animals for creating models of human genetic disease and for treating such disease by gene transfer will be described. Insertional mutagenesis, an important use of transgenic technology in understanding birth defects, will be reviewed, followed by the use of transgenics in revealing the significance of inheriting genetic information from both the mother and the father. Finally, genetic engineering of new and valuable traits in animals will be described.

B. Fundamentals of Gene Expression Elucidated by Transgenic Mice

An important feature of transgenic mice is that donor genes, if designed properly before insertion into the embryo, are efficiently expressed [14]. Moreover, the developmental regulation and tissue distribution of expression of transferred genes ("transgenes") have answered critical questions about the regulatory signals that allow differential gene expression. Thus, these animals constitute attainment of goals sought for decades by biologists who performed reciprocal transplantation experiments in amphibians and birds and by scientists who produced chimeric mice.

In 1981, Brinster et al. [14] introduced a recombinant molecule into transgenic mice, which consisted of the herpesvirus thymidine kinase (*TK*) gene linked to the promoter region for mouse metallothionein-1 (*MT-1*). Metallothionein genes are ubiquitously expressed, but are most active in liver and kidney. Moreover, these genes, the products of which serve to detoxify heavy metals, are induced by heavy metals such as cadmium or zinc.

When the recombinant *MT–TK* gene was transferred into mice, high quantities of the herpes gene product were elaborated from liver and kidney, the normal sites of high *MT* expression. Moreover, when cadmium was administered to the transgenic animals, even higher levels of herpes *TK* gene expression were observed. Integration of foreign genes in transgenic mice is random [13] and was subsequently shown in the MT–TK mice not to have occurred into the endogenous mouse *MT* locus [32]. This indicated that small regions of DNA, within or near genes, carried signals for tissue-specific gene expression. These regulatory regions are often situated near the gene promoter, but may also be distant enhancers. For convenience, regulatory elements near the gene promoter will be referred to simply as the promoter in this discussion. The ability to target heterologous gene expression to other tissues by use of the MT promoter was then reinforced by a spectacular experiment in which rat growth hormone expression was induced in mouse liver and caused gigantism [33]. Subsequently, a variety of similar experiments were performed in which genes were inserted into transgenic mice. When the recombinant clones carried within them the discrete regions of DNA required for tissue-specific expression, such expression was seen in transgenic mice [reviewed in Refs. 34,35].

As experiments progressed, it became increasingly clear that these discrete regulatory elements were not necessarily located near the promoter (5′

flanking) region of genes, but instead, could be relatively distant enhancers [36]. Moreover, it became apparent that even genes expressed in a single differentiated cell type may have many enhancers. For example, when small regions that encompassed the human or rabbit adult β-globin genes were introduced into mice, expression was weak and shown to be profoundly inhibited by the presence of DNA, such as that of bacterial plasmid or bacteriophage, used for gene cloning [37–39]. Additional study showed that two areas in the 3′ region of human globin, one in the third exon and another in the 3′ flanking area, acted cooperatively to stimulate expression of the gene in adult bone marrow [40]. Finally, DNaseI hypersensitive regions of the genome as far as 50 kilobases (kb) 5′ or 20 kb 3′ to the globin gene, when grafted to the globin-coding sequence by recombinant DNA techniques, were shown to markedly enhance expression [41]. Thus, adult β-globin, expressed in only one tissue, may be regulated by at least five regions within, or relatively close to, the coding sequence.

When genes are expressed in more than one tissue, each expression site may have its own enhancer. For example, mouse *Thy-1*, a protein of the immunoglobulin gene family, is normally expressed in thymus, splenic T cells, fibroblasts, and throughout the brain. When an 8-kb clone of the gene was put into mice, expression was reproducibly lacking in splenic T cells and specific regions of the brain [42]. This indicates that the 8-kb clone does not contain all of the enhancers needed for the full repertoire of *Thy-1* expression. Similarly, major histocompatibility complex (MHC) genes contain enhancers that are specific for B cells and distinct from those that direct expression in thymus and splenic T cells [43,44].

Another general principle elicited from these and similar experiments is that when donor genes lack all relevant signals for tissue-specific expression, they are more subject to inhibition by the presence of extraneous bacterial or viral DNA and to the vicissitudes of integration site. For example, a 7.8-kb adult β-globin fragment inserted into mice was quite variably expressed, depending on integration site, and was strongly inhibited by the presence of plasmid DNA [37]. However, when the DNaseI hypersensitive regions located 50-kb 5′ and 20-kb 3′ to the gene were included on the donor clone, expression was far higher, was in direct proportion to the number of copies of the gene integrated, and was unaffected by integration site [41]. However, it should be recognized that in almost all series of transgenic animals produced, some integration sites are not permissive for expression.

Another factor in determining expression of transgenes (and undoubtedly reflective of endogenous regulatory mechanisms) is the elaboration of *trans*-acting factors that stimulate gene expression. Each *trans*-activator of gene expression is presumed to act on several genes, but is clearly also tissue-specific and developmentally regulated. An example of the importance of these factors to gene regulation is exemplified by experiments in which the human *Thy-1* gene was introduced into mice. Unlike mouse *Thy-1*, the human gene is expressed in kidney, vascular endothelium, and peripheral nerve. When the human gene was placed into transgenic mice, its expression followed a pattern identical with that seen in humans [42]. This shows that *trans*-activators of *Thy-1* expression exist in the same tissues in mice as in humans, and indicates that mouse *Thy-1* is not expressed in kidney, vascular endothelium, and nerve because the mouse gene has lost *cis*-acting responder elements that can interact with the *trans*-acting factors. Experiments wherein the human fetal globin gene was introduced into mice also show that species-specific differences in *trans*-acting factor reproduction can alter the developmental regulation of genes. When the human fetal gene is introduced into mice, it follows a pattern of expression typical of the mouse *embryonic* gene; that is, it is expressed in the yolk sac [38,45]. Thus, the mouse elaborates in its embryonic erythropoietic tissues a factor similar to the human *trans*-activator of fetal globin expression, which is produced in human fetal liver [45].

In summary, experiments with transgenic mice show that differential gene expression is somehow dictated by discrete regulatory elements within or close to genes, and that the chromosomal integration site is relatively unimportant. For expression to occur, highly conserved *trans*-acting factors must also be elaborated in an appropriate developmental pattern. Because genes usually assume their expected pattern of expression after insertion into other mammals, and because it is possible to graft regulatory elements from one gene to coding sequences of another and obtain targeted expression of a transferred gene [14], it has been possible to penetrate the most formidable problems in mammalian development at the molecular level.

C. Transgenic Mice in Immunology

The immune system, which must aggressively recognize and expel foreign antigens, while at the same time tolerating the myriad "self"-proteins present in the host, is regulated by mechanisms as complex as its functions. Immunoglobulin (Ig) molecules, composed of heavy- and light-chain polypeptides, are encoded by three sets of heavy-chain genes (V_H, D_H, and J_H) and two sets of light-chain genes (V_L and J_L). Molecular mechanisms regulating the production of Ig are characterized by several unique features. For Ig to be expressed, the genes must be physically rearranged, with the result that the functional gene's somatic configuration differs substantially from the unexpressed, germline structure. Furthermore, expression of one Ig allele is associated with blockage of rearrangement and expression of the other parental allele, a process known as *allelic* exclusion. The genes are also developmentally regulated: the pre-B cell produces μ heavy chains before light-chain production is initiated. When light chains are produced, isotypic exclusion occurs, such that only κ or λ chains are made in any one B cell. How these and other regulatory functions of Ig gene regulation operate have been greatly clarified with transgenic mice. An example of the usefulness of transgenic animals in elucidating these mechanisms is allelic exclusion, which will be reviewed briefly here. Other aspects of the regulatory apparatus, such as isotypic exclusion and somatic hypermutation, have also been fruitfully probed with transgenic animals. Detailed reviews of transgenic animals in the study of immunology have been presented elsewhere [35,46].

It is a relatively simple task to clone functional Ig genes from cultured myeloma cells. Once cloned, such genes can be introduced into the mouse germline. Transfer of a rearranged, expressible Ig gene thus "tricks" the immune system by inserting into the germline a gene with the structure of a somatically active Ig gene in a mature B cell. With this experiment it is thus possible to probe mechanisms of allelic exclusion.

It was precisely such an experiment that was first performed with Ig genes in transgenic mice. A rearranged κ light-chain gene cloned from the myeloma cell line MOPC-21 was inserted into mice. The gene was found to have a powerful enhancer that directed expression specifically to B cells [47,48]. To study the effects of "prearranged" Ig transgenes in the B cells of mice, hybridomas were produced from transgenic mouse B cells, and both the structure and expression of endogenous and donor Ig genes were analyzed [49]. Hybridomas exhibited varying degrees of transgene processing and expression. In some cell lines, the foreign κ-protein was combined with heavy chains and secreted. In these instances, endogenous κ gene rearrangement and expression were both blocked. In hybridomas that synthesized the transgene protein, but did not secrete it as part of a mature Ig molecule, host κ gene expression was still absent, but rearrangement of host genes could take place. In some hybridoma lines, endogenous genes were rearranged and expressed. If such expression was not associated with Ig secretion, the transgene could still be expressed. But if endogenous gene products were secreted as part of a functional Ig molecule, transgene expression was invariably extinguished. These results show that allelic exclusion of light-chain genes occurs at two levels. When one allele is rearranged and expressed, this is sufficient to block expression, but not rearrangement, of the other allele. However, once one allele contributes to production of a secreted Ig molecule, rearrangement of the other allele is blocked as well. Thus, both parental light-chain genes appear to have the potential for expression until Ig secretion is initiated. Transgenic mice thus allowed determination of the regulatory hierarchy that allows expression of only one parental light-chain allele [49].

Similar experiments have been conducted for μ heavy-chain genes. When rearranged μ heavy-chain genes encoding antibodies against the synthetic hapten 4-hydroxy-3-nitrophenyl (NP) were used to produce transgenic mice, endogenous μ gene rearrangement was blocked in a significant number of immature and mature B cells [50,51]. Additional experiments have demonstrated that it is the membrane-bound form of the heavy-chain gene product that is responsible for allelic exclusion. When genes are modified before microinjection so that they do not contain a membrane anchor, no allelic exclusion occurs [51]; however, when signals for secretion, but not membrane anchoring, are removed, allelic exclusion still occurs [52]. Another interesting feature of μ genes in transgenic mice is their aberrant patterns of expression. For example, it has been demonstrated repeatedly that rearranged μ genes inserted into the pronucleus are expressed in T cells as well as B cells. Therefore, it appears that the factor normally preventing μ gene expression in T cells is the inability of those cells to rearrange the gene [50,52]. In addition, rearranged μ transgenes may be expressed in other unexpected sites such as the heart [50] or brain [53].

When unrearranged transgenes are inserted into mice, they can be rearranged. Rabbit Ig genes are rearranged and can even contribute protein to mouse–rabbit hybrid antibody molecules in transgenic mice [54]. Moreover, rearrangement of chicken genes inserted into mice in the germline configuration shows that mechanisms of gene rearrangement are highly conserved in evolution, even extending across classes of organisms [55]. Surprisingly, T cells, which do not normally rearrange Ig genes, can rearrange Ig transgenes [55].

As mentioned previously, studies of allelic exclusion represent only one use of transgenic animals in the study of Ig gene regulation. The MHC genes are also expressed appropriately in transgenic mice [56,57], but, as noted previously, multiple sites of expression may each rely on a specific enhancer. Exploitation of this latter phenomenon has allowed important investigations into mechanisms of immune tolerance. When MHC genes, modified such that they are not expressed in thymus, are inserted into mice, tolerance to the encoded I-E antigen is observed, indicating that, contrary to previous dogma, thymic expression is not required for self-tolerance. These and a variety of other studies illustrate the enormous impact transgenic technology has had on the understanding of the immune system. Moreover, the ability to confer immunity to specific antigens by transfer of MHC genes [56,57; see models for gene therapy] indicates that such knowledge could be applied to genetic engineering of livestock and, perhaps in the future, human beings as well.

D. Transgenic Mice in Oncology

The aberrant patterns of gene expression that characterize neoplastic cells, and the predisposition to malignancy that accompanies increased exposure to mutagenic compounds or genetic deficiencies in DNA repair, all support the notion that cancer is the result of genetic changes in the cell. What are these changes, and how many are required to convert a normal cell to a potentially lethal malignant one? Transgenic mice provide a creative new approach to the study of malignant disease, because they allow the insertion and targeted expression of genes implicated in neoplastic transformation. Cancer research in transgenic mice can be divided into two major categories: transfer of viral genes suspected to be oncogenic and insertion of cellular protooncogenes or their oncogene counterparts. Both strategies have yielded fundamental, highly important information relevant to our understanding of malignancy. Examples of experiments involving viral and cellular oncogenes will be presented, and the significance of the findings to understanding neoplasia underscored.

1. Viral Oncogenes in Transgenic Mice

The first viral oncogene to be introduced into transgenic mice was the T antigen of SV40. When this gene was inserted with its own promoter element, which is entirely foreign to the mouse (mouse cells are normally not permissive for SV40 replication), tumors of the choroid plexus resulted [58,59]. The result showed that this viral oncogene could transform cells normally not susceptible to SV40 infection, and that choroid plexus cells, for reasons unknown, produced a *trans*-acting factor that was able to stimulate the T-antigen promoter.

Further experiments with SV40 T antigen showed that the protein causes marked hyperplasia of cells and predisposes them to malignant conversion, but that expression of T antigen is not in itself sufficient to cause cancer. For example, when T antigen was directed to express in pancreas by substituting the insulin promoter for the native T-antigen promoter, islet cells of the pancreas proliferated abnormally [60]. Moreover, islet cells, which produce tyrosine hydroxylase as well as insulin, and which normally disappear, are immortalized by T-antigen expression [61]. However, only about 1 : 1000 of the T–antigen-expressing cells actually becomes malignant [60]. Because this frequency is far higher than normal, almost every animal expressing T antigen in the endocrine pancreas develops a malignancy. However, the results indicate that cellular transformation, at least when mediated by this SV40 oncogene, is a “two-hit” phenomenon, and that T-antigen expression constitutes only one of the two insults necessary for neoplastic degeneration.

Continued experiments with SV40 T antigen have illustrated another very important principle of oncogenesis—cells are differentially susceptible to genetic changes that predispose to malignant transformation. When SV40 T antigen was linked to the promoter for atrial natriuretic peptide, the T antigen was expressed equally in both the right and left atria of transgenic mouse hearts. However, although few pathological changes were seen in the left atria, the right atria showed marked hyperplasia, reaching several hundred times their normal size before the animals died of cardiac conduction abnormalities [62]. These results provide insight into the observation that cancers are common in some organ systems (e.g., breast), but rare in others (heart).

An important use of transgenic mice has been to reveal the oncogenic potential of viruses suspected, but not proved, to be underlying causes of cancer. For example, the human virus JC has been associated epidemiologically with the development of glial tumors. When JC viral genes are introduced into transgenic mice, neuroblastomas form, thereby confirming that the virus is oncogenic [63]. One of the best-known examples of this in vivo test of viral oncogenic potential resides in experiments with the human T-lymphotrophic virus type I (HTLV-1). A tenuous but unmistakable relationship between infection by this virus and the development of T-cell leukemia was known. However, fewer than 0.1% of those with anti-HTLV antibodies actually develop leukemia, and the time interval between HTLV-1 infection and the appearance of leukemia often exceeds 20 years. Moreover, leukemic cells in these patients showed no evidence of viral antigen production.

To investigate the oncogenic potential of HTLV-1, Hinrichs et al. [64] inserted the *tat* gene of this virus, a gene known to stimulate the viral promoter, into transgenic mice along with its own HTLV-1 promoter element (the long terminal repeat, or LTR). Transgenic mice developed multiple tumors, later shown to be neurofibromas [64,65]. Thus, the ability of HTLV-1 gene expression to be tumorigenic was established.

2. Cellular Protooncogenes and Oncogenes in Transgenic Mice

A variety of cellular genes have been noted to be mutated in malignant cells. These protooncogenes and their mutated oncogene counterparts have been inserted into mice to determine their roles in malignant transformation.

Because mutations in protooncogenes often lead to products with increased activity, an obvious question concerning these protooncogenes was whether or not abnormal regulation of the unmutated protooncogene could lead to malignancy. Stewart et al. [66] investigated this question by inserting the c-*myc* protooncogene into transgenic mice driven by the promoter for mouse mammary tumor virus (MMTV). The LTR of this virus is hormone-responsive and is inducible by androgens, estrogens, and corticosteroids. When the MMTV-c-*myc* construct was inserted into mice, females developed breast tumors after rearing several litters of pups. Presumably, lactation was associated with high hormone stimulation to breast tissue, which, in turn, caused tumors. Again, tumors arose sporadically, indicating at least a two-hit mechanism. The c-*myc* protooncogene has similar effects in lymphoid tissue, causing lymphomas when linked to Ig enhancers [67]. These experiments provide critically important evidence that aberrant regulation of normal cellular genes can greatly increase the likelihood of malignant transformation. They provide a new perspective on cancer as a developmental genetic disorder. Several similar experiments that underscore these findings have been reported and recently reviewed [35].

Evidence exists that some genes transferred into mice cause cancer by a one-hit mechanism. Such a mechanism is implied if all cells that express the transgene become transformed. The possibility of single-hit tumorigenesis exists for *ras* genes expressed in the exocrine pancreas [68], polyoma middle-T antigen in vascular endothelium [69], and the c-*neu* oncogene driven by the MMTV promoter [70]. In the latter, a two-hit mechanism has also been suggested [71]. In none of these cases has unequivocal evidence for single-hit carcinogenesis been provided, but the possibility that such phenomena can occur must now be seriously considered because of results from transgenic animals.

E. Transgenic Mice as Models for Human Disease

Disease states that closely resemble human pathological conditions can be induced in transgenic mice by gene transfer. Such disease models are extremely valuable because they not only provide insight into mechanisms of disease, they can serve as models for testing therapeutic strategies. Two approaches to creating such models can be used. First, expression of foreign genes can cause disease and, second, insertional disruption of host genes by integration of new DNA can be used to induce host genetic deletions identical with those underlying human disease states. Examples of both approaches will be given.

1. Gene Expression

A variety of genes have been introduced into mice, the expression of which causes abnormalities similar to those seen in human disorders. As a model of Down syndrome, Epstein et al. [72] introduced the human superoxide dismutase (*SOD*) gene into mice. The *SOD* gene maps to chromosome 21, and trisomy for this chromosome is the underlying cause of Down syndrome. No abnormalities were initially observed [70], but subsequent evaluation revealed disturbances of the neuromuscular junctions in the tongues of the animals, an abnormality also seen in humans with Down syndrome

[73]. These data thus indicate that some of the pathological manifestations of this syndrome, in which many genes are present in triplicate, can be attributed to triploidy at a single locus.

The hepatitis B (HBV) surface antigen carrier state is a highly significant pathological condition because it is strongly associated with the development of hepatocellular carcinoma. In areas of the world where hepatitis B is a common disease, hepatocellular carcinoma is a leading killer. When the cloned hepatitis B surface antigen gene is introduced into mice, the antigen is present in liver as well as peripheral blood. Accordingly, these animals constitute a potential model for the surface antigen carrier state [74,75]. Recently, evidence has been supplied that transgenic mice will support the complete replicative cycle of HBV introduced by pronuclear microinjection [76]. If this proves true, transgenic mice can serve as excellent models for drug testing.

Osteogenesis imperfecta is a fatal disorder of collagen formation and is a mutation. Collagen is a triple-helical protein with two α_1-chains and one α_2-chain. In osteogenesis imperfecta, a point mutation in the α_1-chain, which substitutes an arginine for a glycine at position 391 or cysteine for glycine at either positions 748 or 988 of the protein, retards formation of the triple-helix and leads to the disease. Stacey et al. [77] exploited site-directed mutagenesis techniques to introduce similar mutations into the cloned α_1-gene in vitro, then microinjected the mutated genes into pronuclei. The result was a fatal disorder of collagen formation that was similar, in many respects, to osteogenesis imperfecta.

2. Insertional Mutagenesis

Potential models of human disease have also been created by insertional disruption of host genetic loci. Lesch–Nyhan disease results from deletion of the X-linked gene for hypoxanthine phosphoribosyl transferase (HPRT) and is characterized by choreoathetosis, mental retardation, and bizarre, self-mutilation behavior. The ES cell system provides an excellent approach to creating mice with a deletion for HPRT. When cells are grown in 6-thioguanine (6-TG), HPRT activity results in incorporation of the purine analogue into DNA, with resultant death of the cell. Thus, simply growing ES cells in the presence of 6-TG leads to emergence of clones that proliferate by virtue of having spontaneously lost HPRT activity. The identification of such mutants can also be facilitated by inserting DNA into ES cells by retroviral infection or other approaches and by using 6-TG to select for disruptive insertions into the HPRT gene.

It was precisely these approaches that led to production of mouse strains with HPRT deficiency [27,28]. Interestingly, HPRT-deficient mice do not manifest abnormality, apparently because of the presence of an alternate salvage pathway for purine biosynthesis. Nonetheless, the creation of these mutant strains illustrates an important feature unique to the ES cell system of transgenic mouse production.

F. Transgenic Mice as Models for Gene Therapy

Although the treatment of human genetic disease by gene transfer into the germline is now a remote objective, transgenic mice can provide information on the feasibility and efficacy of the approach. Several interesting efforts at gene therapy have been made in the transgenic mouse.

Strains of mice exist that lack the adult β-globin gene, a situation analogous to the human condition β-thalassemia. Costantini et al. [78] transferred the human or mouse β-globin gene into mice to determine if its expression could correct the thalassemia. Transfer of either gene led to production of β-globin protein in bone marrow, with consequent alleviation of the thalassemia. Not only was the human β-globin protein able to combine with the mouse α-chain to produce a functional protein, the human protein was actually superior to the mouse β-globin in ameliorating the thalassemic state.

Some strains of mice lack the *Ia* genes of the E complex of the major histocompatibility MHC and, as a result, are unable to respond to certain antigenic stimuli. One such strain is C57BL/6. When the cloned *Ia* gene was inserted into pronuclei of this strain, tissue-specific expression was observed, and responsiveness to the synthetic antigen (poly)Glu-Lys-Phe, which depends on E complex expression, was observed. Thus, these animals constitute a model for broadening the immune responsiveness of animals by gene transfer [56,57].

Mice deficient in growth hormone (GH) have had GH genes transferred into them with resultant increased somatic growth and restoration of fertility to males, which are sterile in the absence of GH production [79]. This experiment constituted partial correction of the GH-deficient state. However, because the GH transgenes were inserted under *MT* regulation and were expressed constitutively, females, formerly fertile, were rendered infertile. Another example of gene transfer to correct endocrinopathy is insertion of the gonadotropin-releasing hormone (GnRH) into mice deficient for this gene. Insertion of the gene leads to restoration of normal hormone cycling and fertility [80].

In summary, transgenic mice have clearly shown that genetic deficiencies can be permanently corrected by insertion of heritable transgenes. In some instances, as in MHC gene transfer, deficiencies were not corrected, but the potential fitness of animals was improved by gene insertion. These experiments have thus laid the foundation for gene therapy in humans, when and if the hazards of gene insertion are eliminated.

G. Insertional Mutagenesis

Insertional mutations occur in transgenic animals when integration of donor DNA disrupts host genes. The phenomenon probably occurs in about 5% of pedigrees and, usually, manifests as a recessive trait. Spontaneous insertional mutations have been observed after retroviral-mediated gene transfer and microinjection. Insertional mutations of specific loci have been created in ES cells as well [28]. The value of insertional mutations, when they occur in unidentified genes, is that insertion of a defined new DNA provides a hybridization target that can be used for molecular cloning of the disrupted host gene. An example of how a recessive insertional mutation is produced, identified, and cloned in bacteriophage is illustrated in Figure 4. When such genes regulate complex developmental processes, cloning and molecular analysis of their expression can lead to unprecedented understanding of the genetic basis of birth defects.

Molecular cloning of genes by DNA transfer is more readily accomplished when retroviral-mediated gene transfer is used. This is because the retroviral insertion mechanism is characterized by integration of a single proviral DNA copy of the virus, with minimal disruption of host DNA sequences. However, when microinjection is used, long, head-to-tail concatamers of the foreign sequence are often inserted [13], and integration can be associated with perturbations of host DNA structure [81], which may even include translocations [82–84]. Such rearrangements substantially increase the difficulty of cloning. However, it is possible that retroviral DNA integration is not random [85,86]; consequently, the kinds of insertional mutations produced may differ when genes are inserted by microinjection versus retroviral infection. It is accordingly important to explore transgenic mice produced by either method for the appearance of insertional mutations.

The best-characterized insertional mutation arose from insertion of MoMuLV DNA. In one strain of such mice, designated Mov-13, homozygotes for the retrovirus proviral DNA insert died in midgestation. Molecular cloning of the flanking DNA revealed integration of the proviral DNA into the first intervening sequence of the α_1(I)-collagen gene with consequent alteration of chromatin structure around the gene and blockage of transcription [87–90]. That failure to produce α_1(I)-collagen kills the embryo is hardly surprising, and a mutation of this kind is not highly revealing for complex mechanisms of organogenesis. However, transgenic mice produced by microinjection have manifested highly interesting insertional mutations. In one

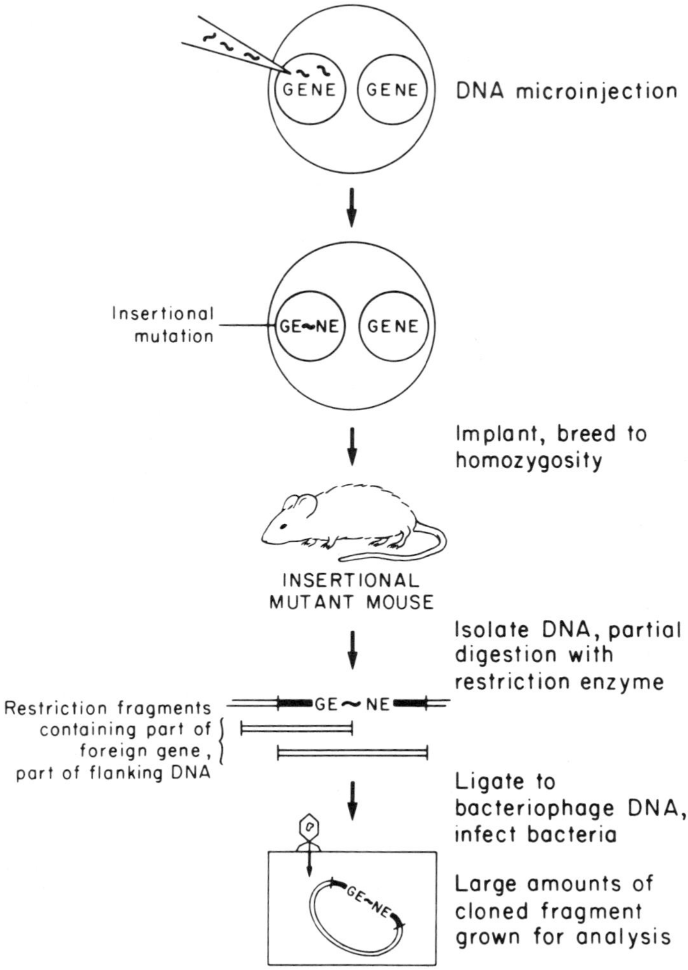

Figure 4 Identification and cloning of genes disrupted by insertional mutation: A hypothetical gene designated "GENE" (top) is disrupted by integration of a microinjected fragment. When transgenic mice are bred to homozygosity, the recessive mutation's phenotypic effects appear. A genomic library is constructed from DNA of the mutant animal (a bacteriophage library is shown) and probed with the microinjected DNA. Large clones, some of which carry flanking material derived from the mutated host gene, are inserted in the clones and expanded in bacteria for analysis (bottom).

instance, mice have been reported to have a specific deformity of the limbs characterized by fusion of the distal bones [91]. In another, the distal portions of the limbs were missing entirely [92]. In a pedigree identified in our laboratory, affected animals lose all cerebellar Purkinje cells by 20 days of postnatal development, and they also manifest marked, precocious degeneration of olfactory bulb mitral cells and retinal photoreceptors [93]. Mutations of this kind, once cloned, could provide new insight into birth defects, information for which access without transgenic technology is exceedingly difficult.

H. Transgenic Mice Reveal Parental Gene Imprinting

Under normal circumstances, individuals receive one copy of every gene from each parent. Nuclear transplantation studies, made possible by micromanipulation, have demonstrated that an embryo cannot develop unless it carries genes from both parents. When mouse embryos are generated which, although diploid, carry genes derived entirely from the male (androgenetic) or female (gynogenetic) parent, they fail to develop to term [94–97]. This result was somewhat surprising, as the regulation of all genes previously examined appeared to be the same, regardless of the parental origin.

Studies of foreign gene expression in transgenic mice have demonstrated, however, that parental origin can affect gene regulation. Expression and/or DNA methylation of c-*myc* [98], hepatitis B surface antigen [99], the bacterial *CAT* gene [100], and the quail troponin gene [101], all have been found, in some pedigrees, to differ when inherited from the father, rather than from the mother. These differences in expression are correlated with different patterns of cytosine methylation [98]. Usually, when such differences are observed, higher methylation and lower expression occurs when the genes are transmitted through the female germline.

Although no endogenous gene has ever been demonstrated to undergo such differential regulation, the failure of androgenetic and gynogenetic embryos to develop indicates that, at least for some critical genes, differential parental imprinting probably takes place. These pedigrees of transgenic mice provide a potential molecular explanation for the requirement that the conceptus receive one package of genes from each parent.

I. Transgenic Mice and Genetic Engineering

Examples alluded to previously in which genetic disease is either induced or cured by gene transfer are obvious examples of genetic engineering in transgenic animals. However, other models have more direct potential impact on agriculture and medicine. The production of transgenic pigs, sheep, and rabbits by microinjection [102] indicates that farm animals can be genetically modified to great benefit. Introduction of growth hormone genes into pigs [102] has established an experimental paradigm for genetic engineering of increased growth rates into animals. If such increases could be successfully engineered, the economics of rearing livestock for food consumption could be dramatically altered. Transgenic livestock might also be used as living factories for production of human proteins with medicinal value. Linkage of the whey acidic protein promoter to tissue plasminogen activator (TPA) has resulted in secretion of TPA in the milk of transgenic mice [102]. If a similar result can be obtained in large farm animals, such animals could supply through their milk biologically active compounds that are difficult to obtain in pure form by any other method. Another example of such proteins would include clotting factors (factor VIII and factor IX) for treatment of clotting disorders.

J. Summary

The previous section has given only a few examples of the importance of transgenic technology to basic and applied research. It is safe to say that this technology has revolutionized the field of mammalian developmental genetics. Not only have our insights into fundamentals of gene regulation been profoundly improved, but the practical potential of transgenic technology is manifest. As we look to the future, we may ask if germline gene insertion will ever be applied to humans. Before this can be done, the gene transfer frequency (currently 15–30% of embryos born after transfer) must be increased, the hazards of insertional mutagenesis eliminated, and the efficient expression of donor genes obtained in every procedure. It is also important to resolve profound ethical issues relating to such methods of medical intervention. Who will be allowed to benefit from a technology that would not only increase the fitness of the individual, but also his or her descendants? Is this form of tampering with the human genetic endowment, even if consented to by some individuals, an unacceptable violation of ethical and moral principles that apply to all? As transgenic technology continues to undergo refinement, these issues will require ever more thorough discussion.

III. MICROMANIPULATION-ASSISTED FERTILIZATION

A. Background

Fertilization of mammalian eggs is accomplished by a single spermatozoon. However, the chance that any given sperm cell will contribute genes to the offspring is exceedingly small. In most mammals, tens of millions of sperm are deposited in the female reproductive tract. Of these, only a few hundred reach the site of fertilization, and only one of those remaining few hundred actually penetrates the oocyte. At several stages preceding the fertilization event itself, sperm are lost. Many spermatozoa fail to traverse the cervical barrier, and loss of sperm number at this step and other points en route to the oocyte accounts for substantial sperm loss. However, even when the oocyte is reached, the sperm must satisfy several important prerequisites to be capable of fusing with the egg. Sperm must undergo capacitation and the acrosome reaction [104] before encountering the surface of the egg. Experiments in the mouse reveal, however, that premature completion of the acrosome reaction (AR) precludes binding to and penetration of the oocyte zona pellucida [105]. Consequently, any sperm that undergo a spontaneous AR before zona contact are excluded from fertilization [106]. Sperm must also be vigorously motile to bore their way through the zona; hence, it is not surprising that even minor perturbations of spermatogenesis can lead to male infertility. Individuals with sperm counts below 20×10^6 are considered to have oligospermia, with an increased likelihood of infertility. Moreover, when oligospermia is accompanied by teratospermia or asthenospermia, as often happens in humans, the chances that those few sperm that do reach the egg will have all the characteristics necessary for penetration of the zona are relatively reduced.

In vitro fertilization (IVF) can overcome the problem of low sperm numbers by placing hundreds of thousands of sperm in the immediate vicinity of the oocyte. However, although sometimes successful in reversing the consequences of male infertility [107], IVF often fails to overcome deficiencies of spermatogenesis. This failure is partly due to the relative inefficiency of sperm–egg interaction in vitro, and partly to the increased proportion of sperm with abnormal motility and structure in ejaculates of infertile males.

Advances in gamete micromanipulation offer a new approach to the problem of male infertility. Eggs can be retrieved and manipulated to facilitate access to the oolemma. Figure 5 illustrates the currently considered approaches to oocyte manipulation. These include microinjection of sperm directly into the ooplasm (procedure 1), insertion of sperm under the zona pellucida (procedure 2), and introduction of a small gap in the zona pellucida through which sperm may pass directly to the egg (zona drilling, procedure 3). Zona drilling, the least aggressive of these methods will be considered first, after which the state of advancement of subzonal sperm insertion and microinjection will be reviewed.

B. Zona Pellucida Drilling (Zona Drilling)

The rationale for zona drilling is that the zona pellucida of the oocyte poses a substantial obstacle to sperm penetration. When this structure is present, it is difficult to obtain fertilization in vitro with fewer than 1000 sperm per egg.

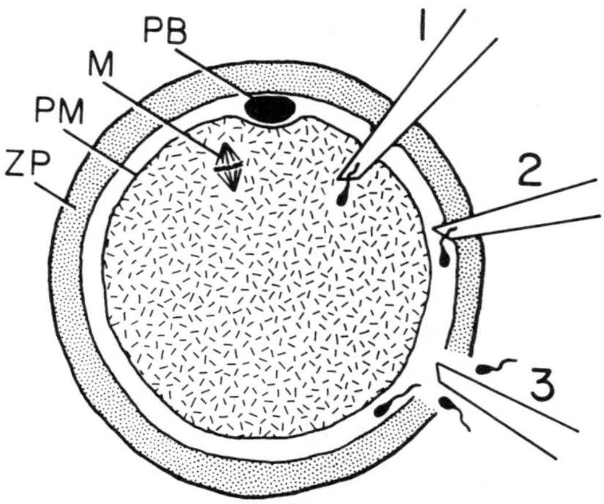

Figure 5 Methods for assisting fertilization by micromanipulation: Sperm are microinjected directly into the cytoplasm (1), inserted under the zona pellucida (2), or allowed to swim through a hole drilled in the zona (3). Abbreviations: PB, first polar body; M, metaphase plate of the second meiotic division; PM, plasma membrane of the oocyte; ZP, zona pellucida.

However, when the zona is removed in its entirety, as few as three sperm may be enough to fertilize the mouse oocyte [108]. Complete removal of the zona, though, is undesirable. This structure often contributes to the block to polyspermic fertilization, and its absence is also associated with a reduced rate of implantation [106].

In an attempt to increase the efficiency of zona penetration, but still retain the important biological characteristics of the zona, Gordon and Talansky devised the approach of zona drilling. This procedure entails introduction of a small hole in the zona through which sperm may freely swim [109]. In experiments with the mouse, acid Tyrode's solution, which solubilizes the zona, was loaded in a microneedle and delivered as a fine stream against a localized region of the zona of an egg fixed in position by suction of a holding pipet. This maneuver (illustrated diagrammatically in Fig. 6), produced a visible gap in the zona and was associated with a marked improvement in IVF: Fertilization rates of zona-drilled eggs were increased by a factor of approximately 100 by zona drilling. The block to polyspermy was not severely affected by introduction of a hole in the zona, and the development of zygotes fertilized after the procedure was no different than that from control eggs fertilized without drilling [109]. This study thereby indicated that zona drilling could be used as a clinical remedy for infertility.

The first clinical trial of zona drilling reinforced this notion [110]. In this study, patients were brought to zona drilling because they had totally failed

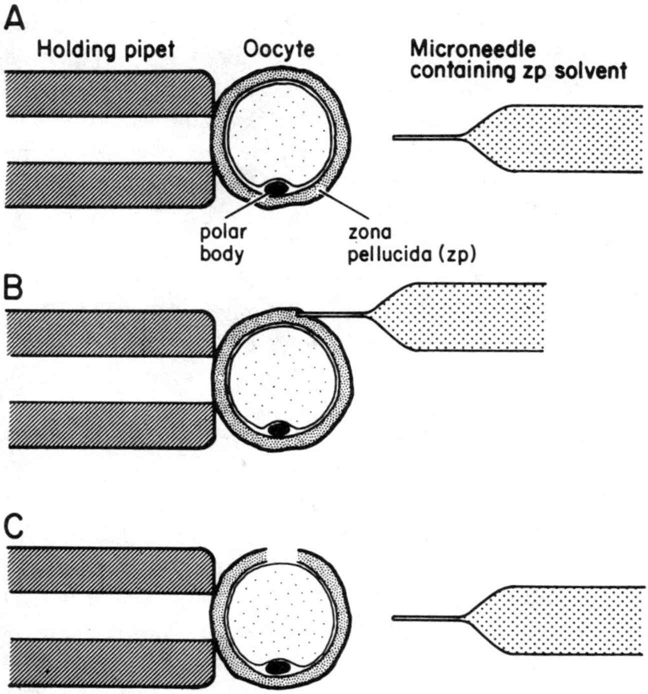

Figure 6 Diagrammatic representation of zona drilling [105]. An oocyte is fixed in position by a holding pipet, and a microneedle is loaded with zona solvent (A). The microneedle is placed tangentially against the zona and solvent is expelled (B). A small gap is introduced into the zona (C).

to achieve fertilization of zona-intact eggs in one or more previous IVF cycles. A fertilization rate of 32% was observed in the same group after zona drilling, and three embryo transfers were performed. Although no pregnancies were established in this trial, results indicated that further use of the procedure would allow couples with previously refractory cases of male infertility to conceive.

Other important findings were made in this study [110]. First, it was found that unlike the mouse egg, human oocytes were sensitive to toxic effects of acid Tyrode's solution. Second, high rates and degrees of polyspermy were observed in the zona-drilled human eggs, this despite the fact that they were inseminated with markedly abnormal sperm. Both a high rate (the percentage of fertilized eggs that are polyspermic) and degree (the number of sperm that penetrate polyspermic eggs) of polyspermy indicate that the human oocyte relies heavily on its zona pellucida for blocking polyspermy. The human oocyte therefore appears similar to that of the hamster, which has a significantly delayed plasma membrane block to polyspermy and which, when denuded of its zona, is subject to high degrees of polyspermic fertilization. These results indicated that zona drilling, in addition to providing a new clinical approach to infertility, can be used to reveal fundamental characteristics of sperm–egg interaction [110].

A modified approach to zona drilling termed partial zona dissection (PZD) was subsequently used to establish human pregnancies [111]. The PZD technique appears superior to standard zona drilling in two major ways. First, acid Tyrode's solution is not used, and, second, the egg is dehydrated in hypertonic sucrose before the procedure so that the perivitelline space can be more safely entered [111]. It appears that sucrose exposure changes the character of the oolemma, making it less likely to sustain polyspermic fertilization. The pregnancies reported after PZD occurred in couples with less severe infertility problems than those previously studied by zona drilling. However, the establishment of normal pregnancies indicates that manipulation of the human zona does not prevent normal development, and indicates that this procedure will soon become a widely used approach to infertility. The procedure has additional value in animal studies, for which it can confer fertility on sterile males [112] and increase the penetration rate of eggs matured in vitro [113].

C. Subzonal Sperm Insertion

Subzonal sperm insertion is more aggressive than zona drilling because the sperm need not be motile to gain access to the oocyte surface. Consequently, the procedure is less selective, and the chance is increased that sperm that are incompetent to fertilize will be approximated to the egg. Initial efforts in our laboratory to obtain newborn mice by subzonal insertion were unsuccessful, probably because manipulation of the spermatozoon with the microneedle altered its surface characteristics and rendered it unable to interact appropriately with the egg [114]. However, Mann [115] subsequently used methylcellulose as a lubricating agent to protect mouse sperm from contact with the microneedle, and this procedure resulted in fertilization and birth of newborn mice after embryo transfer. Previously, experiments with the human in whom embryo transfers were not permitted had shown that sperm of this species are more amenable to manipulation with the microneedle. When these sperm were inserted under the zonae of hamster or human eggs without the use of a protective agent, oocyte penetration and cleavage (of the human zygotes) was observed [116,117]. These results now set the stage for attempts at human subzonal insertion.

Recently, this procedure has been accompanied by embryo transfer and has led to a pregnancy [118]. However, only one of five eggs subjected to the procedure was fertilized normally, a rate of success that indicates that further improvements in the technique are desirable. An important area in need of progress is sperm selection. Unlike zona-intact eggs, which can be fertilized only by sperm with intact acrosomes (zona binding requires an intact acrosome [105]), sperm exposed directly to the egg must have completed the acrosome reaction to fuse. Consequently, selection procedures to produce pure populations of acrosome-reacted sperm could reduce the number of sperm required to assure fertilization after subzonal insertion and, thereby, reduce instances during which too many competent sperm are inserted and polyspermic fertilization results.

D. Sperm Microinjection

Sperm microinjection is the most aggressive of all fertilization enhancement procedures. In this instance, sperm need not be motile, and even a single spermatozoon is sufficient for fertilization. Experiments with animals have indicated that sperm decondensation and formation of metaphase chromosomes are possible after this procedure [119,120]. However, it is exceedingly difficult to avoid oocyte damage [114,121]. Moreover, the oocytes of mice as test animals do not respond to sperm microinjection in a manner resembling a normal response to fertilization. Frequently, the procedure does not activate the oocyte to complete meiosis [114]. This problem appears less severe when hamster oocytes are used [119]; hence, in the future, this may be applicable to the most severe forms of human male infertility. However, until animals can routinely be conceived by this methodology, it will not acquire an important role in IVF.

E. Summary

Gamete micromanipulation, which includes the procedures of zona drilling, subzonal sperm insertion and sperm microinjection, offers new promise for treatment in cases of male infertility. It is also possible, although not yet proved, that eggs may have zona defects that render them incapable of allowing sperm passage under normal circumstances. In such instances, gamete micromanipulation might also be effective. In addition to their potential clinical value, these procedures can provide important data on the nature of sperm–egg interactions and the biological prerequisites that must be satisfied for sperm to be capable of participating not only in fertilization, but in the support of normal embryonic and fetal development.

IV. PRONUCLEUS REMOVAL

It is possible to remove pronuclei from mouse eggs and replace them, either with a pronucleus of the same parental origin as that which was removed, or with one of the opposite paternal origin. Although generation of gynogenetic and androgenetic embryos results in embryonic death [96,97], replacement of a male pronucleus with another male pronucleus from another embryo can be performed with minimal harm to the zygote and is associated with subsequent normal development [96]. Moreover, when embryos rendered triploid by suppression of second polar body release are restored to diploidy by removal of one of the supernumerary female pronuclei, mice can develop normally [95].

These results raise the possibility that removal of supernumerary male pronuclei from polyspermic eggs could increase the number of normal embryos available for transfer in cases of human IVF. In human IVF, the frequency of polyspermic fertilization is 5–7% [122]. Rawlins et al. [123] first attempted the procedure in humans: Three eggs were manipulated, but none cleaved. We [124] and Malter and Cohen [125] have performed a similar manipulation, except that the method of pronuclear extraction differed slightly [124]. We found that when cytochalasin and demecolcine (Colcemid) were used to weaken the cytoskeletal structure of the zygote, 100% of manipulated embryos survived pronuclear extraction and cleaved. These data suggest that pronucleus removal might be used routinely in the future when eggs are penetrated by two sperm.

However, before this can be done, some technical and biological questions must be addressed. From the technical standpoint, it must be possible to unequivocally distinguish the two male pronuclei from the single female pronucleus so that the female is not inadvertently removed. Such an error could lead to embryo demise or to development of a hydatidiform mole, a tumor previously shown to be diploid and androgenetic [126]. Although male and female pronuclei can be readily distinguished in the mouse, this is not so in the human [124,125]. From the biological perspective, it has not yet been demonstrated that supernumerary male pronuclei are actually normal. Although this appears to be true for female pronuclei [95], the substance that induces male pronucleus formation is clearly distinct from that which forms the female pronucleus and is limited in supply [127]. Therefore, animal studies must be conducted to show that eggs can undergo polyspermic fertilization, followed by withdrawal of a male pronucleus, and still develop into normal young. However, success of such experiments is anticipated, and, hence, pronucleus removal may soon play a role in human IVF management.

V. PRENATAL DIAGNOSIS BY EMBRYO BIOPSY

One of the most disappointing possible outcomes in human IVF is establishment of a pregnancy, only to find that the embryo carries a serious genetic disorder. This prospect causes major concern because IVF couples, owing to the advanced maternal age typical of this patient group, are at especially high risk for genetic disorders such as Down syndrome. Because eggs fertilized in vitro are maintained in the culture dish for about 48 hr, they are accessible to manipulation. Logically, one goal of developing micromanipulation technology is for prenatal diagnosis.

The ability to culture mammalian preimplantation embryos quickly led to the discovery that the embryo can undergo extensive manipulation without loss of viability. For example, it is quite feasible to split two-cell or four-cell stage embryos, reimplant the parts, and obtain identical twins or quadruplets [128]. Thus, given the appropriate circumstances, micromanipulation capability, and diagnostic technology, it should be possible to perform genetic diagnosis on blastomeres removed from cleaving embryos. Strategies for performing such diagnoses by enzymatic assay or DNA analysis are illustrated in Figure 7.

Efforts at performing such diagnoses began with enzymatic assays. Monk et al. [129] exploited the availability of HPRT-deficient mice [27] to attempt diagnosis of this deficiency in biopsied blastomeres. Single cells were removed from the cleaving embryo, or several cells were removed from the trophoblast region of the blastocyst (see Fig. 5), and subjected to a microassay for HPRT. The corresponding embryos were then reimplanted and evaluated after developing, to confirm the diagnosis made during cleavage. In all cases, the deficiency was successfully identified [129].

The same system was then used in an effort to determine embryo sex. Of all genetic traits, sex is probably the most important to determine, because a variety of genetic disease, such as Lesch-Nyhan disease, Duchenne's muscular dystrophy, and hemophilia A, are sex-linked. In addition, identification of sex is of enormous value in agriculture, for which an excess of male or female livestock may be desired.

The HPRT assay can potentially be used for embryo sexing because of the developmental regulation of X-chromosome inactivation. X-chromosome inactivation occurs as a compensatory mechanism to equalize the dosage of X-linked genes between females (XX) and males (XY) [130]. However, X inactivation does not occur until late preimplantation development. Consequently, X-linked genes that are active during early cleavage produce more enzyme in XX than XY embryos.

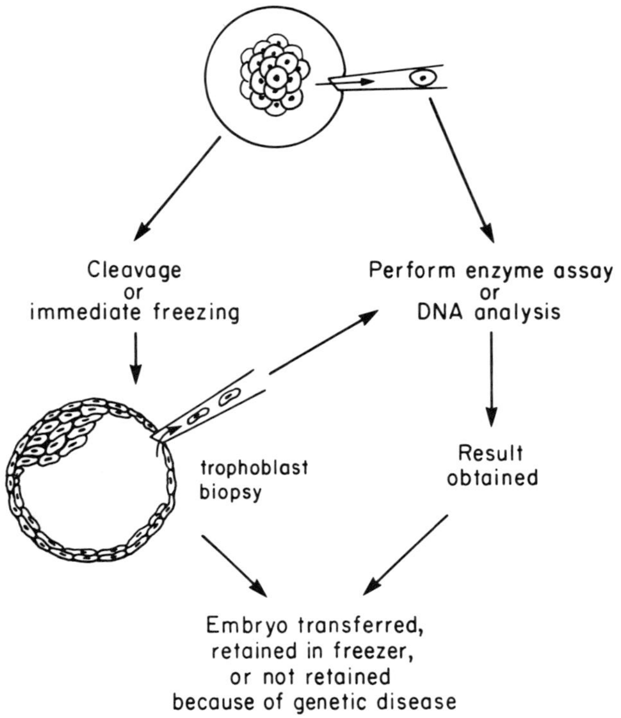

Figure 7 Strategies for prenatal diagnosis by embryo biopsy: A morula (top) can have a blastomere removed, or several cells can be sampled from the trophoblast region of the blastocyst (lower left). The biopsy material is placed in an enzymatic or DNA assay (middle right) and the result obtained (lower right) is used to determine if the embryo has genetic disease and, if so, its disposition (bottom).

In mice, comparisons of cytoplasmic HPRT in biopsied blastomeres, with an autosomal enzyme (adenine phosphoribosyl transferase) used as a reference, has resulted in successful determination of embryo sex [131]. However, drawbacks of prenatal diagnosis by enzymatic assay remain. First, the biopsied cell(s) must remain intact until inserted into the microassay, or cytoplasmic enzyme could be lost. Second, and most importantly, such diagnoses can only be made for enzymes that are produced during cleavage. In humans, it has not been established that HPRT is activated before implantation, and for all genetic traits that affect genes activated later (e.g., Huntington disease, sickle cell anemia), such assays are not possible. Moreover, some genetic diseases alter proteins for which a suitable enzymatic assay is lacking.

A potential solution to these problems is prenatal diagnosis by DNA analysis. The Southern hybridization technique [132] can be used to evaluate DNA structure directly and does not require gene expression. However, it is difficult to perform Southern analysis on DNA purified from fewer than about 150,000 cells.

However, the new technique of enzymatic amplification (also known as the polymerase chain reaction; PCR) can be used to exponentially amplify a defined region of genomic DNA for detailed analysis [133]. This procedure entails synthesis of oligonucleotide primers that correspond to short regions of DNA situated on opposite ends of a small intervening region (100–3000 base pairs). One oligonucleotide is homologous to the coding DNA strand and, the other, to the noncoding strand. When small amounts of DNA are heat denatured and annealed with a large molar excess of the oligonucleotides, the primers intercede in the genomic DNA by base pairing with the opposite strand. DNA polymerase and substrates can then be used to extend strand synthesis from the primers, thereby duplicating the intervening region of DNA. The oligonucleotide priming site is also replicated, so that a repeated round of annealing and synthesis results in an exponential amplification of the defined DNA segment (Fig. 8). Because the PCR reaction requires only a few minutes, it is possible to amplify a region of DNA several hundred thousand-fold in a few hours. PCR is so sensitive that it has even been successful in studying genes amplified from single spermatozoa [134]. However, when single cells are used for the assay, a significant failure rate is encountered. In part, this is due to spurious amplification of extraneous sequences in the highly complex mammalian genome [135].

Because PCR of single genes from individual cells is not yet without error, efforts to diagnose traits in biopsied blastomeres have focused on attempts to amplify repeated elements in the genome. The Y chromosome carries repeated sequences found on no other chromosome; consequently, PCR can be used in an attempt to amplify such repeats. A successful reaction indicates that the cell carries a Y chromosome, whereas failure indicates that the embryo is of the XX karyotype.

Partially successful attempts to make such diagnoses have been reported for human [136] and mouse [137] embryos. In mice, successful diagnosis could be confirmed by reimplanting the embryos and ascertaining the phenotypic sex of fully developed fetuses [137].

Although additional refinement of PCR technology is required before medical or agricultural applications can be made, its potential is staggering. Because the assay can be quantitative, it may be possible to detect trisomy 21 in embryos, a disorder for which IVF couples are at especially high risk. In addition, several gene segments can be amplified simultaneously, making possible the diagnosis of multiple traits from a single blastomere. Once

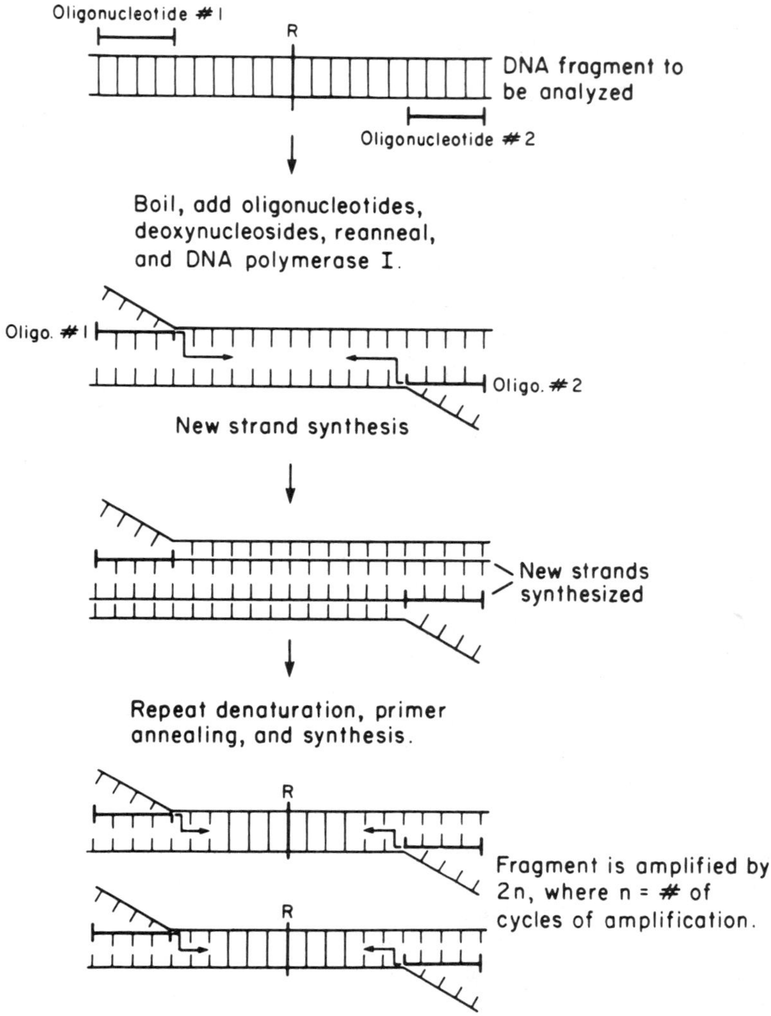

Figure 8 The polymerase chain reaction (PCR; see text and Ref. 126 for details): R indicates a restriction endonuclease site that might be used to diagnose a mutation within the amplified region. Note that the procedure exponentially amplifies a selected region of DNA because the oligonucleotide priming sites are replicated in each PCR cycle (bottom).

perfected, embryo biopsy and PCR, by ruling out scores of genetic disturbances, could make IVF embryos even more genetically healthy than those conceived by natural means. In addition, because many IVF embryos are now frozen for donation to infertile couples, genetic disease can be eliminated in the setting of embryo donation as well.

VI. CONCLUSIONS

The experiments described in this chapter indicate that a new era of reproductive technology has arrived. The accomplishments made by gene microinjection are formidable, and the success of gamete micromanipulation procedures also are impressive. Other techniques of gamete and embryo manipulation, in conjunction with recombinant DNA technology, promise to further improve reproductive management in both humans and livestock.

The practical applications of these advances are numerous and highly important. However, equally important is the fundamental information imparted by these experiments. We are unlocking the greatest secrets of mammalian development, which stands as one of the most complex processes ever observed in nature. As we gain this profound knowledge, we will certainly arrive at a new perspective of the meaning of life as a biological as well as a physical process.

ACKNOWLEDGMENTS

Work in this manuscript was supported by NIH grants HD20484 and CA42103, and March of Dimes Grant 1-1026. This is manuscript No. 21 of the Brookdale Center for Molecular Biology, Mt. Sinai Medical Center.

REFERENCES

1. Spemann, H. (1918). Uber die determination der ersten Organanlagen des Amphibienembryo I-VI. *Roux Arch. Entwmech. 43*:448–555.
2. Tarkowski, A. K. (1961). Mouse chimeras developed from fused eggs. *Nature 190*:857–860.
3. Mintz, B. (1965). Genetic mosaicism in adult mice of quadriparental lineage. *Science 148*:1232–1233.
4. Mintz, B. (1974). Gene control of mammalian differentiation. *Annu. Rev. Genet. 8*:411–470.
5. Graham, F. L., and A. J. Van der Eb (1973). A new technique for the assay of infectivity of human adenovirus 5 DNA. *Virology 52*:456–467.
6. Wigler, M., S. Silverstein, L. S. Lee, A. Pellicer, T. Cheng, and R. Axel (1977). Transfer of purified herpes virus thymidine kinase gene to cultured mouse cells. *Cell 11*:223–232.
7. Maitland, H., and J. McDougall (1977). Biochemical transformation of mouse cells by fragments of herpes simplex virus DNA. *Cell 11*:233–241.
8. Bachetti, S., and F. L. Graham (1977). Transfer of the gene for thymidine kinase to thymidine kinase deficient human cells by means of purified herpes simplex viral DNA. *Proc. Natl. Acad. Sci. USA 74*:1590–1594.
9. Gordon, J. W., G. A. Scangos, D. J. Plotkin, J. A. Barbosa, and F. H. Ruddle (1980). Genetic transformation of mouse embryos by microinjection of purified DNA. *Proc. Natl. Acad. Sci. USA 77*:7380–7384.

10. Graessmann, A., M. Graessmann, W. C. Topp, and M. Botchan (1979). Retransformation of a simian virus 40 revertant cell line, which is resistant to viral and DNA infections, by microinjection of viral DNA. *J. Virol. 32*:989–994.
11. Capecchi, M. R. (1980). High efficiency transformation by direct microinjection of DNA into cultured mammalian cells. *Cell 22*:479–488.
12. Gordon, J. W., and F. H. Ruddle (1981). Integration and stable germ line transmission of genes injected into mouse pronuclei. *Science 214*:1244–1246.
13. Costantini, F., and E. Lacy (1981). Introduction of a rabbit β-globin gene into the mouse line. *Nature 294*:92–94.
14. Brinster, R. L., H. Y. Chen, M. E. Trumbauer, A. W. Senear, R. Warren, and R. D. Palmiter (1981). Somatic expression of herpes thymidine kinase in mice following injection of a fusion gene into eggs. *Cell 27*:223–231.
15. Gordon, J. W., and F. H. Ruddle (1983). Gene transfer into mouse embryos. Production of transgenic mice by pronuclear injection. *Methods Enzymol. Recombinant DNA, Part C 101*:411–433.
16. Jaenisch, R. (1976). Germ line integration and mendelian transmission of the exogenous Moloney leukemia virus. *Proc. Natl. Acad. Sci. USA 73*:1260–1264.
17. Mann, J. R., R. C. Mulligan, and D. Baltimore (1983). Construction of retrovirus packaging mutant and its use to produce helper-free defective retrovirus. *Cell 33*:153–159.
18. Williams, D. A., I. R. Lemischka, D. G. Nathan, and R. C. Mulligan (1984). Introduction of new genetic material into pluripotent haematopoietic stem cells of the mouse. *Nature 310*:476–480.
19. Van der Putten, H., H. Botteri, F. M. Miller, A. D. Rosenfeld, M. G. Fan, R. M. Evans, and I. Verma (1985). Efficient insertion of genes into the mouse germ line via retroviral vectors. *Proc. Natl. Acad. Sci. USA 82*:6148–6152.
20. Jahner, D., K. Haase, R. Mulligan, and R. Jaenisch (1985). Insertion of the bacterial *gpt* gene into the germ line of mice by retroviral infection. *Proc. Natl. Acad. Sci. USA 82*:6927–6931.
21. Mintz, B., and K. Illmensee (1975). Normal genetically mosaic mice produced from malignant teratocarcinoma cells. *Proc. Natl. Acad. Sci. USA 72*:3585–3589.
22. Papaioannou, B. E., M. W. McBurney, R. L. Gardner, and M. J. Evans (1975). Fate of teratocarcinoma cells injected into early mouse embryos. *Nature 258*:69–73.
23. Pellicer, A., E. F. Wagner, A. El Kareh, M. J. Dewey, A. J. Reuser, S. Silverstein, R. Axel, and B. Mintz (1980). Introduction of a viral thymidine kinase gene and the human β-globin gene into developmentally multipotential teratocarcinoma cells. *Proc. Natl. Acad. Sci. USA 77*:2098–2101.
24. Dewey, M. J., D. W. Martin, Jr., G. R. Martin, and B. Mintz (1977). Mosaic mice with teratocarcinoma-derived mutant cells deficient in hypoxanthine phosphoribosyl transferase. *Proc. Natl. Acad. Sci. USA 74*:5564–5568.
25. Martin, G. R. (1981). Isolation of a pluripotent cell line from early mouse embryos cultured in medium conditioned by teratocarcinoma stem cells. *Proc. Natl. Acad. Sci. USA 78*:7634–7638.

26. Evans, M. J., and M. H. Kaufman (1981). Establishment in culture of pluripotential cells from mouse embryos. *Nature 292*:154–156.
27. Hooper, M., K. Hardy, A. Handyside, S. Hunter, and M. Monk (1987). HPRT-deficient (Lesch-Nyhan) mouse embryos derived from germline colonization by cultured cells. *Nature 326*:292–295.
28. Kuehn, M. R., A. Bradley, E. J. Robertson, and M. J. Evans (1987). A potential animal model for Lesch-Nyhan syndrome through introduction of HPRT mutations into mice. *Nature 326*:295–298.
29. Smithies, O., R. G. Grett, S. S. Boggs, M. A. Koralewski, and R. S. Kucherlapati (1985). Insertion of DNA sequences into the human chromosomal β-globin locus by homologous recombination. *Nature 317*:230–234.
30. Thompson, S., A. R. Clarke, A. M. Pow, M. L. Hooper, and D. W. Melton (1989). Germ line transmission and expression of corrected HPRT gene produced by gene targeting in embryonic stem cells. *Cell 56*:313–321.
31. Lavitrano, M., A. Camaioni, V. M. Fazio, S. Dolcki, M. G. Farace, and C. Spadafora (1989). Sperm cells as vectors for introducing foreign DNA into eggs: Genetic transformation of mice. *Cell 57*:717–723.
32. Palmiter, R. D., H. Y. Chen, and R. L. Brinster (1982). Differential regulation of metallothionein–thymidine kinase fusion genes in transgenic mice and their offspring. *Cell 29*:701–710.
33. Palmiter, R. D., R. L. Brinster, R. E. Hammer, M. E. Trumbauer, M. G. Rosenfeld, N. C. Birnberg, and R. M. Evans (1982). Dramatic growth of mice that develop from eggs microinjected with metallothionein–growth hormone fusion genes. *Nature 300*:611–615.
34. Palmiter, R. D., and R. L. Brinster (1986). Germ-line transformation of mice. *Annu. Rev. Genet. 20*:465–499.
35. Gordon, J. W. (1989). Transgenic animals. *Int. Rev. Cytol. 115*:171–230.
36. Pinkert, C. A., D. M. Ornitz, R. L. Brinster, and R. D. Palmiter (1987). An albumin enhancer located 10 kb upstream functions along with its promoter to direct efficient, liver-specific expression in transgenic mice. *Genes Dev. 1*:268–276.
37. Chada, K., J. Magram, K. Raphael, G. Radice, E. Lacy, and F. Costantini (1985). Specific expression of a foreign β-globin gene in erythroid cells of transgenic mice. *Nature 314*:377–380.
38. Townes, T. M., J. B. Lingrel, H. Y. Chen, R. L. Brinster, and R. D. Palmiter (1985). Erythroid-specific expression of human β-globin genes in transgenic mice. *EMBO J. 4*:1715–1723.
39. Lacy, E., S. Roberts, E. P. Evans, M. D. Burtenshaw, and F. D. Costantini (1983). A foreign β-globin gene in transgenic mice: Integration at abnormal chromosomal position and expression in inappropriate tissues. *Cell 34*:343–358.
40. Behringer, R. R., R. E. Hammer, R. L. Brinster, R. D. Palmiter, and T. M. Townes (1987). Two 3′ sequences direct adult erythroid-specific expression of human β-globin genes in transgenic mice. *Proc. Natl. Acad. Sci. USA 84*:7056–7060.
41. Grosveld, F., G. B. Van Assendelft, D. R. Greaves, and G. Kollias (1987). Position-independent, high-level expression of the human β-globin gene in transgenic mice. *Cell 51*:975–985.

42. Gordon, J. W., P. G. Chesa, H. Nishimura, W. J. Rettig, J. E. Maccari, T. Endo, E. Seravalli, T. Seki, and J. Silver (1987). Regulation of *Thy-1* gene expression in transgenic mice. *Cell 50*:445–452.
43. Widera, G., L. C. Burkly, C. A. Pinkert, E. C. Bottger, C. Cowing, R. D. Palmiter, R. L. Brinster, and R. A. Flavell (1987). Transgenic mice selectively lacking MHC class II (I-E) antigen expression on B cells: An in vivo approach to investigate *Ia* gene function. *Cell 51*:175–187.
44. van Ewik, W., Y. Ron, J. Monaco, J. Kappler, P. Marrack, M. LeMeur, P. Gerlinger, B. Durand, C. Benoist, and D. Mathis. Compartmentalization of MHC class II gene expression in transgenic mice. *Cell 53*:357–370.
45. Chada, K., J. Magram, and F. Costantini (1986). An embryonic pattern of expression of a human fetal globin gene in transgenic mice. *Nature 319*:685–689.
46. Storb, U. (1987). Transgenic mice with immunoglobulin genes. *Annu. Rev. Immunol. 5*:151–174.
47. Brinster, R. L., K. A. Ritchie, R. E. Hammer, R. L. O'Brien, B. Arp, and U. Storb (1983). Expression of a microinjected immunoglobulin gene in the spleen of transgenic mice. *Nature 306*:332–336.
48. Storb, U., R. L. O'Brien, M. D. McMullen, K. A. Gollahon, and R. L. Brinster (1984). High expression of a cloned immunoglobulin κ gene in transgenic mice is restricted to B lymphocytes. *Nature 310*:238–241.
49. Ritchie, K. A., R. L. Brinster, and U. Storb (1984). Allelic exclusion and control of endogenous immunoglobulin gene rearrangement in κ transgenic mice. *Nature 312*:517–520.
50. Grosschedl, R., D. Weaver, D. Baltimore, and F. Costantini (1984). Introduction of a μ immunoglobulin gene into the mouse germ line: Specific expression in lymphoid cells and synthesis of functional antibody. *Cell 38*:647–658.
51. Weaver, D., F. Costantini, T. Imanishi-Kari, and D. Baltimore (1985). A transgenic immunoglobulin μ gene prevents rearrangement of endogenous genes. *Cell 42*:117–127.
52. Storb, U., C. Pinkert, B. Arp, P. Engler, K. Gollahon, J. Manz, W. Brady, and R. L. Brinster (1986). Transgenic mice with μ and κ genes encoding antiphosphorylcholine antibodies. *J. Exp. Med. 164*:627–641.
53. Nussenzweig, M. C., A. C. Shaw, E. Sinn, D. B. Danner, K. L. Holmes, H. C. Morse III, and P. Leder (1987). Allelic exclusion in transgenic mice that express the membrane form of immunoglobulin μ. *Science 236*:816–819.
54. Goodhardt, M., P. Cavelier, M. A. Akimenko, G. Lutfallam, C. Babinet, and F. Rougeon (1987). Rearrangement and expression of rabbit immunoglobulin κ light chain gene in transgenic mice. *Proc. Natl. Acad. Sci. USA 84*:4429–4433.
55. Bucchini, D., C.-A. Reynaud, M.-A. Ripoche, H. Grimal, J. Jami, and J.-C. Weill (1987). Rearrangement of a chicken immunoglobulin gene occurs in the lymphoid lineage of transgenic mice. *Nature 326*:409–411.
56. Le Meur, M., P. Gerlinger, C. Benoist, and D. Mathis (1985). Correcting an immune-response deficiency by creating *Ea* gene transgenic mice. *Nature 316*:38–42.

57. Yammamura, K., H. Kikutani, V. Folsom, L. K. Clayton, M. Kimoto, S. Akira, S. Kashiwamura, S. Tonegawa, and T. Kishimoto (1985). Functional expression of a microinjected *Ead* gene in C57BL/6 transgenic mice. *Nature 316*:67–69.
58. Brinster, R. L., H. Y. Chen, A. Messing, T. van Dyke, A. J. Levine, and R. D. Palmiter (1984). Transgenic mice harboring SV40 T-antigen genes develop characteristic brain tumors. *Cell 37*:367–379.
59. Small, J. A., D. G. Blair, S. D. Showalter, and G. A. Scangos (1985). Analysis of a transgenic mouse containing simian virus 40 and v-*myc* sequences. *Mol. Cell. Biol. 5*:642–648.
60. Hanahan, D. (1985). Heritable formation of pancreatic B-cell tumors in transgenic mice expressing recombinant insulin/SV40 oncogenes. *Nature 315*:115–122.
61. Teitelman, G., S. Alpert, and D. Hanahan (1988). Proliferation, senescence, and neoplastic progression of B cells in hyperplasic islets. *Cell 52*:97–105.
62. Field, L. J. (1988). Atrial natriuretic factor–SV40–T antigen transgenes produce tumors and cardiac arrhythmias in mice. *Science 239*:1029–1032.
63. Small, J., G. Scangos, L. Cork, G. Jay, and G. Khoury (1986). The early region of human papovavirus JC induces dysmyelination in transgenic mice. *Cell 46*:13–18.
64. Hinrichs, S. H., M. Nerenberg, R. K. Reynolds, G. Khoury, and G. Jay (1987). A transgenic mouse model for human neurofibromatosis. *Science 237*:1340–1344.
65. Nerenberg, M., S. H. Hinrichs, R. K. Reynolds, G. Khoury, and G. Jay (1987). The *tat* gene of human T-lymphotrophic virus type 1 induces tumors in transgenic mice. *Science 237*:1324–1329.
66. Stewart, T. A., P. K. Pattengale, and P. Leder (1984). Spontaneous mammary adenocarcinomas in transgenic mice that carry and express *MTV/myc* fusion genes. *Cell 38*:627–637.
67. Adams, J. M., A. W. Harris, C. A. Pinkert, L. M. Corcoran, W. S. Alexander, S. Cory, R. D. Palmiter, and R. L. Brinster (1985). The c-*myc* oncogene driven by immunoglobulin enhancers induces lymphoid malignancy in transgenic mice. *Nature 318*:533–536.
68. Quaife, C. J., C. A. Pinkert, D. N. M. Ornitz, R. D. Palmiter, and R. L. Brinster (1987). Pancreatic neoplasia induced by *ras* expression in acinar cells of transgenic mice. *Cell 48*:1023–1034.
69. Williams, R. L., S. A. Courtneidge, and E. Wagner (1987). Embryonic lethalities and endothelial tumors in chimeric mice expressing polyoma virus middle T oncogene. *Cell 52*:121–131.
70. Muller, W. J., E. Sihn, P. K. Pattengale, R. Wallace, and P. Leder (1988). Single-step induction of mammary adenocarcinoma in transgenic mice bearing the activated c-*neu* oncogene. *Cell 54*:105–115.
71. Bouchard, L., L. Lamarre, P. J. Tremblay, and P. Jolicoeur (1989). Stochastic appearance of mammary tumors in transgenic mice carrying the *MMTV*/c-*neu* oncogene. *Cell 57*:931–936.

72. Epstein, C. J., K. B. Avraham, M. Lovett, S. Smith, O. Elroy-Stein, G. Rotman, C. Bry, and Y. Groner (1987). Transgenic mice increased Cu/Zn-superoxide dismutase activity: Animal model of dosage effects in Down syndrome. *Proc. Natl. Acad. Sci. USA 84*:8044–8048.
73. Avraham, K. B., M. Schickler, D. Sapoznikov, R. Yarom, and Y. Groner (1988). Down's syndrome: Abnormal neuromuscular junction in tongue of transgenic mice with elevated levels of human Cu/Zn-superoxide dismutase. *Cell 54*:823–829.
74. Chisari, F. V., C. A. Pinkert, D. R. Milich, P. Filippi, A. Lachlan, R. D. Palmiter, and R. L. Brinster (1985). A transgenic mouse model of the chronic hepatitis B surface antigen carrier state. *Science 230*:1157–1160.
75. Babinet, C., H. Farza, D. Morello, M. Hadchouel, and C. Pourcel (1985). Specific expression of hepatitis B surface antigen (HBsAg) in transgenic mice. *Science 230*:1160–1163.
76. Araki, K., J. Miyazaki, O. Hino, N. Tomita, O. Chisaka, K. Matsubara, and K. Yamamura (1989). Expression and replication of hepatitis B virus genome in transgenic mice. *Proc. Natl. Acad. Sci. USA 86*:207–211.
77. Stacey, A., J. Bateman, T. Choi, T. Mascara, W. Cole, and R. Jaenisch (1988). Perinatal lethal osteogenesis imperfecta in transgenic mice bearing an engineered mutant pro-α_1(I) collagen gene. *Nature 332*:131–136.
78. Costantini, F., K. Chada, and J. Magram (1986). Correction of murine β-thalassemia by gene transfer into the germ line. *Science 233*:1192–1194.
79. Hammer, R. E., R. D. Palmiter, and R. L. Brinster (1984). Partial correction of murine hereditary disorder by germline incorporation of a new gene. *Nature 311*:65–67.
80. Mason, A. J., S. L. Pitts, K. Nikolics, E. Szonyi, J. N. Wilcox, P. H. Seeburg, and T. A. Stewart (1986). The hypogonadal mouse: Reproductive functions restored by gene therapy. *Science 234*:1372–1378.
81. Covarrubias, L., Y. Hishida, and B. Mintz (1986). Early postimplantation embryo lethality due to DNA rearrangements in a transgenic mouse strain. *Proc. Natl. Acad. Sci. USA 83*:6020–6024.
82. Overbeek, P. A., S. P. Lai, K. R. Van Quill, and H. Westphal (1986). Tissue-specific expression in transgenic mice of a fused gene containing RSV terminal sequences. *Science 231*:1574–1577.
83. Mahon, K. A., P. A. Overbeek, and H. Westphal (1988). Prenatal lethality in a transgenic mouse line is the result of a chromosomal translocation. *Proc. Natl. Acad. Sci. USA 85*:1165–1168.
84. Gordon, J. W., D. Pravtcheva, P. A. Poorman, M. J. Moses, W. A. Brock, and F. H. Ruddle (1989). Association of a foreign DNA sequence with male sterility and a translocation in a line of transgenic mice. *Somat. Cell Mol. Genet.* 15:569–578.
85. Vijaya, S., D. L. Steffen, and H. L. Robinson (1986). Acceptor sites for retroviral integration map near DNaseI-hypersensitive sites in chromatin. *J. Virol. 60*:683–692.
86. Rohdewohld, H., H. Weiher, W. Reik, R. Jaenisch, and M. Breindl (1987). Retrovirus integration and chromatin structure: Moloney murine leukemia

proviral integration sites map near DNaseI-hypersensitive sites. *J. Virol. 61*:336–343.

87. Jaenisch, R., K. Harbers, A. Schnieke, J. Lohler, I. Chumakov, D. Jahner, D. Grotkopp, and E. Hofmann (1983). Germline integration of the Moloney murine leukemia virus at the *Mov 13* locus leads to recessive lethal mutation and early embryonic death. *Cell 32*:209–216.
88. Harbers, K., M. Kuehn, H. Delius, and R. Jaenisch (1984). Insertion of retrovirus into the first intron of α_1(I) collagen gene leads to embryonic lethal mutation in mice. *Proc. Natl. Acad. Sci. USA 81*:1504–1508.
89. Breindl, M., K. Harbers, and R. Jaenisch (1984). Retrovirus-induced lethal mutation in collagen I gene of mice is associated with an altered chromatin structure. *Cell 38*:9–16.
90. Hartung, S., R. Jaenisch, and M. Breindl (1986). Retrovirus insertion inactivates mouse α_1(I) collagen gene by blocking initiation of transcription. *Nature 320*:365–367.
91. Woychik, R. P., T. A. Stewart, L. G. Davis, P. D. Eustachio, and P. Leder (1985). An inherited limb deformity created by insertional mutagenesis in a transgenic mouse. *Nature 318*:36–40.
92. McNeish, J. D., W. J. Scott, and S. S. Potter (1988). Legless, a novel mutation found in PHT-1 transgenic mice. *Science 241*:837–839.
93. Krulewski, T. F., P. E. Neumann, and J. W. Gordon (1989). Insertional mutation in a transgenic mouse pedigree allelic with *pcd. Proc. Natl. Acad. Sci. USA 86*:3709-3712.
94. Markert, C. L., and R. M. Petters (1977). Homozygous diploid embryos produced by microsurgery. *J. Exp. Zool. 201*:295–302.
95. Surani, M. A. H., and S. C. Barton (1983). Development of gynogenetic eggs in the mouse: Implications for parthenogenetic embryos. *Science 222*:1034–1036.
96. McGrath, J., and D. Solter (1983). Completion of mouse embryogenesis requires both the maternal and paternal genomes. *Cell 37*:179–183.
97. Barton, S. C., M. A. H. Surani, and M. L. Norris (1984). Role of paternal and maternal genomes in mouse development. *Nature 311*:374–376.
98. Swain, J. L., T. A. Stewart, and P. Leder (1987). Parental legacy determines methylation and expression of an autosomal transgene: A molecular mechanism for parental imprinting. *Cell 50*:719–727.
99. Hadchouel, M., H. Farza, D. Simon, P. Tiollais, and C. Pourcel (1987). Maternal inhibition of hepatitis B surface antigen gene expression in transgenic mice correlates with de novo methylation. *Nature 329*:454–456.
100. Reik, W., A. Collick, M. L. Norris, S. C. Barton, and M. A. Surani (1987). Genomic imprinting determines methylation of parental alleles in transgenic mice. *Nature 328*:248–251.
101. Sapienza, C., A. C. Peterson, J. Rossant, and R. Balling (1987). Degree of methylation of transgenes is dependent on gamete of origin. *Nature 328*:251–254.
102. Hammer, R. E., V. G. Pursel, C. E. Rexsroad, Jr., R. J. Wall, D. J. Bolt, K. M. Ebert, R. D. Palmiter, and R. L. Brinster (1985). Production of transgenic rabbits, sheep and pigs by microinjection. *Nature 315*:680–683.

103. Gordon, K., E. Lee, J. Vitale, A. Smith, H. Westphal, and L. Henninghausen (1987). Production of human tissue plasminogen activator in transgenic mouse milk. *Biotechnology 5*:1183–1187.
104. Yanagimachi, R. (1981). Mechanisms of fertilization in mammals. In *Fertilization and Embryonic Development In Vitro*. L. Mastroianni, Jr., and J. D. Biggers (eds.) Plenum Publishing, New York, pp. 81–182.
105. Bleil, J. D., and P. M. Wassarman (1983). Sperm-egg interactions in the mouse: Sequence of events and induction of the acrosome reaction by a zona pellucida glycoprotein. *Dev. Biol. 95*:317–324.
106. Talansky, B. E., P. E. Barg, and J. W. Gordon (1987). Membrane ion pump inhibitors block the fertilization of zone-free mouse oocytes by acrosome reacted sperm. *J. Reprod. Fertil. 79*:447–455.
107. Cohen, J., R. Edwards, C. Fehilly, S. Fishel, J. H. Hewitt, J. Purdy, G. Rowland, P. Steptoe, and J. Webster (1985). In vitro fertilization: A treatment of male infertility. *Fertil. Steril. 43*:422–432.
108. Thadani, V. (1982). Mice produced from eggs fertilized in vitro at a very low sperm : egg ratio. *J. Exp. Zool. 219*:277–283.
109. Gordon, J. W., and B. E. Talansky (1986). Assisted fertilization by zona drilling: A mouse model for correction of oligospermia. *J. Exp. Zool. 239*:347–354.
110. Gordon, J. W., L. Grunfeld, G. J. Garrisi, B. E. Talansky, C. Richards, H. Wiczyk, D. Rappaport, and N. Laufer (1988). Fertilization of human oocytes by sperm from infertile males after zona pellucida drilling. *Fertil. Steril. 50*:68–73.
111. Cohen, J., H. Malter, C. Fehilly, G. Wright, C. Elsner, H. Kort, and J. Massey (1988). Implantation of embryos after partial opening of oocyte zona pellucida to facilitate sperm penetration. *Lancet 2*:162.
112. Gordon, J. W. (1988). Use of micromanipulation for increasing the efficiency of mammalian fertilization in vitro. *Ann. N.Y. Acad. Sci. 541*:601–613.
113. Vanderhyden, B. C., K. J. McLaughlin, J. M. Rutledge, and D. T. Armstrong (1989). Zona drilling increases the penetrability of rat oocytes matured in vitro. *Biol. Reprod. 40*:953–960.
114. Barg, P. E., M. Z. Wahrman, B. E. Talansky, and J. W. Gordon (1986). Capacitated, acrosome reacted, but immotile spermatozoa, when microinjected under the mouse zona pellucida, will not fertilize the oocyte. *J. Exp. Zool. 237*:365–374.
115. Mann, J. R. (1988). Full term development of mouse eggs fertilized by a spermatozoon microinjected under the zona pellucida. *Biol. Reprod. 38*: 1077–1083.
116. Lassalle, B., A. M. Courtot, and J. Testart (1987). In vitro fertilization of hamster and human oocytes by microinjection of human sperm. *Gamete Res. 16*:69–78.
117. Laws-King, A., A. Trounson, A. H. Sathananthan, and I. Kola (1987). Fertilization of human oocytes by microinjection of a single spermatozoon under the zona pellucida. *Fertil. Steril. 48*:637–642.

118. Ng, S.-C., A. Bongso, S. S. Ratnam, H. Sathananthan, C. L. K. Chan, P. C. Wong, L. Hagglund, C. Anandakumar, Y. C. Wong, and M. H. H. Goh (1988). Pregnancy after transfer of sperm under zona. *Lancet 2*:790.
119. Uehara, T., and R. Yanagimachi (1973). Microsurgical injection of spermatozoa into hamster eggs with subsequent transformation of sperm nuclei into male pronuclei. *Biol. Reprod. 15*:467–470.
120. Thadani, V. M. (1980). A study of heterospecies sperm-egg interactions in the rat, mouse and deer mouse using in vitro fertilization and sperm injection. *J. Exp. Zool. 212*:435–453.
121. Markert, C. L. (1983). Fertilization of mammalian eggs by sperm injection. *J. Exp. Zool. 228*:195–201.
122. Wentz, A. C., J. E. Repp, W. S. Maxson, D. E. Pittaway, and C. A. Torbit (1983). The problem of polyspermy in in vitro fertilization. *Fertil. Steril. 40*:748–754.
123. Rawlins, R. G., Z. Binor, E. Radwanska, and W. P. Dmowski (1988). Microsurgical enucleation of tripronuclear human zygotes. *Fertil. Steril. 50*:266–272.
124. Gordon, J. W., L. Grunfeld, G. J. Garrisi, D. Navot, and N. Laufer (1989). Successful microsurgical removal of a pronucleus from tripronuclear human zygotes. *Fertil. Steril.* 52:367–372.
125. Malter, H. W., and J. Cohen (1989). Embryonic development after microsurgical repair of polyspermic human zygotes. *Fertil. Steril.* 52:373–380.
126. Kajii, T., and K. Ohama (1977). Androgenetic origin of hydatidiform mole. *Nature 268*:633–634.
127. Tarkowski, A. K., and J. Wroblewska (1967). Development of blastomeres of mouse eggs isolated at the four and eight-cell stage. *J. Embryol. Exp. Morphol. 18*:155–180.
128. Hirao, Y., and R. Yanagimachi (1979). Development of pronuclei in polyspermic eggs of the golden hamster: Is there any limit to the number of sperm heads that are capable of developing into male pronuclei? *Zool. Mag. (Tokyo) 88*:24–33.
129. Monk, M., A. Handyside, K. Harky, and D. Whittingham (1987). Preimplantation diagnosis of deficiency of hypoxanthine phosphoribosyl transferase in a mouse model for Lesch-Nyhan syndrome. *Lancet 2*:423–425.
130. Lyon, M. F. (1974). Mechanisms and evolutionary origins of variable X-chromosome activity in mammals. *Proc. R. Soc. Lond. Ser. B 187*:243–268.
131. Monk, M., and A. H. Handyside (1988). Sexing of preimplantation mouse embryos by measurement of X-linked gene dosage in a single blastomere. *J. Reprod. Fertil. 82*:365–368.
132. Southern, E. M. (1975). Detection of specific sequences among DNA fragments separated by gel electrophoresis. *J. Mol. Biol. 98*:503–517.
133. Saiki, R. K., S. Scharf, F. Faloona, K. B. Mullis, G. T. Horn, H. A. Erlich, and N. Arnheim (1985). Enzymatic amplification of β-globin genomic sequences and restriction site analysis for diagnosis of sickle cell anemia. *Science 230*:1350–1354.

134. Li, H., U. B. Gyllensten, X. Cui, R. K. Saiki, H. A. Erlich, and N. Arnheim (1988). Amplification and analysis of DNA sequences in single human sperm and diploid cells. *Nature 335*:414–417.
135. Saiki, R., D. H. Gelfand, S. Stoffel, S. S. Scharf, R. Higuchi, G. T. Horn, K. B. Mullis, and H. A. Erlich (1988). Primer-directed enzymatic amplification of DNA with a thermostable DNA polymerase. *Science 239*:487–491.
136. Handyside, A. H., R. J. A. Penketh, R. M. L. Winston, J. K. Pattinson, J. D. A. Delhanty, and E. G. D. Tuddenham (1989). Biopsy of human preimplantation embryos and sexing by DNA amplification. *Lancet 2*:347–349.
137. Gordon, J. W., L. M. Isola, and M. W. Bradbury (1989). Rapid sexing of preimplantation mouse embryos by blastomere biopsy and enzymatic amplification [Abstract]. *Proc. Soc. Gynecol. Invest. 1.*

16

Mammalian Cells for Testing Genotoxicity of Chemical Mutagens

T. C. Hsu

University of Texas M.D. Anderson Cancer Center, Houston, Texas

Frederick F. B. Elder

University of Texas Medical School at Houston, Houston, Texas

I. INTRODUCTION

Drosophila was the first test organism used for evaluating mutations induced by x-rays. Soon after H. J. Muller's landmark discovery that x-rays can induce mutations, cytologists, using insect and plant materials, began to study chromosomal changes induced by radiation. Most of the basic cytogenetic information, including the classification of chromosomal aberrations, their origins, and their genetic consequences, had been known for decades. Since the discovery of chemical mutagens and the relationship between mutagen exposure and cancer, environmental mutagenesis has become a major biomedical concern and an active area of research. Numerous new test materials have been employed, and highly sophisticated procedures have been developed to evaluate the genotoxic effects of an ever-increasing array of chemicals.

Dependent upon the test system and the endpoint employed, genotoxic agents may be referred to as *mutagens* (those that can induce mutations), *clastogens* (those that can induce chromosomal aberrations), *carcinogens* (those that can induce cancer), and *teratogens* (those that can induce developmental abnormalities in embryos). The overwhelming majority of genotoxic agents are those that cause damage to cellular DNA. Thus, to merely identify genotoxic agents, mutation tests that use prokaryote systems (such as the Ames' test) are effective and economical. Other test systems that use eukaryote cells, both in vitro and in vivo, can then be used to confirm and modify preliminary findings. One of the testing systems is cytogenetic analysis. Most agents that can cause DNA damage can also induce chromosomal damage; but a number of chemicals, collectively known as mitotic poisons, that interfere with the mitotic mechanisms and cause aneuploid cell progenies, do not necessarily induce chromosomal damage. Because prokaryotes obviously do not employ a mitotic mechanism for nuclear division, the action of mitotic poisons must be tested in eukaryote cells. Compounds found to be negative in prokaryote mutation test systems may still be genotoxic. Therefore, it may be advantageous to use eukaryote cells to identify both mutagens (clastogens) and mitotic poisons.

Improvements in tissue and cell culture methods during the past few decades have made the in vitro system highly amenable to genotoxicity experiments. There are many human and mammalian cell lines available, and one can always initiate new lines.

Other than a few new phenomena, such as sister chromatid exchange, prematurely condensed chromosomes, and fragile sites, basic cytogenetic evaluation of mutagens has changed little since the pioneer days, but the ease in performing experiments with mammalian cells has made experimentation more precise and the data derived from these experiments more meaningful. In the present chapter, we shall limit our presentation to human and mammalian material and protocols that can be applied to evaluate the effects of genotoxic agents. The details of classification of the various types of chromosomal aberration as well as the interpretations of their origins will not be elaborated upon here. They can be found in the literature, including several cytogenetics textbooks [1,2]. We also refrain from using extensive citations except key references.

II. MATERIALS

For most cytogenetic investigations of potentially genotoxic agents, cells in vitro offer distinct advantages over cells in vivo. First, there are a variety of cell lines, of both human and animal origins, available. Second, for preliminary data gathering, such as to determine whether or not the agent to be tested is genotoxic, what its effects might be on cellular growth or on the mitotic apparatus, its general cytotoxicity, and other phenomena, cell cultures offer quicker results and lower costs than can be obtained from animal experimentation. In our opinion, mutagen and clastogen screening programs, as long as the experiments include metabolic conversions, should employ microorganisms or cultured cells and avoid the use of animals until in vivo experiments are deemed necessary and relevant. Third, pulse treatments can be performed with ease and precision by using cultured cells.

A. Cells In Vitro

Unless animal cells offer special advantages, most experiments should be performed with human cells, because the information gained will be more relevant to human welfare. The low diploid numbers of some animal species, such as the Chinese hamster and the Indian muntjac, do not offer a particular advantage in these studies, because their chromosomes are much longer than human chromosomes, so that multiple breaks along both chromatids of each chromosome are more common. This often causes difficulty in analyzing aberrations. Also, without special reasons, heteroploid human cell lines, including most neoplastic cell lines, should be avoided because the chromosomes of these cell lines are often unstable, both numerically and structurally. In analyzing an agent's effects on chromosomes or on mitotic division, heteroploid cell lines cause more uncertainty than diploid cell lines.

1. Human Cells

Three types of cultured human cells may be employed:

Short-Term Lymphocyte Cultures. Short-term lymphocyte cultures from peripheral blood samples have been particularly useful in mutagen assays [3]. Despite some disadvantages, they also appear to be the most sensible material to test an individual's cytogenetic stability and their responses to mutagens. The success rate is usually high when standard blood culture procedure is followed, and few persons will object to donating a small amount of blood. The main disadvantage of this cell system is the difficulty and inconvenience of obtaining the same material repeatedly, for consent of the blood donors must be obtained each time.

Fibroblast Lines. Fibroblast lines give luxuriant cellular growth in early passages and are good material for a variety of investigations. One drawback is that sooner or later they reach a senescence stage. Thus, to maintain a continual supply of the same cellular material, it is necessary to cryopreserve large quantities of cells in the early passages.

Lymphoblastoid Lines. Lymphoblastoid (or lymphoid) lines are usually transformed by Epstein-Barr virus (EBV) infection or other transformants. Once the lymphocytes are transformed, they become "immortal" and will continue to multiply without reaching a senescence stage. Unlike monolayer fibroblasts, lymphoid cells grow as suspension cultures, which may make them more amenable to certain test procedures. Nevertheless, cryopreservation of early passages is still recommended, in the event of contamination or possible genomic alterations of the cell population.

2. Animal Cells

Numerous cell lines derived from a variety of domestic, laboratory, and wild animals are available, and one may initiate cell lines from biopsies to obtain diploid or near-diploid cell populations for mutagen studies. For certain research problems, these lines are indispensable. However, for cytogenetic studies on mutagenesis, animal cell lines are not always desirable. Rodent cell lines, for example, are relatively unstable, often exhibiting aneuploidy and spontaneous chromosomal aberrations, even in early passages. These characteristics make evaluation of genetic damage less precise. Moreover, most long-term cell lines are heteroploid with stemline chromosome number in the triploid or tetraploid range, and that number may vary from cell to cell. Thus, when estimating the frequency of chromosomal aberrations on a per cell or per chromosome basis, the data are not as accurate as those recorded in diploid cell populations.

B. Cells In Vivo

Although one may apply the technique to induce premature chromosome condensation (PCC) to observe interphase chromosomes [4,5], most cytogenetic investigations rely heavily on cells in divisional stages, especially metaphase. To obtain mitotic figures of cells in vivo, it is necessary to use tissues that undergo constant cellular renewal. In adult mammals, only a few tissues meet this requirement: bone marrow, lymphatic organs, intestinal mucosa, cornea, and testis.

Obviously, one cannot use human subjects for mutagen experiments in vivo; but in patients receiving chemotherapeutic treatments, one may examine tissues, such as bone marrow, or initiate short-term lymphocyte cultures to estimate the genotoxic effects of drugs [6,7]. However, contemporary chemotherapy employs a combination of drugs that may act synergistically, making it infeasible to identify the effects of a single clastogen.

When tissues are subjected to in vitro cultivation, the chromosomal constitution of their first mitosis represents that of the cells in vivo. Consequently, cultured lymphocytes, especially those that enter mitosis within 24 hr, can be used as a replacement for cells in vivo. However, the frequency of lymphocytes undergoing mitosis in 24 hr is extremely low, whereas many mitotic cells observed at 72 hr after culture initiation are in the second or third cell generation in vitro. A compromise system is to harvest the cultures at approximately 48 hr, at which time the frequency of mitosis is acceptably higher, but the majority of metaphases still represent the first generation in vitro [8].

The testis is the only organ in which the effects of genotoxic agents on the germline can be effectively analyzed. In humans, testicular biopsy or autopsy specimens may be obtained from patients; but such a procedure certainly cannot be routinely employed. With experimental animals, spermatogonia and spermatocytes of all stages are abundantly available from young males, unless the germinal cells have been destroyed by the genotoxic agent under investigation.

III. HARVEST PROCEDURES

Practically any medium formula used for cell culture should be acceptable. For peripheral blood cultures, most laboratories now employ the RPMI 1640 medium, supplemented with phytohemagglutinin and 10–20% fetal bovine serum. For initiating blood cultures, a proportion of one part peripheral blood and nine parts growth medium appears to be most suitable. For growing lymphoid, bone marrow, and fibroblast cultures, phytohemagglutinin is deleted from the medium. For blood cultures, either culture tubes or T-25 plastic flasks are acceptable as culture vessels. The cultures are incubated at 37°C with a 5% CO_2 atmosphere. If CO_2 incubators are not available, flushing the air space of the culture vessel with 10% CO_2 in air is acceptable.

Actually, the cytogenetic harvest procedures for cells in vivo and in vitro are virtually the same, except for the methods for procuring the cellular materials and for achieving cell suspension. We describe in the following the principal steps for cells in vitro and add necessary modifications for in vivo tissues.

In all cell cultures, the conventional harvest procedure for cytogenetic preparations may be employed. During the last 30 min to 1 hr before cell harvest, a mitotic arrestant [colchicine, 0.5 μg/mL (1.25 μM), or demecolcine (Colcemid), 0.04 μg/mL (0.1 μM)] is added to the culture medium to accumulate metaphase figures. Normally, a short exposure to a mitotic arrestant gives a sufficient accumulation of mitotic figures for analysis, unless the mitotic activity of the cell population is extremely low. Prolonged treatment with a mitotic arrestant will increase the number of metaphases, but the chromosomes of many metaphases will become highly condensed. Extreme condensation of chromosomes is undesirable for observation of chromatid aberrations. For recording of chromatid aberrations, the metaphase chromosomes should have their sister chromatids separated to some extent. This is the opposite of studies on chromosome banding, in which it is most desirable to have the two sister chromatids parallel each other. Thus, for clastogen studies a 1-hr mitotic arrest is recommended.

For blood cultures, the suspended cells are centrifuged and the medium decanted. A hypotonic solution, usually 0.075 M KCl, is introduced, and the cells are gently resuspended in it with a pipette. A treatment time of 10–20 min should be more than sufficient. The hypotonic treatment may be performed at room temperature or at 37°C. Each laboratory should develop a routine protocol to ensure consistency. The cells are then fixed with a freshly prepared Carnoy's mixture (3 : 1 methanol/glacial acetic acid) for at least 10 min, preferably longer. It should be cautioned that cells swollen by a hypotonic treatment are fragile. Abruptly replacing the hypotonic solution with a fixative can cause many cells to rupture, and the chromosomes to suspend in the supernatant. A prudent method is to add about 1 mL of the fixative to 10–15 mL of the cell suspension in the hypotonic solution and gently tilt the centrifuge tube to give the cells a weak fixation before centrifugation.

After centrifugation, remove the supernatant, but leave approximately 1 mL of the supernatant and the cell pellet in the tube. Resuspend the cells gently by tapping the tube with fingers, and then slowly introduce the fixative to the cell suspension to restore the volume to approximately 10 mL, again using the tapping method to achieve a homogeneous cell suspension. Leave cells in the fixative for at least 10 min before recentrifugation. In blood cultures, the first fixation should give a black color because of denatured hemoglobin from lysed red blood cells. When the cells are well fixed, the cellular membrane no longer ruptures as easily. Centrifuge the cells and repeat the fixation process twice (this step is known as washing). In blood cultures, all the hemoglobin should have been removed after washing and a small, whitish cell pellet should be visible after centrifugation.

To prepare slides, remove the fixative, leaving approximately 0.5 mL, then gently suspend the cells with a pipette. The slides to be used should be precleaned with acetone or acid–alcohol, to provide a scrupulously clean surface on which to spread the metaphase chromosomes. Flood a slide with deionized or distilled water (either room temperature, cold, or prewarmed, depending on the protocol established for your laboratory) and drain off the excess until only a film of water remains. Splash a few drops of the cell suspension onto the slide surface at different spots. After the slide has air-dried, inspect it under low-power phase-contrast optics to determine the cell density, frequency of mitoses, and the quality of metaphase spreads. If no metaphases can be found, make one more slide to verify disappointment and discard the sample. It is usually easier to repeat the experiment than to search endlessly for a few available cells. If metaphases are found, but the cell density is too thin or too thick, adjust the amount of the suspension accordingly on the next slide. If the chromosomes are not well spread, try the next slide by dropping the cell suspension from a higher position, by changing the temperature of the slide, or by igniting the fixative immediately after the cell suspension is dropped on the slide [9].

The number of slides to be made depends largely upon the information sought. When mitotic figures are plentiful and the spreads are of good quality, two slides should be sufficient for routine recording of chromosomal aberrations. When banding procedures are to be applied (e.g., for fragile site recording) more slides will be needed.

The harvest procedure for monolayer cell culture differs from that of suspension culture only by the initial step. The cells attached to the flasks or dishes may be dislodged by the use of a trypsin–EDTA solution or by simply scraping with a scraper (rubber policeman). We have found the scraping method effective and time-saving.

For human material (such as bone marrow and biopsy specimens of lymph nodes), a short-term culture of 24 hr or less may be desirable. Because the chromosomes of bone marrow cells are usually condensed, the duration of mitotic arrest should be reduced (no more than 40 min), and the concentration of the arrestant should be reduced.

For small animals (mice, rats, hamsters, and such), the yeast stimulation method for bone marrow may be effectively employed [10]. In animals treated with yeast, the mitotic rate of the bone marrow dramatically increases. Therefore, for short-term experiments, such as to test a drug effect for a treatment of 8 hr or less, the animals can be used 48 hr after yeast injection. The easiest route of administration is to inject the drug intraperitoneally, but other routes of administration may be considered, depending on the drug under investigation.

Under normal conditions, the testis contains division stages of spermatogonia (mitotic) as well as spermatocytes (meiotic), but the frequencies of useful stages (spermatogonial metaphase, diakinesis, metaphase I, and metaphase II) are usually low if the animals are not treated with a mitotic arrestant. Although it is feasible to harvest the cells without mitotic arrest, the procedure of Liang et al. [11], using a sublethal dose of colchicine (40 mg/kg) for intraperitoneal injection into a young adult laboratory mouse for 2 hr, can yield a sevenfold and more than 40-fold increase of metaphase I and metaphase II figures, respectively.

IV. TREATMENT PROCEDURES

A. General

It is a good policy for each laboratory to decide on the test cells and a positive control before assaying unknown compounds. We recommend the use of diploid human cell lines, preferably lymphoid lines, as preliminary test material. For chromosome breakage, several positive control agents, such as triethylene melamine (TEM), *N*-methyl-*N*-nitro-*N*′-nitrosoguanidine (MNNG), and bleomycin, may be used. For sister chromatid exchange, mitomycin C serves well as a desirable positive control agent.

If the molecular weight of the test compound is known, molar concentrations, instead of weight/volume ratios, are recommended because molar concentrations facilitate comparative analyses with other chemicals. If the molecular weight is not known, then it will be necessary to use the weight/volume measurement.

In the following descriptions, we shall use diploid human lymphoid lines derived from normal individuals as an example. Both stock cultures and experimental cultures are maintained in plastic T-25 flasks with 5% CO_2 in the atmosphere. The flasks should stand on end instead of lying flat. The stock cultures are usually subcultured two or three times a week. Experimental cultures should contain 10 mL of growth medium and should be seeded at approximately 5×10^5 healthy cells (dye exclusion test) per milliliter. This initial cell density usually ensures an active mitotic rate after 24–36 hr of incubation.

B. Preliminary Determinations

In assessing the effect of a new or unknown compound, a preliminary experiment to determine growth inhibition and general cytotoxicity often provides a worthwhile guide for future genotoxicity experiments.

The growth curve is one of the best toxicity determinants and is the easiest to quantitate. Most investigators run growth curves in triplicate, or at least duplicate, sets. We believe that running all experiments two or three separate times at different intervals would give a better average picture than running them simultaneously. This is because the conditions of cell culture may vary from one time to another. Moreover, the experimental sets may become too large when they are run simultaneously. The following drug concentrations are recommended for the initial test: untreated control, solvent control, and five drug doses to cover a 5-log range, each 10X as concentrated as the last. Each concentration should have a companion culture with a rat liver microsomal fraction (S-9 mix) to determine if metabolic conversion of the drug occurs. Thus, the initial experiment should consist of 36 cultures. It is not necessary to run a positive control at this time.

The best timetable is to set cultures in the morning, add the drug or solvent in the afternoon, and perform cell counts 24, 48, and 72 hr thereafter. To construct a growth curve, use a small aliquot of each culture for dye exclusion cell count and the remainder for conventional cytogenetic harvest.

In addition to possible growth inhibition, which can be assayed by cell counts, the slides should provide information on several possible cytotoxic effects: (1) a reduction of mitotic rate, (2) presence of distorted mitotic figures and chromosome structure, and (3) an increase of pyknotic nuclei.

If one or all of these abnormalities are found, the agent is most likely cytotoxic and very probably genotoxic as well.

The experiment just described should provide information concerning the range of toxicity, so that in future experiments the range may be narrowed down, whereas intermediate concentrations (e.g., one concentration between the two original concentrations) may be inserted.

Genotoxic agents can be broadly classified into two major types: mitotic poisons (those that interfere with the mitotic apparatus) and clastogens (those that cause chromosomal aberrations). Most mitotic poisons interact with cellular tubulin and cause mitotic arrest or anomalies, or both, in division mechanisms (such as nondisjunction, multipolarity, and lagging chromosomes, all resulting in aneuploid daughter cells). Clastogens can induce all types of chromosomal rearrangements resulting from chromatid or chromosome breakage.

1. Mitotic Poisons

Mitosis is a complex process that ensures proper distribution of chromosomes from a parent cell to the daughter cells. The process involves several cellular organelles and the precisely coordinated timing of their activities. These organelles and their behavior during cell division include aster development and movement of centrosomes, depolymerization of cellular microtubules and repolymerization of tubulin to form a spindle apparatus, condensation of chromosomes, breakdown of the nuclear envelope, arrangement of chromosomes to form a metaphase plate, orientation of kinetochores, separation of sister chromatids (karyokinesis), and, finally, cytokinesis [for detailed information, see Ref. 12]. An error in any of these steps can cause misdivision and result in abnormal distribution of chromosomes into daughter cells, giving rise to polyploid or aneuploid progenies.

The most easily detectable mitotic poisons are mitotic arrestants. These chemicals interfere with the formation of the mitotic spindle, causing an arrest of cell division at metaphase. Many mitotic arrestants are plant alkaloids (e.g., colchicine, vinca alkaloids, taxol), but a number of seemingly unrelated chemicals have also shown mitotic-arresting effects, such as benzene, toluene, halothane, diazepam (Valium), and diethylstibestrol.

If arrested mitoses continue to be arrested, the cells will eventually return to interphase and become polyploid. However, when arrested metaphases recover, as the drug effect subsides, and proceed to reorganize a spindle and undergo anaphase, several abnormalities, including multipolar division, nondisjunction, and lagging chromosomes, may occur. It appears that the frequency of abnormal anaphases is correlated with the duration of arrest, namely, the longer the period of arrest, the higher the frequency of abnormal anaphases.

When daughter cells receive a highly abnormal chromosome complement, they usually die because of their severe genomic imbalance. Cells with minor imbalances, such as "mild" aneuploidy, may survive. Considering that most neoplasms are aneuploid, agents that generate aneuploid progeny should receive as much attention as clastogens in studies of environmental mutagenesis.

Most mitotic poisons arrest mitotic figures at metaphase. Determination of mitotic indices and anaphase/metaphase ratios should indicate the action of a mitotic poison when the treated cell populations are compared with the untreated negative control and colchicine-treated positive control.

2. Clastogenic Effects

Clastogen is a generic term applied to any agent that can induce chromosome breakage. Many contemporary cytogeneticists prefer to use sister chromatid exchange (SCE) as a criterion for measuring genotoxic effects on chromosomes. However, SCE should not be used as the only method because a number of S-independent clastogens, such as ionizing radiations, bleomycin, and zinostatin (formerly, neocarzinostatin), do not significantly increase SCE rates. Alkylating agents, on the other hand, are excellent SCE inducers.

Basically, all chromosomal aberrations result from simple breaks of chromatin fibers. Depending upon the stage of the cell cycle at which breaks are induced and the subsequent rearrangements that may occur before the ensuing metaphase, the chromosomal configurations produced may be complex. This may be one of the reasons that many investigators avoid using chromosomal aberrations as a measure of genotoxicity. Indeed, an investigator should be able to recognize and classify the various chromatid- and chromosome-type aberrations and to understand the mechanisms that cause these aberrations; but for assay purposes, the important information to be obtained is (1) whether or not the agent is a clastogen, and if it is, (2) how potent are its effects on chromosomes. The simplest way of measuring an agent's clastogenic potential, therefore, is to estimate the number of breaks induced, not the configurations. This principle was expressed in an article by Hsu [13], who also advocated a simple method for scoring. If the agent has a clastogenic potential, an increase in the number of chromosomal aberrations, especially chromatid-type aberrations, can be observed in one or more test concentrations (see next section for details). Most likely, not all concentrations will exhibit clastogenic effects. Comparison between the experimental series and the control should reveal the threshold concentration levels necessary to induce chromosome breakage. If none of the concentrations shows an increased frequency of chromosome breakage, both with and without metabolic activation, the conclusion that the agent is nonclastogenic may be made.

It must be borne in mind that clastogens can be classified into two major types, S-dependent and S-independent. S-Dependent compounds can express chromosomal aberrations only after DNA replication. For example, alkylating agents cross-link DNA or DNA and chromosomal proteins, but do not sever DNA molecules. Thus, chromosome breakage can be detected only after DNA replication when the new strands cannot be completed. Cells in the G_2 phase, after exposure to S-dependent agents, will not show chromosomal breaks in the next metaphase, even though cross-links already exist. Many base analogues, which must be incorporated into the new DNA strands, and DNA synthesis inhibitors also belong to the S-dependent category.

S-Independent agents directly cause DNA breakage. If the DNA breaks are not repaired, they will be expressed as chromatid breaks (or chromatid exchanges) in the ensuing metaphase. The best examples of these are ionizing radiations and bleomycin. Such agents can cause chromatid lesions, even after a very short exposure time.

If an unknown agent or a mixture of agents is to be tested, one must consider that the agent may be S-dependent. Therefore, preliminary tests should cover one or more cell cycles.

3. Sister Chromatid Exchange

The phenomenon of SCE has been widely used for estimating genotoxicity of possible chemical mutagens for two major reasons: (1) SCE is a sensitive indicator. Many mutagens can cause an increase of SCE at very low concentrations; (2) Compared with chromosome breakage analysis, the SCE procedure is much easier to learn and requires little interpretation. Detailed descriptions of the procedures may be found in Latt et al. [14].

It should be cautioned here that information derived from SCE experiments cannot be used to replace data on chromosomal aberrations, although SCE data provide important supplementary information. An increase of SCE rate above the control values is a good indicator of an agent's genotoxic potential, but lack of SCE increase does not warrant the conclusion that the agent is harmless.

4. Fragile Sites

Fragile sites denote specific chromosome loci that exhibit breakage at a higher frequency than expected on a random basis. In some cases, fragile sites can be detected in control samples, but usually, the frequencies must be enhanced by various treatment procedures. Fragile sites on human chromosomes have received much attention in the last decade, primarily because of the association of the fragile site at the Xq27 locus and a common form of X-linked mental retardation known as the Martin–Bell syndrome [see Ref. 15 for review]. Although the significance of other chromosomal fragile sites remains largely unknown, they have been implicated in a variety of cytogenetic phenomena, including breakpoints of specific chromosome rearrangements in cancer [16,17], as sequences potentially associated with oncogene loci [18], and as "hot spots" for SCE [19,20]. That fragile sites have some special evolutionary importance has been suggested by some studies that indicate a high degree of evolutionary conservatism among several species [21,22].

Almost all of the data concerning chromosomal fragile sites have been derived from studies of human lymphocytes in culture. In a comparative sense, both among tissue types and among different species, studies of fragile sites are few. With the exception of the fragile X, which, because of its clinical significance, has been studied in some detail in monolayer cultures such as amniotic fluid cells and chorionic villus cells [23,24], little information exists about human fragile site expression in other cell types. Several recent studies of humans, as well as rodents, indicate that some fragile sites are tissue-specific [25–27].

Comparison of fragile sites among species other than primates has proved difficult. This, in part, stems from the problem that most fragile site studies in nonprimates have been carried out on fibroblast cultures, with comparisons then being attempted with human lymphocyte fragile site data. Such comparisons must be considered tenuous, not only because of the use of different cell types, but also because of the underlying cytogenetic differences between the organisms compared. Recent studies of the expression and distribution of chromosomal fragile sites in lymphocyte cultures of the laboratory rat and mouse [28,29] indicate a notable lack of evolutionary conservatism when sites on homologous chromosomes from the same tissue type are compared. Likewise, comparison of fragile site distribution in fibroblasts [22] and lymphocytes [29] in the laboratory mouse suggest the possibility of a high percentage of tissue-specific sites.

For genotoxicity testing, a growing body of evidence indicates that fragile sites may be targets for a wide variety of mutagens and carcinogens, irrespective of the particular mode of action of these agents [18]. Several earlier studies have shown that increases in the frequency of micronucleus formation can be observed in some groups of individuals whose cells are cultured under appropriate fragile site-induction conditions. These findings provide evidence that fragile sites may predispose to chromosomal breakage [see Ref. 15 for review]. Most genotoxicity testing does not take into account site specificity in scoring chromosome breakage because most analysis is done on unbanded chromosomal preparations. Nevertheless, we feel that researchers examining genotoxicity through chromosome breakage analysis should be aware that some or many genotoxic agents may induce site-specific lesions and, given the data currently at hand, these lesions may vary in location and frequency from one tissue type to another and one species to another. If site- and tissue-specificity is suspected, it may become important in the evaluation of the genotoxic agent to do further analysis to determine if a particular agent functions as a specific fragile site inducer.

A detailed discussion of the detection of fragile sites on human chromosomes is given in Chapter 9.

C. Follow-Up Experiments

If the initial experiments show a possible mitotic poison, a repeated experiment should be performed with the test agent *without* using the colchicine arrest before cell harvest. Instead, a colchicine control with the same treatment duration should be run for comparison. The treatment time can be as short as 3–5 hr.

To study possible clastogenic effects, one can select two or three concentrations from the initial toxicity experiments that showed some degree of toxicity and, then, design the next experiment. If so desired, intermediate concentrations of the agent may also be examined. These experiments should cover both S-dependent and S-independent possibilities of the agent's action. Therefore, an 18-hr or a 24-hr treatment schedule is preferred. Set up the cultures as before and harvest the cells the next day, with a 1-hr colchicine arrest before harvest. When positive clastogenic effects are noted in the 18-hr or 24-hr experiment, an 8-hr exposure (to cover S + G_2 phase) and a 2-hr exposure (G_2 only) will simplify the conclusion on whether the clastogenic agent in question is S-dependent or S-independent, respectively. The necessity for further investigation of metabolic activation of the agent will depend upon the results obtained from the initial experiment.

V. DATA COLLECTION

A. Mitotic Index and Anaphase/Metaphase Ratio

Usually, when constructing a mitotic index (percentage of mitotic cells in the entire cell population), all mitotic stages from prophase to telophase are scored. This method has certain advantages, but it is relatively cumbersome because one must be able to recognize early stages of prophase and late stages of telophase. Sometimes it is extremely difficult to differentiate very early prophase nuclei and interphase nuclei with an appearance of coarse chromatin, as well as to differentiate late telophase from early daughter cells. Moreover, in mammalian cells, cytokinesis occurs soon after the completion of anaphase. Sister telophase nuclei are not always associated with each other. This phenomenon causes difficulty in enumerating the frequency of telophases. Thus, some errors are inevitable. However, it is relatively objective to record the frequency of metaphase and anaphase figures. In cell populations treated with a mitotic arrestant, it is good policy to record only metaphases and anaphases. With an exposure to colchicine at a concentration capable of a complete metaphase arrest, there will be no anaphase or telophase cells. Accordingly, if the mitotic index is replaced by a metaphase and anaphase index, the procedure can be much simplified. We suggest a random sampling of at least 2000 cells per sample.

Because all mitotic arrestants have a threshold dose below which the ability to arrest cell division at metaphase is either absent or incomplete, some anaphases may be observed at weaker concentrations. An anaphase/metaphase (A/M) ratio may be calculated by dividing the number of anaphases by the number of metaphases (at least 100 such figures). An A/M ratio of zero signifies a complete mitotic arrest, and ratios smaller than that of the untreated control sample suggest partial mitotic block.

When mitotically arrested cell populations are incubated in a drug-free medium, most metaphase cells will recover and eventually enter anaphase. In the recovering cell populations, a number of abnormal anaphases (nondisjunction, lagging chromosomes, and multipolar spindles) may be noted. Figure 1 shows examples of anaphases of a male Indian muntjac (2n = 7) cell line that was treated with demecolcine (Colcemid; 2.5×10^{-8} M) for 5 hr and reincubated in demecolcine-free medium for 30 min. An obvious nondisjunction is illustrated in Figure 1c, and a lagging chromosome can be noted in Figure 1d. Figure 2 depicts a nondisjunction during anaphase of a Chinese hamster cell (2n = 22) after mitotic arrest and reversal.

To observe the frequencies of anaphase abnormalities, a modification of the harvest procedure should be made. We recommend harvests 30, 60, and 90 min after reincubation in the drug-free medium. The cells are treated with a weak hypotonic solution (3 : 1 drug-free medium/distilled water) before fixation. The classic acetic orcein squash method gives better preparations than air-dried slides for this type of study.

B. Micronucleus Test

When a chromatid break or a chromosome break occurs, the resulting acentric fragment will not be able to move poleward during anaphase. Usually, it lags near the equatorial plate and, subsequently, forms a micronucleus, which may be incorporated into one of the daughter cells. More than one chromosomal fragment in the parent cell may form multiple micronuclei, which may be incorporated into either or both daughter cells. Record-

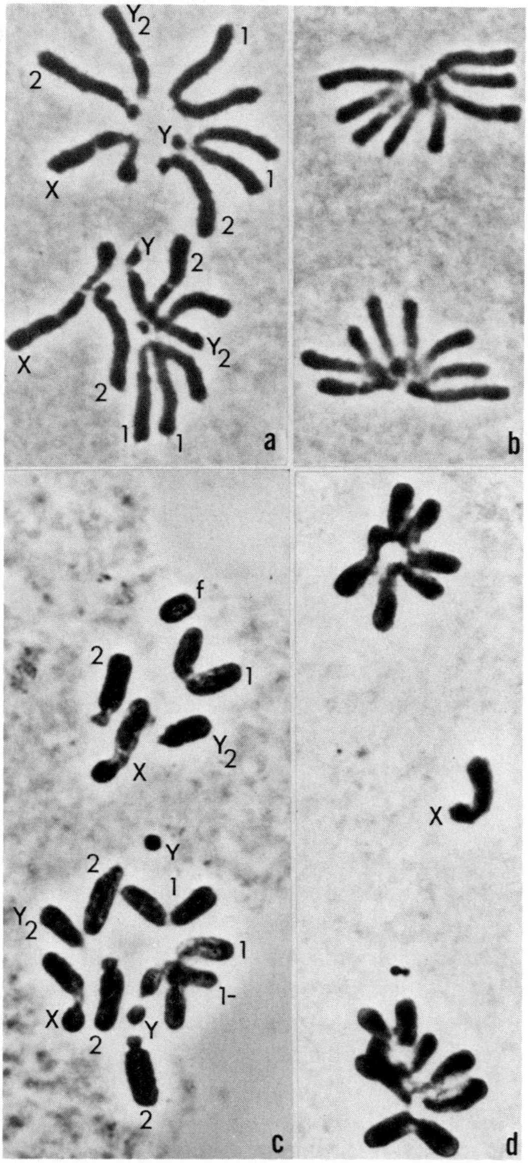

Figure 1 Anaphases from a male Indian muntjac cell line (2n = 7) after treatment with demecolcine (Colcemid; 10^{-8} M) for 5 hr and return to demecolcine-free medium for 30 min. (a) normal anaphase: each chromosome can be identified at both poles; (b) normal late anaphase with eight chromosome arms; (c) nondisjunction (f, acentric fragment); (d) nondisjunction plus a lagging chromosome.

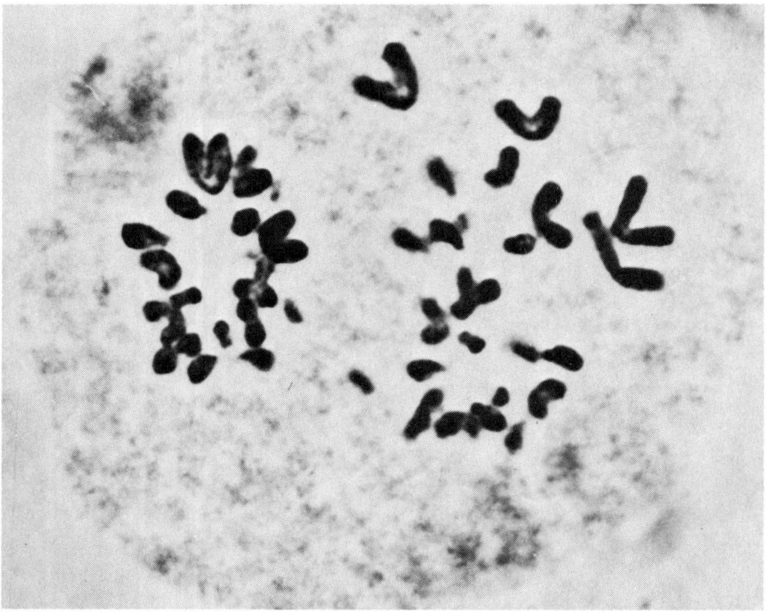

Figure 2 A nondisjunction anaphasic figure from a diploid Chinese hamster cell line (Don) after treatment with demecolcine (Colcemid; 10^{-6} M) for 5 hr and return to demecolcine-free medium for 20 min.

ing micronuclei may give an indirect estimate of the frequency of chromosome breaks in the cell population being assayed. The micronucleus test is most effective for bone marrow samples from animals treated with mutagens in vivo [30]. If an agent can induce chromosomal breaks, it can induce micronuclei in all dividing cells, regardless of the cell type. However, after the last mitosis of the erythroblasts, the micronuclei may remain in the polychromatic erythrocytes after the main nuclei are extruded. Thus, by examining the frequency of micronuclei in the polychromatic erythrocytes, one can obtain a quantitative assessment of the clastogenic action.

Scoring micronuclei in erythrocytes to estimate the genetic damage of a potential clastogen is indeed a simple and elegant technique, but, in cell cultures, nuclear extrusion does not occur. The current procedure takes advantage of the property of cytochalasin B, which prevents cytokinesis of animal cells, but does not interfere with karyokinesis. In the presence of cytochalasin B, a normal mitosis will result in a binucleated cell. If a chromosomal fragment or a lagging chromosome is present, the binucleate cell will contain a micronucleus. By recording the number of micronuclei present in the binucleated cells, one can again indirectly measure the effect of a genotoxic agent. This procedure [31] has been used mostly by radiologists, but it should also be applicable to chemical clastogens and mitotic poisons.

C. Chromosomal Aberrations

One problem related to quantitative assessment of chromosomal damage induced by genotoxic chemicals is that a number of DNA intercalators and aklylating agents cause chromosome stickiness [32]. Chromatin fibers of various dimensions may tangle, and the chromosomes involved cannot be separated into distinct entities. This characteristic causes difficulty in scoring chromosomal aberrations. Also, several chemicals may cause chromosome stretching (actually failure to condense at metaphase) in certain chromosomal regions. These attenuated regions may be considered gaps or breaks by some cytogeneticists, but whether or not they represent real chromatid breaks is not certain at the microscopic level. Other clastogens, such as bleomycin and cytarabine (cytosine arabinoside), do not induce chromatin stickiness, so that chromatid breaks are relatively easy to enumerate. Nevertheless, even with chromosome tangling, obvious chromatid breaks can still be observed and recorded, although quantitation may not be as precise.

The kaleidoscopic array of chromosomal aberrations are the consequences of rearrangements occurring after chromatin breaks. It should be remembered that, other than simple chromatid breaks or gaps, most simple rearrangements (translocations, inversions, interstitial deletions) represent two breaks and a restitution. The chromosomes may exhibit a multitude of seemingly confusing configurations, but it is not exceedingly difficult to view them as a number of separate events of two breaks, followed by a restitution. Therefore, a simple dicentric (or a ring) and an acentric fragment can be recorded as two breaks, and a tricentric, four breaks. Chromatid exchanges, regardless of the number present, can be counted as two breaks each. Accordingly, three chromatid exchanges involving two, three, or four chromosomes are counted as six original breaks. In clastogen assays, the primary purpose is to estimate the potency of genotoxic compounds by cytogenetic criteria; therefore, how the breaks of chromatin fibers restitute is of secondary importance when compared with how many breaks have been induced. Converting each restituted configuration into two breaks greatly simplifies the recording procedure.

The more controversial aspect of cytogenetic recording of chromosomal damage is the differentiation between chromatid breaks and gaps. There are two problems: (1) the criteria for identifying breaks and gaps, and (2) the recommended method of recording.

Chromatid lesions, such as achromatic disruptions of the continuity of a chromatid, may be barely perceptible, up to extremely conspicuous. The very obvious ones, especially when the terminal segment distal to the break is displaced to the other side of the sister chromatid, represent no problem. Most minor lesions may indeed represent true chromatid breaks, but it is also possible that some do not. The immediate question is: Should all of them be recorded as chromatid breaks, or should gaps and breaks be recorded separately, or should gaps not be recorded at all? The next question is: What criterion is generally used to differentiate gaps and breaks?

There are two recording systems, both arbitrary, to differentiate gaps and breaks: (1) The Chatham Barrs Inn Conference recommendation [33], which suggests the use of the length of the lesion as a criterion for distinguishing gaps and breaks; namely, when the length of the achromatic lesion is equal to, or longer than, the width of the chromatid, it is called a chromatid break, whereas when the length of the lesion is shorter than the width of the chromatid, it is called a gap. (2) The system recommended by the International System of Cytogenetic Nomenclature (ISCN) [34], namely, a chromatid break is identified only when the chromatid fragment is displaced; if the fragment is still aligned with its sister at the original position, it is referred to as a chromatid gap regardless of the length of the lesion. Hsu [35] presented a discussion on these nomenclature systems and recommended the Chatham Barrs Conference system as the lesser of two evils.

Probably the best way of enumerating chromatid lesions is to record gaps and breaks separately, but this procedure is rather laborious, especially in samples for which numerous lesions are present. Hsu [35] compared ten bleomycin-treated samples in which chromatid gaps and breaks were recorded separately. He found that when gaps and breaks are combined, the increased value of breaks plus gaps per cell over breaks per cell alone does not significantly alter the overall picture. It should be borne in mind that assay systems for genotoxicity are designed to obtain a glimpse of the actions of agents, not the total output of genetic lesions. With cytogenetic protocols, chromosomal lesions not detected by microscopic observation are necessarily omitted. We are aware that cytogenetic estimation of genotoxic action represents only one spot of the entire leopard. Following this concept, the ignoring of minor gaps should not change the conclusion that an agent is a clastogen and that the potency of the clastogen can still be estimated by recording obvious breaks alone. The time saved, on the other hand, may be substantial. Muller's original mutation test on *Drosophila* was designed to reveal mutation rates only within the inversion loop of the X chromosomes, not the mutation load of the entire genome. In cytogenetic evaluation, a similar principle can be followed. The absolute number of induced chromosome lesions is less important than the relative values, as long as the data are objectively collected and the criteria used are consistently applied.

REFERENCES

1. Swanson, C. P., T. Merz, and W. J. Young (1981). *Cytogenetics*. Prentice-Hall, Englewood Cliffs, N. J.
2. Schulz-Schaeffer, J. (1980). *Cytogenetics: Plants, Animals, Humans*. Springer-Verlag, New York.
3. Buckton, K. E., and H. J. Evans (1982). Human peripheral blood lymphocyte cultures: An in vitro assay for the cytogenetic effects of environmental mutagens. In *Cytogenetic Assays of Environmental Mutagens*. T. C. Hsu (ed.). Allenheld, Osmun, Publishers, Totowa, N. J., pp. 182–202.
4. Rao, P. N., and R. T. Johnson (1974). Induction of chromosome condensation in interphase cells. In *Advances in Cell and Molecular Biology*. E. J. DuPraw (ed.). Academic Press, New York, pp. 135–189.
5. Hittelman, W. M. (1982). Premature chromosome condensation for the detection of mutagenic activity. In *Cytogenetic Assays of Environmental Mutagens*. T. C. Hsu (ed.). Allenheld, Osmun, Publishers, Totowa, N. J., pp. 353–384.
6. Gebhart, E. R. E. (1982). The epidemiological approach: Chromosome aberrations in persons exposed to mutagens. In *Cytogenetic Assays of Environmental Mutagens*. T. C. Hsu (ed.). Allenheld, Osmun, Publishers, Totowa, N. J., pp. 385–408.
7. Gebhart, E. (1987). Clastogenic action of chemotherapy in humans. *In Vivo* *1*:15–22.
8. Bloom, A. (1972). Induced chromosomal aberrations in man. In *Advances in Human Genetics*. H. Harris, and K. Hirschhorn (eds.). Plenum Press, New York, pp. 99–122.
9. Scherz, R. G. (1962). Blaze drying, by igniting the fixative, for improved spreads of chromosomes in leucocytes. *Stain Technol. 37*:386.
10. Lee, M. R., and F. F. B. Elder (1980). Yeast stimulation of bone marrow mitosis for cytogenetic investigations. *Cytogenet. Cell Genet. 26*:36–40.
11. Liang, J. C., T. C. Hsu, and M. Gay (1985). Response of murine spermatocytes to the metaphase-arresting effects of mitotic arrestants. *Cytobios 43*:101–114.
12. Hyams, J. S., and B. R. Brinkley (eds.) (1989). *Mitosis: Molecules and Mechanisms*. Academic Press, New York, pp. 1–350.
13. Hsu, T. C. (1982). Introduction. In *Cytogenetic Assays of Environmental Mutagens*. T. C. Hsu (ed.). Allenheld, Osmun, Publishers, Totowa, N. J., pp. 1–9.
14. Latt, S. A., R. R. Schreck, K. S. Lovejoy, and C. F. Shuler (1982). Sister chromatid exchange analysis: Methodology, application, and interpretations. In *Cytogenetic Assays of Environmental Mutagens*. T. C. Hsu (ed.). Allenheld, Osmun, Publishers, Totowa, N. J., pp. 29–80.
15. Sutherland, G. R., and F. Hecht (1985). Fragile sites on human chromosomes. *Oxford Monographs on Medical Genetics No. 13*. Oxford University Press, New York.

16. LeBeau, M. M. (1986). Chromosomal fragile sites and cancer-specific rearrangements. *Blood 67*:849–858.
17. Yunis, J. J., and A. L. Soreng (1984). Constitutive fragile sites and cancer. *Science 226*:1199–1204.
18. Yunis, J. J., A. L. Soreng, and A. E. Bowe (1987). Fragile sites are the targets of diverse mutagens and carcinogens. *Oncogene 1*:59–69.
19. Glover, T. W., and C. K. Stein (1987). Induction of sister chromatid exchanges at common fragile sites. *Am. J. Hum. Genet. 41*:882–890.
20. Glover, T. W., and C. K. Stein (1988). Chromosome breakage and recombination at fragile sites. *Am. J. Hum. Genet. 43*:265–273.
21. Miro, R., I. C. Clemente, C. Fuster, and J. Egozcue (1987). Fragile sites, chromosome evolution, and human neoplasia. *Hum. Genet. 75*:345–349.
22. Djalali, M., S. Adolph, P. Steinbach, H. Winking, and H. Hameister (1987). A comparative mapping study of fragile sites in the human and murine genomes. *Hum. Genet. 77*:157–162.
23. Sutherland, G. R., E. Baker, S. Purvis-Smith, A. Hockey, E. Krumins, and S. Z. Eichenbaum (1987). Prenatal diagnosis of the fragile X using thymidine induction. *Prenat. Diagn. 7*:197–202.
24. Tommerup, N. (1988). Induction of the fra(X) in amniotic fluid cells by excess thymidine. *Am. J. Med. Genet. 30*:451–453.
25. Hecht, F., B. K. Hecht, and I. R. Kirsch (1987). Fragile sites limited to lymphocytes: Molecular recombination and malignancy. *Cancer Genet. Cytogenet. 26*:95–104.
26. Hunt, P. A., and P. S. Burgoyne (1987). A tissue-specific fragile site associated with the sex reversed (*Sxr*) mutation in the mouse. *Chromosoma 96*:67–71.
27. Murano, I., A. Kuwano, and T. Kajii (1989). Fibroblast-specific common fragile sites induced by aphidicolin. *Hum. Genet. 83*:45–48.
28. Robinson, T. J., and F. F. B. Elder (1987). Multiple common fragile sites are expressed in the genome of the laboratory rat. *Chromosoma 96*:45–59.
29. Elder, F. F. B., and T. J. Robinson (1989). Rodent common fragile sites: Are they conserved? Evidence from the mouse and rat. *Chromosoma 97*:459–464.
30. Schmid, W. (1982). The micronucleus test: An in vitro bone marrow method. In *Cytogenetic Assays of Environmental Mutagens*. T. C. Hsu (ed.). Allenheld, Osmun, Publishers, Totowa, N. J., pp. 221–230.
31. Fenech, M., and A. A. Morley (1986). Cytokinesis-block micronucleus method in human lymphocytes: Effect of in vivo ageing and low dose x-irradiation. *Mutat. Res. 161*:193–198.
32. McGill, M., S. Pathak, and T. C. Hsu (1974). Effects of ethidium bromide on mitosis and chromosomes: A possible material basis for chromosome stickiness. *Chromosoma 47*:157–167.
33. Chatham Workshop Conference on Karyological Monitoring of Normal Cell Populations (1971). *Int. Assoc. Biol. Stand.*, pp. 65–161.
34. ISCN (1985). An International System for Human Cytogenetic Nomenclature. *Cytogenet. Cell Genet. 31*:1–23.
35. Hsu, T. C. (1987). Genetic predisposition to cancer with special reference to mutagen sensitivity. *In Vitro Cell. Dev. Biol. 23*:591–603.

17

Primary Culture of Solid Lung Tumors for Chromosomal Analysis

Jill M. Siegfried

University of Pittsburgh, School of Medicine, Pittsburgh, Pennsylvania

James H. Resau

University of Maryland School of Medicine, Baltimore, Maryland

Ikuo Miura* and Joseph R. Testa

Fox Chase Cancer Center, Philadelphia, Pennsylvania

I. INTRODUCTION

Emerging knowledge of the molecular basis for malignancy has increased our understanding of the mechanistic relationship between chromosomal abnormalities and cancer. In some tumor types for which extensive, detailed karyotyping has been possible, specific translocations and other rearrangements have been detected, which result in alterations in the expression of individual oncogenes. Such molecular aberrations occur repeatedly in specific tumor types. Examples of this kind of abnormality are the 9;22 translocation observed in chronic myelogenous leukemia, resulting in an *abl* oncogene protein with altered tyrosine kinase activity [1–3]; a similar translocation in acute lymphoblastic leukemia, which also causes produc-

**Current affiliation:* Akita University School of Medicine, Akita, Japan

tion of an altered *abl* protein [4,5]; and the 8;14, 2;8, and 8;22 translocations detected in Burkitt's lymphoma, which are associated with unregulated c-*myc* expression [6–8].

Solid tumors, which represent a far greater proportion of cancer cases than the hematologic malignancies, probably also are associated with specific chromosomal abnormalities, which lead to changes in genes that are causally related to the process of tumorigenesis [9]. Because of the relative difficulty in obtaining adequate karyotypes from solid fresh tumor tissue, identification of chromosomal alterations associated with particular tumor types has proved difficult. When enough cytogenetic and molecular information is available, such as for neuroblastoma [10,11], small-cell lung cancer [12], and retinoblastoma [13,14], changes in gene structure and chromosome structure have been linked. A large catalogue of known karyotypic abnormalities has been compiled for other types of solid tumors, but many of these were described in long-term cultured cell lines, effusions, or in other sporadic cases of advanced disease. There is still considerable uncertainty whether the wide range of chromosomal abnormalities seen in solid tumors is the result of genetic instability introduced in the tumor by virtue of the malignant state, or whether any of the observed changes contributed to the malignant state. In leukemias [15], consistent karyotypic alterations have been found that are related to diagnosis and prognosis. If the data were available, such relationships might also be found in solid non–small-cell lung tumors, an important site of human cancer, which has a generally poor prognosis and which will probably continue to be a major disease site because of the popularity of tobacco products [16].

Chromosomal analysis of solid non–small-cell lung tumors on a routine basis has been much more difficult than analysis of leukemias and lymphomas. There are several reasons for this: (1) the low rate of cell growth in these tumors; (2) the presence of cells in many stages of differentiation, many of which will not enter mitosis; (3) the areas of necrosis and hypoxia in which cell division does not occur; (4) the difficulty in adapting these tumor cells to in vitro culture; and (5) the difficulty in obtaining fresh viable surgical specimens for research purposes. Therefore, new approaches are necessary to establish solid tumors in short-term culture and to induce the malignant stem cell population to divide.

We have undertaken several new approaches in an attempt to improve the success rate in obtaining metaphase spreads from primary solid tumors in short-term culture, namely, (1) the use of different substrata to induce tumor cells to attach, and (2) the supplementation of culture medium with "autocrine" growth factors secreted by an established lung tumor cell line. The results from culturing fresh tumor specimens by these techniques and from karyotypic analysis will be discussed.

II. METHODS

A. Use of Murine Fibroblast 3T3 Cells Combined with Tumor Cell-Conditioned Medium

For many normal epithelial cell types, fibroblast feeder cells enhance attachment and growth [17,18]. Primary epithelial cells derived from normal human bronchus have been shown to attach to and grow readily on murine 3T3 fibroblasts inactivated either by irradiation or exposure to mitomycinC [19,20]. Epithelial cells grown on feeder layers often survive longer, have a higher growth index, and can be induced to develop differentiated characteristics in culture more readily than those cultured on plastic or other substrates. The effect of feeder cells appears to be primarily that of enhancing attachment and maintaining function, rather than a direct growth-promoting effect, since medium conditioned by 3T3 cells does not provide the same benefits as the physical presence of the 3T3 cells [20].

B. Inactivation of 3T3 Cells

Murine Swiss 3T3 cells were obtained from Dr. Howard Green, Harvard University. The 3T3 cells are cultured in Dulbecco's modified Eagles' medium (DMEM) with the addition of 10% iron-supplemented bovine calf serum (Hyclone Laboratories) containing antibiotics. Irradiation is accomplished by exposing trypsinized cells to a ^{137}Ce source for 2.5 min (5000-rad total dose) in medium containing serum. Cells are then replated at a density of 2×10^4 cells/cm^2 for feeder layers. Alternatively, confluent cultures are exposed to 4 μg/mL mitomycinC (Sigma Chemical Co.) for 2 hr, rinsed with Hanks' buffered salt solution, trypsinized, and replated. Feeder cells are plated at least 24 hr before addition of cells from solid tumors and can be used up to 2 weeks after inactivation if medium is replenished. The DMEM medium is removed and replaced with tumor cell growth medium (described in Sect. II.C) before plating of lung tumor cells. Inactivated 3T3 cells continue many metabolic functions, but are unable to undergo cell division; therefore, they do not interfere with karyotypic analysis of cells with which they are cocultured.

C. Preparation of Conditioned Medium

Conditioned medium is prepared using the human lung tumor cell line A549, which was derived from a bronchioloalveolar carcinoma [21]. These cells secrete autocrine growth factors into the culture medium that are stimulatory for the growth of normal and malignant primary lung epithelial cells [22,23]. Among the proteins secreted are forms of transforming growth factor-α (TGFα) and insulin-like growth factor I (IGF-I). An unidentified factor, which promotes anchorage-independent growth of primary lung carcinoma cells, is also secreted [23]. Conditioned medium is collected from cell cultures grown in basal medium Eagle's (BME), supplemented with 1% fetal bovine serum (Hyclone Laboratories). Cells are grown in roller bottles, and three 48-hr collections are taken from each bottle once cultures are three-fourths confluent. Medium is acidified to pH 6.8 and stored at 4°C for up to 2 weeks. To prepare tumor culture medium, conditioned medium is centrifuged at 1000 rpm for 10 min to remove any cell debris and combined 1 : 1 with fresh Ham's F12 medium supplemented with 1% fetal bovine serum. Antibiotics and amphotericin B (Fungizone; Gibco) are also added to prevent contamination when culturing primary tissue. The medium is then filtered slowly under positive pressure through a 0.22-μm sterile filter to ensure that no viable A549 cells are introduced into primary tumor cell cultures. This medium is used within 1 week.

D. Culture of Primary Tumor Tissue on 3T3 Feeder Layers

Specimens are cultured by a mince technique as follows: tissue is rinsed, obvious areas of necrosis or connective tissue are cut away, and the tissue is minced between two scalpels. The resulting aggregates and small tissue pieces are triturated in a 10-mL pipet and briefly vortexed at low speed. These fragments and aggregates are then plated onto confluent flasks of inactivated 3T3 cells and fed with conditioned medium as described above. After 48 hr, flasks are checked for attachment of tumor tissue. Floating cell aggregates are removed to additional flasks periodically to increase the number of cells attaching, and all flasks are monitored for colonies of tumor cells, which usually develop in 7–14 days. Flasks containing colonies are usually used between day 14 and day 30 for chromosomal analysis, unless, because of a slow growth rate, more time is needed to obtain enough cells for adequate analysis.

E. Use of Matrix Proteins as Tissue Culture Substrates

To prepare flasks coated with matrix proteins for culturing fresh tumor specimens, fibronectin, laminin, and type IV collagen (Sigma Chemical Co.) are dissolved in calcium- and magnesium-free phosphate-buffered saline (PBS). The concentrations used to coat plastic culture dishes are: 30 μg/mL fibronectin, 100 μg/mL collagen, and 10 μg/mL laminin. A mixture of these three proteins is poured into 75-cm^2 flasks and swirled to cover the bottoms of the flasks. The excess is aspirated off and flasks are air-dried before use. Explants of lung tumor tissue are placed in flasks with serum-free KGM medium (Clonetics), which was found to be superior to serum-supplemented medium (see Sect. III). Colony formation on matrix proteins is assessed after fixation and staining with crystal violet, and colonies are counted and tabulated as modified from Cottrell and Resau [24].

F. Harvest of Cells for Chromosomal Analysis

Cultures are exposed to demecolcine (Colcemid; 0.01 μg/mL) overnight, usually for 16–18 hr. For cell cultures exhibiting more active growth, a shorter exposure to demecolcine (2 hr at 0.02 μg/mL) is used. To inhibit chromosome contraction during prolonged treatment with mitotic arrestant, we sometimes expose cells to ethidium bromide (10 μg/mL) during this same 2-hr period, before termination of the culture. Cells are collected by trypsinization and centrifugation at 1000 rpm. Cells growing on matrix proteins tend to form particularly adherent monolayers, so that we usually trypsinize and scrape cells free with a cell scraper. Floating tumor cells present in culture medium are combined with detached monolayer cells. Trypsin activity is inhibited by resuspension in medium containing 10% fetal bovine serum. The cells are pelleted by centrifugation, and the supernatant is discarded. The cells are then resuspended in hypotonic medium (4 : 1 0.075 M KCl/1% sodium citrate) for 30 min; after centrifugation, the cell pellet is resuspended gradually in fixative (3 : 1 methanol/acetic acid) with shaking. After 30 min, the cells are again centrifuged, the supernatant discarded, and the cells are resuspended in fresh fixative. The cells are placed at −20°C overnight, followed by one to two additional fixative changes. Slide preparations are performed by air-drying or flame ignition techniques. Chromosomes are banded by standard trypsin G-banding (using Wright stain) and quinacrine (Q)-banding methods. Metaphase cells are photographed and karyotypes expressed according to the ISCN [25].

III. RESULTS

A. Tumor Cell Culture

Figure 1 shows primary cultures of two solid lung tumors (an adenosquamous carcinoma and an adenocarcinoma) placed on 3T3 feeder layers and growing in the presence of A549-conditioned medium. The tumor cells can be seen as colonies of tightly joined cells growing on a background of 3T3 fibroblasts. In Figure 1A, a clump of tissue can be seen from which the tumor cells are emerging. In Figure 1B, several small colonies of tumor cells have formed after 1 week in culture. The presence of a confluent layer of 3T3 cells inhibits the attachment of connective tissue cells from the tumor specimen and also promotes the attachment of carcinoma cells. Previous experience over 4 years with fresh solid lung tumor specimens showed a success rate of 90.5% in obtaining such primary colonies (Table 1). These

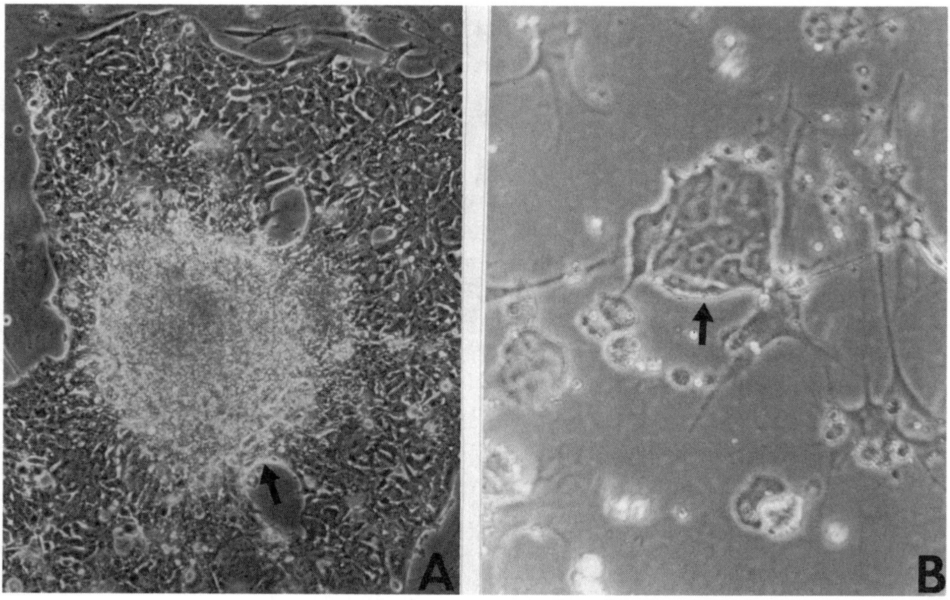

Figure 1 (A) Adenosquamous carcinoma of the lung cultured on fibroblast feeder cells: Arrow indicates portion of fresh tumor tissue (phase-contrast microscopy, ×250). (B) Adenocarcinoma cultured on fibroblast feeder cells: Arrow indicates colony of tumor cells. 3T3 fibroblasts are seen around tumor colony (phase-contrast microscopy, ×1000).

colonies have been confirmed to be composed of epithelial cells by immunostaining with a monoclonal antibody for human keratin (Fig. 2A,B). Colonies were confirmed to be malignant by several criteria: (1) expression of the tumor marker protein pregnancy-specific glycoprotein-1 (SP-1 [26]; see Fig. 2C and Table 2); (2) expression of the muscle (anaerobic) forms of lactate dehydrogenase (LDH [27]; see Table 2); (3) anchorage-independent

Table 1 Establishment of Primary Cultures from Fresh Solid Lung Tumor Specimens

Histology	Percentage successful
Adenocarcinoma	80 (4/5)
Adenosquamous	100 (3/3)
Squamous cell	87.5 (7/8)
Large Cell	100 (5/5)
Total	90.5 (19/21)

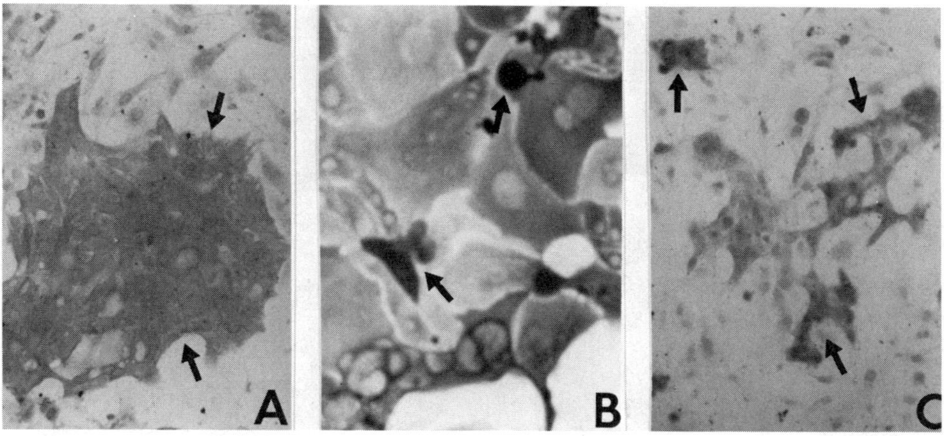

Figure 2 Visualization of keratin and SP-1 glycoprotein by immunoperoxidase staining of colonies of lung tumor cells: Reagents were obtained from Dako, Inc. (Santa Barbara, Calif.). A monoclonal antibody to human keratin or SP-1 glycoprotein was employed as the primary antibody. Counterstained with hematoxylin. (A) Colony of adenosquamous tumor cells (arrows) is keratin-positive; surrounding 3T3 feeder cells are negative. (B) Higher magnification of a squamous cell carcinoma showing cytoplasmic staining of keratin. Some highly positive bundles of residual keratin are also visible (arrows). (C) Cells derived from a large-cell carcinoma stain positively for SP-1 glycoprotein (arrows); 3T3 fibroblasts are negative.

growth (see Table 2); and (4) formation of invasive tumors in immunodeficient mice (see Table 2). The SP-1 expression was determined by immunostaining; LDH expression was measured by extraction of cells harvested from secondary cultures and electrophoretic separation of the extract [23]. Anchorage-independent growth was measured by culture in semisolid agar; between 10^4 and 10^5 cells were plated per agar dish and colony formation was observed over a 4-week period [23]. Formation of tumors in immunodeficient mice was analyzed by intraperitoneal injection of 10^5–10^6 cells, depending on the number available, and observation of tumor growth over a 4-month period. Resulting tumors were fixed, stained with hematoxylin and eosin, and their histology determined. The resulting tumors either displayed the histologic appearance of the original fresh specimen, or an undifferentiated, anaplastic tissue structure. Although every culture was not positive by all the criteria examined (see Table 2), each culture displayed some transformed characteristics. In the present collaborative study, there was a 1-day delay in receiving tumor tissue because of overnight shipping, and the success rate of obtaining primary colonies was reduced to 50% (5/10 tissue specimens).

Table 2 Malignant Characteristics of Cells Cultured from Fresh Solid Lung Tumor Specimens

Specimen no.	Histology	Expression of SP-1	Elevated expression of muscle LDH subunit	Growth in immuno-deficient mice	Anchorage-independent growth
81-86T	Large cell	−	+	+	+
91-86T	Adenosquamous	++	−	+	+
101-87T	Squamous	+++	+	+	+
105-87T	Large cell	+++	++	+	++
108-87T	Adenocarcinoma	+	Not done	−	+
109-87T	Squamous	+++	+	−	+
110-87T	Adenosquamous	+++	+	Not done	Not done
114-87T	Squamous	+	++	Not done	+

Tumor specimens were also cultured on plastic-coated plates with extracellular matrix proteins using KGM serum-free medium (Clonetics Corp.; San Diego, Calif.). The KGM medium was superior to serum-supplemented MEM in producing outgrowths from tumor explants (Table 3). Matrix proteins did not give substantially better results than plastic alone, however.

B. Karyology

During the last several years, we have attempted cytogenetic studies on solid tumor tissue or, in two cases, effusions from 60 patients with non–small-cell

Table 3 Explant Growth of Primary Lung Tumor Cells on Extracellular Matrix Proteins

Medium	Collagen	Fibronectin	Laminin	Plastic substrate
KGM	n = 25	n = 24	n = 15	n = 13
	27.357 ± 53.794	37.096 ± 81.531	35.314 ± 55.374	29.635 ± 51.625
MEM	n = 26	n = 30	n = 20	n = 16
	8.674 ± 7.443	9.597 ± 7.663	7.103 ± 8.916	6.631 ± 4.490
Totals	n = 51	n = 54	n = 35	n = 29
	17.833 ± 38.804	21.819 ± 55.741	19.194 ± 38.828	16.943 ± 35.897

Outgrowth is in square millimeters.
n, number of explants.

lung cancer (NSCLC). With standard culture methods (cell growth on plastic), we found suitable numbers of cytogenetically abnormal metaphase cells in 23 cases (38%). Among the remaining 37 specimens, 6 were culture failures because of contamination, 13 had only normal karyotypes (presumably from fibroblasts or other normal tumor-infiltrating cells), and 18 had no analyzable mitoses.

In an initial investigation of 19 NSCLC specimens, we harvested metaphase cells after a brief 2- to 4-day culture period [28]. In 12 of these specimens, we were successful in obtaining at least 10 analyzable, karyotypically abnormal mitoses, whereas in seven cases, three or fewer mitotic cells were available (in five of the latter cases, no mitoses or no abnormal cells were observed). In subsequent studies, also involving growth on plastic, we attempted to improve our yield of mitotic cells by harvesting after longer periods of in vitro culture (7 to 35 days). The results of these studies were not encouraging, as we began to observe overgrowth by cytogenetically normal cells, even when cultured in medium containing a low percentage (2%) of fetal calf serum. Therefore, we have now returned to short-term (3–4 days) cultures, and we have begun to use biological substrates (feeder layers and matrix proteins) to improve our cytogenetic success rate and to attempt to develop established cell lines. Such lines will be useful for future cell and molecular biological investigations. Karyotypes of lung carcinomas are often extremely complex and can have many additional chromosomes, making it difficult to find enough adequately spread metaphase cells in a single cytogenetic harvest. The availability of long-term cultures and cell lines can improve the quality of metaphase preparations (Fig. 3). Moreover, such lines allow for more detailed karyological assessment by affording multiple opportunities to optimize harvest conditions.

We have also begun to compare the karyotypic pattern of cells proliferating under these different culture conditions to determine if different methods select for phenotypically and cytogenetically different cell populations within a tumor. Continued studies of this type will enable us to determine which biological substrate assay selects for a population of cells dividing in culture that are more representative of those from the original tumor. Thus far, we have completed detailed cytogenetic analyses on two solid lung tumor specimens cultured on each of multiple substrates: (1) plastic, (2) feeder layers (using tumor-conditioned medium), and (3) matrix protein substrates. In one specimen (an adenocarcinoma), all metaphase cells harvested after 9–12 days of growth on biological substrates were cytogenetically abnormal. On the other hand, monolayers grown 9 and 27 days on plastic alone exhibited a mixture of abnormal and normal karyotypes; only a single analyzable metaphase cell, which had an abnormal karyotype, was found at 9 days. By 27 days most (four of six) dividing cells had a normal karyotype. The

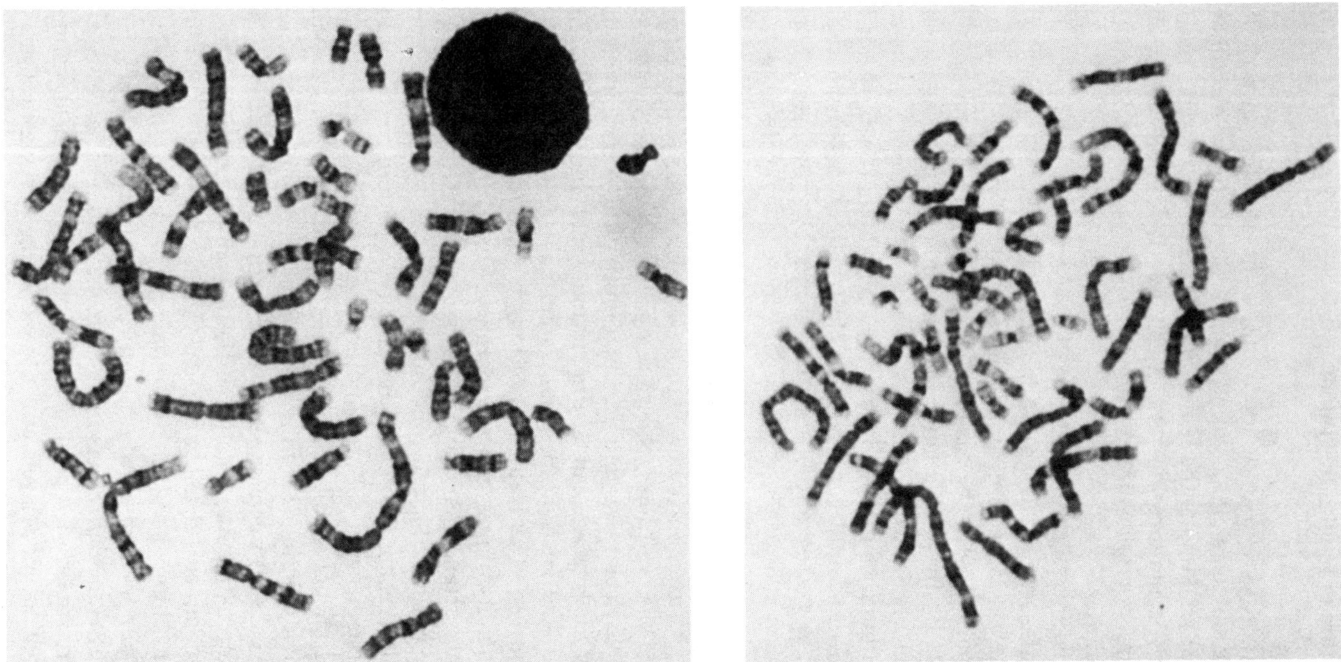

Figure 3 Giemsa (G)-banded hyperdiploid early metaphase cells from established cell line derived from a lung adenocarcinoma.

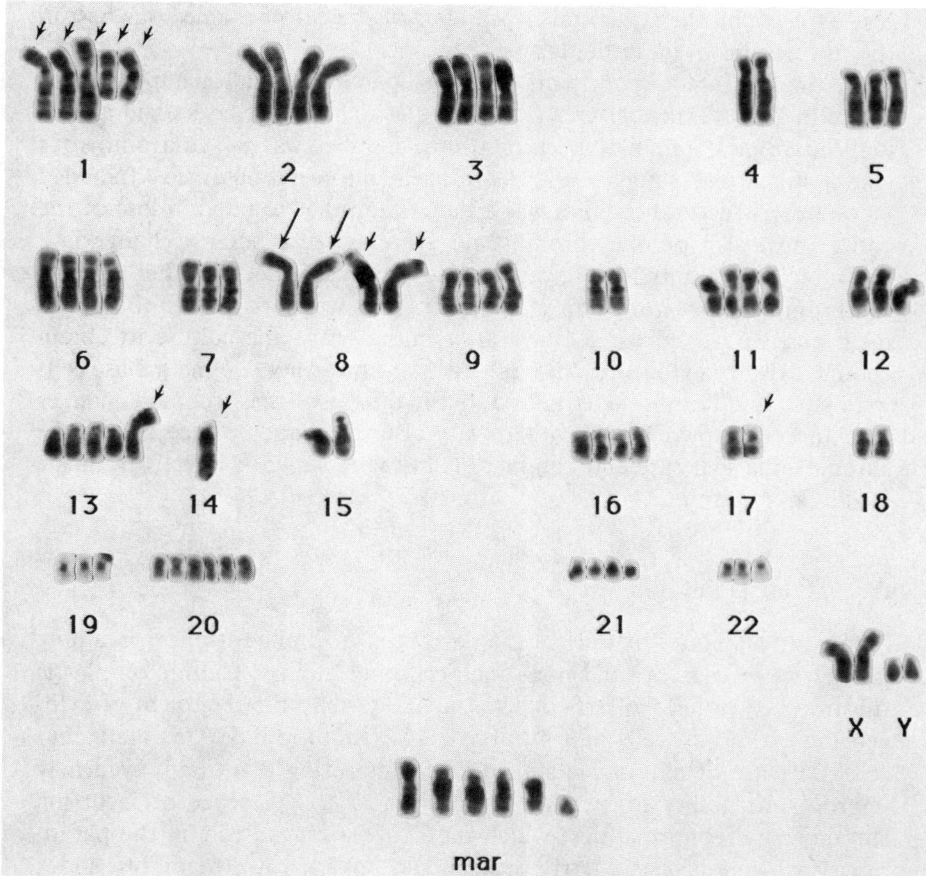

Figure 4 Karyotype of G-banded highly aneuploid metaphase cell from another adenocarcinoma specimen cultured on feeder layers: This cell contains 84 chromosomes with 12 rearranged chromosomes (arrows) and six unidentified markers (mar). The long arrow indicates two copies of an isochromosome of the long arm of chromosome 8.

cytogenetically abnormal cells had highly complex karyotypes with a mode of 82 to 85 chromosomes per cell (Fig. 4). There were numerous structural rearrangements, including two copies of an isochromosome for the long arm of chromosome 8, an abnormality that has recently been reported as a recurring change in adenocarcinoma of the lung [29]. Karyotypes of the cytogenetically abnormal cells did not appear to differ among the three methods used. Thus, there was no evidence for selection of a particular subset of abnormal cells with any method used. However, the mitotic index was highest and the structural appearance of the chromosomes was best in the cultures grown on feeder layers.

In another tumor (poorly differentiated carcinoma), all metaphase cells found in 7- to 12-day cultures grown on plastic or feeder layers had abnormal karyotypes. The modal chromosome number was 84, and numerous clonal numerical changes and rearrangements were observed. Included among the structural changes was an alteration that resulted in loss of the entire short arm of one chromosome 3 homologue. Such a change is a common abnormality in lung carcinomas as well as in several other types of solid tumors [30]. Growth on protein substrates was very slow in this specimen, and we waited for 60 days after initiation of the culture to obtain enough cells for cytogenetic harvest. At that time, only two metaphase cells were seen, and both were tetraploid, but had none of the structural changes seen in cells grown by the other two culture methods. Once again, the chromosome structure and banding quality were superior in cells from the feeder layer culture.

IV. CONCLUSIONS

Standard techniques of cell culture have been of limited use for obtaining metaphase spreads of solid non–small-cell lung tumors. Culture on plastic substrates, even in low levels of fetal bovine serum, allows growth of cytogenetically normal cells and results in a low mitotic index for malignant cells. The use of alternative substrates for culturing this tumor type may improve our ability to analyze karyotypes for the presence of recurring chromosomal abnormalities, which may have a crucial role in the pathogenesis of lung cancer. Matrix proteins did not improve the mitotic index compared with plastic alone. The use of a fibroblast feeder layer for cocultivation of fresh lung tumor specimens, however, increased the number of analyzable mitotic figures obtained compared with plastic alone, and, to date, has allowed more precise identification of structural abnormalities in two cases. This method also discourages normal fibroblast growth and, thereby, may reduce contaminating normal karyotypes in the preparations and also allow establishment of malignant cell lines. We are in the process of producing karyotypes from a series of lung tumor cell cultures that have been established with use of feeder cell cocultivation.

ACKNOWLEDGMENTS

This work was supported by grants to J. M. S. from the American Cancer Society (IN-58-28) and from the Pharmaceutical Manufacturer's Association Foundation, by a grant from the National Cancer Institute (CA45745) to J. R. T. and J. H. R., and by a contract (NOI-CP-51000) to J. H. R. J. R. T. is a Scholar of the Leukemia Society of America. J. M. S. is the recipient of an American Cancer Society Junior Faculty Award.

REFERENCES

1. Collins, S. J., I. Kulsoniski, I. Mijioslin, and M. T. Groudine (1984). Altered transcription of the c-*abl* oncogene in K652 and other chronic myelogenous leukemia cells. *Science 225*:72–74.
2. Shtivelman, E., B. Lifshitz, R. P. Gale, and E. Canaani (1985). Fused transcript of *abl* and *bcr* genes in chronic myelogenous leukemia. *Nature 315*:550–554.
3. Heisterkamp, N., K. Stam, J. Groffen, A. deKlein, and G. Grosveld (1985). Structural organization of the *bcr* gene and its role in the Ph translocation. *Nature 315*:758–761.
4. Erikson, J., C. Griffin, A. ar-Rushdi, M. Valtieiri, J. Hoxie, J. Finnan, B. S. Emanuel, G. Rovera, P. C. Nowell, and C. M. Croce (1986). Heterogeneity of chromosome 22 breakpoint in Philadelphia-positive (Ph^+) acute lymphocytic leukemia. *Proc. Natl. Acad. Sci. USA 83*:1807–1811.
5. Clark, S. S., J. McLaughlin, W. M. Crist, R. Champlin, and O. N. Witte (1987). Unique forms of the *abl* tyrosine kinase distinguish Ph^1-positive CML from Ph^1 positive ALL. *Science 235*:85–87.
6. Erikson, J., A. ar-Rushdi, H. L. Drwinga, P. C. Nowell, and C. M. Croce (1983). Transcriptional activation of the c-*myc* oncogene in Burkitt lymphoma. *Proc. Natl. Acad. Sci. USA 80*:820–824.
7. Taub, R., K. Kelly, J. Battey, S. Latt, G. M. Lenoir, U. Tantravahi, Z. Tu, and P. Leder (1984). A novel alteration in the structure of an activated c-*myc* gene in a variant t(2;8) Burkitt lymphoma. *Cell 37*:511–520.
8. Croce, C., W. Thierfelder, J. Erikson, K. Nishikura, J. Finan, G. M. Lenoir, and P. C. Nowell (1983). Transcriptional activation of an unrearranged and untranslocated c-*myc* oncogene by translocation of a C-lambda locus in Burkitt lymphoma cells. *Proc. Natl. Acad. Sci. USA 80*:6922–6926.
9. Weston, A., J. C. Willey, R. Modali, H. Sugimura, E. M. McDowell, J. Resau, B. Light, A. Haugen, D. L. Mann, B. F. Trump, and C. C. Harris (1989). Differential DNA sequence deletions from chromosomes 3, 11, 13, and 17 in squamous-cell carcinoma, large-cell carcinoma, and adenocarcinoma of the human lung. *Proc. Natl. Acad. Sci. USA 86*:5099–5103.
10. Schwab, M., K. Alitalo, K. H. Klempnauer, H. W. Varmus, J. M. Bishop, F. Gilbert, G. Brodeur, M. Goldstein, and J. Trent (1983). Amplified DNA with limited homology to *myc* cellular oncogene is shared by human neuroblastoma cell lines and a neuroblastoma tumour. *Nature 305*:245–248.
11. Schwab, M., H. E. Varmus, and J. M. Bishop (1985). Human N-*myc* gene contributes to neoplastic transformation of mammalian cells in culture. *Nature 316*:160–162.

12. Whang-Peng, J., C. S. Kao-Shan, E. C. Lee, P. A. Bunn, D. N. Carney, A. F. Gazdar, and J. D. Minna (1982). Specific chromosome defect associated with human small-cell lung cancer: Deletion 3p (14–23). *Science 215*:181–182.
13. Yunis, J. J., and N. Ramsay (1978). Retinoblastoma and subband deletion of chromosome 13. *Am. J. Dis. Child. 132*:161–163.
14. Friend, S. H., R. Bernards, S. Rogelj, R. A. Weinberg, J. M. Rapaport, D. M. Albert, and T. P. Dryja (1986). A human DNA segment with properties of the gene that predisposes to retinoblastoma and osteosarcoma. *Nature 323*:643–646.
15. Heim, S., and F. Mitelman (1987). *Cancer Cytogenetics*, Alan R. Liss, New York.
16. Marcus, A. C., D. R. Shopland, L. A. Crane, and W. R. Lynn (1989). Prevalence of cigarette smoking in the United States: Estimates from the 1986 current population survey. *J. Natl. Cancer Inst. 81*:409–414.
17. Rheinwald, J. G., and H. Green (1975). Serial cultivation of strain of human epidermal keratinocytes: The formation of keratinizing colonies from single cells. *Cell 6*:331–344.
18. Stoner, G. D., Y Katoh, J. M. Foidart, G. A. Myers, and C. C. Harris (1980). Identification and culture of human bronchial epithelial cells. *Methods Cell Biol.* 21A:15–35.
19. Lechner, J. F., A. Haugen, H. Autrup, I. A. McClendon, E. W. Pettis, B. F. Trump, and C. C. Harris (1981). Clonal growth of epithelial cells from normal adult human bronchus. *Cancer Res. 41*:2294–2304.
20. Siegfried, J. M., and S. Nesnow (1984). Cytotoxicity of chemical carcinogens towards human bronchial epithelial cells evaluated in a clonal assay. *Carcinogenesis 5*:1317–1322.
21. Lieber, M., B. Smith, A. Szakal, W. Nelson-Rees, and G. A. Todaro (1976). A continuous tumor-cell line from a human lung carcinoma with properties of type II alveolar epithelial cells. *Int. J. Cancer 17*:62–70.
22. Siegfried, J. M. (1987). Detection of human lung epithelial cell growth factors produced by a lung carcinoma cell line: Use in culture of primary solid lung tumors. *Cancer Res. 47*:2903–2910.
23. Siegfried, J. M., and S. E. Owens (1980). Response of primary human lung carcinoma to autocrine growth factors produced by a lung carcinoma cell line. *Cancer Res. 48*:4976–4981.
24. Cottrell, J. R., and J. H. Resau (1985). A method to determine the viability of explant tissue in organ culture. *J. Tissue Cult. Methods 9*:191–192.
25. ISCN (1985). *An International System for Human Cytogenetic Nomenclature*. D. G. Harnden, and H. P. Klinger (eds.). Published in collaboration with *Cytogenet. Cell Genet*. Karger, Basel, 1985.
26. Wachner, R., C. Wittekind, and S. Von Kleist (1984). Localisation of CEA, β-HCG, SP1, and keratin in the tissue of lung carcinomas. *Virchows Arch. Pathol. Anat. 402*:415–423.
27. Langvad, E. (1968). Lactate dehydrogenase isoenzyme patterns in bronchogenic carcinoma. *Eur. J. Cancer 4*:107–115.

28. Tanaka, K., and J. R. Testa (1987). Assessment of methods for the cytogenetic analysis of human solid tumors. *J. Natl. Cancer Inst. 79*:1287–1293.
29. Jin, Y.-S., N. Mandahl, S. Heim, H. Schuller, and F. Mitelman (1988). Isochromosomes i(8q) or i(9q) in three adenocarcinomas of the lung. *Cancer Genet. Cytogenet. 33*:11–17.
30. Mitelman, F. (1988). *Catalog of Chromosome Aberrations in Cancer*, 3rd ed. Alan R. Liss, New York.

18

The Practical Use of Colchicine and Colcemid to Reversibly Block Microtubule Assembly in Living Cells

Greenfield Sluder

Worcester Foundation for Experimental Biology, Shrewsbury, Massachusetts

I. INTRODUCTION

A. Background

For over 60 years colchicine has been an important experimental tool for cell biologists. It was one of the first plant alkaloids known to inhibit the assembly of the mitotic spindle and prevent cell division [1]. Even before its mechanism of action was fully known, colchicine was widely used to arrest cells in "metaphase" to facilitate the study of chromosome structure for both clinical and research purposes. The demonstration in the late 1960s and early 1970s that colchicine prevents microtubule assembly by specifically binding to the tubulin dimer established this drug as the agent of choice for assaying the role of microtubules in a wide variety of cellular processes. The usefulness of colchicine was extended by the discovery that the *N*-methyl-*N*-desacetyl derivative of colchicine, demecolcine (widely known by its trade name Colcemid, the term used throughout this chapter), binds with higher affinity to the tubulin dimer than does colchicine. This has provided the investigator with an agent that has the same properties as colchicine, but is effective at lower external concentrations [2].

The power of these drugs as experimental tools stems from unique circumstances, not shared by any other microtubule-inhibiting drugs. Colchicine is the one that is most extensively characterized in terms of its mode of action and possible side effects. It also has the unique property that its ability to inhibit microtubule assembly can be abolished by exposing it to light. This property allows the investigator to rapidly reverse the effects of this drug both in vivo and in vitro. Importantly, light-inactivated colchicine has the same side effects as the native drug. Thus, by treating cells with inactivated colchicine the researcher can empirically determine the extent to which the native colchicine has side effects under the particular experimental conditions used.

B. Scope of This Chapter

This chapter is meant to provide a practical guide for the use of colchicine and Colcemid on living cells. We will briefly review the mechanism of action and the side effects of these drugs to give the reader an awareness of the pitfalls attendant with the improper use of these agents and also to provide an understanding of the rationale for the methods we describe. Then, drawing upon our experience with sea urchin eggs, we will describe methods to precisely control the dosage of drug, test for side effects, and provide for optimal photochemical reversal of drug effects in vivo. Throughout, our aim is to stress the principles of the methods, without burdening the reader with the experimental details, which can be found in our original publications. For each experimental material, the exact drug dosages and the methods for handling the cells will have to be determined empirically. Our hope is that the principles outlined in this article will make this task straightforward and will enhance the investigator's confidence in the results he or she obtains.

II. COLCHICINE AND COLCEMID

A. Mechanism of Action

Stated in general terms, colchicine and Colcemid act in the same way by binding the tubulin dimers, thereby preventing them from self-assembling into microtubules both in vivo and in vitro. These drugs bind to a single site on the tubulin dimer, which is inaccessible to these agents when the tubulin is incorporated into microtubules [3–7]. This means that colchicine and Colcemid prevent new microtubule assembly by restricting the availability of the assembly competent tubulin, rather than by binding to microtubules once they have formed [6]. These drugs cause the loss of existing microtubules by preventing their regrowth during normal microtubule turnover reactions. With the exception of very stable microtubules, such as those found in axonemes or centrioles, individual microtubules in living cells are normally in a constant state of growth and shortening. This behavior, termed *dynamic instability*, produces the turnover of microtubules that is observed in both interphase and mitotic cells [8,9]. If a significant fraction of the free intracellular pool of tubulin is complexed with drug, microtubules can shorten but not regrow. Thus, the speed with which these drugs disassemble existing microtubules is primarily determined by the time constants for microtubule turnover during mitosis or interphase and the rate at which the drugs enter the cell. The fact that axonemal microtubules do not turn over explains why they are not sensitive to colchicine or Colcemid. However, flagellar regrowth or the assembly of new daughter centrioles can be inhibited by high doses of Colcemid [10,11].

Although the exact molecular details of the colchicine–tubulin interaction are not fully known, in vitro studies have revealed the general features of the reactions involved. Colchicine binds stoichiometrically to a specific site on the α–β tubulin dimer by a two-step mechanism involving a fast reversible binding, followed by a slower conformational change in the dimer [3,12]. The binding of colchicine to tubulin occurs best at physiological temperatures and is less favored at 0°C. Once formed, colchicine–tubulin exists as a stable complex that is slow to dissociate. The half-time for spontaneous dissociation of colchicine from tubulin in vitro is approximately 36 hr [13]. However, the relatively rapid (approximately 1-hr) recovery of some, but not all, cell lines from a colchicine or Colcemid block of spindle microtubule assembly [14] suggests that the in vivo rates of colchicine–tubulin dissociation may sometimes be faster than those observed in vitro. Colcemid binds more tightly to the tubulin dimer than colchicine does [15] and, therefore, is effective at lower external concentrations than its parent compound [2,16,17].

The exact mechanism by which colchicine–tubulin complexes inhibit microtubule growth has been the subject of much research and remains somewhat controversial [18]. Although colchicine–tubulin complexes will self-associate in vitro, they form amorphous aggregates, sheets of protofilaments, and spiral structures [19–21]. This suggests that the binding of colchicine to the tubulin dimer prevents normal microtubule assembly by distorting the geometry of the bonds formed between tubulin protomers [19]. At low concentrations, colchicine–tubulin complexes will copolymerize with native tubulin into microtubules, to a limited extent [22]. However, colchicine–tubulin complexes also can form a "cap" on the growing end of the microtubule, producing a potent substoichiometric inhibition to the addition of native tubulin, without detectable incorporation of these complexes into the microtubule lattice [18,23].

B. Side Effects

Although colchicine and Colcemid bind to the tubulin dimer with high specificity, they can exhibit side effects when used continuously at high concentrations. Sadly, some workers have subscribed to the philosophy that "if a little drug works well, more will be better." Continuous exposure of cells to high (generally millimolar) concentrations of colchicine for more than a few hours can reduce macromolecular synthesis [24–28] and slow the entry of cells into mitosis [24,29–32]. Colchicine also binds to membranes and alters their functional properties [33–37]. Furthermore, living cells will accumulate bound colchicine in excess of that needed to prevent spindle microtubule assembly. Taylor [38] showed that cultured cells continue to accumulate bound colchicine for at least 20 hr, yet the binding of colchicine to just 3–5% of the total cellular binding sites is sufficient to prevent spindle assembly. Hence, treating cells with colchicine in excess of that needed to minimally inhibit microtubule assembly only increases the probability of producing toxic side effects that complicate the interpretation of the experimental results.

Regardless of how carefully colchicine or Colcemid are used, it is important to know what nonspecific side effects the drugs can have on the cells being studied and also have some experimental measure for the extent of these effects. To this end, one can use an important characteristic of these drugs. Light in the near-ultraviolet wavelengths converts both colchicine and Colcemid into "lumi" forms (Fig. 1), which do not bind to tubulin and do not inhibit microtubule assembly either in vivo or in vitro [39–41]. However, the lumi forms of these drugs show the same nonspecific side effects as their parent compounds for all parameters that have been tested [33,35]. Consequently, treating cells with lumicolchicine or lumi-Colcemid at the same or higher doses than those used in the experiment allows one to empirically differentiate between the specific and nonspecific effects of the native compounds.

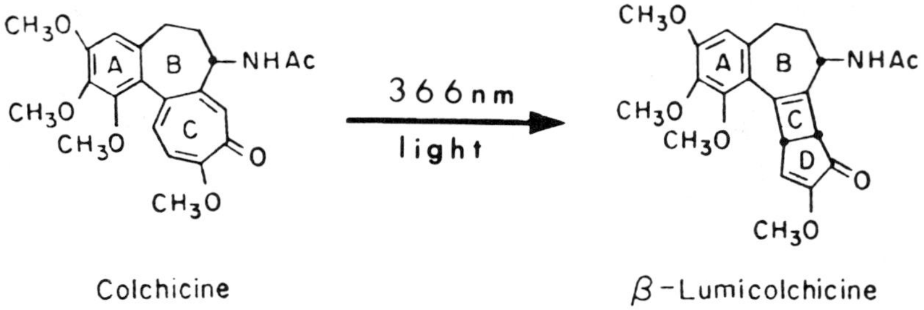

Figure 1 Structure of colchicine and lumicolchicine: The structural rearrangement induced by 366-nm light abolishes the ability of the drug to bind to the tubulin dimer.

III. PRACTICAL METHODS

A. Use of Colchicine and Colcemid

We will describe a simple way to use these drugs that has several practical advantages. Even though our experience is drawn from work with sea urchin eggs, the method should have general applicability. From the start, we chose to work with Colcemid rather than colchicine, because Colcemid is effective at lower external concentrations. Also, Colcemid is reported not to show the toxic side effects of colchicine when both are used at doses just sufficient to block microtubule assembly [14,16,17,24,42–45].

Traditionally, microtubule assembly has been inhibited by continuously exposing cells to a given concentration of drug. Our approach is to treat cells for a limited period with a higher concentration of Colcemid than that required to inhibit microtubule assembly under conditions of continuous exposure. In practice, we treat sea urchin eggs before mitosis for 4–8 min with $5–10 \times 10^{-6}$ M Colcemid. This is at least five- to tenfold the continuously applied concentration that minimally prevents spindle assembly. At the end of the treatment, the free extracellular and intracellular drug is removed by three washes of standard medium (here, sea water). These numbers are given only as guidelines; the exact combination of treatment duration and drug concentration will depend on the experimental material being used.

In determining the exact treatment parameters, we start by choosing a convenient exposure duration and then vary the drug concentration to determine a value that blocks microtubule assembly. The inhibition is then fine-tuned by slightly varying the treatment duration. Assays that can be used to test for complete inhibition of microtubule assembly include: immunofluorescence, with antibodies to tubulin, and various forms of light microscopy on living or fixed material. For example, polarization, differential interference contrast, or phase-contrast microscopy can be used to directly assess the extent of spindle microtubule assembly. We also find that chromosomal orientation is a useful assay for the assembly of relatively few spindle microtubules. When no microtubules are present, the chromosomes have a random distribution. The alignment of chromosomes on a metaphase plate or the formation of a rosette of chromosomes indicates the presence of residual spindle microtubules (Fig. 2).

We find that a single application of Colcemid before mitosis provides a stable inhibition of spindle microtubule assembly for 2–3 hr, or about three to four cell cycles in sea urchin eggs. Thereafter, the eggs show slow recovery from the block. The exact point during interphase at which the Colcemid is applied is not critical. Eggs treated even before fertilization do not assemble spindles upon entering mitosis over an hour later. However, slightly longer

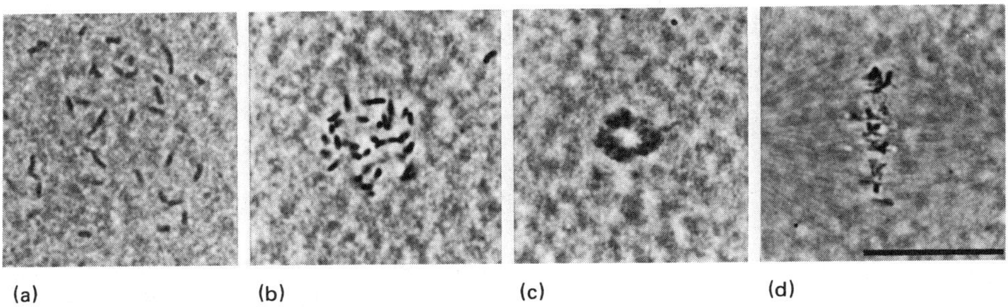

Figure 2 Comparison of chromosome distribution during mitosis in Colcemid-treated sea urchin eggs: Eggs were fixed in acid–alcohol and later stained with 1% orcein in 75% acetic acid. Only the nuclear areas are shown here. (a, b) Complete inhibition of spindle microtubule assembly. Chromosomes are randomly distributed in a spread-out or compact configuration. (c) Incomplete inhibition of microtubule assembly. Chromosomes form a rosette around the centrosomes, which have come together. (d) Slight inhibition of microtubule assembly. Chromosomes are aligned on a metaphase plate (phase-contrast microscopy, bar = 10-μm).

drug treatments are required in these experiments [46]. For rapidly cycling cells that have short mitoses, colchicine or Colcemid treatments during mitosis are not practical if one wishes to completely inhibit microtubule assembly. When externally applied to living cells, these drugs take too long to completely disassemble the spindle during mitosis, unless high concentrations are used [47]. Rapid, complete disassembly of spindle microtubules during mitosis requires that the drug be microinjected [48]. However, drug treatments during mitosis may be used to partially inhibit spindle microtubule assembly, especially in cells that take longer than sea urchin eggs to traverse prometaphase.

The rationale for our approach is that a defined exposure of cells to Colcemid inactivates a finite portion of the tubulin pool before spindle microtubules are assembled. In effect, the amount of tubulin available for microtubule assembly is "set" by the drug treatment before the cell enters mitosis. An important practical advantage of this approach is that one can precisely and repeatedly set the dosage of Colcemid delivered to the cells. This becomes important when one seeks to just minimally inhibit spindle microtubule assembly for the in vivo photoreversals of Colcemid that we describe later. Furthermore, by varying the treatment duration, it is possible to easily titrate the quantity of spindle microtubules that are assembled at mitosis, as is shown in Figure 3 [46]. In this way, spindle assembly can be diminished in a controlled fashion, without concern that the spindle will continue to disassemble, as could be the case with continuous drug treatment. Importantly, removal of the unbound Colcemid ensures that potential side effects are minimized; there is no free drug to bind to other cellular components.

Because the dissociation of colchicine or Colcemid from tubulin is slow, most cells should spontaneously recover slowly from a single exposure to Colcemid. However, it is important to remember that the recovery rate will vary from cell type to cell type. For example, HeLa cells recover poorly, if at all, from a Colcemid block to microtubule assembly, whereas cells from rodent cell lines recover more rapidly [14]. For long-term experiments with cultured cells that recover relatively quickly, the investigator can try regimens whereby cells are briefly treated with a high concentration of drug and then continuously maintained at a lower drug concentration, just sufficient to maintain the degree of microtubule inhibition required. In such cases, there may be some binding of the drug to nontubulin components of the cell and the possibility of side effects. Accordingly, lumicolchicine or lumi-Colcemid control experiments should be conducted.

B. Controlling for Side Effects

To test for possible side effects of Colcemid, we treated fertilized sea urchin eggs with lumi-Colcemid in the same way as the native drug. Lumi-Colcemid was prepared by irradiating Colcemid solutions, at working concentrations, for several hours with unfiltered light from a high-pressure mercury arc lamp. We found that treatment of sea urchin eggs with lumi-Colcemid at double the doses used for the native Colcemid had no effect on the timing of mitotic events or the overall duration of the cell cycle [45]. Also, the time course and extent of DNA synthesis were not affected [49].

However, high doses of lumicolchicine (and probably lumi-Colcemid) clearly have toxic effects, without directly inhibiting microtubule assembly. For sea urchin eggs, continuous application of 1 mM β-lumicolchicine (Sigma Inc., St. Louis, Mo.) reduces spindle size and birefringence (Fig. 4). This inactivated drug also slows both the mitosis and interphase portions of the cell cycle when applied in this fashion [49].

C. Photoreversal of Colchicine and Colcemid

Although the slow spontaneous recovery of cells from colchicine or Colcemid is useful if one wishes to prevent microtubule assembly by briefly treating cells with these drugs, it poses obvious problems for experiments in which the rapid reversal of the drug effect is required. For many years colchicine and Colcemid have been known to be sensitive to light. When exposed to sunlight or light from fluorescent lamps, colchicine undergoes a photochemical rearrangement into α-, β-, and γ-lumicolchicines [39,40],

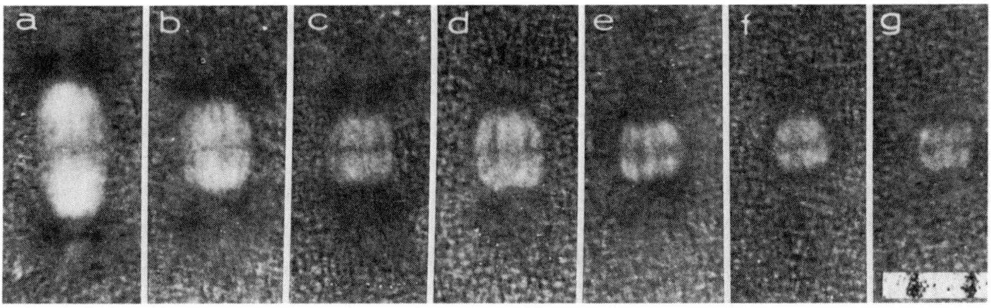

Figure 3 Metaphase spindles of sea urchin eggs treated for various durations with 1×10^{-6} M Colcemid: The eggs were treated with Colcemid 20 min before first nuclear envelope breakdown for (a) 0 min, (b) 1 min, (c) 2 min, (d) 3 min, (e) 4 min, (f) 5 min, and (g) 6 min. These eggs are all from the same female (polarization micrographs, 10-μm/scale division).

which do not prevent microtubule assembly in vivo or in vitro [41]. Aronson and Inoué [50] capitalized upon this property of these compounds to devise a way to inactivate Colcemid in vivo. Through the optical system of the microscope they irradiated individual Colcemid-treated sea urchin eggs with 366-nm light from a mercury arc lamp and found that the eggs assembled spindle microtubules. Although these eggs divided, stable recovery of microtubules was not achieved owing to the background levels of Colcemid in the sea water.

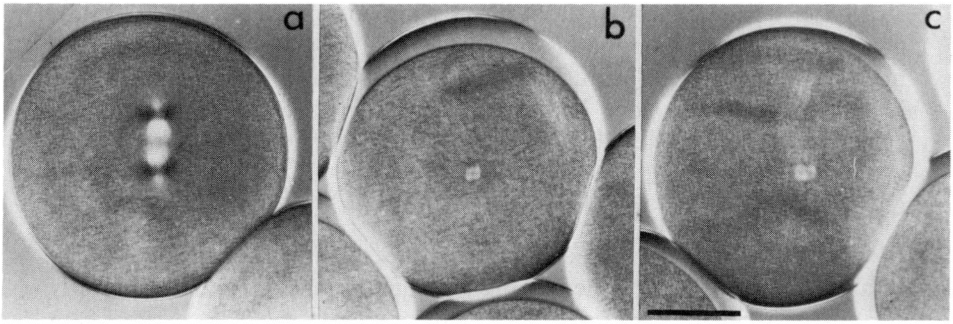

Figure 4 Comparison of prometaphase spindle morphology for (a) control and (b) 1×10^{-3} M lumicolchicine-treated sea urchin eggs: These eggs are from the same female. They were continuously treated with lumicolchicine starting 15 min after fertilization (polarization micrographs; bar, 50 μm).

Our approach to in vivo photoreversals of Colcemid has been to integrate our method for drug treatment with an optical system that optimizes the intensity of 366-nm light at the specimen. To achieve complete, lasting reversals of a Colcemid block to microtubule assembly, it is important to consider the following factors. First, the drug treatment should provide a minimal, but complete, inhibition of microtubule assembly. Overly high doses of Colcemid are difficult or impossible to completely reverse with 366-nm light. Simply increasing the duration of the irradiation is not practical because 366-nm light itself has direct effects on the cell, which we discuss later. The advantage of briefly treating cells with drug is that after the irradiation, there is no free pool of Colcemid outside the cells to reinactivate the intracellular tubulin pool, thereby limiting continued microtubule assembly. For experimental systems that require a defined background level of drug to maintain an inhibition of microtubule assembly, photochemical reversals will be optimized by washing the drug out just before the irradiation. Second, the microscope optics should provide intense 366-nm light at the specimen. We used a single 15-sec irradiation of each cell to completely reverse total blocks to microtubule assembly. We wanted to use short irradiations because they allowed us to observe microtubule assembly after the irradiation and microtubule-dependent processes did not occur to any significant extent during the irradiation. As will become evident from our discussion of the effects of 366-nm light on sea urchin eggs, it is important to empirically determine what effects the irradiations alone have on cells over the expected time course of the experiment. For short-term experiments, these 366-nm light effects are less important than for experiments in which cells are followed for several cell cycles.

An example of a photoreversal of a Colcemid block to spindle microtubule assembly in sea urchin eggs is shown in Figure 5. Recovery of the capacity to assemble microtubules is complete, and cell division is normal. Because the free Colcemid was washed out of the cell at the end of the drug treatment, subsequent mitoses are also normal.

D. Specialized Applications

The foregoing conditions work well when one wants to reverse the effects of Colcemid in whole cells. We have also tried experiments in which we irradiated limited areas of drug-treated sea urchin eggs in hopes of allowing microtubule assembly only at predetermined intracellular sites. We obtained limited microtubule assembly in the irradiated region and at later times observed that some microtubules were assembled at microtubule-organizing centers outside of the irradiated region. In retrospect, this was to be expected because we were not irradiating just a single point in the cell. Irradiation of

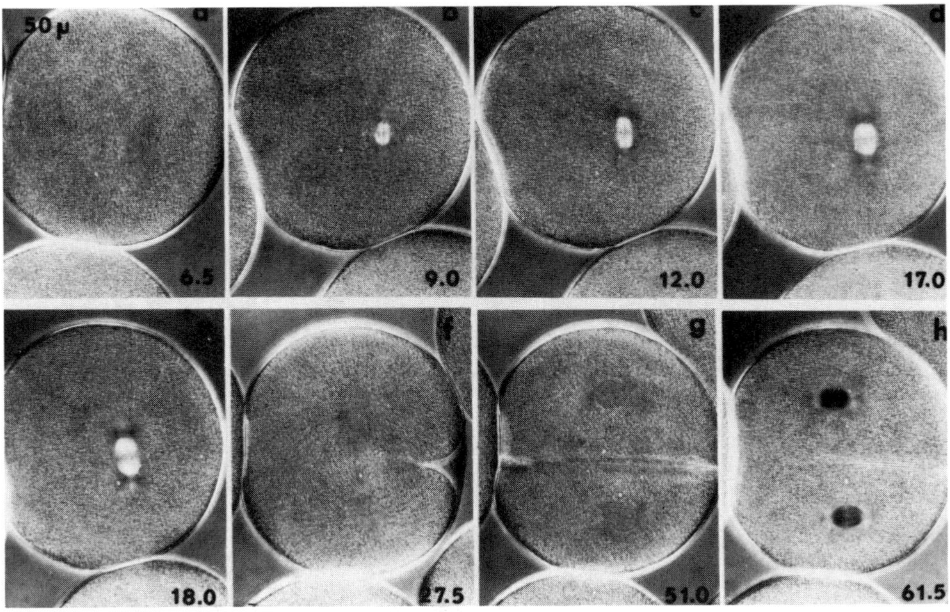

Figure 5 Mitosis in a Colcemid-treated sea urchin egg after irradiation with 366-nm light: This egg was treated for 3.5 min with 5 × 10^{-6} M Colcemid in early prophase of the first division; 8 min after nuclear envelope breakdown, it was irradiated for 15 sec with 366-nm light. (a) After nuclear envelope breakdown, but before irradiation. Spindle assembly is completely inhibited; (b–d) recovery of the birefringent spindle after irradiation; (e) early anaphase; (f) cleavage; (g) nuclear envelopes reformed; (h) second prometaphase. Minutes after first nuclear envelope breakdown are shown in the lower corner of each frame. (polarization micrographs, 50-μm/scale division).

drug-treated cells produces assembly-competent tubulin at locations above and below the focal point; this tubulin is then free to diffuse to other cellular microtubule-organizing centers [51].

These comments notwithstanding, one elegant approach deserves comment. Hamaguchi and Hiramoto [52] devised conditions under which they were able to balance continuous 366-nm light irradiation of defined areas against a background level of Colcemid in the medium to produce stable microtubule assembly only within the irradiated areas. Assembly-competent tubulin diffusing out of the irradiated region was inactivated by the free intracellular Colcemid. This approach requires close attention to the balance between 366-nm light intensity and the background concentration of Colcemid. Attention must also be paid to the effects of 366-nm light alone on the cellular events being investigated.

E. Effects of 366-nm Light

We describe the effects of 366-nm light on sea urchin eggs to show what it can do to living cells and to provide an indication of what to look for in other experimental systems. Other types of cells could show different responses and sensitivities to this light. We irradiated individual eggs with 366-nm light at approximately 1.8×10^{11} quanta · sec^{-1} · μm^{-2} for various durations at different points in the cell cycle. A 15-second irradiation of eggs before nuclear envelope breakdown delayed this event by 1–3 min, yet that mitosis was nonetheless normal. A 15-sec irradiation during prometaphase, such as that used to reverse a Colcemid block to microtubule assembly, had no effect on spindle birefringence, spindle structure, or the timing of mitotic events. However, the following nuclear envelope breakdown was delayed by approximately 1–3 min. Despite this delay, that mitosis was always normal. Irradiations of prometaphase eggs for up to 2.5 min diminished spindle birefringence and size in a dose-dependent fashion. The progression of mitotic events after the irradiation was normal, with the exception that the time from nuclear envelope breakdown to anaphase onset was slightly prolonged. This is consistent with a previous finding that eggs with small spindles traverse prometaphase slower than equivalent eggs with normal-sized spindles [45]. Also, such irradiations delayed the subsequent nuclear envelope breakdown in a dose-dependent fashion. For 2-min irradiations this delay could be as long as 1 hr. However, the timing and morphology of subsequent mitotic events were normal regardless of the delay. Irradiations longer than 3–4 min, administered at any point in the cell cycle, caused the cytoplasm to become granular, and cell death occurred at a variable time thereafter.

The cellular target(s) of 366-nm light which lead to delays in nuclear envelope breakdown are not known. A series of experiments in which we locally irradiated nonnuclear volumes of cytoplasm indicate that 366-nm light affects the cytoplasm, not the nucleus. Furthermore, the effects on the cytoplasm remain spatially localized to the irradiated area. Thus, if a nucleus moves into irradiated cytoplasm, nuclear envelope breakdown will be delayed. Conversely, if an irradiated nucleus is moved by micromanipulation out of the irradiated region, nuclear envelope breakdown occurs in synchrony with the unirradiated nuclei.

IV. OPTICAL SYSTEMS

A. Irradiation of Whole Cell Cultures

Reversal of a colchicine or Colcemid block of microtubule assembly can be achieved by irradiating individual cells on the microscope or by irradiating entire cell cultures at one time. Irradiation of whole cultures can be performed on long-wave ultraviolet light boxes available from a number of manufacturers. In our experience, it is important to prevent heating of the culture by performing the irradiation in a cold room, raising the culture approximately 1-cm above the light box with spacers, and using a small fan to circulate air under the culture dish. As standard plastic culture dishes transmit 366-nm light, special glassware is not required. The duration of the irradiation will necessarily have to be determined empirically.

B. Irradiation of Individual Cells

1. Layout of the Optical System

To irradiate individual cells on the microscope we have used the optical system described in the following. As long as the basic principles that we describe are borne in mind, other optical systems, such as epi-illumination configurations, should work well. Basically, one wants to transmit as much energy in the 366-nm wavelength as possible and have a convenient way to limit the area of the specimen that is irradiated.

A schematic diagram of the optical system we use on a polarizing microscope is shown in Figure 6. To facilitate irradiations, we modify our microscopes to provide an easily reached field diaphragm, a two-position filter slider, and a swing-out polarizer. We use objective lenses as condensers to optimize the extinction of our microscopes. The use of an objective lens in this capacity is not mandatory to perform irradiations; standard condensers should also work as well. In principle, irradiations could be done with microscopes equipped for any of the commonly available contrast methods, as long as the following desired characteristics of the illuminating optical path are maintained:

1. Optical components should transmit 366-nm light without fluorescing strongly.
2. Optical components that absorb 366-nm light, such as polarizers, should be easily removable.
3. Apertures that restrict light, for example phase-contrast annuli, should be removed.
4. The number of relay lenses in the illuminating pathway should be kept to a minimum.
5. The highest condenser numerical aperture possible should be used.
6. A flat first-surface mirror should be used to reflect the illuminating beam into the microscope.

In practice, cells can be irradiated using preparations made from conventional glass slides and cover slips; quartz slides are not required.

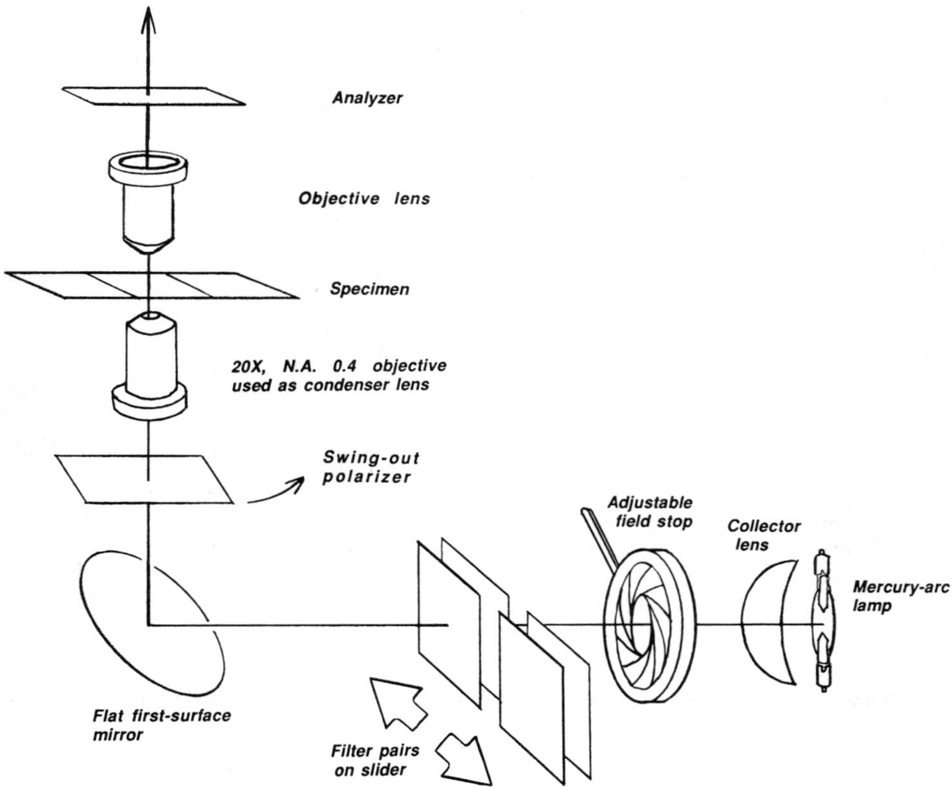

Figure 6 Schematic diagram of the optical system we have used to irradiate individual sea urchin eggs with 366-nm light and observe them with polarized light.

Importantly, one should not look into the microscope during irradiations, unless there is a barrier filter for 366-nm light or a polarizer in the optical path before the oculars. *Intense 366-nm light can cause eye damage.*

2. Microscope Illuminator

The light source must be capable of providing a high intensity of 366-nm light in addition to the visible wavelengths used for observing the cells. For practical purposes incandescent sources, even tungsten halogen lamps, are not suitable. We have used 100-W and 200-W high-pressure mercury arc lamps because they are line emitters with intense emission lines at 365–366 nm (Fig. 7). Most recently, we have used 100-W high-pressure mercury arc lamps because the flux density within the arc is higher than that of 200-W lamps. Although not intuitively obvious, light intensity at the specimen

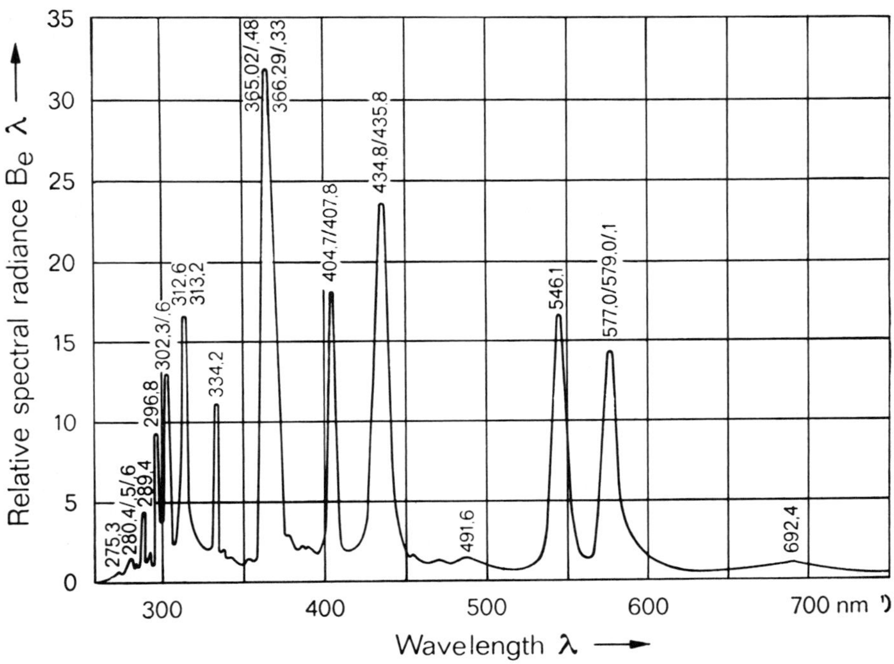

Figure 7 Emission spectrum of the 200-W high-pressure mercury arc lamp.

plane is a function of flux density, not total wattage [53]. However, 100-W lamps are more demanding to use than the 200-W lamps because of the smaller arc size. Thus, a good quality collector lens in the lamp housing is required to ensure even illumination at the specimen plane and project a reasonably sized image of the arc into the back focal plane condenser. Because most glasses used in collector lenses transmit 366-nm light well, quartz collector lenses in the lamphouse are not required. However, such an ultraviolet-transmitting collector, if available, may afford a slight increase in the intensity of 366-nm light at the specimen.

Although we have not used short-arc 75-W xenon lamps, they might offer a reasonable alternative to the mercury arc lamps. Xenon lamps do not emit as much 366-nm light as mercury lamps and produce a higher intensity "floor" of wavelengths across the visible spectrum. Thus, the use of xenon lamps in combination with the simple absorbing filters we describe in the next section will yield irradiations that are relatively richer in wavelengths other than 366-nm than those performed with a mercury arc lamp as a light source. More information on high-intensity light sources can be found in Inoue [53] and the Oriel Corporation catalog Vol. II "Light Sources, Monochromators, Detection Systems," Oriel Corp., 250 Long Beach Blvd., PO Box 872, Stratford, CT, 06497.

3. Filters

Isolation of the 366-nm line from a mercury arc lamp emission spectrum can be achieved by filtering the raw output of the lamp, first through a heat-absorbing filter to diminish the infrared portion of the spectrum, followed by an absorbing glass filter that transmits maximally at 366 nm. As mercury arc lamps are line emitters, a suitable broad-band pass filter of the absorbing type will effectively transmit primarily the 366-nm line. We use a Corning 1-75 (glass number 4605) heat-cut filter followed by a Corning 7-60 (glass number 5840) filter. We have obtained these filters from VWR Scientific Inc. The filters are relatively inexpensive and have a high transmittance for 366-nm light. The percentage transmittance across the spectrum for these filters are shown separately and in combination in Figure 8. With this filter combination there is, however, minor transmittance of the 334-nm and 404- to 407-nm lines. Alternatively, a two- or three-cavity interference filter could be used to isolate the 366-nm line. Although they work well, interference filters generally have a lower peak transmittance at 366-nm than the absorbing filter combination and cost more. If an epifluorescence microscope is used to irradiate individual cells, one should use the filter cube for DAPI fluorescence.

4. 366-nm Light Microbeam Applications

If one wishes to precisely limit the area of a cell exposed to 366-nm light, several factors should be considered. Glass achromatic objective and condenser lenses are generally not well corrected for 366-nm light. Thus, off-axis imaging of the field aperture is not recommended. For whole-cell irradiations, however, condenser aberrations at this wavelength should not be a problem.

A more important practical problem is that these lenses project the image of the field aperture at a slightly different plane and at a slightly different size for 366-nm light and the green light commonly used to observe cells in vivo. Hence, the best focus of the field aperture set with green light must be corrected to yield a properly focused field aperture in 366-nm light. This can be accomplished by placing a second removable field aperture, to be used only for irradiations, at the appropriate distance from the microscope field diaphragm or by refocusing the condenser by a predetermined amount and direction before the irradiation. In addition, Forer [54] describes an elegant

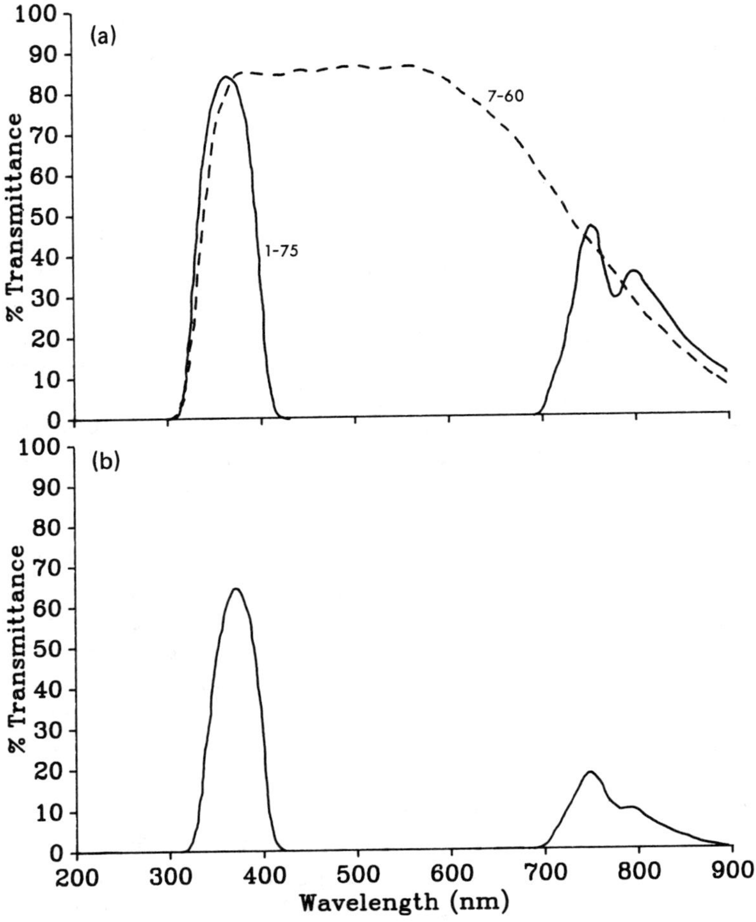

Figure 8 Spectral characteristics of the 1–75 and 7–60 filters: (a) Percentage transmittance as a function of wavelength for these filters individually. (b) Percentage transmittance as a function of wavelength of the filters in combination. Since high-pressure mercury arc lamps do not emit much energy in the far-red portion of the spectrum, the secondary transmittance of the filter combination in the 700- to 900-nm region of the spectrum is of little practical consequence.

method to correct for focal shift by using a calibrated optical flat or a fluid-filled cell.

To visualize the image of the microbeam aperture in 366-nm light for these calibrations, one can prepare a thin slide/coverslip preparation of a dye that fluoresces when irradiated with 366-nm light, for example 8-anilino-1-

napthalene sulfonate in *n*-butanol. To establish a focal plane in such preparations, we use a diamond scribe to scratch the surface of the coverslip that will be in contact with the dye.

When irradiating specific portions of a cell, one should bear in mind that the volume of irradiated cytoplasm has the appearance of an hourglass. The image of a field aperture defines the shape and area of the crossover point; the numerical aperture of the condenser defines the solid angle of the light coming to and from the focal plane. Obviously, the amount of irradiated cytoplasm outside the focal plane will depend on the thickness of the cell. For critical applications, one must also consider the extent to which 366-nm light is scattered by flare in the condenser, imperfections in the slide (or coverslip), and cellular organelles. Such scattering will deliver some 366-nm light to portions of the cell outside the volume of cytoplasm nominally irradiated.

V. CONCLUDING REMARKS

We have sought to provide a practical guide for the use of colchicine and Colcemid on living cells. Although our methods have been designed for work on sea urchin eggs, the principles underlying the specific techniques described here should be applicable to other types of cells. The effort involved in determining optimal drug doses, irradiation durations, and performing the lumicolchicine control experiments should greatly enhance the confidence the researcher has in his or her results. When used intelligently, colchicine and Colcemid are still among the most powerful tools available for the study of microtubule function in living cells.

ACKNOWLEDGMENTS

I am indebted to Frederick J. Miller for his skilled work in the preparation of the figures and Melissa Melan for helpful suggestions on the text. The efforts of Sandy Johnson and Rebecca Bell in the typing of the manuscript are most appreciated. In addition, I thank the Rockefeller University Press for permission to use figures from *J. Cell Biol. 70*:75–85 and *J. Cell Biol. 80*:674–691. Also, I thank Alan R. Liss Inc. for permission to use figures from *J. Exp. Zool. 238*:325–336. Some of the work described in this article was performed in the laboratories of Shinya Inoue and Daniel Mazia; I thank them for their financial support, use of their facilities, and their encouragement. I also acknowledge the support of National Institutes of Health grant GM 30758 (to G.S.), and National Cancer Institute grant PO 3012708 to the Worcester Foundation for Experimental Biology.

REFERENCES

1. Eigsti, O. J., and P. Dustin (1955). *Colchicine in Agriculture Medicine Biology and Chemistry*. Iowa State College Press, Ames, Iowa.
2. Dustin, P. (1978). In *Microtubules*. Springer-Verlag, New York, pp. 167–225.
3. Andreu, J. M., and S. N. Timasheff (1986). Tubulin-colchicine interactions and polymerization of the complex. *Ann. N.Y. Acad. Sci. 466*:676–689.
4. Sherline, P., J. T. Leung, and D. M. Kipnis (1975). Binding of colchicine to purified microtubule protein. *J. Biol. Chem. 250*:5481–5486.
5. Wilson, L., and M. Friedkin (1967). The biochemical events of mitosis. II. The in vivo and in vitro binding of colchicine in grasshopper embryos and its possible relation to inhibition of mitosis. *Biochemistry 6*:3126–3135.
6. Wilson, L., and I. Meza (1973). The mechanism of action of colchicine. Colchicine binding properties of sea urchin sperm tail outer doublet tubulin. *J. Cell Biol. 58*:709–719.
7. Borisy, G. G., J. B. Olmsted, and R. A. Klugman (1972). In vitro aggregation of cytoplasmic microtubule subunits. *Proc. Natl. Acad. Sci. USA 69*:2890–2894.
8. Mitchison, T., and M. Kirschner (1984). Dynamic instability of microtubule growth. *Nature 312*:237–242.
9. Cassimeris, L., N. K. Pryer, and E. D. Salmon (1988). Real-time observations of microtubule dynamic instability in living cells. *J. Cell Biol. 107*:2223–2231.
10. Rosenbaum, J. L., and K. Carlson (1969). Cilia regeneration in *Tetrahymena* and its inhibition by colchicine. *J. Cell Biol. 40*:415–425.
11. Kuriyama, R. (1982). Effect of colcemid on the centriole cycle in Chinese hamster ovary cells. *J. Cell Sci. 53*:155–172.
12. Garland, D. L. (1978). Kinetics and mechanism of colchicine binding to tubulin: Evidence for ligand-induced conformational change. *Biochemistry 17*:4266–4272.
13. Garland, D., and D. C. Teller (1975). A reexamination of the reaction between colchicine and tubulin. *Ann. N.Y. Acad. Sci. 253*:232–238.
14. Stubblefield, E., and R. Klevecz (1965). Synchronization of Chinese hamster cells by reversal of Colcemid inhibition. *Exp. Cell Res. 49*:660–664.
15. Ray, K., B. Bhattacharyya, and B. B. Biswas (1980). Colcemid and colchicine binding to tubulin: Similarity and dissimilarity. *Eur. J. Cell Biol. 22*:288.
16. Dustin, P. (1963). New aspects of the pharmacology of antimitotic agents. *Pharmacol. Rev. 15*:449–480.
17. Kleinfeld, R. G., and J. E. Sisken (1966). Morphological and kinetic aspects of mitotic arrest by and recovery from Colcemid. *J. Cell Biol. 31*:369–379.
18. Wilson, L., and K. W. Farrell (1986). Kinetics and steady state dynamics of tubulin addition and loss at opposite microtubule ends: The mechanism of action of colchicine. *Ann. N.Y. Acad. Sci. 466*:690–708.
19. Andreu, J. M., and S. M. Timasheff (1982). Conformational states of tubulin liganded to colchicine, tropolone methyl ether, and podophyllotoxin. *Biochemistry 21*:6465–6476.

20. Andreu, J. M., and S. M. Timasheff (1982). Tubulin bound to colchicine forms polymers different from microtubules. *Proc. Natl. Acad. Sci. USA 79*:6753–6756.
21. Saltarelli, D., and D. Pantaloni (1982). Polymerization of the tubulin–colchicine complex and guanosine 5′-triphosphate hydrolysis. *Biochemistry 21*:2996–3006.
22. Sternlicht, H., and I. Ringel (1979). Colchicine inhibition of microtubule assembly via copolymer formation. *J. Biol. Chem. 254*:10540–10550.
23. Margolis, R. L., and L. Wilson (1977). Addition of colchicine–tubulin complex to microtubule ends: The mechanism of substoichiometric colchicine poisoning. *Proc. Natl. Acad. Sci. USA 74*:3466–3470.
24. Hell, E., and D. G. Cox (1963). Effects of colchicine and Colcemid on synthesis of deoxyribonucleic acid in the skin of guinea pig's ear in vitro. *Nature 197*:287–288.
25. Creasey, W. A., and M. E. Markiw (1964). Biochemical effects of the vinca alkaloids. II. A comparison of the effects of colchicine, vinblastine, and vincristine on the synthesis of ribonucleic acids in Ehrlich ascites carcinoma cells. *Biochim. Biophys. Acta 87*:601–609.
26. Creasey, W. A., and M. E. Markiw (1965). Biochemical effects of the vinca alkaloids. III. The synthesis of ribonucleic acid and the incorporation of amino acids in Ehrlich ascites cells in vitro. *Biochim. Biophys. Acta 103*:635–645.
27. Fitzgerald, P. H., and L. A. Brehaut (1970). Depression of DNA synthesis and mitotic index by colchicine in cultured human lymphocytes. *Exp. Cell Res. 59*:27–31.
28. Suzuki, N., and Y. Mano (1973). Dependence of periodic DNA synthesis on protein synthesis: Delay of DNA synthesis in the early cleavage stage of sea urchin embryos treated with cycloheximide and colchicine. *Dev. Growth Differ. 15*:113–125.
29. Gustafsson, M. K. (1968). Effect of colchicine on DNA synthesis in plerocerioids of *Diphyllobothrium dendriticum* (Cestoda). *Exp. Cell Res. 50*:1–8.
30. Webster, P. L., and D. Davidson (1969). Changes in the duration of the mitotic cycle induced by colchicine and indol-3yl-acetic acid in *Vicia faba* roots. *J. Exp. Bot. 20*:671–685.
31. Mueller, G. A., M. E. Gaulden, and W. Drane (1971). The effects of varying concentrations of colchicine on the progression of grasshopper neuroblasts into metaphase. *J. Cell Biol. 48*:253–268.
32. MacLeod, R. D. (1971). The response of apical meristems of primary roots of *Vicia faba* L. to colchicine treatments. *Chromosoma 35*:217–232.
33. Mizel, S. B., and L. Wilson (1972). Nucleoside transport in mammalian cells. Inhibition by colchicine. *Biochemistry 11*:2573–2578.
34. Wunderlich, F., R. Muller, and V. Speth (1973). Direct evidence for a colchicine-induced impairment in the mobility of membrane components. *Science 182*:1136–1138.

35. Stadler, J., and W. W. Franke (1974). Characterization of the colchicine binding of membrane fractions from rat and mouse liver. *J. Cell Biol. 60*:297–303.
36. Furcht, L. T., and R. E. Scott (1975). Effect of vinblastine sulfate, colchicine and lumicolchicine on membrane organization of normal and transformed cells. *Exp. Cell Res. 96*:271–282.
37. Hirsimaki, Y., A. U. Arstila, and B. F. Trump (1975). Autophagocytosis: In vitro induction by microtubule poisons. *Exp. Cell Res. 92*:11–14.
38. Taylor, E. W. (1965). The mechanism of colchicine inhibition of mitosis. I. Kinetics of inhibition and the binding of ^{3}H-colchicine. *J. Cell Biol. 25*:145–160.
39. Chapman, O. L., H. G. Smith, and R. W. King (1963). The structure of β-lumicolchicine. *J. Am. Chem. Soc. 85*:803–806.
40. Chapman, O. L., H. G. Smith, and R. W. King (1963). The structure of α-lumicolchicine: Some examples of diamagnetic shielding by the carbon–oxygen double bond. *J. Am. Chem. Soc. 85*:806–812.
41. Borisy, G. G. (1971). Mechanism of UV reversal of colchicine inhibition of microtubule aggregation in vitro. *Proc. Annu. Meet. Am. Soc. Cell Biol. 27*:35a.
42. Kobayashi, N., and K. Nakamura (1957). The cleavage inhibition and abnormal cleavage of sea urchin eggs induced by demecolcine, I. *Zool. Mag. (Tokyo) 66*:258.
43. Kobayashi, N., and K. Nakamura (1957). The cleavage inhibition and abnormal cleavage of sea urchin eggs induced by demecolcine, II. *Zool. Mag. (Tokyo) 66*:377.
44. Zimmerman, A. M., and S. Zimmerman (1967). Action of Colcemid in sea urchin eggs. *J. Cell Biol. 34*:483–488.
45. Sluder, G. (1979). Role of spindle microtubules in the control of cell cycle timing. *J. Cell Biol. 80*:674–691.
46. Sluder, G. (1976). Experimental manipulation of the amount of tubulin available for assembly into the spindle of dividing sea urchin eggs. *J. Cell Biol. 70*:75–85.
47. Inoué, S. (1952). The effect of colchicine on the microscopic and submicroscopic structure of the mitotic spindle. *Exp. Cell Res. Suppl. 2*:305–314.
48. Salmon, E. D., M. McKeel, and T. Hays (1984). Rapid rate of tubulin dissociation from microtubules in the mitotic spindle in vivo measured by blocking polymerization with colchicine. *J. Cell Biol. 99*:1066–1075.
49. Sluder, G., F. J. Miller, and K. Spanjian (1986). The role of spindle microtubules in the timing of the cell cycle in echinoderm eggs. *J. Exp. Zool. 238*:325–336.
50. Aronson, J., and S. Inoué (1970). Reversal by light of the action of *N*-methyl-*N*-desacetylcolchicine on mitosis. *J. Cell Biol. 45*:470–477.
51. Salmon, E. D., W. M. Saxton, R. J. Leslie, M. L. Karow, and J. R. McIntosh (1984). Diffusion coefficient of fluorescein-labeled tubulin in the cytoplasm of embryonic cells of a sea urchin: Video image analysis of fluorescence redistribution after photobleaching. *J. Cell Biol. 99*:2157–2164.

52. Hamaguchi, M. S., and Y. Hiramoto (1986). Analysis of the role of astral rays in pronuclear migration in sand-dollar *Clypeaster japonicus* eggs by the Colcemid–UV method. *Dev. Growth Differ. 28*:143–156.
53. Inoué, S. (1986). *Video Microscopy*. Plenum Press, New York, pp. 126–128.
54. Forer, A. (1965). Local reduction of spindle fiber birefringence in living *Nephorotoma suturalis* (Loew) spermatocytes induced by ultraviolet microbeam irradiation. *J. Cell Biol. 25*:95–117.

Index